Additive Manufacturing Technology

Additive Manufacturing Technology

Design, Optimization, and Modeling

Edited by
Kun Zhou

WILEY-VCH

Editor

Prof. Kun Zhou
Nanyang Technological University
Mechanical & Aerospace Engineering
50 Nanyang Avenue
639798 Singapore
Singapore

Cover Image: © Nordroden/Shutterstock

Library of Congress Card No.: applied for

British Library Cataloguing-in-Publication Data
A catalogue record for this book is available from the British Library.

Bibliographic information published by the Deutsche Nationalbibliothek
The Deutsche Nationalbibliothek lists this publication in the Deutsche Nationalbibliografie; detailed bibliographic data are available on the Internet at <http://dnb.d-nb.de>.

Print ISBN: 978-3-527-34952-4
ePDF ISBN: 978-3-527-83391-7
ePub ISBN: 978-3-527-83392-4
oBook ISBN: 978-3-527-83393-1

Typesetting Straive, Chennai, India
Printing and Binding CPI Group (UK) Ltd, Croydon, CR0 4YY

C9783527349524_040624

Contents

Preface

Additive manufacturing (AM) covers a wide range of techniques capable of transforming digital models into solid objects through layer-by-layer accumulation of feedstock. The AM approach offers great flexibility in part design and enables the fabrication of complex parts without additional machining or tooling. Since the invention of the first AM technique in the 1980s, AM has undergone extensive development and become a disruptive technology capable of manufacturing products from a wide spectrum of materials, including polymers, metals, ceramics, composites, and biological materials. With the design freedom, high efficiency, ease of customization, and industrial-grade part quality that it offers, the AM technology has gained the attention of numerous industries and has been increasingly employed to replace conventional manufacturing methods in commercial applications.

AM technology, with its tremendous untapped potential, is one of the fastest growing fields today. Extensive research has been dedicated to pursuing new discoveries and attaining new insights into various AM techniques to supplement our existing knowledge and further the advancement of the technology. In particular, the scientific community has placed significant emphasis on researching design, optimization, and modeling for AM. The development of novel material systems expands the selection of AM materials, which is advantageous for uncovering new scientific and industrial applications. Meanwhile, process optimization and modeling are also crucial research domains that provide a deeper understanding of the process mechanisms involved in AM techniques. This book reviews and discusses the latest developments in AM technology with a focus on material design, process optimization, and modeling.

Chapter 1 provides a brief introduction to four-dimensional (4D) printing that utilizes programmable polymeric materials and advanced structural methodologies. The four major types of programmable polymers for 4D printing, namely, shape memory polymers, hydrogel composites, liquid crystal elastomers, and magnetoactive materials, are presented. The design methods and principles of controlling shape change in 4D printing are summarized, and the trends, prospects, and potential applications of 4D printing are discussed.

Chapter 2 covers the recent progress in four common 3D cell-laden bioprinting technologies, namely, inkjet printing, laser printing, support bath-enabled extrusion printing, and digital light processing printing. Various strategies implemented in these printing technologies that control and optimize the precision of cell deposition while ensuring cell viability and functionality are also discussed.

Chapter 3 grants an in-depth understanding of the mechanisms for metallurgical defect formation of alloys and their mechanical properties in fusion-based AM with an emphasis on powder-bed fusion techniques. The development of methodologies to evaluate the suitability of designed alloys for fusion-based AM is delineated.

Chapter 4 introduces multi-material AM (MMAM) technologies and elucidates the design of MMAM-processed components. The working principles of direct energy deposition (laser- and arc-based) and laser powder-bed fusion MMAM are explained, and the multi-material feeding mechanisms and material properties of metal–metal, metal–ceramic, and metal–polymer interfaces are described. Specific challenges in materials science and engineering pertaining to different MMAM technologies are stated, followed by a summary and evaluation of the potential research directions in this area.

Chapter 5 presents a modified inherent strain method for the fast prediction of residual stress and distortion in metal parts fabricated by AM. Inherent strains are applied to a layerwise quasi-static finite element analysis model to simulate residual stress and deformation evolution during the fabrication of the part by a heat source. Applications such as crack prevention, support structure optimization, and laser scanning path designs are described in several examples.

Chapter 6 introduces the high-fidelity modeling of key aspects of metal AM, including powder spreading, powder pre-sintering, powder melting, spattering, denudation, and thermal stress evolution. The underlying physical mechanisms at different stages of a printing process, as well as strategies for mitigating defect formation, are highlighted and discussed.

Chapter 7 reports the recent progress in modeling polymer powder-based AM processes, in particular, powder recoating, melting, and cooling stages. A discrete element model for simulating powder recoating processes is introduced, and the effects of the recoating process parameters on powder-bed quality are analyzed. Additionally, a finite-element model for optimizing printing process parameters and predicting the temperature field, degree of recrystallization, and strain and stress fields within printed parts is presented.

I would like to thank the contributing authors of all the chapters and Lifen Yang and Katrina Maceda at Wiley-VCH. I would also like to thank the members of my research group, with whom my discussions have always been fruitful. In particular, I appreciate the efforts of Changjun Han, Wei Zhu, Wei Shian Tey, Devesh Kripalani, Yujia Tian, Raj Kiran, Priyanka Viveganananthan, Han Zheng, Liming You, and Asker Jarlöv.

Kun Zhou

1

Introduction to 4D Printing: Concepts and Material Systems

Xiao Kuang, Liang Yue, and H. Jerry Qi

Georgia Institute of Technology, The George W. Woodruff School of Mechanical Engineering, 801 Ferst Dr., Atlanta, GA 30332, USA

1.1 Background

In 2013, Tibbits introduced the concept of "four-dimensional printing" (or 4D printing) [1]. He used an inkjet printer to print active (or programmable) polymer composites, whose shape could evolve with time under the water. Soon after, Qi and coworkers presented the concept of printed active composites (PACs), which can shift into various complex configurations by leveraging shape-memory polymers (SMPs) [2]. These pioneering works open the new field of 4D printing, which attracts increasing attention in research communities and industry.

In Figure 1.1a, the schematic illustration shows the difference of structures from one-dimensional (1D) to 4D. The structures show improved complexity in dimensions from 1D to three-dimensional (3D). 4D printing is initially defined as "3D-printed shape + time," where the fourth dimension is time [2–6]. Currently, the definition of 4D printing has been broadened to include not only the shape but also the property and functionality of 3D-printed structures that would change with time under a predetermined stimulus after printing. In 4D printing technique, stimuli-responsive materials and proper structural design are usually used for 3D printing of shape-programmable structures. Compared with conventional 3D-printed static objects, 4D-printed structures are able to change their shape, size, color, or other functional properties with time under external stimuli. The on-demand shape and property changes after printing enable several prominent advantages. First, it allows the direct manufacturing of smart and intelligent structures/devices. Second, it saves printing time and materials, particularly, for thin-wall structure fabrication. Third, it can save storage and transportation space for printed parts [7]. Due to these reasons, 4D printing has become a fast-growing research field in various disciplines, such as smart materials and advanced manufacturing.

4D printing has witnessed significant progress during the past years, thanks to the tremendous advances in 3D printing techniques, stimuli-responsive materials, and design and modeling-based methods [3, 8–11]. 3D printing, or additive manufacturing (AM), becomes a disruptive technology for advanced manufacturing [12, 13].

Additive Manufacturing Technology: Design, Optimization, and Modeling, First Edition. Edited by Kun Zhou.

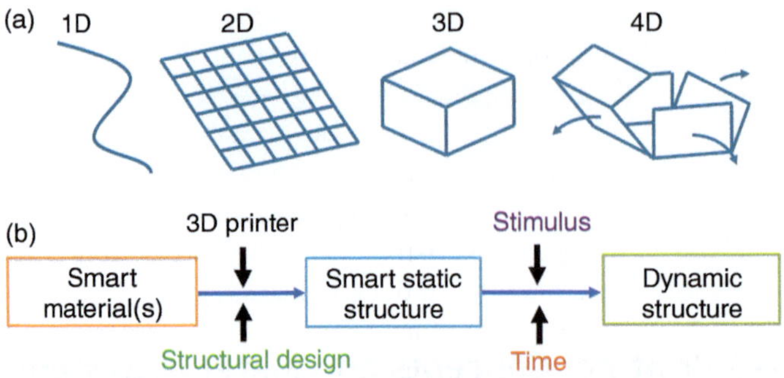

Figure 1.1 (a) Schematic illustration of 1D to 4D. (b) Flowchart showing the basic elements of 4D printing. Source: Momeni et al. [3]/with permission of Elsevier.

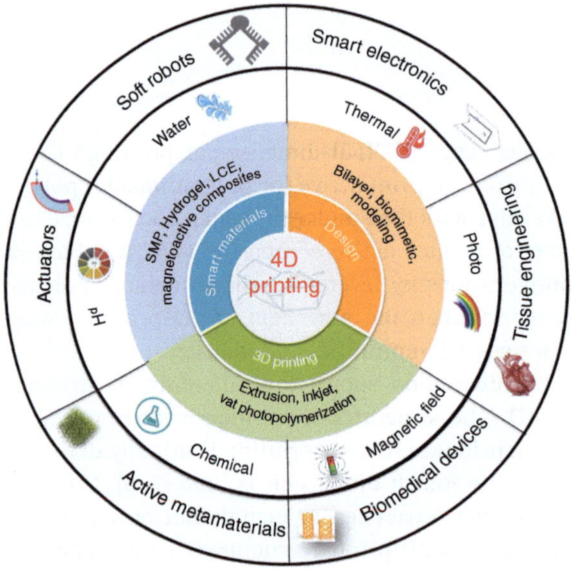

Figure 1.2 Schematic summary of the primary 3D printing techniques, various shape-programmable materials, and commonly used design approaches for 4D printing and their applications.

According to predetermined design models (such as computer-aided design [CAD]), raw materials in various forms (powder, filament, or liquid) can be deposited by different printing modalities, such as extrusion-based printing, inkjet-based printing, and vat photopolymerization-based printing (Figure 1.2). Notably, multi-material 3D printing by constructing heterogeneous structures with spatial control of material distributions becomes a leading direction for 4D printing. Metals [14], ceramics [15, 16], and polymers [2, 4, 17] have been used for 4D printing. However, polymers are still the primary materials for 4D printing due to their cost-effectiveness, diverse stimuli-responding, and large deformability [14]. For example, shape-programmable polymers can be engineered to respond to a wide range of physical and/or chemical stimuli, such as temperature, chemical, light, and magnetic field. The most extensively studied shape-programmable smart polymers for 4D printing include SMPs, hydrogels, liquid crystal elastomers (LCEs), and magnetoactive soft materials (MSMs). The 4D-printed structures with on-demand complex geometry and stimuli-responsive capability find applications in various fields, including actuators, soft robots, active metamaterials, smart electronics, self-folding structures, biomedical devices, and tissue engineering [18–24]. The shape-shifting performance of 4D-printed structures is highly dependent on both the properties of shape-programmable materials and structures. Several review papers in 4D printing can be found with different focuses on stimuli-responsive materials [7, 25, 26], shape-changing mechanisms [3, 27, 28], and applications [18, 29]. This chapter provides a fundamental understanding and recent advances in programmable polymeric material systems for 4D printing. It is organized by starting with an overview of 3D printing techniques, followed by the advances in smart material systems for 4D printing. Particular attention is paid to the use of the single-material and multi-material design of smart materials with different shape-changing mechanisms for shape-changing. In addition, we briefly introduce the modeling-guided design for 4D printing to predict the shape-shifting. To close, we discuss the major challenges and perspectives in 4D printing.

1.2 Overview of 3D Printing Techniques

According to International Organization for Standardization (ISO Technical Committee 261) and American Society for Testing and Materials (ASTM) International (Committee F42), AM technologies are classified into seven groups, including binder jetting, directed energy deposition, material extrusion, material jetting, powder bed fusion, sheet lamination, and vat photopolymerization [30]. In the following section, we introduce the 3D printing methods for polymers and composites most relevant to 4D printing. In most cases, 3D printers are only compatible with single-material processing, meaning a single material is printed by one printing technique in one printing job. By contrast, multi-material 3D printers have been developed to fabricate heterogeneous structures with precisely controlled materials distribution to manipulate more complex shape evolution and pathway.

1.2.1 Single-Material 3D Printing Techniques

Extrusion-based printing, inkjet printing, and vat photopolymerization-based printing are extensively used 3D printing techniques for 4D printing (Figure 1.3). Among them, the most versatile printing technique is extrusion-based printing, which constructs the 3D object in a line-by-line and then a layer-by-layer manner. Depending on the ink, extrusion-based printing is divided into fused filament fabrication (FFF) (or fused deposition modeling [FDM]) and direct-ink-writing (DIW). In the FFF process, a solid filament is fed into a heating nozzle to form a polymer melt, which is continuously deposited onto a building tray with a controlled temperature and then solidified by environmental cooling. For example, engineering thermoplastics and composites, such as acrylonitrile butadiene styrene (ABS), polylactic acid (PLA), polyurethane (PU), as well as fiber or particle-embedded

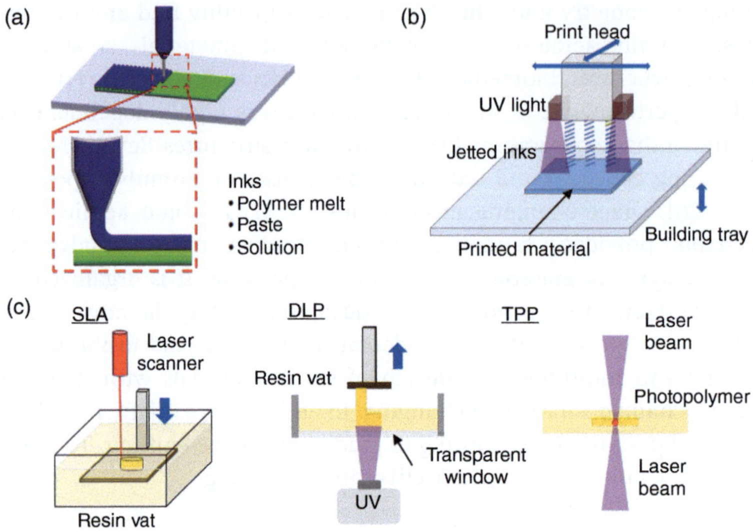

Figure 1.3 Schematics of three major 3D printing techniques for polymers: extrusion-based printing, inkjet-based printing, and vat photopolymerization-based printing. Source: Kuang [31]/with permission of Elsevier.

composites, can be printed by FFF [32]. In FFF, residual stress usually exists due to the shearing and moving nozzle's temperature gradient and polymer chain alignment [33, 34]. This residual stress can be utilized to induce shape-shifting by stress releasing after printing [35, 36]. For FFF-based 4D printing, the shape memory effect (SME) of printed plastics is also frequently used to accommodate shape change [37]. By contrast, DIW uses a mechanical extruder or pressure to deposit paste or solution ink, solidified later by solvent evaporation, UV curing, or thermal curing. The prominent advantage of extrusion-based methods is their applicability to a broad width of inks with different curing chemistry and viscosity. Assisted by post-curing, DIW has been used to print different engineering thermosetting polymers [38–42], and functional materials, such as MSMs [43–47] and LCEs [48–53]. Compared with FFF printing with weak interfaces between filaments [54], DIW-printed materials with post-curing generally present a strong interface. The resolution of FFF and DIW is typically in the range of 100–200 μm, limited by the nozzle diameter. In addition, the extrusion-based methods usually show a low printing speed due to the line-by-line fabrication manner.

Inkjet 3D printing is another popular method for photopolymer printing. During inkjet printing, photocurable ink drops (in the size of 20–40 μm) are deposited onto the building bed from printing nozzles controlled by the piezoelectric device, heating modules, or electromagnetic unit, followed by UV curing to form a solid layer [55]. The process is repeated until the whole structure is built. Various photocurable acrylate-based monomers and crosslinkers are formulated as low-viscosity inks to print materials with different properties [56]. Inkjet printing has high resolution with a typical in-plane resolution of 30–40 μm [57]. Usually, a support material made of weak and water-soluble polymer is used to allow complex structure printing. Despite its many advantages, inkjet printing suffers from the high equipment cost and the stringent requirement on ink viscosity [58].

Vat photopolymerization-based printing is a fast-growing method for 3D printing polymers. This printing technique includes several different printing modalities, including stereolithography (SLA), digital light processing (DLP), and two-photon polymerization (TPP), which relies on the light to selectively cure photo-sensitive polymer resins to build 3D structures. SLA is the earliest 3D printing method, which employs a UV laser beam to rapidly solidify the photopolymers inks [56, 57]. However, the line-by-line scanning-based SLA printing is usually slow, especially in large part. By contrast, DLP exploits dynamic light patterns by digital micromirror devices (DMDs) to cure a layer of resin at a time, enabling fast printing [59]. The emergence of continuous liquid interface production (CLIP) is a breakthrough for DLP, which further enhances the printing speed and is among the fastest 3D printing technology suitable for large-scale industrial production [60, 61]. CLIP uses an oxygen-permeable film to form a dead zone between the transparent window and liquid, allowing continuous build stage elevation without ink renewal step. The in-plane resolution of DLP falls in the range of 50–100 μm, determined by the DMDs and focusing lens used in the projector. By coupling a proper lens system with a DLP projector, micro-DLP (μDLP; or projection micro-stereolithography, PμSLA) has been developed to realize higher resolution up to 0.6 μm [59, 62]. Nanoscale printing resolution is achieved by using TPP-based printing. The simultaneous absorption of two photons at the high-intensity focal point of two lasers allows printing structures with a submicron resolution typically around 40–100 nm [63].

1.2.2 Multi-Material 3D Printing

Driven by the need to fabricate heterogeneous functional structures and devices, multi-material 3D printing has been developed and attracted great attention [64–69]. The direct approach to multi-material printing uses multiple nozzle extruders [67, 69–73]. Microfluidics-extrusion [74] and micromixer-based co-extruders [66, 75–77] have also been developed for extrusion-based multi-material printing. Espalin et al. [72] developed a multi-material FFF system to print parts either using discrete multi-materials or varying building processes. Recently, Lewis and coworkers developed a multi-material multi-nozzle DIW printing approach to realize high-resolution, high-speed multi-material printing [67]. Besides, multi-material DIW printing was achieved using photothermal dual-cure resins [78].

The PolyJet technology by Stratasys enables inkjet printing of digital materials and multi-materials using multiple print heads and inks [56]. To achieve this, multiple inks were deposited in a predetermined ratio and cured together, generating so-called digital materials with digitalized mechanical properties within a certain range. This technique allows for the facile fabrication of parts composed of materials with widely tunable mechanical and thermomechanical properties [56, 57]. However, the resolution of multi-material printing by PolyJet printer is reduced to 200–400 μm [79].

Multi-material printing can be achieved by switchable multiple resin vats [80–82]. Wicker and coworker [80] utilized the scanning SLA system with multiple vat carousel designs to print 3D multi-material objects. The process requires a cleaning step (in a cleaning vat) before being immersed in a different material vat to reduce contamination. Recently, Ge and coworkers developed multi-material PμSLA printing by switching between different ink droplets and using high-pressure air for surface cleaning [81, 82]. Instead of using multiple resin vats, single-vat DLP printing of multi-materials has been achieved by selective wavelength photocuring [83, 84], and grayscale printing [85, 86]. Particularly, grayscale DLP printing using grayscale images for spatially controlled light dose correlated with different crosslinking densities and mechanical properties. The monochromic images with grayscale red, green, and blue (RGB) values ranging from 255 to 0 (brightness from 1 to 0) indicate full light intensity to completely dark. By leveraging the material properties gradients by the grayscale DLP, various shape-shifting materials using different mechanisms can be created for 4D printing [87–89].

Besides, integrated multiple printing technologies (or hybridizing) can provide or improve spatial control of material, geometry, and functionality. Early in 2005, Wicker's group [90] integrated commercial SLA machines with DIW for automated and efficient hybrid manufacturing of complex electrical devices. Recently, Peng et al. [91] reported a hybrid printing system consisting of top-down DLP printing and DIW printing for multi-material 3D printing (Figure 1.4a). The DLP module enabled high-resolution, high-speed printing of geometrical complex structures using various photopolymer inks. The DIW printing module was used for printing functional materials, such as LCEs and conductive silver ink. With this technique, multi-material soft robots and actuators could be directly printed in one print job. Roach et al. [92] developed a multi-material multi-method (m4) AM platform by integrating four different printing modules (inkjet printing, FFF, DIW, and aerosol jetting), and robotic arms for pick-and-place (PnP), as well as intense pulsed light (IPL) sintering system (Figure 1.4b). With this platform, complex structures and functional devices, including actuators, soft robotics, and electronics, could be readily printed.

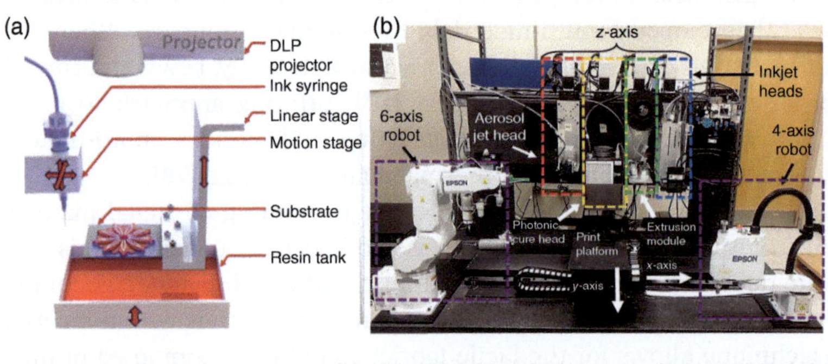

Figure 1.4 Typical examples of multi-material printing techniques. (a) Hybrid 3D printer integrating top-down DLP printing and DIW printing techniques. Source: Peng et al. [91]/with permission of Elsevier. (b) Picture of a multi-material multi-method 3D printing platform. Source: Roach et al. [92]/with permission of Elsevier B.V.

1.3 Shape-Programmable Materials for 4D Printing

Shape-programmable materials are widely used as stimuli-responsive materials for 4D printing. Four different types of shape-programmable materials, namely SMPs, hydrogels, LCEs, and MSMs and their composites, are extensively studied [93]. These materials have different shape-shifting mechanisms. SMPs are mechanically programmed in a temporary shape, which is triggered by external stimuli for shape recovery due to the entropic elasticity of the polymer network [94, 95]. Shape-changing of hydrogels originates from water uptake/release by swelling and shrinkage under external stimuli [96]. LCEs enable reversible shape change via the phase transition between the nematic (N) state and isotropic (I) state [97, 98]. MSMs composed of magnetic particles embedding soft polymer matrix can undergo rapid and reversible shape actuation, owing to the interactions between magnetic particles and magnetic fields [99, 100]. Although there are some other materials and shape-morphing mechanisms, such as polymerization-induced volume shrinkage [87], evaporation-induced shrinkage [101, 102], and thermal expansion mismatch [103], we will focus on using the aforementioned shape-programmable materials for 4D printing in this chapter. This section will summarize advances in these shape-programmable materials, and typical examples will be provided.

1.3.1 Shape-Memory Polymers and Composites

SMPs are smart materials showing the SME, which has been extensively used for 4D printing. For SMPs, temporary shapes can first be programmed, and then the initial shape can be recovered in response to external stimuli, such as heat [104–109], solvent [110–114], and light [115–118]. Vernon first discovered the SME in 1941, when he used e-beam or peroxide crosslinked high-density polyethylene to make heat-shrinkable tubes and films [119].

SME can be understood by the concept of "fixing phase" and "switching phase" [94, 95]. In the form of chemical or physical crosslinking points, the fixing phase can hold the permanent shape of the material. By contrast, the switching phase can be either phase level (amorphous and crystalline) or molecular level (supramolecular entities and reversible molecular units). The switching phase can transform between an active state and a frozen state featuring different chain mobility. The shape fixing and recovery of SMPs are realized by altering the chain mobility via applying and eliminating external stimuli. Heat is the most common trigger for SMPs, which can be achieved by direct heating or indirect heating [120]. Phase transition temperature (T_t), including glass transition temperature (T_g) and melting temperature (T_m), is employed by heating to trigger the shape change (or shape recovery) in SMPs. In thermally activated SMPs, the initial shape is in the thermodynamically stable conformation with the highest entropy. Chain mobility is significantly enhanced upon heating above T_t, allowing for deformation under external force. The programmed shapes possess lower entropy. When SMPs are cooled down to a temperature lower than T_t and undergo the phase transition (glass transition or crystallization), a metastable state with relatively low energy is obtained, leading to the temporary shape fixing of SMP. Upon reheating above T_t, SMPs regain high chain mobility for shape recovery driven by entropic elasticity.

1.3.1.1 Single SMP

Various semicrystalline polymers (such as PLA and PCL) [121, 122] and amorphous polymers (such as polyacrylate and epoxy) [123] with tunable thermomechanical properties have been used as thermally triggered SMPs, which are good candidates for [124] 4D printing. FFF can print physically crosslinked semicrystalline polymers for 4D printing, while chemically crosslinked amorphous SMPs with better shape memory performance are mainly printed via vat photopolymerization-based techniques using various photopolymer inks. The T_g and stretchability of polyacrylate-based photopolymers can be tuned by the rigidity of acrylate monomers and crosslinker content [124, 125]. For example, Ge et al. [68] used benzyl methacrylate (BMA) and different difunctional (meth)acrylates for high-resolution PμSL printing of SMPs with tunable mechanical and thermomechanical properties (Figure 1.5a). The rubbery modulus increased from 1 to 100 MPa by increasing the crosslinker content. In addition, by using crosslinkers with different chain rigidity, the T_g of the polymer was tuned from −50 to 180 °C. However, the aforementioned methacrylate-based resins required very high exposure energy (or light dose) to accommodate adequate curing due to slow crosslinking. Figure 1.5a also shows the shape recovery of 3D-printed Eiffel Tower model from a programmed bent shape.

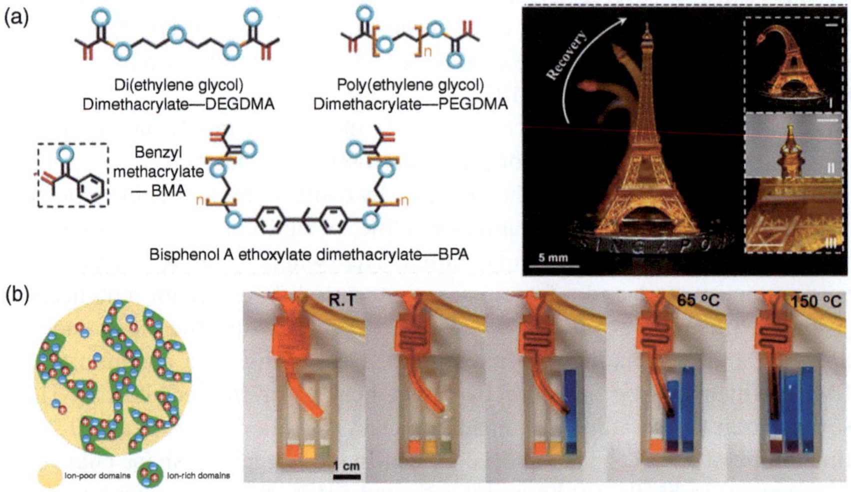

Figure 1.5 Design examples of SMP for 4D printing of SMP. (a) The chemical structures of a rigid mono-functional monomer of several and difunctional oligomers as crosslinkers for PμSL printing of SMPs. The printed high-resolution Eiffel Tower with shape memory effect. Source: Ge et al. [68]/with permission of Springer Nature. (b) A microphase separation SMP can be printed by DLP via photopolymerization-induced microphase separation. The use of the printed SMP with two different T_g for selective feeding process using triple SME devices. Source: Peng et al. [126]/with permission of American Chemical Society.

Multiple shape-shifting by multi-SME in single-material SMPs has also been achieved by 3D printing. For example, polymers with a broad phase transition temperature can realize multi-SME [127–129]. Yu et al. [130] used Objet Connex printer material library to print a digital material Gray 60 with a broad T_g (a range of ~40 °C) with multi-SME. The printed SMP was programmed into different temporary shapes at different temperatures to obtain sequential shape-changing. Recently, Peng et al. [126] reported the DLP printing of a microphase-separated material for a triple SME. An ion–pair comonomer was introduced and served as a physical crosslinker in ink. The photopolymerization-induced microphase separation could rapidly occur during printing, and two separated glass transition temperatures were obtained in the material. It was demonstrated that the printed microfluidic devices could be used as selective feeding devices using temperature-dependent shape-shifting (Figure 1.5b).

To enhance the properties of printed SMPs, inks can be rationally designed to modulate the material network architectures. To this end, the molecular engineering of interpenetrating polymer network (IPN) and semi-interpenetrating polymer network (semi-IPN) has been widely used. Kuang et al. [131] developed a new photopolymer ink containing aliphatic urethane diacrylate (AUD), *n*-butyl acrylate, and semicrystalline polycaprolactone (PCL) as well as fumed silica (rheology modifier) for UV-assisted DIW printing (Figure 1.6a). The composite ink was deposited at around 70 °C and then crosslinked by photocuring. The obtained semi-IPN polymer composite contains a percolating network of nanoscale crystals and a covalent network. The loosely crosslinked chemical

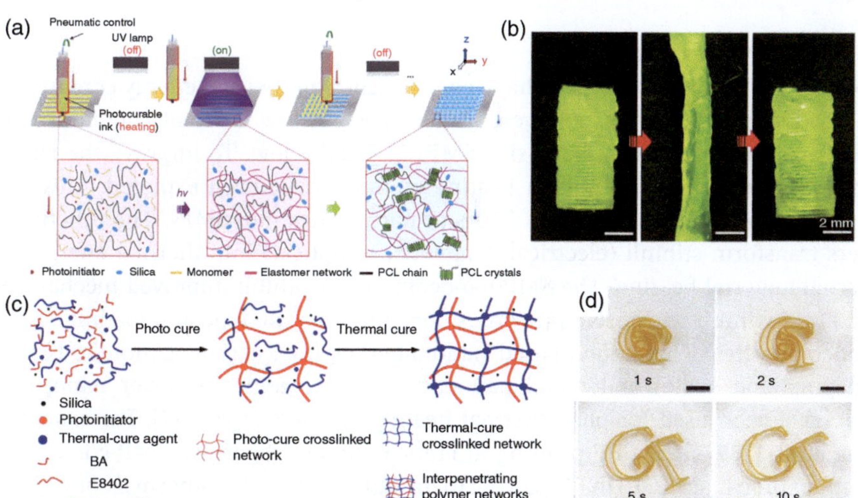

Figure 1.6 SMP composites with hybrid polymer network design. (a) Semi-interpenetrating polymer network SMP was printed by UV-assisted DIW printing [131]. (b) DIW-printed tubes with shape memory effect [131]. (c) Interpenetrating polymer network SMP was printed by DIW via a photothermal dual-cure approach [39]. (d) The results shape the recovery of printed epoxy SMP in a "GT" logo in an oil bath [39]. Source: (a) and (b) Kuang et al. [131]/with permission of American Chemical Society. (c) and (d) Chen et al. [39]/with permission of Royal Society of Chemistry.

network and PCL crystals acted as fixing and switching phases for the SME with large deformation, respectively. Besides, the PCL nanocrystals played the role of a healing agent for material self-healing. The SME could facilitate crack closing and assist the material healing process. It was shown shape memory tubular structures can be printed with on-demand size for vascular repair applications (Figure 1.6b). Zhang et al. [132] used DLP for high-resolution 4D printing using similar material systems. Similarly, PCL is incorporated into a methacrylate photopolymer ink for high-resolution (up to 30 μm) printing of SMPs.

IPN can be obtained by dual-curing or photothermal two-stage curing approaches. Yu et al. [133] reported the use of epoxy-acrylate hybrid photopolymer to print SMP via the SLA 3D printing technique. A polyacrylate-*co*-epoxy particle was first performed and dispersed in epoxy resin to enhance the toughness. Then, both free radical and cationic initiators are used to simultaneously cure the acrylates and epoxy resin by UV irradiation during printing. Finally, the printed SMP shows good thermal stability, high strength, and good toughness. To further enhance the mechanical properties, dual-stage curing was adopted to fabricate IPN [39, 134, 135]. Chen et al. [39] developed a hybrid ink containing photocurable acrylates and thermal-curable epoxy for UV-assisted DIW printing. The UV curing leads to a first polymer network in the green part, which is further treated with the second-stage (thermal-curing) to boost the mechanical properties (Figure 1.6c). The modulus and T_g of SMPs can be tuned by altering the ratio of the photocurable and thermal-curable components. With this approach, IPN-based SMPs with high toughness and high resolution can be printed (Figure 1.6d).

1.3.1.2 SMP Nanocomposites

Thermal activation of SMPs for shape change relies on thermal energy conduction for the material. As pure polymers usually possess low thermal conductivities, the shape recovery rate of large-size SMPs is usually low. To improve the shape recovery speed of SMPs, indirect heating methods have been explored. This can be achieved by adding functional fillers or particulates into SMPs. The functional fillers transform stimuli (electrical, magnetic, or optical) into thermal energy for internal material heating. The SMP nanocomposites exhibit improved mechanical and multi-stimuli-responsive properties depending on the type and loading of the nanocomposites. The carbon-based nanomaterials (i.e., carbon nanotube (CNT) and graphene oxide) and metal nanoparticles (i.e., gold and silver nanoparticles) have been used for photothermal heating of SMPs [136, 137]. For example, Yang et al. [137] used FFF printing to fabricate photoresponsive SMP containing /9 wt% carbon black (CB) in PU. Owing to the outstanding photothermal conversion efficiency of CB (92 ± 3%) [138], PU/CB composites showed high photothermal conversion efficiency. It was demonstrated the 3D-printed sunflower with photoresponsive shape memory petals can be triggered to recover its original state with light illumination of 87 mW cm^{-2}. Iron oxides (i.e., Fe_2O_3 and Fe_3O_4) and CNT have been used for inductive heating of SMPs by alternating magnetic fields [139–144]. As an example, Wei et al. [145] reported the DIW printing of UV-curable PLA/Fe_3O_4 composite for thermally and remotely actuated SMP. The composite ink viscosity

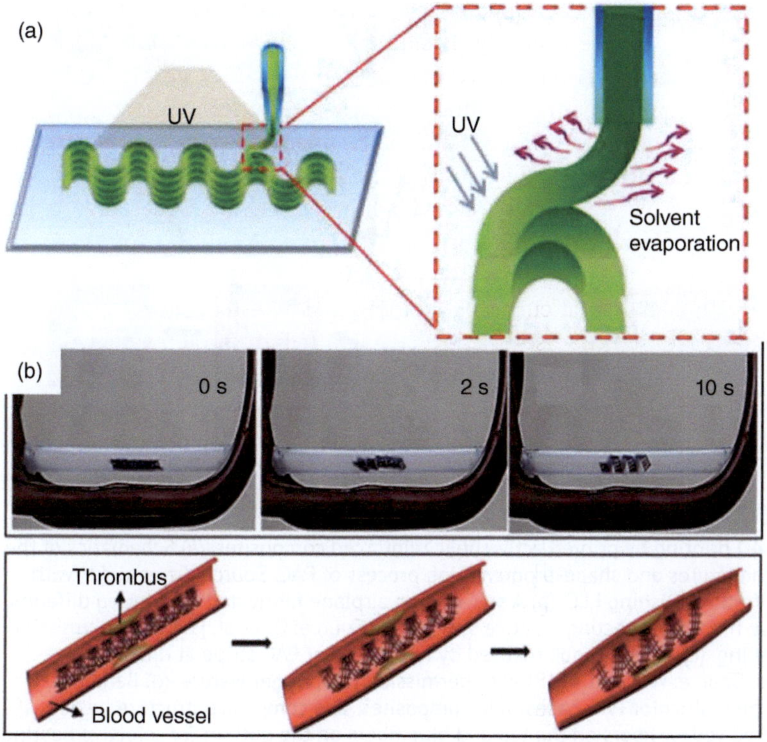

Figure 1.7 4D printing of SMP nanocomposites. (a) Schematics of extrusion of PLA/iron oxide composite ink with fast solvent evaporation and *in situ* UV crosslinking [145]. (b) Application demonstration of the 4D-printed scaffold as an intravascular stent. Source: Wei et al. [145]/with permission of American Chemical Society.

with added highly volatile solvent was extruded, followed by fast solvent removal and *in situ* UV curing (Figure 1.7a). The printed scaffolds with remote activation find potential biomedical applications as intravascular stents (Figure 1.7b). Similarly, Hua et al. [146] used FFF to print photoresponsive PLA/CNT SMP composites on paper substrates as photothermal-responsive actuators. It is also noted that, instead of utilizing nanoparticles, selectively coating colorful inks on SMPs can also realize effective photothermal heating for the activation of SMPs [117, 118].

1.3.1.3 Printed Active Fiber-Reinforced Composites

The SMP-based PACs have been extensively studied for 4D printing [2]. The bilayered PACs, fabricated by a PolyJet printer, consist of parallel glassy polymer fibers (i.e., VeroWhite) embedded in a rubbery matrix (i.e., TangoBlack). The glassy polymer fiber is an SMP with a T_g of 60 °C. After printing, the bilayer lamina was trained for shape change by heating, stretching, cooling, and releasing sequence. Finally, the composite laminates turn to curled shapes due to the strain mismatch between the rubbery matrix and the glassy polymer fibers. By regulating fiber architecture and orientation, complex 3D configurations can be obtained, including bent, coiled, twisted, and folded shapes. The SMP composites laminates were used

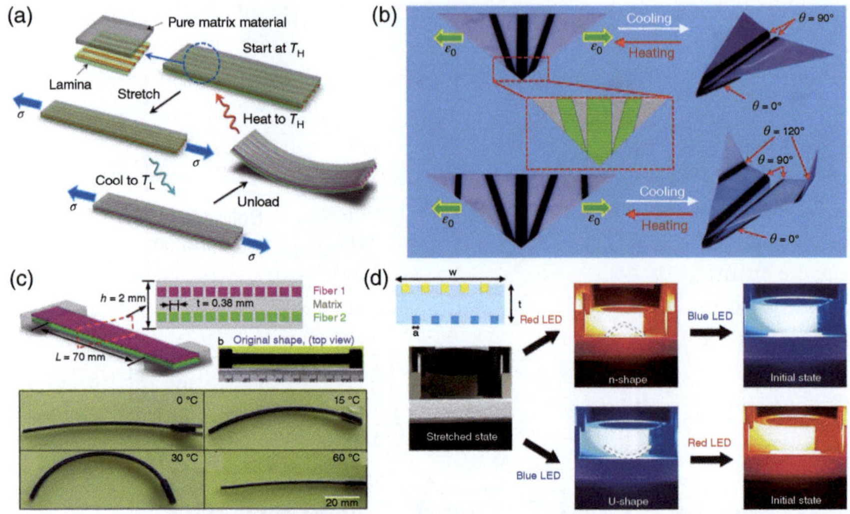

Figure 1.8 4D printing by printed active fiber-reinforced composites. (a) Schematics of the laminate architectures and shape-programming process of PAC. Source: Ge et al. [2]/with permission of AIP Publishing LLC. (b) A self-folding airplane fabricated by printing different PACs as active hinges connecting inactive stiff plates. Source: Qi et al. [17]/with permission of IOP Publishing. (c) A smart hook realized by multi-SME of PAC stripe at different temperatures. Source: Wu et al. [148]/with permission of Springer Nature. (d) Bending behavior of the multicolor PAC-based SMP composites. The composite structure has bias-distributed yellow fibers on the top and blue fibers on the bottom for selective heating under light with different wavelengths. Source: Jeong et al. [149]/with permission of Springer Nature.

as smart hinges for active origami, morphing 2D structures into 3D architectures [17, 147]. Figure 1.8b shows a 2D origami airplane with PAC-based smart hinges that can be morphed into a 3D-shaped plane [17].

Similarly, Wu et al. [148] reported the use of PACs SMP composites that shows multiple shapes change via digital SMPs. Digital SMPs with tunable T_g were printed by digital material printing on an Objet Connex printer. To do so, two materials, including a rigid plastic (VeroWhite) and a soft rubber (TangoBlack), were jetted at specific ratios, followed by UV curing to print digital materials with different T_g. With this method, SMP fibers with different T_g were printed into laminated composite structures to achieve multiple and sequential shape-shifting after a one-step shape programming (Figure 1.8c). In addition, PACs with proper arrangements can be used for metamaterial lattice, tubular grippers, and stents for self-expanding and self-shrinking shape changes [150, 151]. More recently, Jeong et al. reported 4D printing of photoresponsive PACs using selective light absorption/heating in multi-color PACs for remote-controlled shape-shifting [149]. The multi-material 3D printer (Stratasys, J750) was used to print yellow fibers (VeroYellow) and blue fibers (VeroCyan) in a rubber-like transparent material (Tango +), and the colored fibers were biasedly distributed on the top and bottom. The yellow fibers can absorb blue light, and the blue fiber absorbs red light for selective heating in the composites. As a result, the laminated composites bent to an *n*-shape under the red light illumination and then recovered to an initial flat state by blue light illumination. The illumination blue-red light sequence enabled the structure to bend to a U-shape and recover to the initial state (Figure 1.8d).

1.3.1.4 Bilayer SMPs

Bilayer SMP composites made of SMPs as an active layer and passive rubbery layers are simple and versatile designs for shape-shifting [152]. Besides, the bilayer can be used to realize direct 4D printing, where the residual stress can be incorporated during the printing or heating step without additional shape programming. Ding et al. [153] presented the direct 4D printing of bilayer laminate by an Objet 500 Connex3 printer (Stratasys). The bilayer strip was printed by VeroClear as the SMP and TangoBlack+ (or Tango+) as the elastomer. During printing, the elastomer layer contained a compressive strain regulated by printing parameters (such as the layer printing time). The release of the residual stress led to laminate bending into new permanent shapes (Figure 1.9a). Figure 1.9b shows some typical examples of direct 4D-printed structures. The lattice with an unsymmetrical bilayer design was printed to achieve both expanding and bending upon heating. For example, upon heating, a printed flat star-shaped laminated structure and multi-layer flower could be deployed into a 3D dome shape and a blossom configuration, respectively.

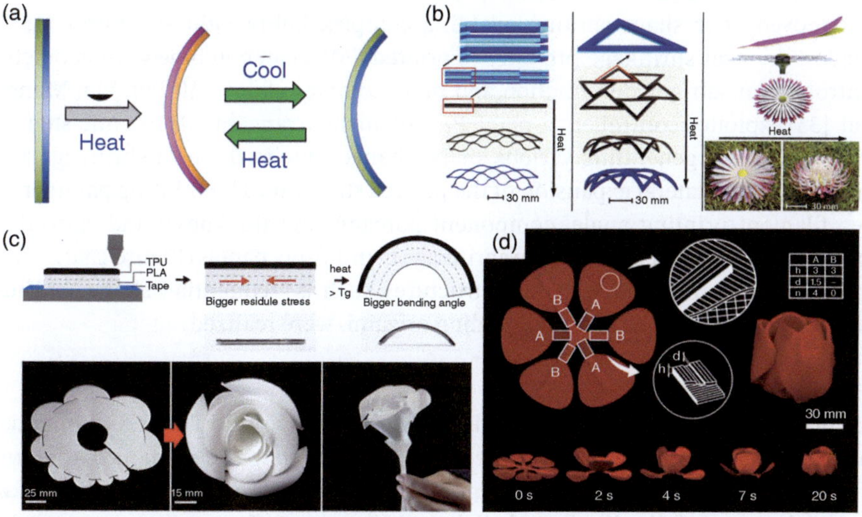

Figure 1.9 4D printing of SMP-based bilayer composites. (a) Schematics of direct 4D printing of multi-material bilayer composites. (b) Similar lattice structure with both expanding and bending upon heating with the unsymmetrical architecture. A flat star-shaped structure transforms into a 3D dome. Printed flower blooms into a blossom shape upon heating. Source: Ding et al. [153]/with permission of American Association for the Advancement of Science – AAAS. (c) FFF printing Thermorph material causes bending using an active PLA layer and passive TPU layer. A printed flat sheet self-folded into a 3D rose. Source: An et al. [154]/with permission of ACM SIGCHI. (d) FFF-printed PLA filaments with different orientations in layers can simultaneously shrink in length and thickness upon heating to a temperature above T_g. The two-step folding of the initially flat petals creates a tulip. Source: van Manen et al. [35]/from Royal Society of Chemistry/CC BY 3.0.

Ding et al. [155] further exploited the bilayer composite-based rods for direct 4D printing. Besides, the thermal expansion mismatch between the rubber and SMP layer in printed bilayer composites could also be harnessed for shape change [156]. The morphed bilayer structures were stable even after being cooled down. These examples suggested a great advantage of the 4D printing technique, which can tremendously reduce the printing time and material consumption up to 70–90% for thin-wall 3D structures.

Heat shrinkage of SMPs has also been utilized for direct 4D printing of bilayer composites without shape programming. The SMP chains (such as PLA) can be stretched and aligned along the direction of extrusion during the FFF printing process, storing internal residual stress in the polymer. Upon heating to a temperature above its T_g, the printed parts can show macroscopic contraction along the chain alignment direction due to residual stress release. For example, An et al. [154] reported the concept of Thermorph by FFF printing of bilayers consisting of an active PLA layer and a passive thermoplastic polyurethane (TPU) layer to form an actuator (Figure 1.9c). The printed 2D bilayer composites could self-fold into 3D configurations upon heating. With this Thermorph approach, complex geometries, including 3D Stanford Bunny and 3D rose, were designed, printed, and self-folded from 2D flat sheets. Zhang et al. [36] reported heat-triggered shape transformation of printed 2D lattice patterns. When the paper was used as the passive material, it could be removed after shape-shifting, yielding complex lightweight structures [157]. Besides, the heat-shrinkage property of printed SMPs can combine with geometry control for anisotropic deformation and more complex shape-shifting. Van Manen et al. [35] exploited controlled in-plane PLA filament arrangements in a one-step 3D printing process, generating various shape changes due to the heat shrinkage and the concomitant lateral expansion of the printed structure. The printing parameters (i.e., filament printing angle, component porosity, and thickness) and activation temperatures were used to tune the shrinkage of multi-ply panels (Figure 1.9d). With this method, complex shape-shifting structures, such as "sequential" shape-shifting, instability-driven pop-up, and self-folding origami, were realized.

1.3.1.5 Multi-material SMPs

3D printing of multi-material SMP is utilized for multiple and sequential shape-shifting more complex target shapes and shape evolution pathways. This was first achieved by using digital SMPs via PolyJet printing. Mao et al. [158] printed spatially distributed digital SMPs with different T_g to control the shape-shifting sequences in 3D-printed structures. Hinges composed of SMPs with higher T_g recovered slower at the same recovery temperature, leading to sequential folding. This was illustrated by self-closing and locking a box from a flat shape (Figure 1.10a). Liu et al. [160] also exploited the digital SMPs to realize programmable deployment of tensegrity structures. The shape-shifting structures find application potential as a deployable antenna, reflectors on satellites, and biomedical devices as a self-deployable stent.

Multi-material SMPs with location-specific properties can be printed by DLP with multiple switchable resin vats. Ge et al. [68] formulated different inks with tunable

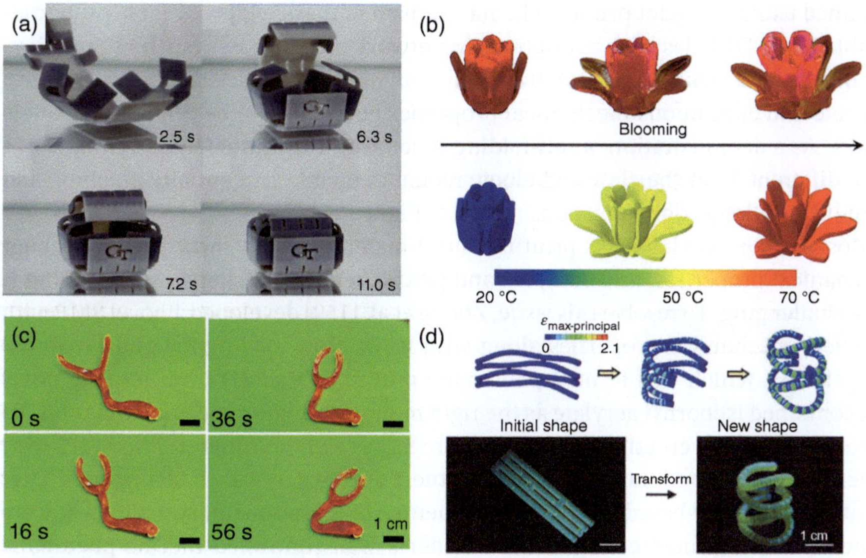

Figure 1.10 4D printing by multi-material SMPs. (a) Pictures showing the folding of a 3D-printed sheet to a box with self-locking. Source: Mao et al. [158]/with permission of Springer Nature. (b) The experiment and simulation of shape recovery of a PμSL-printed multi-material flower with different T_g for the two layers of petals. Source: Ge et al. [68]/ with permission of Springer Nature. (c) Multi-material SMP for artificial arms with sequential shape recovery by grayscale DLP printing. Scale bars: 1 cm. Source: Kuang et al. [89]/with permission of American Association for the Advancement of Science – AAAS. (d) Grayscale DLP-printed shape memory balloon-based planar raft consisting of three types of hoop sequences transformed into a 3D cylindrical tent-like structure actuated by pneumatic blowing. Scale bars: 10 mm. Source: Zhang et al. [159]/with permission of John Wiley & Sons, Inc.

T_g for multi-vat PμSL printing of SMP multi-materials. As shown in Figure 1.10b, a flower was printed with two layers of petals featuring different T_g. In the programmed temporary shape of the bud state, the outer petals with lower T_g opened first after heating to 50 °C, and the inner petals with higher T_g opened at 70 °C, which can be captured by corresponding finite element (FE) simulations.

Alternatively, multi-material SMPs can be printed by grayscale DLP using a single vat [89, 159]. Compared with regular one-step grayscale DLP printing with limited mechanical properties gradients, Kuang et al. [89] reported the grayscale DLP printing of digital materials with widely tunable mechanical properties and T_g using a customized photothermal two-stage curing ink. The hybrid resin comprises a diacrylate crosslinker, a dual-functional monomer of glycidyl methacrylate (GMA), and a diamine as a thermal crosslinker. The grayscale light patterns were used during the printing stage to print green parts with locally distinct degrees of reaction conversion associated with different crosslinking densities. After printing, the green parts were further subjected to thermal curing between different functional groups. This thermal curing substantially boosted the mechanical strength and enlarged the mechanical properties contrast (Young's modulus from 1.4 MPa –1.2 GPa). In addition, the T_g can be tuned from 14 to 68 °C, which resembles the digital SMPs obtained using a PolyJet printer. The mechanism of these widely tunable properties enabled by g-DLP lies in the variations in network architectures (such as crosslinking density and chain branches). By simply assigning grayscale levels by design, both discrete and continuous mechanical properties gradients can be printed in a single object. As a demonstration, a self-folding robot arm was printed by assigning SMPs with different T_g at the claw and elbow region to enable sequentially grabbing and then lifting a long stick in response to heat (Figure 1.10c).

Most of the previous 3D printing multi-material SMPs show relatively poor mechanical properties, and complex and precise temporary shape programming is very challenging. To resolve this issue, Zhang et al. [159] developed a novel SMP with excellent mechanical properties along with facile blow loading for multi-material 4D printing, which was termed pneumatic multi-material 4D printing. The novel ink contained isobornyl acrylate as the rigid monomer, high-molecular-weight AUD (Ebecryl 8413) as a crosslinker, and Irgacure 819 as a photoinitiator. Two SMPs with different T_g were modulated by light brightness during grayscale DLP printing. After printing, the SMP-based hollow components (or shape-memory balloons) were blown into predefined configurations under a well-controlled thermo-pneumatic condition. The shape actuation was performed at the temperature between the T_g of two SMPs so that the SMP with lower T_g was in a soft and rubbery state (Young's modulus: 0.37 MPa), while the SMP with higher T_g was still in the viscoelastic state (Young's modulus: 22.5 MPa). The significant modulus difference between these two SMPs was leveraged to realize selectively blowing of the printed parts into various shapes, which could be fixed after being cooled below the T_g of both SMPs. They showed the shape memory balloon-based planar raft transformed into a 3D cylindrical tent-like structure actuated by pneumatic blowing (Figure 1.10d).

1.3.2 Hydrogels and Composites for 4D Printing

Hydrogels are made of three-dimensional hydrophilic polymer networks that can absorb a large amount of water or fluid [161]. The hydrophilicity of hydrogels is attributed to the existence of hydrophilic functional groups (i.e., $-NH_2$, $-COOH$, $-OH$, $-CONH_2$, $-CONH-$, and $-SO_3H$). The water uptake and release contribute to the expansion or shrinkage of hydrogels, leading to the shape actuation. Hydrogels enable significant volume-phase transition triggered by physical and chemical stimuli (i.e., temperature, electricity, solvent composition, light, pH, and ions) [162]. Basically, the swelling equilibrium of hydrogel is determined by the crosslinking density, temperature, and interaction between the polymer chain and aqueous solution (Flory–Huggins solubility parameter). The mechanism of the stimuli-responsive reversible swelling/shrinkage of hydrogels lies in the change of polymer–water interactions. The external factors that alter the intermolecular interactions can influence the swelling state of hydrogels. Usually, single-material hydrogel shows simple expansion and contraction. To realize more complex and high-order deformation, alternating the stimulus field or using heterogeneous material structures is adopted to realize bending, twisting, and snap-buckling [163–167]. For this reason, the printing of hydrogel composites and multi-materials is mainly used for 4D printing. This section will summarize the advances of hydrogel composites and multi-material hydrogel structures for 4D printing. More progress in the 4D printing of hydrogels can be found in the review paper [162].

1.3.2.1 Single-Material Hydrogels and Composites

Stimuli-responsive smart hydrogels and composites have been printed using different 3D printing techniques for 4D printing. Among them, the poly(*N*-isopropyl acrylamide) (PNIPAm) and its copolymers have been widely studied as temperature-sensitive hydrogels, exhibiting large and reversible volume transition at the lowest critical solution temperature (LCST, $T_C \approx 32$–$35\,°C$) [168]. When increasing the temperature above T_c, the polymer network strands experience a coil–globule transition resulting in a large volume change due to the decreases in the water content [169]. Bakarich et al. [170] presented the DIW printing of ionic covalent entanglement hydrogels alginate/PNIPAm with enhanced mechanical properties and thermal actuation. A printed smart valve can realize the control of the water flow at 20 and 60 °C because the thermal-responsive hydrogel enabled the change in the size of the channel with temperature.

Polyacrylic acid (PAA) is another widely used hydrogel for 4D printing [171–173]. The presence of the carboxyl acid group enables multiple stimuli-reposing capability, including chemical, electric field, and pH [172]. Lee and coworkers [171] used PμSL to print PAA–PEG hydrogels that exhibited large deformation under electric field in an electrolyte medium (phosphate-buffered saline [PBS]). This shape-shifting was attributed to the difference in mobile ion concentrations between the inside and outside of the ionic gels. A printed human-like structure showed untethered and bidirectional locomotion under an electric field (Figure 1.11). Jin et al. [172] used TPP to print PAA-based structures with a sub-micrometer resolution for pH-triggered complex shape change.

Usually, a single hydrogel gives a limited shape deformation mode. Inspired by the complex shape change of some natural plants due to microstructural anisotropy of cell walls, composite hydrogels showing anisotropic swelling are also developed for 4D printing with complex shape-shifting behavior [174, 175]. To achieve anisotropic

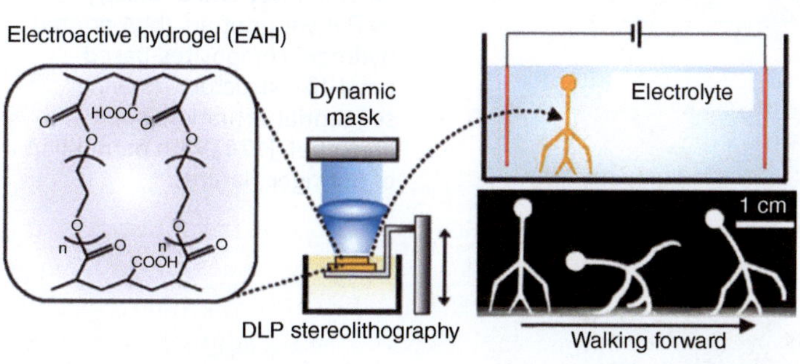

Figure 1.11 A typical example of 4D printing of single hydrogel and composite. PμSL printing PAAc-based electric active hydrogels for bidirectional locomotion human-like electroactive hydrogels (EAH) structure. Source: [171]/with permission of American Chemical Society.

swelling in hydrogel composites, micro-particulates/fillers are usually added and aligned by shearing forces [174] or external fields, such as an electric field or magnetic field [175]. For example, Gladman et al. [174] utilized DIW to print hydrogel composite using nanofibrillated cellulose (NFC) composite inks. NFC was aligned during the ink deposition process. By controlling printing pathways, the orientation of fibers in each layer could be encoded in the bilayer hydrogel structures. As a result, localized, anisotropic swelling behavior could be modulated to precisely tune the curvature of bilayer hydrogel composite structure for complex shape-shifting.

1.3.2.2 Multi-Material Hydrogels

Multi-material hydrogels with a swelling mismatch of different materials are the earliest and most extensively studied systems for complex shape change. The first 4D printing example by Tibbits is the printing of hinges composed of a rigid plastic base and a hydrogel layer using an Objet 260 Connex printer [4, 5]. The hydrogels expand upon immersion in water triggering local bending deformation. The folding angle of the hinges was precisely controlled by varying the distribution of the two materials, enabling controlled shape change. It was shown the 3D-printed multi-material strand gradually transformed into a three-dimensional cube in water with time. Using the same type of 3D printer, Mao et al. [176] further combined the SME of the SMP layer with the expansion of a hydrogel layer for 3D-printed reversible shape-changing components with enhanced mechanical properties. The multi-material printing enabled the fabrication of bilayer structures containing an SMP layer, elastomer columns, and hydrogel sandwiched between the two layers (Figure 1.12a). To allow water to flow in and out rapidly, tiny holes were positioned in the elastomer layer. When putting the structures in cold water, the deformation was constrained by the stiff SMP despite the hydrogels' swelling. Upon heating, the

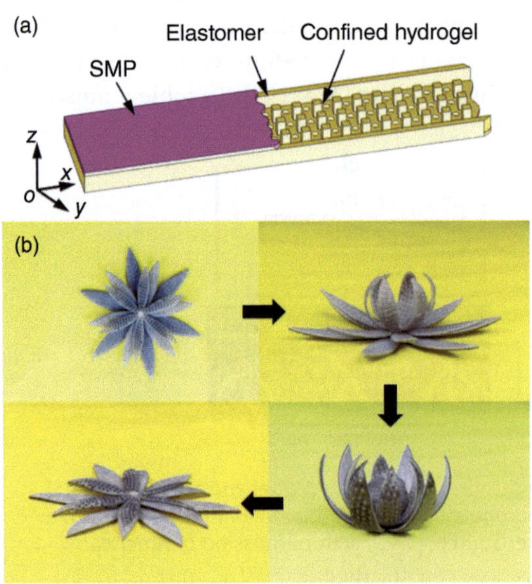

Figure 1.12 4D printing of multi-material hydrogels. (a) Schematics of 3D-printed hydrogel composite consisting of an SMP layer and a rubbery layer by PolyJet printing. (b) A printed hydrogel composites-based petal-like structure showing sequential actuation. Source: Mao et al. [176]/with permission of Springer Nature.

release of internal stress led to the structure bending due to the decreasing stiffness of SMP. The cooling and drying of the structure locked the bent shape, which recovered to the original shape after reheating. By adjusting the dimensions of the component and ratios of the two materials, the actuation speed can be tuned. Consequently, actuation of the printed flower with programmed sequences was achieved (Figure 1.12b).

DIW printing of multi-material hydrogels for 4D printing is also widely investigated. Naficy et al. [177] exploited dual-nozzle DIW printing of bilayer structures containing a thermoresponsive hydrogel (PNIPAm) and non-thermoresponsive hydrogels to realize bidirectional bending upon temperature change. Similarly, Liu et al. [178] used PNIPAm as an active layer and polyacrylamide (PAAm) as a passive layer for dual-nozzle DIW printing of biomimetic tube structures. Complex shape actuation, including radial expansion, bending, uniaxial elongation, and gripping, was realized using heat as a trigger.

DLP was used for rapid printing of multi-material hydrogels for 4D printing. Zhao et al. [166] patterned photomasks to fabricate bilayer structures with distinct physical properties and complex geometries. Hydrophilic poly(ethylene glycol) (PEGDA) was cured to form highly crosslinked hydrogels. A hydrophobic poly(propylene glycol) dimethacrylate (PPGDMA) layer was cured as a passive elastomeric layer. The resultant bilayer 2D hydrogel structures showed high actuation speed and large actuation force. In another work, Xie and coworkers [179] controlled the reaction conversion of multi-material hydrogels by adjusting the local exposure time. Shorter exposure times (or light dose) generated lower reaction conversions, leading to larger swelling ratios. With this approach, the ratio between the highest and lowest swelling ratios was as high as 40, allowing the fabrication of shapes to change with a sharp transition. Consequently, complex shape change with complex swelling patterns, such as a 3D cartoon face, was realized.

1.3.3 Liquid Crystal Elastomers

As soft-active materials, LCEs can reversibly actuate under external stimuli, such as heat or light [180–186]. LCEs comprise crosslinked polymer networks with covalently linked liquid crystals (or mesogens) in chain backbones or as side chains. Particularly, the actuation of main-chain thermal-responsive LCEs is achieved by the phase transition (nematic to isotropic) across the transition temperature (T_{NI}), which has been extensively investigated for 4D printing. By contrast, photoresponsive azobenzene-based LCEs are realized by photochemistry of *trans–cis* configuration transition under UV light illumination. The azobenzene-based LCE is less reported for 3D printing, partially attributed to the high cost of raw materials. In addition, thermal-responsive LCEs can be engineered with photo-actuation by leveraging the photothermal effect of nanoparticles [187, 188]. To facilitate large reversible shape changes, LCE mesogens are required to be aligned into a mono-domain state before use. Several alignment methods, including mechanical stretching [189–193], polarized light [186], shearing by extrusion [48–52], and magnetic fields [194], have been exploited to align the LCE mesogens. As DIW printing enables effective mesogen alignment by the material extrusion-induced shear, DIW printing is the most widely used printing technique for 4D printing of LCEs.

1.3.3.1 Single-Material LCEs

Based on the tremendous progress in the fabrication of LCEs using aza-Michael addition chemistry [186, 195, 196], LCE ink containing viscous oligomers was printed by Ware and coworkers for the first time [48]. They synthesized a main-chain LC macromer by the Michael addition of a liquid crystal monomer, 1,4-bis-[4-(6-acryloyloxhexyloxy)benzoyloxy]-2-methylbenzene (RM82), and a chain extender, *n*-butylamine (1.1 : 1 M ratio) at 75 °C for 12 hours. The obtained LCE ink with high viscosity (~8 Pa·s at 50 s^{-1} and 85 °C) and a T_{NI} of 105 °C was DIW-printed at 85 °C. At a temperature below T_{NI}, the mesogens in the nematic regime were aligned by the shearing force during extrusion. The subsequent UV curing allowed to lock the mesogen alignment. Upon heating/cooling, the printed LCE filament exhibited reversible contraction as high as 40% strain along the printing direction. By controlling the printing path, the mesogen alignment was locally programmed. Complex shape transformations, including volumetric contractions and repetitive snap-through transitions, were realized. Soon after, Kotikian et al. [50] exploited the same strategy for 4D printing of LCE actuators with spatially programmed nematic order in arbitrary form factors for more intricate shape manipulation. A main-chain LCE ink with a lower T_{NI} of 95 °C but higher viscosity was obtained by altering the ratio of RM82 and *n*-butylamine to 1 : 1. Therefore, much higher pressure was required to operate the LCE ink printing at 50 °C. Actuation contraction of 43.6% in the printing direction and expansion of 29.8% normal to printing direction were obtained. The printed LCE-based architectures showed reversible 2D-to-3D and 3D-to-3D shape transformations. However, the actuation temperature of the above LCEs was as high as 200 °C for full actuation, limiting their actual applications.

To this end, new LCE oligomer inks have also been prepared to reduce the printing temperature or tune the actuation temperature. Researchers used soft thiol-spacers to tune the actuation temperature and realized room temperature printing of LCE by modifying the synthesis chemistry [49, 197]. In a typical example, Roach et al. [49] reported an LCE oligomer with much lower T_{NI} (42 °C) for room temperature DIW printing. The new LC ink was synthesized between a diacrylate mesogen of RM 257 and a flexible dithiol spacer 2,2′-(ethylenedioxy) diethanethiol (1.15 : 1 M ratio) in acetone with a proper catalyst. Before use, the excess acetone in the LC oligomers was evaporated under a vacuum. The extruded LC ink with aligned mesogens was cured by UV light, yielding reversible contraction actuation strain of 48% between −20 and 75 °C. Likewise, Saed et al. [51] combined the aza-Michael addition and thiol–ene reaction scheme for LCE oligomers synthesis and crosslinking. By altering the difunctional thiol's chain length and stiffness as a chain spacer, tunable actuation temperature (T_{NI}) was achieved from 28 to 105 °C. The obtained LC inks could be 3D-printed into multi-material-based structures, exhibiting sequentially and reversibly shape changes.

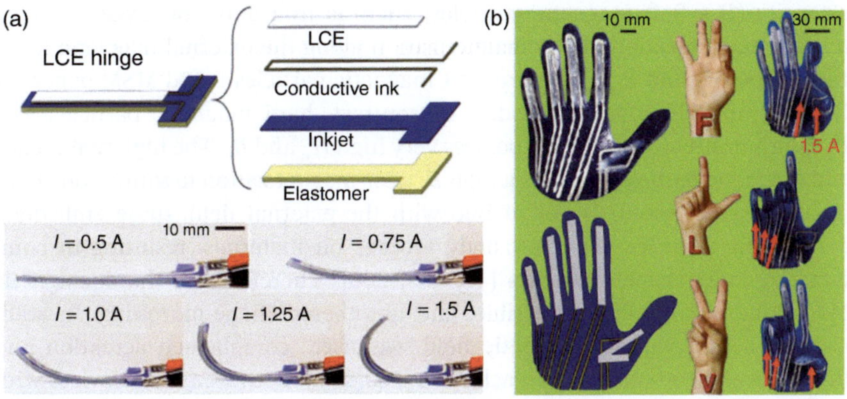

Figure 1.13 Liquid crystal elastomer (LCE)-based multi-material 4D printing. (a) Hybrid-printed multiple-layered LCE hinges connected to a power supply can be actuated to bend by Joule heating under various applied currents. (b) A fully 3D-printed LCE-based hand with individually controlled fingers to show the American sign language letters. Source: (a) and (b) Roach et al. [49]/with permission of IOP Publishing.

1.3.3.2 LCE-Based Multi-Materials

LCE-based multi-material structures containing active LCE and passive inert materials enable more complex shape change behavior. For example, Kotikian et al. [52] used DIW to print multi-LCE bilayers with orthogonal director alignment and locally different T_{NI}s to form active hinges in soft robotic structures. A self-twisting origami polyhedron and a soft robot with programmed folding sequences were realized. Roach et al. [49] also presented the printing of LCE-based multi-material electronics and metamaterials. As shown in Figure 1.13a, a rubbery substrate (1 mm) was first printed by DIW printing, and the coating layer (30 μm) was printed by inkjet printing to create a smooth surface. Then, conductive wire traces (1 mm wide, 20 μm thick) were printed and cured, followed by covering an LCE layer (0.5 mm thick) by DIW printing. When the current was applied, the conductive wire enabled Joule heating, leading to LCE contraction and bilayer bending due to the deformation constrain by the elastomer substrate. With this approach, a printed hand with individually controlled fingers could be realized, which could produce American sign language letters (Figure 1.13b). 4D-printed LCE-based structures and devices with reversible large-strain actuation, find applications as novel actuators and soft robotics [52, 198].

1.3.4 Magnetoactive Soft Materials

MSMs consisting of magnetic particles and a soft polymer matrix allow for rapid and large actuation in response to magnetic fields. The magnetically actuated shape-morphing derives from the interactions between the magnetic particles and the applied magnetic field. Two types of magnetic particles, i.e., soft and hard magnetic particles, are divided in terms of coercivity (H_c) and residual magnetic flux density (B_r). Soft magnetic particles, such as iron and iron oxide, develop prominent magnetization and demagnetization along the external magnetic fields owning to low H_c and B_r. Therefore, soft magnetic particles-based MSM generally exhibited simple deformation modes. In contrast, hard magnetic particles, i.e., neodymium-iron-boron (NdFeB), possess very high H_c and B_r. The high remanence of hard magnetic particles enables a high B_r after magnetization to saturation [199]. When the magnetization is not in line with the external field, these embedded hard magnetic particles can exert body torques on materials, resulting in complex macroscopic shape actuation [47, 199, 200]. Therefore, the hard magnetic MSMs enable remote, fast, reversible, and untethered shape-morphing. In addition, when a time-varying magnetic field was used, complicated actuation and multi-modes of locomotion can be achieved [200, 201]. The MSMs and devices with remote-controlled actuation can find application potentials in soft robotics, biomedical devices, and active metamaterials [47, 200, 202–207]. Recent progress in the properties and applications of MSMs can be found in these review papers [99, 100]. This section discusses the progress of 3D printing of MSMs using single-material and multi-materials design.

1.3.4.1 Single Magnetoactive Soft Material Composite

Early works in the printing of MSMs mainly involve using soft magnetic materials [208]. Several recent works reported the large and complex magnetic structures using DIW printing [45, 202]. For example, Roh et al. [202] presented 3D-printed architectures with programmed magneto-capillary reconfiguration. The magnetically driven soft actuators consisting of soft silicone and carbonyl iron particles could deform easily under magnetic fields and by lateral capillary forces. Hard magnetic soft materials were also printed to enable a more complex shape-shifting mode. Kim et al. developed the magnetic field-assisted DIW printing to pattern the local magnetization by applying a magnetic field on the printing nozzle [47]. The viscous ink consisting of commercial silicone rubber (polydimethylsiloxane [PDMS]), magnetized NdFeB particles, and fumed silica as a rheological modifier was used for extrusion printing. After printing and curing, complex 2D and 3D structures with programmed magnetization enabled fast and reversible shape actuation. For example, flat sheets can repeatedly actuate into origami and recover rapidly by applying and removal of a magnetic field.

Besides, photocuring-based fabrication has been used to manufacture MSMs in small sizes [209–211]. Xu et al. [209] used UV lithography and magnetic field to manipulate the orientation of NdFeB in planar materials. Millimeter-scale flexible robots with sophisticated magnetization patterns and intricate shapes can be fabricated. It was shown that the printed multi-legged paddle-crawling robots exhibited different bending angles under magnetic actuation. The alternating magnetization profile in each leg enabled the robot's locomotion in a microchannel. In another work, Peters et al. [210] printed degradable magnetic hydrogel composites as swimming microrobots by TPP printing. The obtained microdevices absorbed and released biologically relevant substances and degraded in aqueous solutions. The magnetically actuated microdevices can be used as *in vivo* noninvasive manual/surgical devices. Likewise, Lee et al. [211] utilized TPP to print size-controllable superparamagnetic hydrogel for the fabrication of microscrews and microrollers that could respond to diverse stimuli, including magnetic field, pH, temperature, and cations.

1.3.4.2 Multi-material MSMs

Multi-material 4D printing has been achieved to further enhance the functionality of MSMs. Li et al. [212] used switchable resin vat DLP printing for DLP printing of multi-materials and devices. A magnetic field actuated flexible gripper with integrated magnetic-responsive, and inert materials can be printed into a single part. The magnetic composite photopolymer ink contained Fe_3O_4 nanoparticles. A simple match-like actuator contains the top head printed by magnetic resin and a thin load-bearing piece. The dual-material DLP printing enables good surface quality and interfacial bonding for effective actuation. A printed flexible gripper could pick up and release a cotton ball in response to magnetic fields.

Recently, Ma et al. [43] used multi-material DIW printing to fabricate complex structures consisting of photocurable magnetic shape memory polymers (M-SMPs) and MSMs (Figure 1.14). The M-SMP is made of an AUD-based photopolymer and NdFeB microparticles, which combines the advantages of MSMs and SMPs [144]. Fumed silica nanoparticles as a rheological modifier were also added to impart

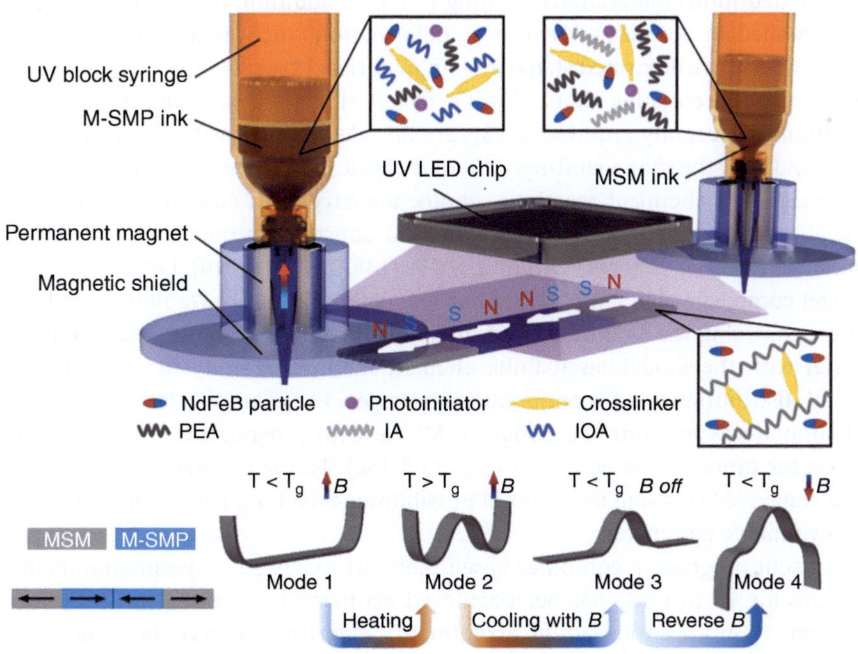

Figure 1.14 4D printing by MSMs-based multi-material. Schematic of the magnetic multi-material DIW system to print two magnetoactive materials with different glass transition temperatures. Cooperatively controlling the temperature and magnetic field led to different deformation modes. Source: Ma et al. [43]/with permission of American Chemical Society.

shearing shining for extrusion. To make MSM with low glass transition temperature, a rigid monomer (isobornyl acrylate) was replaced by a flexible monomer (isodecyl acrylate). 2D lattice structures exhibited multimodal shape actuation and tunable physical properties through cooperative heating and magnetic field control. For example, only the MSM parts (T_g below room temperature) could be actuated at room temperature by the magnetic field, and the stiff M-SMP parts could be kept undeformed state. Upon heating above T_g, both materials were activated for the magnetic actuation. In addition, the M-SMP parts could lock the deformed shape after cooling down, which was stable without magnetic field. The locked shape could still recover upon heating due to the SME of the M-SMP.

1.4 Modeling-Guided Design for 4D Printing

Mechanics modeling and multiphysics simulation-based approaches play a vital role in predicting the shape-shifting and guiding the design in 4D printing. For bilayer structure-based 4D printing involving simple expansion and shrinkage, the beam theory and its modified models have been extensively employed to predict the shape-shifting [102, 103, 166, 213]. For example, Boley et al. reported the precise and complex shape manipulation of multiplex bilayer lattices via the mechanics model-guided multi-material 4D printing [103]. In addition, the beam theory has been extended to voxelized 3D object's deformation, such as a lattice frame, to capture the object's deformed shape in bulk materials [214].

To predict the complex shape-shifting in smart materials, multiphysics-based simulations are usually required to capture all relevant physical processes before, during, and after the shape-shifting. Different physical processes, including thermal, electromagnetic, chemical reactions, photochemistry, or a combination of these physical processes, may be involved. Different constitutive models of smart programmable material systems, including SMPs [148, 153, 215–219], LCEs [220–222], hydrogel composites [165, 223–225], and magnetic materials [47, 199], have been established. Using the well-developed materials models, a designer can design the material with the model-based finite element method to study the influence of material distributions and programming strategies. For example, Zhao et al. [199] used simulations to guide the design of MSMs with programmed magnetization patterns for target shape actuation (Figure 1.15a). However, this intuition-based design can be tedious, and large sets of possibilities exist for a particular design due to many tunable parameters.

The recent progress in computer-based methods to design shape-programmable structures for 4D printing has been achieved via machine learning (ML). Genetic algorithms (GAs), also known as evolutionary algorithms (EAs), are typical ML methods. EA is population-based gradient-free optimization, which is well suitable for searching problems-based optimization. For the design problem in 4D printing, EA can be used to search the complex design of active material distribution. Based on a predefined criterion, the ideal material distribution can be found by selecting the best-performing designs to continue reproducing and discarding the most inferior performing designs in an iteration manner. For example, Sossou et al. [228]

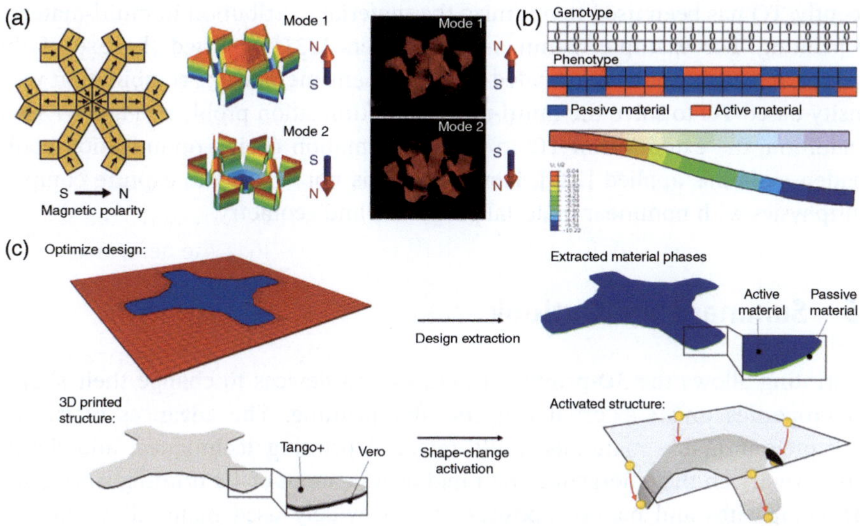

Figure 1.15 Simulation-based design for 4D printing. (a) An FE-based model-guided design of shape-shifting structures with programmed patterns of ferromagnetic domains for 3D-printed MSMs [199]. (b) Schematic illustration of genetic algorithm representing active materials (red) and passive materials (blue) in the phenotype to realize a target deformation [226]. (c) A level-set topology optimization approach for 4D printing using digital materials. Source: Geiss et al. [227]/with permission of American Society of Mechanical Engineers (ASME).

reported an EA-based modeling framework to simulate the materials in a voxelized manner. The initial shape was first voxelized and meshed with a chosen resolution. Secondly, the voxelized source shape and target deformation mode were compared to get the necessary underlying frame nodes' rotational degrees of freedom for the target shape. Lastly, the inverse design problem was solved by MATLAB toolbox, yielding the optimal material distribution design. The simulations with reasonable simulation speed can be used to predict the shape change of complex 3D structures. In another work, Hamel et al. [226] utilized the EA to guide the active material composite design in 4D printing. EA was combined with the finite element method to solve multi-material design optimization problems for target shape-shifting. To do so, the discrete passive or active materials were voxelized. As indicated by the "phenotype," their distribution was optimized to reach predetermined configuration by multiple-generation iterations (Figure 1.15b). When the active composite configuration matched the target shape as the best results, the corresponding material distribution indicated the optimized active material design. More recently, Wu et al. [46] reported EA-assisted voxel-encoding for DIW printing to control both the magnetic density and polarity direction distributions in MSMs. The simulation-guided design for 4D printing of MSMs enabled complicated and on-demand shape-morphing.

Besides, topology optimization (TO) provides another robust tool for material distribution optimization in 4D printing design. In a TO process, a continuous function is used to represent the distribution of discrete material [229, 230].

Recently, TO has been used to optimize the material distribution in multi-material 4D printing. For example, Dunn and coworkers [231] reported the use of the level set approach and the extended finite element method in combination with density-based TO to solve the multi-material optimization problem (Figure 1.15c). In addition, the extension of TO to large-deformation design optimization could broaden materials applied [227]. However, TO is not suitable to capture complex multiphysics with nonlinear material properties and geometry.

1.5 Summary and Outlook

4D printing allows the 3D-printed structures and devices to change their shapes and properties under external triggers after printing. The advances of various shape-programmable materials, multi-material printing techniques, and design tools have driven the emergence and rapid development of 4D printing since 2013. SMP composites and nanocomposites are the widely used material systems for 4D printing. Multi-material SMPs further expand the shape-shifting mode and complexity. However, 4D-printed SMPs usually showed irreversible shape-shifting. By contrast, hydrogel composites with stimuli-responsive swelling and contraction enable reversible shape-shifting by the water uptake/release. The major limitations of hydrogels are their slow actuation due to the diffusion-induced volume change and the weak mechanical properties. LCEs exhibit reversible actuation and enhanced mechanical properties compared with hydrogels. MSMs are emerging smart materials with remote-controlled, untethered, and fast actuation. 4D printing of SMPs, hydrogel, and MSMs is compatible with various single-material and multi-material printing modalities. However, 4D printing of LCEs requires the mesogens to be aligned, which is currently limited to extrusion-based printing. 4D printing using smart materials exhibits several unique features complementing advanced manufacturing and also enables functional properties. Therefore, 4D printing has become one of the fasting-growing techniques, attracting broad attention to multidiscipline.

For future materials in 4D printing, printable smart materials with robust physical properties for extreme conditions, multi-stimuli-responsive, sensing-actuation-integrated intelligence, and excellent bioactivity can be developed to satisfy different applications. In aerospace, 4D-printed programmable materials usually possess large sizes, outstanding mechanical/thermomechanical properties, long-service lifespans, and tolerance to extreme space conditions. To this end, mechanically robust SMPs exhibiting high service temperature and excellent environmental stability are desired. One interesting trend in material design is the integrated multifunctions of smart programmable materials. For instance, the common elastomeric matrix of MSMs can be replaced by SMPs, hydrogels, or LCEs, which can expand the functional properties of MSMs [144, 232]. Another emerging functional programmable material is vitrimer (or covalent adaptive networks), which is capable of rearranging the network via dynamic chemical reactions [232–236]. The stimuli-induced plasticity enables sustainable applications via repeatable programmability, recycling, and self-healing for 4D printing [237–241].

More robust design methods, particularly the simulation-based design tools, can empower the design for 4D printing. The biomimetic concept can provide great inspiration for the design in 4D printing [242–244]. In terms of simulation-guided design, ML offers a powerful tool to design structures with multiphysics involving geometric and material nonlinearity that is usually challenging through traditional approaches. The simulation/modeling approaches also provide a nonintuitive solution to inverse design problems [201, 214]. In addition, TO has been used to optimize the design of smart structures with reduced weight.

Lastly, the rapid development of new 3D printing techniques will no doubt advance the 4D printing technique. Multiscale-geometry, high-speed, and multimaterial printing will play a critical role in future 4D printing techniques. Recent progress of volumetric 3D printing can tremendously improve the printing speed [245–248], while high-area rapid printing can increase the printing size enormously [249]. In addition, external fields (such as electric fields and magnetic fields) have been combined with 3D printing, allowing the fabrication of anisotropic stimuli-responsive smart materials for 4D printing [250, 251]. With this interdisciplinary research progress, 4D printing can transform into a practical technique, finding applications in broad fields such as soft robots, aerospace, biomedical devices, and more.

Acknowledgments

H.J.Q. acknowledges the support of AFOSR grants (FA9550-19-1-0151 and FA9550-20-1-0306; Dr. B.-L. “Les” Lee, Program Manager) and a gift fund from HP, Inc.

References

1 Tibbits, S. (2013). The emergence of “4D printing”. https://www.ted.com/talks/skylar_tibbits_the_emergence_of_4d_printing (accessed 26 June 2022).
2 Ge, Q., Qi, H.J., and Dunn, M.L. (2013). Active materials by four-dimension printing. *Applied Physics Letters* 103 (13): 131901.
3 Momeni, F., Mehdi Hassani, S.M., Liu, X. et al. (2017). A review of 4D printing. *Materials & Design* 122: 42–79.
4 Tibbits, S. (2014). 4D printing: multi-material shape change. *Architectural Design* 84 (1): 116–121.
5 Raviv, D., Zhao, W., McKnelly, C. et al. (2014). Active printed materials for complex self-evolving deformations. *Scientific Reports* 4: 7422.
6 Pei, E. (2014). 4D printing: dawn of an emerging technology cycle. *Assembly Automation* 34 (4): 310–314.
7 Kuang, X., Roach, D.J., Wu, J. et al. (2018). Advances in 4D printing: materials and applications. *Advanced Functional Materials* 29 (2): 1805290.
8 Choi, J., Kwon, O.C., Jo, W. et al. (2015). 4D printing technology: a review. *3D Printing and Additive Manufacturing* 2 (4): 159–167.

9 Wu, J.-J., Huang, L.-M., Zhao, Q. et al. (2017). 4D printing: history and recent progress. *Chinese Journal of Polymer Science* 36 (5): 563–575.

10 Shin, D.-G., Kim, T.-H., and Kim, D.-E. (2017). Review of 4D printing materials and their properties. *International Journal of Precision Engineering and Manufacturing-Green Technology* 4 (3): 349–357.

11 Khoo, Z.X., Teoh, J.E.M., Liu, Y. et al. (2015). 3D printing of smart materials: a review on recent progresses in 4D printing. *Virtual and Physical Prototyping* 10 (3): 103–122.

12 Stansbury, J.W. and Idacavage, M.J. (2016). 3D printing with polymers: challenges among expanding options and opportunities. *Dental Materials* 32 (1): 54–64.

13 Huang, S.H., Liu, P., Mokasdar, A. et al. (2013). Additive manufacturing and its societal impact: a literature review. *The International Journal of Advanced Manufacturing Technology* 67 (5): 1191–1203.

14 Ma, J., Franco, B., Tapia, G. et al. (2017). Spatial control of functional response in 4D-printed active metallic structures. *Scientific Reports* 7: 46707.

15 Bargardi, F.L., Le Ferrand, H., Libanori, R. et al. (2016). Bio-inspired self-shaping ceramics. *Nature Communications* 7: 13912.

16 Liu, G., Zhao, Y., Wu, G. et al. (2018). Origami and 4D printing of elastomer-derived ceramic structures. *Science Advances* 4 (8).

17 Qi, G., Conner, K.D., Qi, H.J. et al. (2014). Active origami by 4D printing. *Smart Materials and Structures* 23 (9): 094007.

18 Lui, Y.S., Sow, W.T., Tan, L.P. et al. (2019). 4D printing and stimuli-responsive materials in biomedical aspects. *Acta Biomaterialia*.

19 Li, Y.-C. et al. (2016). 4D bioprinting: the next-generation technology for biofabrication enabled by stimuli-responsive materials. *Biofabrication* 9 (1): 012001.

20 Zhang, Z., Demir, K.G., and Gu, G.X. (2019). Developments in 4D-printing: a review on current smart materials, technologies, and applications. *International Journal of Smart and Nano Materials* 10 (3): 205–224.

21 Zolfagharian, A., Kaynak, A., and Kouzani, A. (2020). Closed-loop 4D-printed soft robots. *Materials & Design* 188: 108411.

22 Pei, E. and Loh, G.H. (2018). Technological considerations for 4D printing: an overview. *Progress in Additive Manufacturing* 3 (1): 95–107.

23 Gao, B., Yang, Q., Zhao, X. et al. (2016). 4D bioprinting for biomedical applications. *Trends in Biotechnology* 34 (9): 746–756.

24 Montgomery, S.M., Kuang, X., Armstrong, C.D. et al. (2020). Recent advances in additive manufacturing of active mechanical metamaterials. *Current Opinion in Solid State and Materials Science* 24 (5): 100869.

25 González-Henríquez, C.M., Sarabia-Vallejos, M.A., and Rodriguez-Hernandez, J. (2019). Polymers for additive manufacturing and 4D-printing: materials, methodologies, and biomedical applications. *Progress in Polymer Science* 94: 57–116.

26 Ryan, K.R., Down, M.P., and Banks, C.E. (2020). Future of additive manufacturing: overview of 4D and 3D printed smart and advanced materials and their applications. *Chemical Engineering Journal* 126162.

27 Peng, B., Yang, Y., and Cavicchi, K.A. (2020). Sequential shapeshifting 4D printing: programming the pathway of multi-shape transformation by 3D printing stimuli-responsive polymers. *Multifunctional Materials* 3 (4): 042002.

28 Shen, B., Erol, O., Fang, L. et al. (2020). Programming the time into 3D printing: current advances and future directions in 4D printing. *Multifunctional Materials* 3 (1): 012001.

29 Spiegel, C.A., Hippler, M., Münchinger, A. et al. 4D printing at the microscale. *Advanced Functional Materials* 30 (26): 1907615.

30 International Organization for Standardization/ASTM International (2015). *Additive manufacturing — General principles — Terminology (ISO/ASTM 52900).*

31 Kuang, X. (2021). Chapter 9 – Introduction to 4D printing: methodologies and materials. In: *Additive Manufacturing* (ed. J. Pou, A. Riveiro and J.P. Davim), 303–342. Elsevier.

32 Ligon, S.C., Liska, R., Stampfl, J. et al. (2017). Polymers for 3D printing and customized additive manufacturing. *Chemical Reviews* 117 (15): 10212–10290.

33 Mukherjee, T., Zhang, W., and DebRoy, T. (2017). An improved prediction of residual stresses and distortion in additive manufacturing. *Computational Materials Science* 126: 360–372.

34 Jia, D., Li, F., and Zhang, Y. (2020). 3D-printing process design of lattice compressor impeller based on residual stress and deformation. *Scientific Reports* 10 (1): 600.

35 van Manen, T., Janbaz, S., and Zadpoor, A.A. (2017). Programming 2D/3D shape-shifting with hobbyist 3D printers. *Materials Horizons* 4 (6): 1064–1069.

36 Zhang, Q., Yan, D., Zhang, K. et al. (2015). Pattern transformation of heat-shrinkable polymer by three-dimensional (3D) printing technique. *Scientific Reports* 5: 8936.

37 Tao, Y., J. Gu, B. An, et al. (2018). Demonstrating thermorph: democratizing 4D printing of self-folding materials and interfaces. In: *Extended Abstracts of the 2018 CHI Conference on Human Factors in Computing Systems*. ACM.

38 Hansen, C.J., Saksena, R., Kolesky, D.B. et al. (2013). High-throughput printing via microvascular multinozzle arrays. *Advanced Materials* 25 (1): 96–102.

39 Chen, K., Kuang, X., Li, V. et al. (2018). Fabrication of tough epoxy with shape memory effects by UV-assisted direct-ink write printing. *Soft Matter* 14 (10): 1879–1886.

40 Compton, B.G. and Lewis, J.A. (2014). 3D-printing of lightweight cellular composites. *Advanced Materials* 26 (34): 5930–5935.

41 Invernizzi, M., Natale, G., Levi, M. et al. (2016). UV-assisted 3D printing of glass and carbon fiber-reinforced dual-cure polymer composites. *Materials* 9 (7): 583.

42 Lebel, L.L., Aissa, B., Khakani, M.A.E. et al. (2010). Ultraviolet-assisted direct-write fabrication of carbon nanotube/polymer nanocomposite microcoils. In: *Advanced Materials*, 592–596. Wiley-VCH Verlag.

43 Ma, C., Wu, S., Ze, Q. et al. (2020). Magnetic multimaterial printing for multimodal shape transformation with tunable properties and shiftable mechanical Behaviors. *ACS Applied Materials & Interfaces* 13 (11): 12639–12648.

44 Zhang, Y., Wang, Q., Yi, S. et al. (2021). 4D printing of magnetoactive soft materials for on-demand magnetic actuation transformation. *ACS Applied Materials & Interfaces* 13 (3): 4174–4184.

45 Zhu, P., Yang, W., Wang, R. et al. (2018). 4D printing of complex structures with a fast response time to magnetic stimulus. *ACS Applied Materials & Interfaces* 10 (42): 36435–36442.

46 Wu, S., Hamel, C.M., Ze, Q. et al. (2020). Evolutionary algorithm-guided voxel-encoding printing of functional hard-magnetic soft active materials. *Advanced Intelligent Systems* 2 (8): 2000060.

47 Kim, Y., Yuk, H., Zhao, R. et al. (2018). Printing ferromagnetic domains for untethered fast-transforming soft materials. *Nature* 558 (7709): 274–279.

48 Ambulo, C.P., Burroughs, J.J., Boothby, J.M. et al. (2017). Four-dimensional printing of liquid crystal elastomers. *ACS Applied Materials & Interfaces* 9 (42): 37332–37339.

49 Roach, D.J., Kuang, X., Yuan, C. et al. (2018). Novel ink for ambient condition printing of liquid crystal elastomers for 4D printing. *Smart Materials and Structures* 27 (12): 125011.

50 Kotikian, A., Truby, R.L., Boley, J.W. et al. (2018). 3D printing of liquid crystal elastomeric actuators with spatially programed nematic order. *Advanced Materials* 30 (10): 1706164.

51 Saed, M.O., Ambulo, C.P., Kim, H. et al. (2019). Molecularly-engineered, 4D-printed liquid crystal elastomer actuators. *Advanced Functional Materials* 29 (3): 1806412.

52 Kotikian, A., McMahan, C., Davidson, E.C. et al. (2019). Untethered soft robotic matter with passive control of shape morphing and propulsion. *Science Robotics* 4: eaax7044.

53 Wang, Z., Wang, Z., Zheng, Y. et al. (2020). Three-dimensional printing of functionally graded liquid crystal elastomer. *Science Advances* 6 (39): eabc0034.

54 Gao, X., Qi, S., Kuang, X. et al. (2020). Fused filament fabrication of polymer materials: a review of interlayer bond. *Additive Manufacturing* 101658.

55 de Gans, B.J., Duineveld, P.C., and Schubert, U.S. (2004). Inkjet printing of polymers: state of the art and future developments. *Advanced Materials* 16 (3): 203–213.

56 Cazón, A., Morer, P., and Matey, L. (2014). PolyJet technology for product prototyping: tensile strength and surface roughness properties. *Proceedings of the Institution of Mechanical Engineers, Part B: Journal of Engineering Manufacture* 228 (12): 1664–1675.

57 Barclift, M.W. and Williams, C.B. (2012). Examining variability in the mechanical properties of parts manufactured via polyjet direct 3D printing.

In: *International Solid Freeform Fabrication Symposium*. University of Texas at Austin Austin, Texas.

58 Jang, D., Kim, D., and Moon, J. (2009). Influence of fluid physical properties on ink-jet printability. *Langmuir* 25 (5): 2629–2635.

59 Sun, C., Fang, N., Wu, D. et al. (2005). Projection micro-stereolithography using digital micro-mirror dynamic mask. *Sensors and Actuators A: Physical* 121 (1): 113–120.

60 Janusziewicz, R., Tumbleston, J.R., Quintanilla, A.L. et al. (2016). Layerless fabrication with continuous liquid interface production. *Proceedings of the National Academy of Sciences of the United States of America* 113 (42): 201605271.

61 Tumbleston, J.R., Shirvanyants, D., Ermoshkin, N. et al. (2015). Continuous liquid interface production of 3D objects. *Science* 347 (6228): 1349–1352.

62 Qi, G., Zhiqin, L., Zhaolong, W. et al. (2020). Projection micro stereolithography based 3D printing and its applications. *International Journal of Extreme Manufacturing*.

63 Mao, M., He, J., Li, X. et al. (2017). The emerging Frontiers and applications of high-resolution 3D printing. *Micromachines* 8 (4): 113.

64 Rafiee, M., Farahani, R.D., and Therriault, D. (2020). Multi-material 3D and 4D printing: a survey. *Advanced Science* 7 (12): 1902307.

65 Zhang, Y.-F., Zhang, N., Hingorani, H. et al. (2019). Fast-response, stiffness-tunable soft actuator by hybrid multimaterial 3D printing. *Advanced Functional Materials* 29 (15): 1806698.

66 Kokkinis, D., Bouville, F., and Studart, A.R. (2018). 3D printing of materials with tunable failure via bioinspired mechanical gradients. *Advanced Materials* 30 (19): 1705808-n/a.

67 Skylar-Scott, M.A., Mueller, J., Visser, C.W. et al. (2019). Voxelated soft matter via multimaterial multinozzle 3D printing. *Nature* 575 (7782): 330–335.

68 Ge, Q., Sakhaei, A.H., Lee, H. et al. (2016). Multimaterial 4D printing with tailorable shape memory polymers. *Scientific Reports* 6: 31110.

69 MacDonald, E. and Wicker, R. (2016). Multiprocess 3D printing for increasing component functionality. *Science* 353 (6307).

70 Bakarich, S.E., Gorkin, R. III, Gately, R. et al. (2017). 3D printing of tough hydrogel composites with spatially varying materials properties. *Additive Manufacturing* 14: 24–30.

71 Li, V.C.-F., Kuang, X., Hamel, C.M. et al. (2019). Cellulose nanocrystals support material for 3D printing complexly shaped structures via multi-materials-multi-methods printing. *Additive Manufacturing* 28: 14–22.

72 Espalin, D., Ramirez, J.A., Medina, F. et al. (2014). Multi-material, multi-technology FDM: exploring build process variations. *Rapid Prototyping Journal* 20 (3): 236–244.

73 Xu, C., Quinn, B., Lebel, L.L. et al. (2019). Multi-material direct ink writing (DIW) for complex 3D metallic structures with removable supports. *ACS Applied Materials & Interfaces* 11 (8): 8499–8506.

74 Hardin, J.O., Ober, T.J., Valentine, A.D. et al. (2015). Microfluidic printheads for multimaterial 3D printing of viscoelastic inks. *Advanced Materials* 27 (21): 3279–3284.

75 Ren, L., Song, Z., Liu, H. et al. (2018). 3D printing of materials with spatially non-linearly varying properties. *Materials & Design* 156: 470–479.

76 Miriyev, A., Xia, B., Joseph, J.C. et al. (2019). Additive manufacturing of silicone composites for soft actuation. *3D Printing and Additive Manufacturing* 6 (6): 309–318.

77 Ortega, J.M., Golobic, M., Sain, J.D. et al. (2019). Active mixing of disparate inks for multimaterial 3D printing. *Advanced Materials Technologies* 4 (7): 1800717.

78 Chen, K., Zhang, L., Kuang, X. et al. (2019. **0**(0)). Dynamic photomask-assisted direct ink writing multimaterial for multilevel triboelectric nanogenerator. *Advanced Functional Materials* 1903568.

79 Jochen, M., Diana, C., Manuel, S. et al. (2017). Mechanical properties of interfaces in inkjet 3D printed single- and multi-material parts. *3D Printing and Additive Manufacturing* 4 (4): 193–199.

80 Choi, J.-W., MacDonald, E., and Wicker, R. (2010). Multi-material microstereolithography. *The International Journal of Advanced Manufacturing Technology* 49 (5-8): 543–551.

81 Kowsari, K., Akbari, S., Wang, D. et al. (2018). High-efficiency high-resolution multimaterial fabrication for digital light processing-based three-dimensional printing. *3D Printing and Additive Manufacturing* 5 (3): 185–193.

82 Ge, Q., Chen, Z., Cheng, J. et al. (2021). 3D printing of highly stretchable hydrogel with diverse UV curable polymers. *Science Advances* 7 (2): eaba4261.

83 Dolinski, N.D., Page, Z.A., Callaway, E.B. et al. (2018). Solution mask liquid lithography (SMaLL) for one-step, multimaterial 3D printing. *Advanced Materials* 30 (31): 1800364.

84 Schwartz, J.J. and Boydston, A.J. (2019). Multimaterial actinic spatial control 3D and 4D printing. *Nature Communications* 10 (1): 791.

85 Peterson, G.I., Schwartz, J.J., Zhang, D. et al. (2016). Production of materials with spatially-controlled cross-link density via vat photopolymerization. *ACS Applied Materials & Interfaces* 8 (42): 29037–29043.

86 Na, J.-H., Bende, N.P., Bae, J. et al. (2016). Grayscale gel lithography for programmed buckling of non-Euclidean hydrogel plates. *Soft Matter* 12 (22): 4985–4990.

87 Zhao, Z., Wu, J., Mu, X. et al. (2017). Origami by frontal photopolymerization. *Science Advances* 3 (4): e1602326.

88 Jiangtao, W., Zeang, Z., Xiao, K. et al. (2018). Reversible shape change structures by grayscale pattern 4D printing. *Multifunctional Materials* 1 (1): 015002.

89 Kuang, X., Wu, J., Chen, K. et al. (2019). Grayscale digital light processing 3D printing for highly functionally graded materials. *Science Advances* 5 (5): eaav5790.

90 Medina, F., A. Lopes, A. Inamdar, et al. (2005). *Hybrid manufacturing: integrating direct-write and stereolithography. Proceedings of the 2005 solid freeform fabrication*, p. 39–49.

91 Peng, X., Kuang, X., Roach, D.J. et al. (2021). Integrating digital light processing with direct ink writing for hybrid 3D printing of functional structures and devices. *Additive Manufacturing* 40: 101911.

92 Roach, D.J., Hamel, C.M., Dunn, C.K. et al. (2019). The m4 3D printer: a multi-material multi-method additive manufacturing platform for future 3D printed structures. *Additive Manufacturing* 29: 100819.

93 Kuang, X., Roach, D.J., Hamel, C.M. et al. (2020). Materials, design, and fabrication of shape programmable polymers. *Multifunctional Materials* 3 (3).

94 Hu, J., Zhu, Y., Huang, H. et al. (2012). Recent advances in shape–memory polymers: structure, mechanism, functionality, modeling and applications. *Progress in Polymer Science* 37 (12): 1720–1763.

95 Xie, H., Yang, K.-K., and Wang, Y.-Z. (2019). Photo-cross-linking: a powerful and versatile strategy to develop shape-memory polymers. *Progress in Polymer Science* 95: 32–64.

96 Erol, O., Pantula, A., Liu, W. et al. (2019). Transformer hydrogels: a review. *Advanced Materials Technologies* 4 (4): 1900043.

97 Ula, S.W., Traugutt, N.A., Volpe, R.H. et al. (2018). Liquid crystal elastomers: an introduction and review of emerging technologies. *Liquid Crystals Reviews* 6 (1): 78–107.

98 White, T.J. and Broer, D.J. (2015). Programmable and adaptive mechanics with liquid crystal polymer networks and elastomers. *Nature Materials* 14 (11): 1087–1098.

99 Nguyen, V.Q., Ahmed, A.S., and Ramanujan, R.V. (2012). Morphing soft magnetic composites. *Advanced Materials* 24 (30): 4041–4054.

100 Wu, S., Hu, W., Ze, Q. et al. (2020). Multifunctional magnetic soft composites: a review. *Multifunctional Materials* 3 (4): 042003.

101 Zhang, Q., Kuang, X., Weng, S. et al. (2020). Rapid volatilization induced mechanically robust shape-morphing structures toward 4D printing. *ACS Applied Materials & Interfaces* 12 (15): 17979–17987.

102 Weng, S., Kuang, X., Zhang, Q. et al. (2021). 4D printing of glass fiber-regulated shape shifting structures with high stiffness. *ACS Applied Materials & Interfaces* 13 (11): 12797–12804.

103 Boley, J.W., van Rees, W.M., Lissandrello, C. et al. (2019). Shape-shifting structured lattices via multimaterial 4D printing. *Proceedings of the National Academy of Sciences of the United States of America* 116 (42): 20856–20862.

104 Behl, M., Razzaq, M.Y., and Lendlein, A. (2010). Multifunctional shape-memory polymers. *Advanced Materials* 22 (31): 3388–3410.

105 Mather, P.T., Luo, X., and Rousseau, I.A. (2009). Shape memory polymer research. *Annual Review of Materials Research* 39: 445–471.

106 Lendlein, A. and Gould, O.E.C. (2019). Reprogrammable recovery and actuation behaviour of shape-memory polymers. *Nature Reviews Materials* 4 (2): 116–133.

107 Hager, M.D., Bode, S., Weber, C. et al. (2015). Shape memory polymers: past, present and future developments. *Progress in Polymer Science* 49–50(0): 3–33.

108 Liu, Y., Genzer, J., and Dickey, M.D. (2016). "2D or not 2D": shape-programming polymer sheets. *Progress in Polymer Science* 52: 79–106.

109 Zhao, Q., Qi, H.J., and Xie, T. (2015). Recent progress in shape memory polymer: new behavior, enabling materials, and mechanistic understanding. *Progress in Polymer Science* 49-50(0): 79–120.

110 Liu, Y., Li, Y., Yang, G. et al. (2015). Multi-stimulus-responsive shape-memory polymer nanocomposite network cross-linked by cellulose nanocrystals. *ACS Applied Materials & Interfaces* 7 (7): 4118–4126.

111 Fang, Y., Ni, Y., Choi, B. et al. (2015). Chromogenic photonic crystals enabled by novel vapor-responsive shape-memory polymers. *Advanced Materials* 27 (24): 3696–3704.

112 Huang, W., Yang, B., An, L. et al. (2005). Water-driven programmable polyurethane shape memory polymer: demonstration and mechanism. *Applied Physics Letters* 86 (11): 114105.

113 Guo, Y., Lv, Z., Huo, Y. et al. (2019). A biodegradable functional water-responsive shape memory polymer for biomedical applications. *Journal of Materials Chemistry B* 7 (1): 123–132.

114 Zhu, Y., Hu, J., Luo, H. et al. (2012). Rapidly switchable water-sensitive shape-memory cellulose/elastomer nano-composites. *Soft Matter* 8 (8): 2509–2517.

115 Mishra, S.R. and Tracy, J.B. (2018). Sequential actuation of shape-memory polymers through wavelength-selective photothermal heating of gold nanospheres and nanorods. *ACS Applied Nano Materials* 1 (7): 3063–3067.

116 Han, D.-D., Zhang, Y.-L., Ma, J.-N. et al. (2016). Light-mediated manufacture and manipulation of actuators. *Advanced Materials* 28 (38): 8328–8343.

117 Liu, Y., Shaw, B., Dickey, M.D. et al. (2017). Sequential self-folding of polymer sheets. *Science Advances* 3 (3): e1602417.

118 Liu, Y., Boyles, J.K., Genzer, J. et al. (2012). Self-folding of polymer sheets using local light absorption. *Soft Matter* 8 (6): 1764–1769.

119 Vernon, L. and Vernon, H. (1941), *Producing molded articles such as dentures from thermoplastic synthetic resins.* US Patent 2234993.

120 Luo, X. and Mather, P.T. (2010). Conductive shape memory nanocomposites for high speed electrical actuation. *Soft Matter* 6 (10): 2146–2149.

121 Liu, R., Kuang, X., Deng, J. et al. (2018). Shape memory polymers for body motion energy harvesting and self-powered mechanosensing. *Advanced Materials* 30 (8): 1705195. -n/a.

122 Zarek, M., Layani, M., Cooperstein, I. et al. (2015). 3D printing of shape memory polymers for flexible electronic devices. *Advanced Materials* 28 (22): 4449–4454.

123 Zheng, N., Fang, G., Cao, Z. et al. (2015). High strain epoxy shape memory polymer. *Polymer Chemistry* 6 (16): 3046–3053.

124 Voit, W., Ware, T., Dasari, R.R. et al. (2010). High-strain shape-memory polymers. *Advanced Functional Materials* 20 (1): 162–171.

125 Yakacki, C.M., Shandas, R., Safranski, D. et al. (2008). Strong, tailored, biocompatible shape-memory polymer networks. *Advanced Functional Materials* 18 (16): 2428–2435.

126 Peng, B., Yang, Y., Gu, K. et al. (2019). Digital light processing 3D printing of triple shape memory polymer for sequential shape shifting. *ACS Materials Letters* 1 (4): 410–417.

127 Zotzmann, J., Behl, M., Feng, Y. et al. (2010). Copolymer networks based on poly(ω-pentadecalactone) and poly(ε-caprolactone)segments as a versatile triple-shape polymer system. *Advanced Functional Materials* 20 (20): 3583–3594.

128 Behl, M. and Lendlein, A. (2010). Triple-shape polymers. *Journal of Materials Chemistry* 20 (17): 3335–3345.

129 Xie, T. (2010). Tunable polymer multi-shape memory effect. *Nature* 464 (7286): 267–270.

130 Yu, K., Dunn, M.L., and Qi, H.J. (2015). Digital manufacture of shape changing components. *Extreme Mechanics Letters* 4: 9–17.

131 Kuang, X., Chen, K., Dunn, C.K. et al. (2018). 3D printing of highly stretchable, shape-memory and self-healing elastomer toward novel 4D printing. *ACS Applied Materials & Interfaces* 10 (8): 7381–7388.

132 Zhang, B., Zhang, W., Zhang, Z. et al. (2019). Self-healing four-dimensional printing with ultraviolet curable double network shape memory polymer system. *ACS Applied Materials & Interfaces* 11 (10): 10328–10336.

133 Yu, R., Yang, X., Zhang, Y. et al. (2017). Three-dimensional printing of shape memory composites with epoxy-acrylate hybrid photopolymer. *ACS Applied Materials & Interfaces* 9 (2): 1820–1829.

134 Kuang, X., Zhao, Z., Chen, K. et al. (2018). High-speed 3D printing of high-performance thermosetting polymers via two-stage curing. *Macromolecular Rapid Communications* 39 (7): 1700809.

135 Weigand, J.J., Miller, C.I., Janisse, A.P. et al. (2020). 3D printing of dual-cure benzoxazine networks. *Polymer* 189: 122193.

136 Han, B., Zhang, Y.L., Chen, Q.D. et al. (2018). Carbon-based photothermal actuators. *Advanced Functional Materials* 28 (40): 1802235.

137 Yang, H., Leow, W.R., Wang, T. et al. (2017). 3D printed photoresponsive devices based on shape memory composites. *Advanced Materials* 29 (33): 1701627.

138 Wang, Z., Tao, P., Liu, Y. et al. (2014). Rapid charging of thermal energy storage materials through plasmonic heating. *Scientific Reports* 4 (1): 1–8.

139 Tong, S., Quinto, C.A., Zhang, L. et al. (2017). Size-dependent heating of magnetic iron oxide nanoparticles. *ACS Nano* 11 (7): 6808–6816.

140 Fuhrer, R., Schumacher, C.M., Zeltner, M. et al. (2013). Soft iron/silicon composite tubes for magnetic peristaltic pumping: frequency-dependent pressure and volume flow. *Advanced Functional Materials* 23 (31): 3845–3849.

141 Razzaq, M., Behl, M., and Lendlein, A. (2012). Magnetic memory effect of nanocomposites. *Advanced Functional Materials* 22 (1): 184–191.

142 Kobayashi, K., Yoon, C., Oh, S.H. et al. (2018). Biodegradable thermomagnetically responsive soft untethered grippers. *ACS Applied Materials & Interfaces* 11 (1): 151–159.

143 He, Z., Satarkar, N., Xie, T. et al. (2011). Remote controlled multishape polymer nanocomposites with selective radiofrequency actuations. *Advanced Materials* 23 (28): 3192–3196.

144 Ze, Q., Kuang, X., Wu, S. et al. (2020). Magnetic shape memory polymers with integrated multifunctional shape manipulation. *Advanced Materials* 32 (4): 1906657.

145 Wei, H., Zhang, Q., Yao, Y. et al. (2017). Direct-write fabrication of 4D active shape-changing structures based on a shape memory polymer and its nanocomposite. *ACS Applied Materials & Interfaces* 9 (1): 876–883.

146 Hua, D., Zhang, X., Ji, Z. et al. (2018). 3D printing of shape changing composites for constructing flexible paper-based photothermal bilayer actuators. *Journal of Materials Chemistry C* 6 (8): 2123–2131.

147 Yuan, C., Wang, T., Dunn, M.L. et al. (2017). 3D printed active origami with complicated folding patterns. *International Journal of Precision Engineering and Manufacturing-Green Technology* 4 (3): 281–289.

148 Wu, J., Yuan, C., Ding, Z. et al. (2016). Multi-shape active composites by 3D printing of digital shape memory polymers. *Scientific Reports* 6: 24224.

149 Jeong, H.Y., Woo, B.H., Kim, N. et al. (2020). Multicolor 4D printing of shape-memory polymers for light-induced selective heating and remote actuation. *Scientific Reports* 10 (1): 6258.

150 Bodaghi, M., Damanpack, A., and Liao, W. (2016). Self-expanding/shrinking structures by 4D printing. *Smart Materials and Structures* 25 (10): 105034.

151 Teoh, J.E.M., An, J., Feng, X. et al. (2018). Design and 4D printing of cross-folded origami structures: a preliminary investigation. *Materials* 11 (3): 376.

152 van Manen, T., Janbaz, S., and Zadpoor, A.A. (2018). Programming the shape-shifting of flat soft matter. *Materials Today* 21 (2): 144–163.

153 Ding, Z., Yuan, C., Peng, X. et al. (2017). Direct 4D printing via active composite materials. *Science Advances* 3 (4): e1602890.

154 An, B., Y. Tao, J. Gu, et al. (2018). Thermorph: democratizing 4D printing of self-folding materials and interfaces. In *Proceedings of the 2018 CHI Conference on Human Factors in Computing Systems*. ACM.

155 Ding, Z., Weeger, O., Qi, H.J. et al. (2018). 4D rods: 3D structures via programmable 1D composite rods. *Materials & Design* 137: 256–265.

156 Yuan, C., Ding, Z., Wang, T. et al. (2017). Shape forming by thermal expansion mismatch and shape memory locking in polymer/elastomer laminates. *Smart Materials and Structures* 26 (10): 105027.

157 Zhang, Q., Zhang, K., and Hu, G. (2016). Smart three-dimensional lightweight structure triggered from a thin composite sheet via 3D printing technique. *Scientific Reports* 6: 22431.

158 Mao, Y., Yu, K., Isakov, M.S. et al. (2015). Sequential self-folding structures by 3D printed digital shape memory polymers. *Scientific Reports* 5: 13616.

159 Zhang, Q., Kuang, X., Weng, S. et al. (2021). Shape-memory balloon structures by pneumatic multi-material 4D printing. *Advanced Functional Materials* 2010872.

160 Liu, K., Wu, J., Paulino, G.H. et al. (2017). Programmable deployment of tensegrity structures by stimulus-responsive polymers. *Scientific Reports* 7 (1): 3511.

161 Wichterle, O. and Lim, D. (1960). Hydrophilic gels for biological use. *Nature* 185 (4706): 117–118.

162 Champeau, M., Heinze, D.A., Viana, T.N. et al. (2020). 4D printing of hydrogels: a review. *Advanced Functional Materials* 30 (31): 1910606.

163 Wu, Z.L., Moshe, M., Greener, J. et al. (2013). Three-dimensional shape transformations of hydrogel sheets induced by small-scale modulation of internal stresses. *Nature Communications* 4: 1586.

164 Kim, J., Hanna, J.A., Byun, M. et al. (2012). Designing responsive buckled surfaces by halftone gel lithography. *Science* 335 (6073): 1201–1205.

165 Han, D., Lu, Z., Chester, S.A. et al. (2018). Micro 3D printing of a temperature-responsive hydrogel using projection micro-stereolithography. *Scientific Reports* 8 (1): 1963.

166 Zhao, Z., Kuang, X., Yuan, C. et al. (2018). Hydrophilic/hydrophobic composite shape-shifting structures. *ACS Applied Materials & Interfaces* 10 (23): 19932–19939.

167 Zhao, Q., Yang, X., Ma, C. et al. (2016). A bioinspired reversible snapping hydrogel assembly. *Materials Horizons* 3 (5): 422–428.

168 Schild, H.G. (1992). Poly (N-isopropylacrylamide): experiment, theory and application. *Progress in Polymer Science* 17 (2): 163–249.

169 Illeperuma, W.R., Sun, J.-Y., Suo, Z. et al. (2013). Force and stroke of a hydrogel actuator. *Soft Matter* 9 (35): 8504–8511.

170 Bakarich, S.E., Gorkin, R., Panhuis, M.i.h. et al. (2015). 4D printing with mechanically robust, thermally actuating hydrogels. *Macromolecular Rapid Communications* 36 (12): 1211–1217.

171 Han, D., Farino, C., Yang, C. et al. (2018). Soft robotic manipulation and locomotion with a 3D printed electroactive hydrogel. *ACS Applied Materials & Interfaces* 10 (21): 17512–17518.

172 Jin, D., Chen, Q., Huang, T.-Y. et al. (2019). Four-dimensional direct laser writing of reconfigurable compound micromachines. *Materials Today* 32: 19–25.

173 Huang, T.-Y., Huang, H.-W., Jin, D.D. et al. (2020). Four-dimensional micro-building blocks. *Science Advances* 6 (3): eaav8219.

174 Sydney Gladman, A., Matsumoto, E.A., Nuzzo, R.G. et al. (2016). Biomimetic 4D printing. *Nature Materials* 15 (4): 413–418.

175 Kokkinis, D., Schaffner, M., and Studart, A.R. (2015). Multimaterial magnetically assisted 3D printing of composite materials. *Nature Communications* 6: 8643.

176 Mao, Y., Ding, Z., Yuan, C. et al. (2016). 3D printed reversible shape changing components with stimuli responsive materials. *Scientific Reports* 6: 24761.

177 Naficy, S., Gately, R., Gorkin, R. et al. (2017). 4D printing of reversible shape morphing hydrogel structures. *Macromolecular Materials and Engineering* 302 (1).

178 Liu, J., Erol, O., Pantula, A. et al. (2019). Dual-gel 4D printing of bioinspired tubes. *ACS Applied Materials & Interfaces* 11 (8): 8492–8498.

179 Huang, L., Jiang, R., Wu, J. et al. (2017). Ultrafast digital printing toward 4D shape changing materials. *Advanced Materials* 29 (7): 1605390. –n/a.

180 Kim, Y., Yeom, B., Arteaga, O. et al. (2016). Reconfigurable chiroptical nanocomposites with chirality transfer from the macro- to the nanoscale. *Nature Materials* 15 (4): 461–468.

181 Palagi, S., Mark, A.G., Reigh, S.Y. et al. (2016). Structured light enables biomimetic swimming and versatile locomotion of photoresponsive soft microrobots. *Nature Materials* 15 (6): 647–653.

182 Pei, Z., Yang, Y., Chen, Q. et al. (2014). Mouldable liquid-crystalline elastomer actuators with exchangeable covalent bonds. *Nature Materials* 13 (1): 36–41.

183 Yu, Y., Nakano, M., and Ikeda, T. (2003). Photomechanics: directed bending of a polymer film by light. *Nature* 425 (6954): 145–145.

184 Guin, T., Settle, M.J., Kowalski, B.A. et al. (2018). Layered liquid crystal elastomer actuators. *Nature Communications* 9 (1): 2531.

185 Wani, O.M., Zeng, H., and Priimagi, A. (2017). A light-driven artificial flytrap. *Nature Communications* 8: 15546.

186 Ware, T.H., McConney, M.E., Wie, J.J. et al. (2015). Voxelated liquid crystal elastomers. *Science* 347 (6225): 982–984.

187 Lu, X., Zhang, H., Fei, G. et al. (2018). Liquid-crystalline dynamic networks doped with gold nanorods showing enhanced photocontrol of actuation. *Advanced Materials* 30 (14): 1706597.

188 Kohlmeyer, R.R. and Chen, J. (2013). Wavelength-selective, IR light-driven hinges based on liquid crystalline elastomer composites. *Angewandte Chemie International Edition* 52 (35): 9234–9237.

189 Roach, D.J., Yuan, C., Kuang, X. et al. (2019). Long liquid crystal elastomer fibers with large reversible actuation strains for smart textiles and artificial muscles. *ACS Applied Materials & Interfaces* 11 (21): 19514–19521.

190 Saed, M.O., Torbati, A.H., Nair, D.P. et al. (2016). Synthesis of programmable main-chain liquid-crystalline elastomers using a two-stage thiol-acrylate reaction. *Journal of Visualized Experiments: JoVE* 0 (107): 53546.

191 Yakacki, C.M., Saed, M., Nair, D.P. et al. (2015). Tailorable and programmable liquid-crystalline elastomers using a two-stage thiol-acrylate reaction. *RSC Advances* 5 (25): 18997–19001.

192 Barnes, M. and Verduzco, R. (2019). Direct shape programming of liquid crystal elastomers. *Soft Matter* 15 (5): 870–879.

193 Li, Y., Zhang, Y., Rios, O. et al. (2017). Liquid crystalline epoxy networks with exchangeable disulfide bonds. *Soft Matter* 13 (29): 5021–5027.

194 Tabrizi, M., Ware, T.H., and Shankar, M.R. (2019). Voxelated molecular patterning in 3-dimensional Freeforms. *ACS Applied Materials & Interfaces* 11 (31): 28236–28245.

195 Kowalski, B.A., Guin, T.C., Auguste, A.D. et al. (2017). Pixelated polymers: directed self assembly of liquid crystalline polymer networks. *ACS Macro Letters* 6 (4): 436–441.

196 Ware, T.H., Perry, Z.P., Middleton, C.M. et al. (2015). Programmable liquid crystal elastomers prepared by thiol–ene photopolymerization. *ACS Macro Letters* 4 (9): 942–946.

197 Yu, L., Shahsavan, H., Rivers, G. et al. (2018). Programmable 3D shape changes in liquid crystal polymer networks of uniaxial orientation. *Advanced Functional Materials* 28 (37): 1802809.

198 Yuan, C., Roach, D.J., Dunn, C.K. et al. (2017). 3D printed reversible shape changing soft actuators assisted by liquid crystal elastomers. *Soft Matter* 13 (33): 5558–5568.

199 Zhao, R., Kim, Y., Chester, S.A. et al. (2019). Mechanics of hard-magnetic soft materials. *Journal of the Mechanics and Physics of Solids* 124: 244–263.

200 Hu, W., Lum, G.Z., Mastrangeli, M. et al. (2018). Small-scale soft-bodied robot with multimodal locomotion. *Nature* 554 (7690): 81.

201 Lum, G.Z., Ye, Z., Dong, X. et al. (2016). Shape-programmable magnetic soft matter. *Proceedings of the National Academy of Sciences of the United States of America* 113 (41): E6007–E6015.

202 Roh, S., Okello, L.B., Golbasi, N. et al. (2019). 3D-printed silicone soft architectures with programmed magneto-capillary reconfiguration. *Advanced Materials Technologies* 4 (4): 1800528.

203 Cui, J., Huang, T.-Y., Luo, Z. et al. (2019). Nanomagnetic encoding of shape-morphing micromachines. *Nature* 575 (7781): 164–168.

204 Rahmer, J., Stehning, C., and Gleich, B. (2017). Spatially selective remote magnetic actuation of identical helical micromachines. *Science Robotics* 2 (3): eaal2845.

205 Kim, Y., Parada, G.A., Liu, S. et al. (2019). Ferromagnetic soft continuum robots. *Science Robotics* 4 (33): eaax7329.

206 Wu, S., Ze, Q., Zhang, R. et al. (2019). Symmetry-breaking actuation mechanism for soft robotics and active metamaterials. *ACS Applied Materials & Interfaces* 11 (44): 41649–41658.

207 Montgomery, S.M., Wu, S., Kuang, X. et al. (2020). Magneto-mechanical metamaterials with widely Tunable mechanical properties and acoustic bandgaps. *Advanced Functional Materials* n/a(n/a): 2005319.

208 Kim, J., Chung, S.E., Choi, S.-E. et al. (2011). Programming magnetic anisotropy in polymeric microactuators. *Nature Materials* 10 (10): 747.

209 Xu, T., Zhang, J., Salehizadeh, M. et al. (2019). Millimeter-scale flexible robots with programmable three-dimensional magnetization and motions. *Science Robotics* 4 (29): eaav4494.

210 Peters, C., Hoop, M., Pané, S. et al. (2016). Degradable magnetic composites for minimally invasive interventions: device fabrication, targeted drug delivery, and cytotoxicity tests. *Advanced Materials* 28 (3): 533–538.

211 Lee, Y.-W., Ceylan, H., Yasa, I.C. et al. (2021). 3D-printed multi-stimuli-responsive mobile micromachines. *ACS Applied Materials & Interfaces* 13 (11): 12759–12766.

212 Ji, Z., Yan, C., Yu, B. et al. (2017). Multimaterials 3D printing for free assembly manufacturing of magnetic driving soft actuator. *Advanced Materials Interfaces* 4 (22): 1700629.

213 van Rees, W.M., Vouga, E., and Mahadevan, L. (2017). Growth patterns for shape-shifting elastic bilayers. *Proceedings of the National Academy of Sciences of the United States of America* 114 (44): 11597–11602.

214 Sossou, G., Demoly, F., Belkebir, H. et al. (2019). Design for 4D printing: a voxel-based modeling and simulation of smart materials. *Materials & Design* 175: 107798.

215 Westbrook, K.K., Parakh, V., Chung, T. et al. (2010). Constitutive modeling of shape memory effects in semicrystalline polymers with stretch induced crystallization. *Journal of Engineering Materials and Technology* 132 (4): 041010-041010-9.

216 Ge, Q., Luo, X., Rodriguez, E.D. et al. (2012). Thermomechanical behavior of shape memory elastomeric composites. *Journal of the Mechanics and Physics of Solids* 60 (1): 67–83.

217 Ge, Q., Luo, X., Iversen, C.B. et al. (2014). A finite deformation thermomechanical constitutive model for triple shape polymeric composites based on dual thermal transitions. *International Journal of Solids and Structures* 51 (15): 2777–2790.

218 Mao, Y., Chen, F., Hou, S. et al. (2019). A viscoelastic model for hydrothermally activated malleable covalent network polymer and its application in shape memory analysis. *Journal of the Mechanics and Physics of Solids* 127: 239–265.

219 Hamel, C.M., Cui, F., and Chester, S.A. (2017). A finite element method for light activated shape-memory polymers. *International Journal for Numerical Methods in Engineering* 111 (5): 447–473.

220 Warner, M. and Terentjev, E.M. (2003). *Liquid Crystal Elastomers*, vol. 120. OUP Oxford.

221 Oates, W. and Wang, H. (2009). A new approach to modeling liquid crystal elastomers using phase field methods. *Modelling and Simulation in Materials Science and Engineering* 17 (6): 064004.

222 Zhang, Y., Xuan, C., Jiang, Y. et al. (2019). Continuum mechanical modeling of liquid crystal elastomers as dissipative ordered solids. *Journal of the Mechanics and Physics of Solids* 126: 285–303.

223 Chester, S.A., Di Leo, C.V., and Anand, L. (2015). A finite element implementation of a coupled diffusion-deformation theory for elastomeric gels. *International Journal of Solids and Structures* 52 (Supplement C): 1–18.

224 Zhao, Z., Wu, J., Mu, X. et al. (2017). Desolvation induced origami of photocurable polymers by digit light processing. *Macromolecular Rapid Communications* 38 (13): 1600625. -n/a.

225 Bosnjak, N., Wang, S., Han, D. et al. (2019). Modeling of fiber-reinforced polymeric gels. *Mechanics Research Communications* 96: 7–18.

226 Craig, H., Devin, R., Kevin, L. et al. (2019). Machine-learning based design of active composite structures for 4D printing. *Smart Materials and Structures* 28: 065005.

227 Geiss, M., Boddeti, N., Weeger, O. et al. (2018). Combined level-set-XFEM-density topology optimization of 4D printed structures undergoing large deformation. *Journal of Mechanical Design* 141.

228 Sossou, G., Demoly, F., Belkebir, H. et al. (2019). Design for 4D printing: modeling and computation of smart materials distributions. *Materials & Design* 181: 108074.

229 Sigmund, O. (2001). Design of multiphysics actuators using topology optimization – part I: one-material structures. *Computer Methods in Applied Mechanics and Engineering* 190 (49): 6577–6604.

230 Sigmund, O. (2001). Design of multiphysics actuators using topology optimization – part II: two-material structures. *Computer Methods in Applied Mechanics and Engineering* 190 (49): 6605–6627.

231 Maute, K., Tkachuk, A., Wu, J. et al. (2015). Level set topology optimization of printed active composites. *Journal of Mechanical Design* 137 (11): 111402.

232 Kuang, X., Wu, S., Ze, Q. et al. (2021). Magnetic dynamic polymers for modular assembling and reconfigurable morphing architectures. *Advanced Materials* 33 (30): e2102113.

233 Montarnal, D., Capelot, M., Tournilhac, F. et al. (2011). Silica-like malleable materials from permanent organic networks. *Science* 334 (6058): 965–968.

234 Podgórski, M., Fairbanks, B.D., Kirkpatrick, B.E. et al. Toward stimuli-responsive dynamic thermosets through continuous development and improvements in covalent adaptable networks (CANs). *Advanced Materials.* **n/a**(n/a) 1906876.

235 Chakma, P. and Konkolewicz, D. (2019). Dynamic covalent bonds in polymeric materials. *Angewandte Chemie International Edition.*

236 Denissen, W., Winne, J.M., and Du Prez, F.E. (2016). Vitrimers: permanent organic networks with glass-like fluidity. *Chemical Science* 7 (1): 30–38.

237 Deng, J., Kuang, X., Liu, R. et al. (2018). Vitrimer elastomer-based jigsaw puzzle-like healable triboelectric nanogenerator for self-powered wearable electronics. *Advanced Materials* 30 (14): 1705918. -n/a.

238 Denissen, W., Droesbeke, M., Nicolaÿ, R. et al. (2017). Chemical control of the viscoelastic properties of vinylogous urethane vitrimers. *Nature Communications* 8 (1): 14857.

239 Rekondo, A., Martin, R., de Luzuriaga, A.R. et al. (2014). Catalyst-free room-temperature self-healing elastomers based on aromatic disulfide metathesis. *Materials Horizons* 1 (2): 237–240.

240 Davidson, E.C., Kotikian, A., Li, S. et al. (2019). 3D printable and reconfigurable liquid crystal elastomers with light-induced shape memory via dynamic bond exchange. *Advanced Materials* 32 (1): e1905682.

241 Zhang, G., Peng, W., Wu, J. et al. (2018). Digital coding of mechanical stress in a dynamic covalent shape memory polymer network. *Nature Communications* 9 (1): 4002.

242 Yang, Y., Chen, Z., Song, X. et al. (2017). Biomimetic anisotropic reinforcement architectures by electrically assisted nanocomposite 3D printing. *Advanced Materials* 29 (11): 1605750. –n/a.

243 Yang, Y., Xuan, S., Xiangjia, L. et al. (2018. **0**(0)). Recent Progress in biomimetic additive manufacturing technology: from materials to functional structures. *Advanced Materials* 1706539.

244 Studart, A.R. (2016). Additive manufacturing of biologically-inspired materials. *Chemical Society Reviews* 45 (2): 359–376.

245 Regehly, M., Garmshausen, Y., Reuter, M. et al. (2020). Xolography for linear volumetric 3D printing. *Nature* 588 (7839): 620–624.

246 Kelly, B.E., Bhattacharya, I., Heidari, H. et al. (2019). Volumetric additive manufacturing via tomographic reconstruction. *Science* 363 (6431): eaau7114.

247 de Beer, M.P., van der Laan, H.L., Cole, M.A. et al. (2019). Rapid, continuous additive manufacturing by volumetric polymerization inhibition patterning. *Science Advances* 5 (1): eaau8723.

248 Shusteff, M., Browar, A.E., Kelly, B.E. et al. (2017). One-step volumetric additive manufacturing of complex polymer structures. *Science Advances* 3 (12): eaao5496.

249 Walker, D.A., Hedrick, J.L., and Mirkin, C.A. (2019). Rapid, large-volume, thermally controlled 3D printing using a mobile liquid interface. *Science* 366 (6463): 360–364.

250 Erb, R.M., Sander, J.S., Grisch, R. et al. (2013). Self-shaping composites with programmable bioinspired microstructures. *Nature Communications* 4: 1712.

251 Martin, J.J., Fiore, B.E., and Erb, R.M. (2015). Designing bioinspired composite reinforcement architectures via 3D magnetic printing. *Nature Communications* 6: 8641.

2

Strategies in 3D Bioprinting of Cell-Laden Bioinks

Jun Yin[1,2], Changxue Xu[3], Zhengyi Zhang[4], Yifei Jin[5], and Yong Huang[6]

[1]*Zhejiang University, The State Key Laboratory of Fluid Power and Mechatronic Systems, School of Mechanical Engineering, Hangzhou 310028, China*
[2]*Zhejiang University, Key Laboratory of 3D Printing Process and Equipment of Zhejiang Province, School of Mechanical Engineering, Hangzhou 310028, China*
[3]*Texas Tech University, Department of Industrial, Manufacturing, and Systems Engineering, Lubbock, TX 79409, USA*
[4]*Huazhong University of Science and Technology, School of Naval Architecture and Ocean Engineering, Wuhan 430074, China*
[5]*University of Nevada Reno, Mechanical Engineering Department, Reno, NV 89557, USA*
[6]*University of Florida, Department of Mechanical and Aerospace Engineering, Gainesville, FL 32611, USA*

2.1 Introduction

With the introduction of additive manufacturing, scientists have been able to generate bioengineered 3D structures to replicate the complex nature of tissues [1]. The 3D printing-based tissue engineering approach can be classified into two groups: scaffold-based and cell-embedded [2, 3]. Traditionally, the scaffold-based or "top-down" approach (Figure 2.1) employs 3D printed porous solid biodegradable scaffolds as the synthetic extracellular matrix (ECM) to seed cells on their surfaces to form 3D cellular architectures. Although the classic scaffold-directed approach has been used to fabricate 2D/3D constructs, it still face some limitations and challenges: (i) effective vascularization of thick tissue constructs, (ii) heterogeneous placement of multiple cell types inside a scaffold [4], and (iii) less control over cell–cell contact and tissue architecture [5]. Because of the aforementioned drawbacks, recent tissue engineering studies tend toward the cell-embedded or "bottom-up" approach (Figure 2.1), which allows cells, biomaterials, and bioactive moieties to be precisely combined and deposited to form 3D complex constructs through additive manufacturing technology [6]. Recently, 3D printing materials including living cells (either suspended or cell aggregates) and ECM materials such as applicable hydrogels, cell media, and/or other additives are referred as "cell-laden bioinks" [4]. Therefore, the cell-embedded or "bottom-up" 3D printing-based tissue engineering approach, which has the ability to print cell-laden bioinks instead of plastic and metal inks using conventional 3D printing, is termed as 3D bioprinting/cell printing [7]. The cell printing approach, which allows skipping the cell seeding

Additive Manufacturing Technology: Design, Optimization, and Modeling, First Edition. Edited by Kun Zhou.

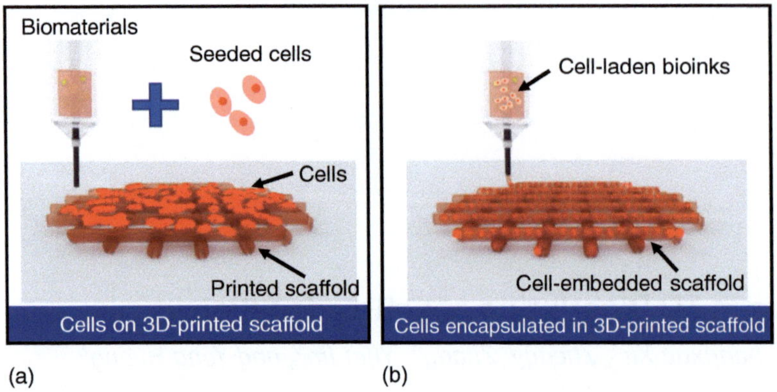

Figure 2.1 Scaffold-based (a) and cell embedded (b) tissue engineering.

procedure, has specific advantages including accurate control of multiply cells and materials distribution, high resolution of scaffold, and cost-effectiveness [8, 9].

Biomedical applications have emerged based on the rapid development of cell printing, including tissue scaffolds and implants, organoids for 3D biological models, micro-organ chips, and cellular machining and biorobots [10]. More importantly, since cells have been understood to be nature's building blocks [10] and cell biology in 2D and 3D culture is significantly different [7], the ultimate goal of cell printing is to hold cells together in a 3D matrix to produce functional biomimetic composite tissues [11]. To achieve the goal, bioinks containing living cells are the key components in 3D bioprinting, and the materials present in bioinks protect the cells from stressors during the printing process [12], where hydrogels are the main type of bioink materials. The implementation of hydrogels in cell printing is mainly based on their biological properties. (i) Hydrogels are attractive candidates to incorporate cells and bioactive compounds due to their high water content [1]. (ii) Hydrogels provide an instructive and aqueous environment, which simulates the ECM due to their innate structural and compositional similarities [13]. (iii) The abundance of chemical singles in the hydrogels leads to high cell viability and proliferation rates for embedded cells [1]. However, it has to be noticed that biological properties provided by hydrogels are not sufficient to recover the functionality of the native tissues, and the printed cell-encapsulated constructs also need to mimic the organizational complexity of native tissues. It is believed that geometry is as important in tissue structure as its function [7]. Therefore, how to print cell-laden hydrogel inks to the designed hierarchical objects through automated cell-friendly printing methods plays a key role in 3D cell printing-based tissue engineering [14].

To process cell-laden bioinks at increased resolutions, several bioprinting fabrication strategies have been explored, which can be classified based on the dimension of building blocks [4, 15]. The basic building blocks are bioink droplets during inkjet and laser printing, bioink filaments during microextrusion, and bioink layer during digital light processing (DLP). Therefore, these four bioprinting technologies can be further classified into three groups: (i) material jetting, which includes inkjet printing and laser-assisted printing, (ii) material extrusion, and (iii) vat polymerization as

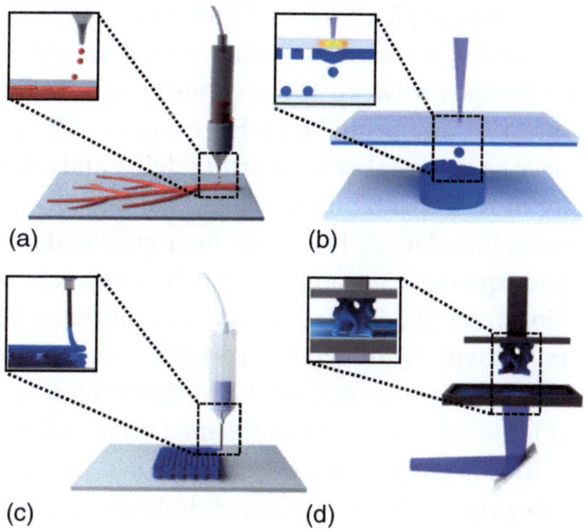

Figure 2.2 Four bioprinting technologies. (a) Inkjet printing; (b) laser printing; (c) extrusion-based printing; and (d) DLP printing.

shown in Figure 2.2. In this chapter, four bioprinting techniques, including inkjet, laser-assisted, micro-extrusion, and DLP are discussed in detail, and the brief comparison of these four bioprinting techniques is shown in Figure 2.2. Although many efforts have been made to enhance the shape fidelity of 3D-printed tissue structures, continuous advances are required to improve the printability of cell-laden bioinks. The bioink printability is characterized in terms of the controllable formation of well-defined droplets/jets/filaments/layers and/or the morphology of the deposited building blocks [4]. Moreover, mechanical integrity and stability also need to be considered during bioink printing, since the printed bioink material should retain its printed shape and bear loading during experimental operations. Figure 2.3 is a

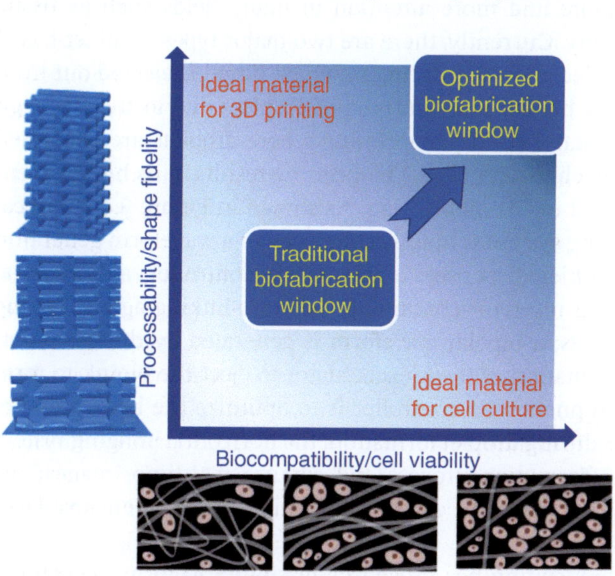

Figure 2.3 Biofabrication window for the rational design of bioinks requiring compromises among different requirements.

typical example of the biofabrication window showing the interplay among different bioink material properties. Traditionally, a compromised decision is often made on bioinks between these opposing requirements of biocompatibility, mechanical integrity, and printability. For example, Figure 2.3 describes the compromises that have been made between print fidelity and cell viability. Generally, a better printability of hydrogel bioinks can be achieved by increasing hydrogel concentration and cross-link density [1], while hydrogels with a lower network density provide a soft and aqueous environment that is beneficial to support cell survival, migration, and bioactivity [6]. Therefore, cell viability may be compromised when cells are being mixed with hydrogels with a high concentration or cross-link density for bioprinting. Strategies to extend the biofabrication window to the advanced bioink region (Figure 2.3) have to be performed on the aforementioned bioprinting technologies.

In this chapter, we try to introduce the recent progress in four widely used 3D cell printing technologies, including inkjet printing, laser printing, support-bath-enabled extrusion printing, and DLP printing.

2.2 Drop-on-Demand (DOD)-Based Inkjet Printing

2.2.1 Introduction to Inkjet Printing

As one of the most promising 3D bioprinting technologies for fabrication of 3D functional tissues/organs to solve the severe organ transplantation problem, inkjet-based bioprinting holds the advantages of high printing resolution, high deposition accuracy, and low cost despite its low allowance on fluid viscosity, and it has recently attracted more and more attention in many fields such as tissue engineering and drug delivery. Currently, there are two major types of inkjet-based bioprinting: continuous inkjet (CIJ) bioprinting in which liquid is ejected out from an orifice and subsequently breaks into continuous droplets due to the Rayleigh instability; and drop-on-demand (DOD) bioprinting where droplets are generated only when required. The mechanism of DOD bioprinting results in its higher accuracy and efficiency compared to CIJ bioprinting. As shown in Figure 2.4, a typical DOD inkjet-based bioprinting system is mainly composed of a waveform generator, a piezoactuator-attached inkjet dispenser, a pneumatic controller, a horizontal imaging system, and a small reservoir containing the cross-linking agent. During the droplet formation process, a bipolar waveform is generated by the waveform generator causing the deformation of the piezoactuator to eject the bioink to form the cell-laden droplets; the pneumatic controller is to optimize the back pressure and ensure a flat meniscus during droplet formation; the horizontal imaging system is to monitor the droplet formation process and capture real-time images; the cross-linking agent is to cross-link the cell-laden droplets into cell-encapsulated microspheres.

Recently, inkjet-based bioprinting has proved its feasibility to form 3D cellular structures and has been widely implemented in a variety of biomedical applications. For example, Negro et al. [17] reported an innovative method to inkjet print

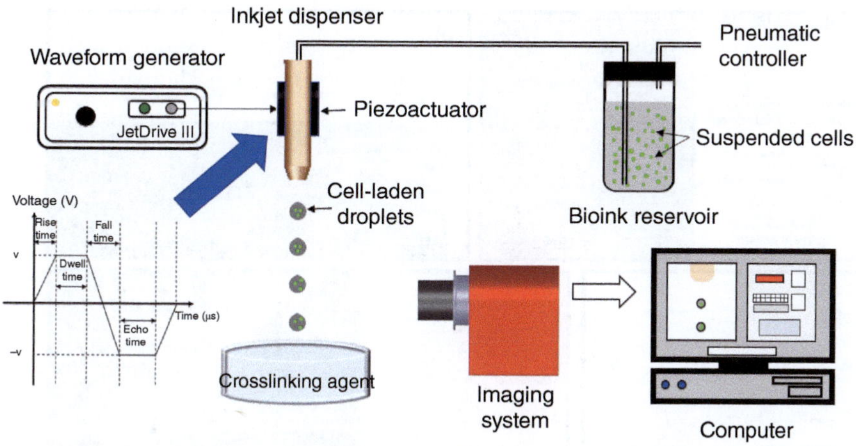

Figure 2.4 Schematic diagram of inkjet-based bioprinting system. Source: Xu et al. [16]/ AIP publishing/CC BY.

microfluidic chips and perfusable channels with a long-term culturing purpose (Figure 2.5a). Hewes et al. [18] fabricated free-standing microvessels (Figure 2.5b) with high precision using inkjet-based bioprinting of a bioink composed of endothelial cells and alginate/fibrinogen. Compaan et al. [19] selected inkjet-based bioprinting and a two-step gelation process to form a stable silk fibroin-based construct used for cell proliferation and spreading (Figure 2.5c). Takagi et al. [20] sequentially deposited two bioinks composed of different cell types to form 3D multilayering constructs with high precision (Figure 2.5d). Ji et al. [21] proposed an innovative inkjet-based bioprinting method using upward printing. Compared to the conventional downward printing structure, the printing resolution and shape fidelity of the upward printing construct (Figure 2.5e) could be significantly improved. Teo et al. [22] successfully used micro-reactive inkjet printing (MRIJP) to print low-viscosity hydrogels with high shape fidelity (Figure 2.5f).

2.2.2 Droplet Formation During DOD Inkjetting of Cell-laden Bioink

Numerous studies have focused on the droplet formation mechanism during DOD printing of inks with hard-particle laden. However, very few studies worked on the droplet formation process during DOD inkjet printing of bioink containing living cells which are considered as soft particles. Since the formation of cell-laden droplets during DOD inkjet printing is the first step toward a viable printing of organs, it is significant to understand the droplet formation mechanism of cell-laden bioink.

2.2.2.1 Bioink Preparation and Experimental Setup

In Xu et al.'s work [23], NIH 3T3 mouse fibroblasts and sodium alginate were used to prepare the bioinks with four different cell concentrations of 0, 10^6, 5×10^6, and 10^7 cells ml^{-1}, and the sodium alginate concentration is 1% (w/v). A MicroFab

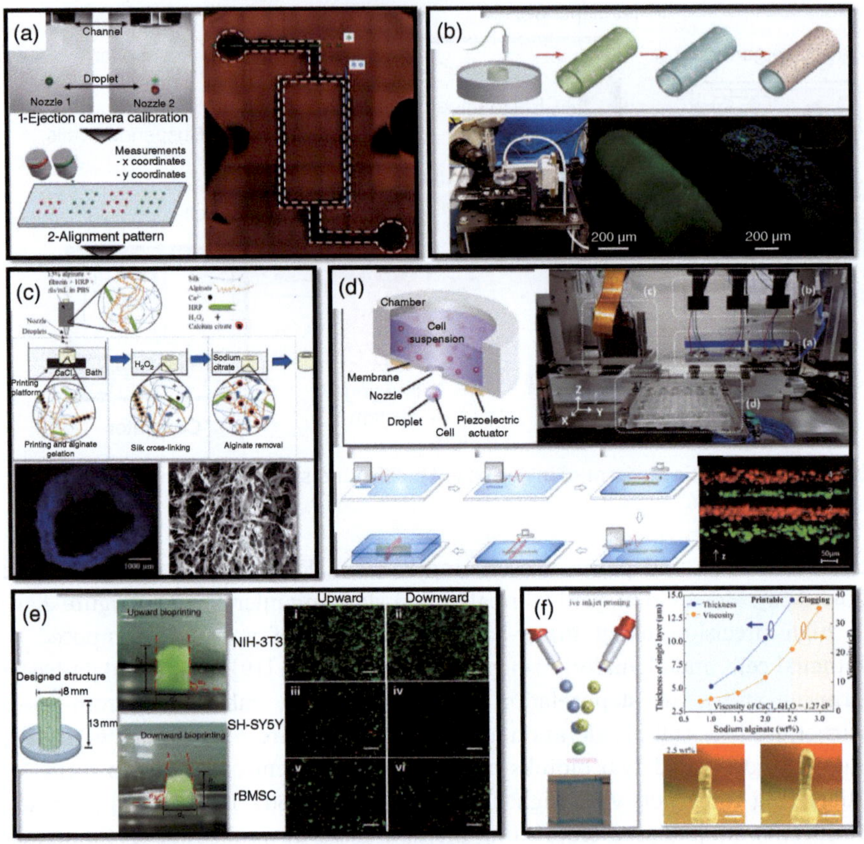

Figure 2.5 Recent progress of inkjet-based bioprinting. (a) Inkjet-printed microfluidic chips. Source: Negro et al. [17]/with permission of Springer Nature; (b) fabrication of microvessels using inkjet-based bioprinting. Source: Hewes et al. [18]/with permission of Elsevier. (c) Cell-laden silk fibroin construct fabricated with inkjet-based bioprinting using a two-step gelation method. Source: Compaan et al. [19]/with permission of American Chemical Society. (d) Inkjet bioprinting of a structure using alternative types of bioink. Source: Takagi et al. [20]/Whioce Publishing Pte. Ltd./CC BY-NC 4.0; (e) Downward and upward inkjet bioprinting. Source: Yang et al. [21]/with permission of American Chemical Society. (f) Sodium alginate-based hydrogel structures using MRIJP. Source: Teo et al. [22]/with permission of American Chemical Society.

nozzle dispenser (MicroFab, Plano, TX) was used to implement the inkjet printing process (Figure 2.4), and the bioink was ejected out of the MicroFab nozzle dispenser with an orifice diameter of 120 μm (MJABL-01-120-6MX) in a DOD mode. The DOD driving pulse was controlled by a MicroFab Jet Driver, and a MicroFab pneumatic controller was used to adjust the backpressure of the fluid reservoir to obtain an ideal meniscus for good droplet formation. The dispensing head was generally driven by a bipolar driving waveform consisting of a succession of two wave pulses, and the second wave pulse was aimed at eliminating some residual acoustic oscillations that remained in the DOD dispensing head after droplet ejection.

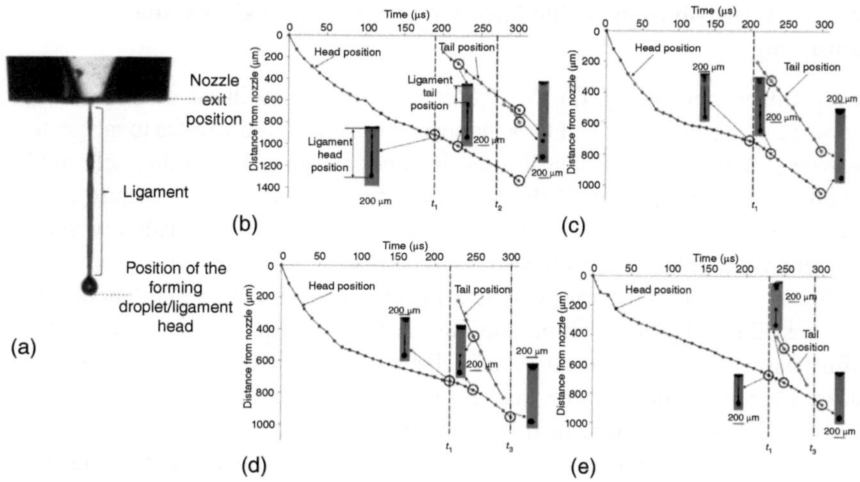

Figure 2.6 (a) Morphological features of a forming droplet. Jet/droplet morphological evolution under different cell concentrations: (b) without cells, (c) 10^6, (d) 5×10^6, and (e) 10^7 cells Ml^{-1}. Source: Xu et al. [23]/with permission of American Chemical Society.

2.2.2.2 Representative Droplet Formation Observations

The bioink was ejected into a 2% (w/v) calcium chloride solution to form alginate microspheres embedded with cells. As shown in Figure 2.6, the jet/droplet morphology was characterized under different cell concentrations, including the head and tail positions and breakup time. Besides, three critical times are illustrated: the breakup time at the nozzle orifice (exit pinching) to form the ligament tail (t_1), the breakup time of the ligament to form new satellites (t_2), and the merging time for the tail and primary droplet without satellites (t_3). It can be concluded that: (i) the position of forming filament/droplet at a given time decreases with the cell concentration, indicating the decrease of the ligament length as well as the droplet velocity; (ii) the breakup time (t_1) increases with the cell concentration; and (iii) the satellite droplets are suppressed with higher cell concentrations.

During a typical piezoactuator-based inkjet process, the piezo-element first enlarges and then compresses the nozzle as driven by the applied waveform (Figure 2.4). During nozzle enlarging, dilation pressure pulses propagating to both ends of the nozzle are generated by the piezo-element. The liquid reservoir, which connects and supplies the dispensing fluid to the nozzle, is assumed to be an open boundary condition, and a compression pulse is generated after being reflected from the liquid reservoir as a result of a 180° phase shift of the reflected pressure pulse. The nozzle orifice represents a large increase in acoustic impedance, and the pulse is reflected with the same sign. Consequently, compression pressure pulses are generated and propagated in both directions as the piezo-element compresses the nozzle, eliminating the dilation pulse reflected from the orifice and reinforcing the compression pressure wave reflected from the liquid reservoir. The resulting reinforced compression pulse drives the fluid inside the nozzle until it is large enough to eject a droplet from the orifice.

2.2.3 Cell Distribution Within Microspheres During Inkjet-Based Bioprinting

During inkjet bioprinting, the cell-laden microspheres utilized as basic building blocks are precisely controlled to be deposited on various substrates to fabricate 3D constructs. A uniform cell distribution is beneficial to post-printing cell viability as well as the functionality of the fabricated tissues and organs. Despite the previous droplet formation studies during inkjet bioprinting of cell-laden bioink, the comprehensive cell distribution within microspheres is still absent.

In Xu et al.'s work [16], the cell-laden droplets ejected from the inkjet dispenser were cross-linked by the calcium chloride solution to form cell-laden microspheres, within which different numbers of cells may be encapsulated. As shown in Figure 2.7, some microspheres had no cells encapsulated, while some had one or more cells encapsulated. In this work, the sodium alginate concentration was 0.5–2% (w/v) with an interval of 0.5%, which significantly affected the bioink rheological properties, further influencing the jetting and pinch-off behaviors during the droplet formation process. The excitation voltage was 40–70 V with an interval of 10 V. The excitation voltage determined the jetting velocity, which further influenced the jet instability for satellite droplet formation. The cell concentrations used in this study were 5×10^5, 2×10^6, 5×10^6, and 8×10^6 cells ml^{-1}. Under each condition, 1000 microspheres were randomly selected to quantify the cell distribution, where normal distribution was utilized to fit the experimental results to obtain the mean and standard deviation of the distribution.

2.2.3.1 Effect of Cell Concentration on Cell Distribution

The effect of cell concentration on the post-printing cell distribution is shown in Figure 2.8, when the other printing conditions are fixed: the excitation voltage of 60 V, the sodium alginate concentration of 1% (w/v), and the calcium chloride concentration of 2% (w/v). It can be seen that at the very low cell concentration of 5×10^5 cells ml^{-1}, most microspheres are without cells, and the microspheres with one cell are only 7.4%. At the low cell concentration of 2×10^6 cells ml^{-1}, most microspheres are still without cells, while the percentage of the microspheres with cells significantly increases. The microspheres with one cell and two cells are 18.6% and

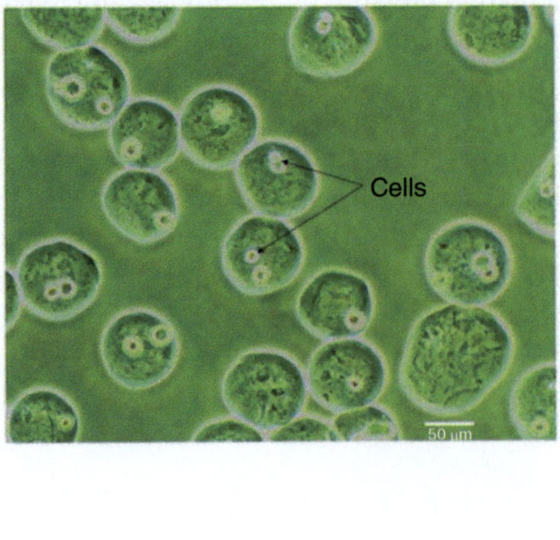

Figure 2.7 Alginate microspheres embedded with living cells. Source: Xu et al. [23]/with permission of American Chemical Society.

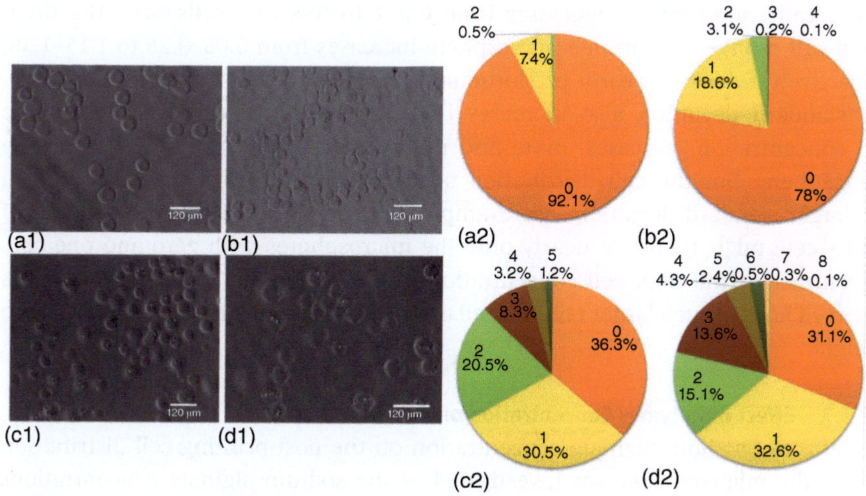

Figure 2.8 Representative images of microspheres and cell distribution in the microspheres using the bioink with different cell concentrations: (a) 5×10^5 cells ml^{-1}, (b) 2×10^6 cells ml^{-1}, (c) 5×10^6 cells ml^{-1}, and (d) 8×10^6 cells ml^{-1}. The other printing conditions are fixed: the excitation voltage 60 V and the sodium alginate concentration 1% (w/v). Source: Xu et al. [16]/AIP Publishing LLC./CC BY 4.0.

3.1%, respectively, and there are even 0.3% microspheres with three and four cells. At the high cell concentration, the microspheres with zero cell, one cell, two cells, and three cells are 31.1%, 32.6%, 15.1%, and 13.6%, respectively. There are more than 20% microspheres with more than three cells.

As shown in Figure 2.9, the normal distribution has been implemented to analyze the mean and the standard deviation of the distribution. It is seen that

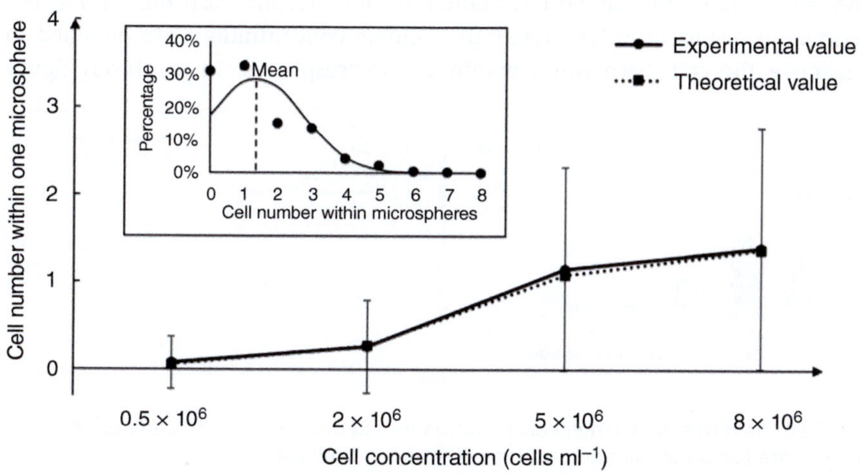

Figure 2.9 Effect of cell concentration on the mean cell number within one microsphere. The insert figure shows the fitted normal distribution based on the experimental results at the cell concentration of 8×10^6 cells ml^{-1}. Source: Xu et al. [16]/AIP publishing/CC BY.

as the cell concentration increases from 0.5–2 to $5–8 \times 10^6$ cells ml^{-1}, the fitted mean cell number within one microsphere increases from 0.08–0.26 to 1.15–1.39, respectively, which is nearly proportional to the cell concentration. Meanwhile, the standard deviation also increases from 0.30–0.53 to 1.16–1.39. When the cell concentration increases, more and more cells are encapsulated within one microsphere. But the cell distribution becomes more nonuniform according to the larger standard deviation. For example, at the very low cell concentration of 5×10^5 cells ml^{-1}, there are nearly only the microspheres with zero and one cell. In contrast, at the high cell concentration of 8×10^6 cells ml^{-1}, the microspheres observed have a much larger range of cell number within one microsphere from zero to eight.

2.2.3.2 Effect of Polymer Concentration on Cell Distribution

The effect of sodium alginate concentration on the post-printing cell distribution within the microspheres was investigated as the sodium alginate concentrations increased from 0.5%, 1%, 1.5%, to 2% (w/v) [19]. The other printing conditions are fixed: the excitation voltage of 60 V, the cell concentration of 3×10^6 cells ml^{-1}, and the calcium chloride concentration of 2% (w/v).

It is seen from Figure 2.10a, as the sodium alginate concentration increases from 0.5% to 2% (w/v), the microspheres without cells decrease significantly from 65.1% to 48.1%. On the contrary, the microspheres with two cells increase significantly from 5.2% to 13.9%, and the microspheres with three cells also increase from 1.1% to 4.9%. The microspheres with one cell are considered as the medium case between a low cell number and a high cell number, which almost have the same percentage at different sodium alginate concentrations. It is also seen from Figure 2.10a that when the sodium alginate concentration increases from 0.5% to 2% (w/v), the maximum cell number within the microspheres increases from 3–4 to 5–6.

The mean and the standard deviation of the normal distribution for the experimental results at different sodium alginate concentrations are obtained to characterize the cell distribution within the microspheres. In the insert figure

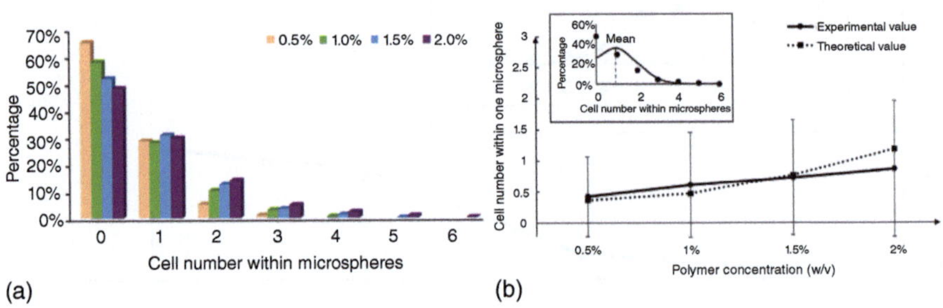

Figure 2.10 (a) Effect of polymer concentration on cell distribution. The other printing conditions are fixed: the excitation voltage 60 V and the cell concentration 3×10^6 cells ml^{-1}. (b) Effect of sodium alginate concentration on the mean cell number within one microsphere. The insert figure shows the fitted normal distribution based on the experimental results at the sodium alginate concentration of 2% (w/v). Source: Xu et al. [16]/AIP publishing/CC BY.

of Figure 2.10b, it shows the representative fitted normal distribution based on the experimental results at the sodium alginate concentration of 2% (w/v), which shows that with the increase in sodium alginate concentration, the mean cell number within one microsphere increases as a result of the increasing diameter of the microspheres. At the low sodium alginate concentration (e.g. 0.5% (w/v)), the droplet formation process is governed by the balance of the inertial stress and capillary stress as the viscoelasticity of the bioink is negligible. The ejected ligament is very long and quickly breaks up into several droplets due to the Rayleigh instability. The size of these droplets is usually similar, which is mainly determined by the ligament diameter. On the contrary, at the high sodium alginate concentration (e.g. 2% (w/v)), the droplet formation process is governed by the balance of the viscoelastic stress and the capillary stress as the viscoelasticity of the bioink is dominant. The ejected ligament is short, and the breakup time is significantly delayed by the bioink viscoelasticity, so the fluid within the ligament has enough time to flow into the primary droplet. As a result, usually, only one large primary droplet is formed.

2.2.3.3 Effect of Excitation Voltage on Cell Distribution

The effect of excitation voltage on the post-printing cell distribution within the microspheres was investigated by increasing the voltage from 40, 50, 60, to 70 V. The other printing conditions are fixed: the sodium alginate concentration is 1% (w/v), the cell concentration is 3×10^6 cells ml^{-1}, and the concentration of calcium chloride remains 2% (w/v). From Figure 2.11, it is seen that at 40 V and 50 V the mean cell number within one microsphere is very similar, and after 50 V the mean cell number within one microsphere increases with the increase in the excitation voltage. As the excitation voltage increases from 50 to 70 V, the microsphere diameter increases from 67.2 to 74.8 μm, which results in an increase of the mean cell number within one microsphere from 0.52 to 0.85. However, at 40 and 50 V, the formed microsphere diameters are 68.6 and 67.2 μm, respectively, since at 40 V only one primary droplet was formed, while one satellite droplet with a similar size was formed correspondingly at 50 V.

During inkjet-based bioprinting, the cell-laden bioink is ejected out from an inkjet dispenser to form cell-laden droplets, which are cross-linked by calcium chloride solution to form cell-laden microspheres. Based on the results, it is seen that the mean cell number within one microsphere is determined by cell concentration and microsphere size, both of which are related to the droplet formation process. During inkjet-based bioprinting, the bioink used is composed of biological materials and living cells, which are considered as suspensions. The cell-laden droplet formation is a very complex process. The living cells play a critical role in ligament thinning and pinch-off during the droplet formation in inkjet printing [4]. On one hand, the addition of living cells substantially changes the bioink macroscopic rheological properties, especially the viscoelasticity, which significantly affects the bioink jetting behaviors. On the other hand, compared to the nozzle orifice size of 120 μm, the cell size of 15 μm is large enough to affect the bioink flow during the droplet formation process. Moreover, the cell–cell interaction and cell motion significantly affect the ligament thinning and pinch-off processes. Therefore, the living cells significantly

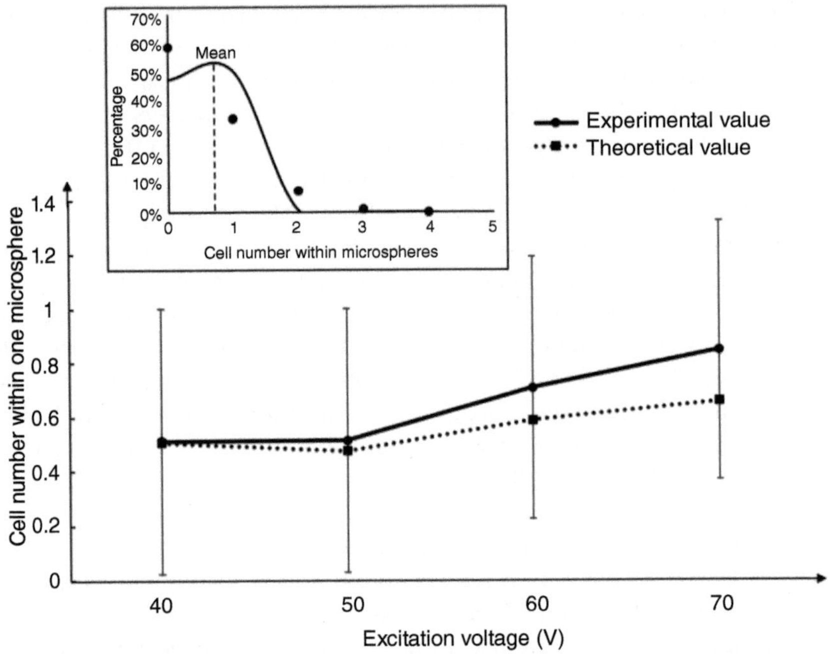

Figure 2.11 Effect of excitation voltage on the mean cell number within one microsphere. The insert figure shows the fitted normal distribution based on the experimental results at the excitation voltage of 50 V. Source: Xu et al. [16]/AIP publishing/CC BY.

affect the droplet formation process as well as the post-printing cell distribution in both macroscopic and microscopic ways.

2.3 Laser Printing

2.3.1 Introduction to Laser Printing

Laser bioprinting, a versatile laser-induced forward transfer-based technique [14, 24, 25], has been emerging as a promising orifice-free direct-write strategy for bioprinting. During a typical laser bioprinting process as shown in Figure 2.12, the bioink to be printed is prepared as a thin-film coating on the bottom side of a UV light transparent quartz support, which are collective to form a "ribbon". Laser beams, such as ultraviolet, are then guided to pass through the top side of the ribbon perpendicularly and focused on the interface of the quartz support and the thin-film coating. The laser–matter interaction generates the localized heat to sublime a small portion of the coating to form a high-temperature, high-pressure vapor bubble [24, 25]. The consequential bubble expansion then propels part of the bioink away from the ribbon, resulting in different types of jets and eventually breaking up into several droplet(s) [26–29]. The generated droplets are the building blocks for biofabrication. The printing resolution is defined by the size and shape of the printed droplets, which are influenced by the coating's rheological properties and thickness,

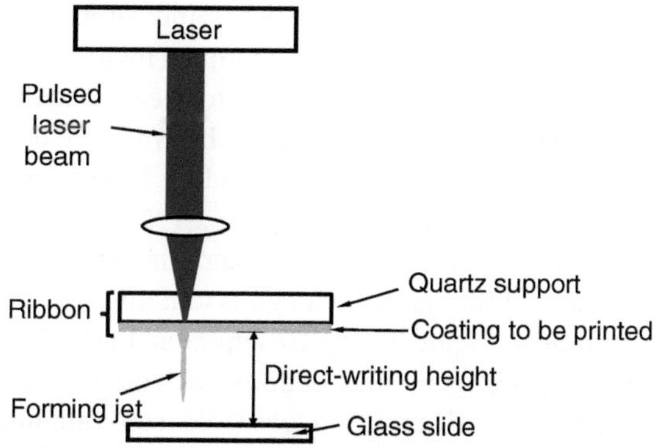

Figure 2.12 Schematics of laser printing experimental setup.

the laser fluence, the direct-writing height, and the wettability of the receiving substrate. Because of its orifice-free nature, laser bioprinting can handle a wide range of bioinks without suffering from the possible clogging and material compatibility issues associated with other nozzle-based bioprinting techniques, such as inkjet printing [30, 31] and microextrusion. Due to these advantages, the laser bioprinting has been adopted to print different biological materials, such as hydrogels [19, 25–27, 32], peptides [33], DNA [34], and even cells [3, 35–39], to name a few.

Despite these advantages, the conventional laser bioprinting experiences two main limitations preventing its broad adoption in 3D bioprinting. (i) It is difficult to achieve good printing quality when printing materials with relatively low-viscosity and high surface tension, such as the 2% alginate solution with a viscosity of around 140 cP and surface tension of 44.6 mN m^{-1}, due to its orifice-free implementation [26]. (ii) The post-printing cell viability may be relatively low (around 70%) [19, 28] compared to other droplet-based printing techniques such as inkjet printing [30, 31], which may be due to the direct interaction between the laser and the printing materials, the mechanical stress coming from the thermomechanical stimuli during the formation of bubbles [40] and jet/droplets, and the impact between cells and their receiving substrate during the landing of cellular droplets [40, 41]. Other minor limitations of laser bioprinting include the high cost of the laser bioprinting system and the time-consuming procedure for the preparation of each ribbon. Therefore, novel implementation is expected to relieve the dilemma introduced by the aforementioned limitations. Recently, energy absorbing layer (EAL)-assisted laser printing has been developed to circumvent these challenges [42, 43]. During EAL-assisted laser printing, the laser pulse energy is absorbed by the EAL causing localized heating and evaporation of a small portion of EAL and forming a small pressurized bubble. The bubble further expands and may eject part of the build material coating as droplets for deposition. Xiong et al. [44] developed a novel EAL-assisted laser printing, which utilizes gelatin as the EAL to transfer cell-lade bioinks. The added gelatin EAL helps to reduce the droplet size and average jet velocity, thereby improving the printing quality and the post-printing cell viability.

2D and 3D constructs with living cells have been successfully fabricated by laser bioprinting. Wu et al. [45] fabricated branch/stem structures of human umbilical vein endothelial cells (HUVECs) and human umbilical vein smooth muscle cells (HUVSMCs) by using laser printing (Figure 2.13a). Gaebel et al. [46] applied laser printing to prepare a cardiac patch seeded with HUVECs in a defined pattern for cardiac regeneration (Figure 2.13b). Guillotin et al. [47] implemented laser printing to fabricate miniaturized tissue-like layouts with *de novo* high cell density and micro scale organization (Figure 2.13c). Koch et al. [48] demonstrated that the 3D arrangement of fibroblasts and keratinocytes embedded in collagen by laser printing as multicellular grafts analogous to native archetype and the formation of tissue by these cells (Figure 2.13d). Gruene et al. [39] printed spots of human adipose-derived stem cells (ASCs) and endothelial colony-forming cells (ECFCs) in a 3D array using laser printing and observed cell–cell interactions in vascular endothelial growth factor-free medium. It has been demonstrated that direct cell–cell contacts trigger the development of stable vascular-like networks. Xiong et al. [49] fabricated both straight and Y-shaped tubes by freeform laser printing (Figure 2.13e). It has been demonstrated that 3D cellular tubes, including constructs with bifurcated overhang structures, can be adequately fabricated under optimal printing conditions.

2.3.2 Effects of Living Cells on the Bioink Printability

High-resolution printed structures can be easily obtained via laser printing due to the orifice-free nature of laser printing, which avoids nozzle clogging during the printing of highly viscous bioinks as well as possible contamination [44]. However, cell-laden bioink contains living cells, which can be considered as soft particles in bioink suspension, resulting in different nature from the Newtonian or viscoelastic solutions. Therefore, the printability of different bioinks must be carefully studied to optimize the laser printing performance [29]. Macroscopically, suspended living cells change the rheological properties of bioinks and consequently the printing dynamics; microscopically, the interaction of suspended living cells may lead to cell aggregation, resulting in nonideal jetting behavior.

2.3.2.1 Representative Observations During Laser Printing of Cell-laden Bioink

Xiong et al. [44] encapsulated NIH 3T3 mouse fibroblasts into sodium alginate to form the model cell-laden bioinks, and Figure 2.14 shows some representative time-resolved images of the jet/droplet formation and deposition processes as well as the corresponding micrographs of printed droplets during the laser printing of the cell-laden bioink under different laser fluences. During the jet/droplet formation process, most of the incident laser energy is converted into the kinetic energy of the initial laser-induced bubble, which is eventually dissipated by viscous flow, or converted into the elastic, surface, and kinetic energies of the forming jet/droplet. The kinetic energy of the forming jet increases with the incident laser fluence, leading to the increase in jet velocity. The jet velocity has a significant influence on the jet morphology and breakup mechanism, since a jet with different velocities may experience the Rayleigh breakup, the first wind-induced breakup, the second wind-induced breakup, or even atomization.

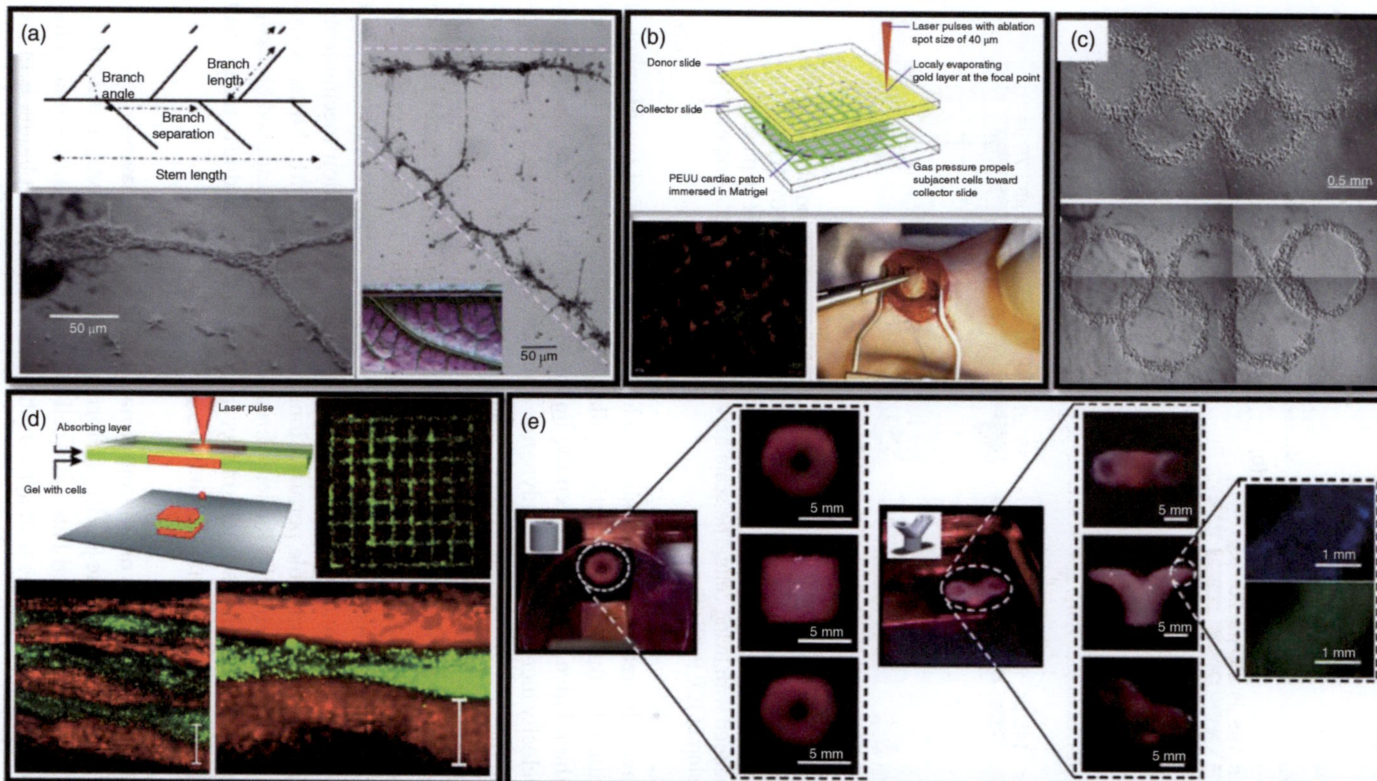

Figure 2.13 (a) Laser printing printed human umbilical vein endothelial cell (HUVEC) and human umbilical vein smooth muscle cell (HUVSMC) branch/stem structures on hydrogel layers. Source: Reproduced with permission from Wu et al. [45]/with permission of IOP Publishing. (b) Laser printing printed cardiac patch seeded with human umbilical vein endothelial cells. Source: Gaebel et al. [46]/with permission of Elsevier. (c) Laser printing of cellularized 2D pattern. Source: Guillotin et al. [47]/with permission of Elsevier. (d) Laser printing printed 3D pattern with fibroblasts (green) and keratinocytes (red). Source: Koch et al. [48]/with permission of John Wiley & Sons, Inc. (e) Laser printing printed straight and Y-shaped cellular tubes with straight and Y-shaped tubes. Source: Xiong et al. [49]/with permission of IOP Publishing.

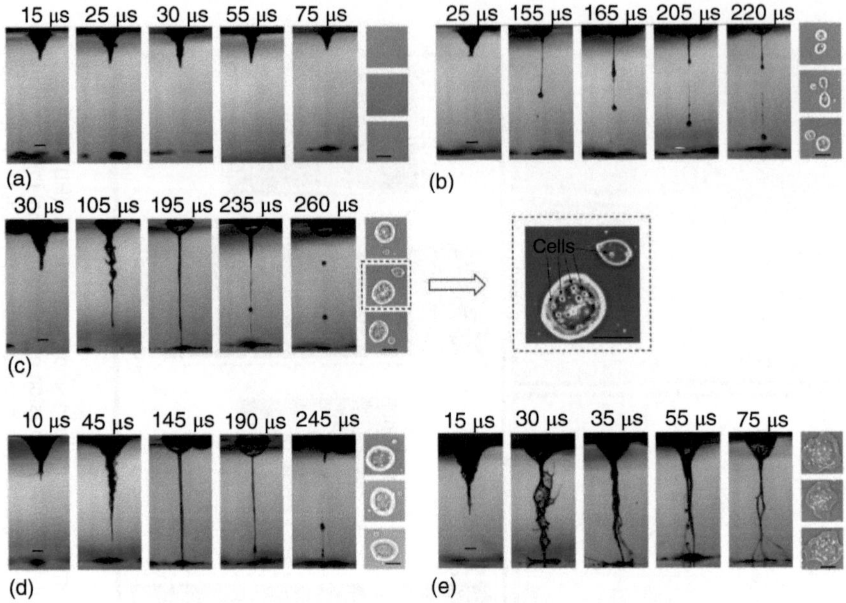

Figure 2.14 Time-resolved representative images during laser printing of cell-laden bioink under laser fluences of: (a) 700 mJ cm^{-2}, (b) 900 mJ cm^{-2}, (c) 1100 mJ cm^{-2}, (d) 1300 mJ cm^{-2}, and (e) 1500 mJ cm^{-2}. (c) Inset shows a representative deposited droplet embedded with living cells. The scale bars are 200 μm. Source: Xiong et al. [44]/ with permission of IOP Publishing.

In the Rayleigh breakup with small jet velocity, the pinch-off of the filament is mainly attributed to the action of inertio-capillary force, while the aerodynamic force caused by the relative motion between the liquid jet and ambient gas is negligible. As the jet velocity increases, the aerodynamic force gradually increases and the perturbation on the filament becomes higher, leading to the first wind-induced breakup of filament occurring during which the aerodynamic force is large enough (up to 10% of the surface tension force) to affect the jet breakup, resulting in a faster breakup process and shorter breakup length. When the jet velocity further increases, the greater aerodynamic force leads to further growth of perturbation. Then, the jet may experience the second wind-induced breakup and atomization mechanisms, during which the most unstable perturbations turn to be those with shorter wavelengths than the ones during the Rayleigh breakup and first wind-induced breakup. Thus, the resulting drop size is much smaller than the jet diameter due to the second wind-induced and atomization mechanisms.

2.3.2.2 Effects of Living Cells on Printing Dynamics and Jetting Behaviors

The effects of living cells on the jet/droplet formation and deposition dynamics were investigated by printing the cell-laden and cell-free bioinks under different laser fluences. It can be found that the transfer threshold of the cell-laden bioink is higher compared to that of the cell-free bioink, but the jet velocity and the printed droplet size of the cell-laden bioink are smaller (Figure 2.15a,b). It was measured that with

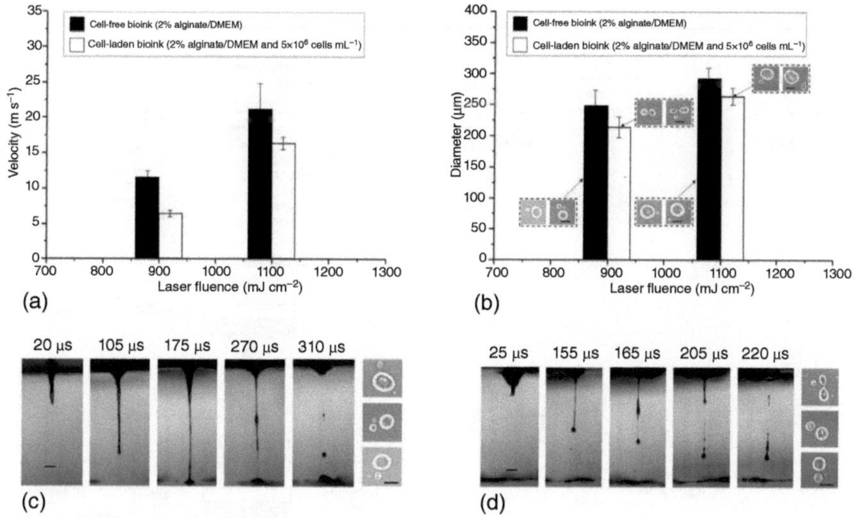

Figure 2.15 (a) Jet velocity and (b) printed droplet diameter of cell-laden and cell-free bioinks under different laser fluences, and time-resolved printing dynamics images under a 900 mJ cm^{-2} laser fluence when using: (c) cell-free and (d) cell-laden bioinks. The scale bars are 200 μm. Source: Xiong et al. [44]/with permission of IOP Publishing.

the increase in the addition of living cells, both the viscous and elastic properties of the cell-laden bioink increased, so more viscous energy is dissipated and more elastic energy is stored in the forming jet, which leads to a higher transfer threshold for the printing of the cell-laden bioink. Therefore, under the same laser fluence, it takes more energy to overcome the higher viscous and elastic effects of the cell-laden bioink, leaving less kinetic and surface energies to the forming jet as well as less material transferred, resulting in a lower jet velocity and a smaller printed droplet diameter. When printing cell-laden bioinks at a fixed direct-writing height, the lower jet velocity may allow a jet to have enough time to break up before it arrives at the receiving substrate. Figure 2.15c,d show that under the same laser fluence of 900 mJ cm^{-2}, the jet breakup length of cell-laden bioink (Figure 2.15d) is shorter than that of the cell-free bioink (Figure 2.15c), and the printing type changes from jet-impingement printing with multiple breakups to droplet-impingement printing.

The bioink material properties, especially the rheological behavior, significantly affect the jetting dynamics during laser printing. The addition of living cells affects the jetting dynamics both macroscopically and microscopically. Macroscopically, the material properties of bulk bioinks, such as the viscosity, storage and loss moduli, and relaxation time, have changed due to the presence of living cells, which indirectly affect the jetting dynamics. Microscopically, the local nonuniformity and nonhomogeneity of cell-laden bioink may be caused by cell aggregation, which may lead to nonideal jetting behaviors during the laser printing of cell-laden bioinks.

Figure 2.16 illustrates the two types of nonideal jetting behaviors observed in the experiments of this study: a nonstraight jet with a nonstraight trajectory (case 1) and a straight jet with a nonstraight trajectory (case 3). It is worth noting that when

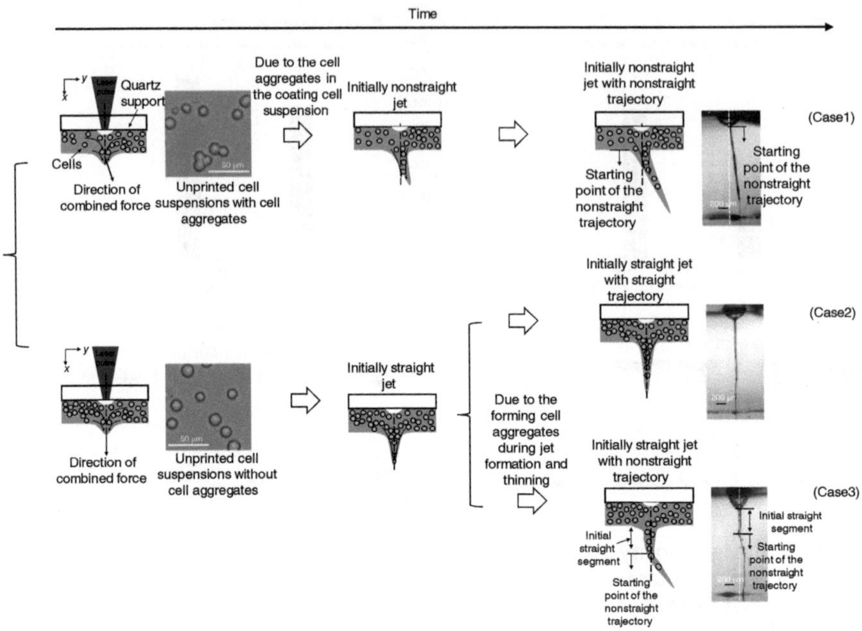

Figure 2.16 Schematics illustrating the formation of different types of jets during the laser printing of the cell-laden bioink. Source: Xiong et al. [44]/with permission of IOP Publishing.

a straight jet is formed, it may develop into two fates: a straight jet with a straight trajectory (case 2) and a straight jet with a nonstraight trajectory (case 3). Among these three jetting behaviors, only the straight jet with a straight trajectory is considered desirable, while the other two are considered nonideal. By observing the morphology of 100 forming jets under an 1100 mJ cm^{-2} laser fluence, the occurrence of these three types of jets is further investigated, and the results show that the proportion of the straight jet with a straight trajectory (case 2), the nonstraight jet (case 1), and the straight jet with a nonstraight trajectory (case 3) is 81%, 11%, and 8%, respectively. During laser printing, the initial laser-induced bubble expansion propels the underneath coating layer downward, which forms a jet. The direction of the resultant ejection force of the inertial, capillary, viscous, and elastic forces acting on the jet determines the jet trajectory. For ideal scenarios, no cell aggregates appeared on the bioink film, and the direction of the resultant ejection force is downward perpendicularly, forming a straight jet as shown in cases 2 or 3 in Figure 2.16. However, the local nonuniformity and nonhomogeneity in the bioink film may be caused once the cell aggregations are formed, thus leading to asymmetric flows during jet formation and resulting in the deviation of resultant ejection force and the jetting direction from the perpendicular centerline, as shown in case 1 in Figure 2.16.

Here, some main conclusions are drawn as follows: (i) higher transfer threshold, as well as lower jet velocity, shorter jet breakup length, and smaller printed droplet size, can be achieved when printing cell-laden bioink instead of cell-free bioink. The printing type changed from jet-impingement printing with multiple breakups to droplet-impingement printing when living cells are embedded in the bioinks and (ii) because of the local nonuniformity and nonhomogeneity of cell-laden bioinks, two nonideal jetting behaviors are identified: a nonstraight jet with a nonstraight trajectory and a straight jet with a nonstraight trajectory.

2.3.3 Freeform Drop-on-Demand Laser Printing of 3D Alginate and Cellular Constructs

Xiong et al. [49] also used laser printing for the freeform fabrication of 3D alginate and cellular tubes. Figure 2.17 shows the printing process of the laser pulse-induced jet traveling over a direct-writing height, which is the stand-off distance between the ribbon coating bottom and receiving substrate. After jet/droplet impingement, the deposited droplets could be chemical or physical cross-linked to spread, coalesce, gel, and maintain their position and shape. In the printing process, countless continuous droplets are deposited on demand to form a layer and then stacked layer-by-layer to obtain a three-dimensional structure. For each layer, the deposited droplets are patterned in a predesigned shape at a given elevation using XY translational stages. Once a layer is printed onto the top of a previously deposited construct,

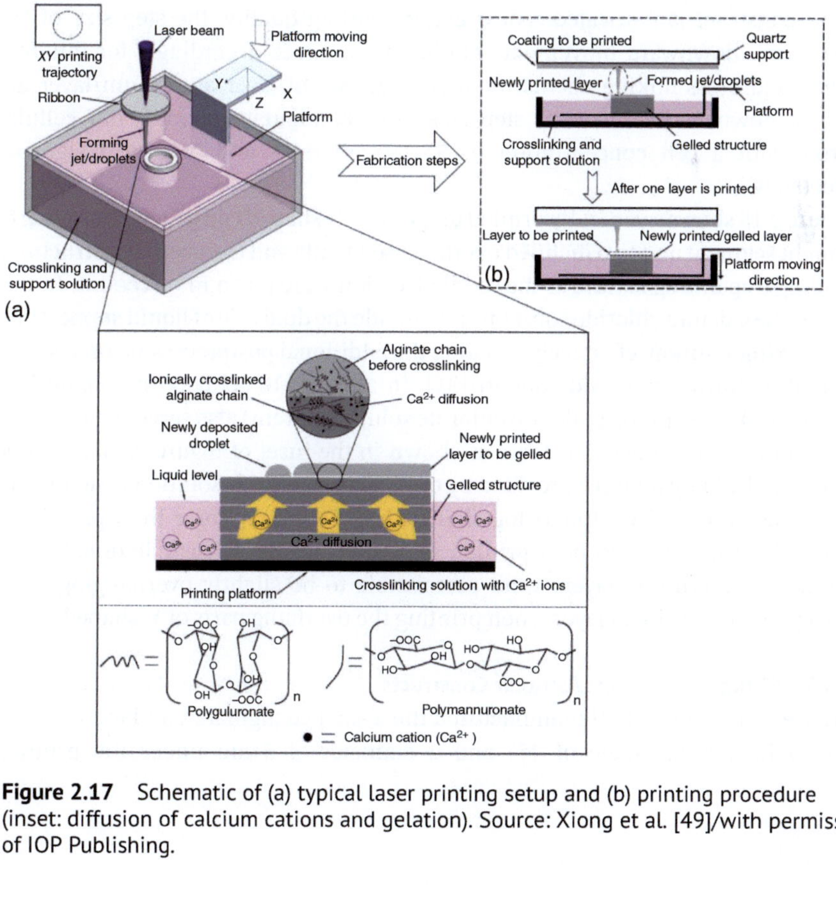

Figure 2.17 Schematic of (a) typical laser printing setup and (b) printing procedure (inset: diffusion of calcium cations and gelation). Source: Xiong et al. [49]/with permission of IOP Publishing.

which is attached to a Z vertical stage, the Z stage should move downward by a specific distance to gel the newly deposited layer (Figure 2.17).

2.3.3.1 Overhang Construct Fabrication

Based on both alginate and fibroblast bioinks, Y-shaped tubes have been constructed by laser printing [49]. The calcium chloride ($CaCl_2$) solution serves as a cross-linking and support medium for spanning and overhang features to achieve in situ gel for each droplet of bioink printed onto the receiving substrate. Here, the gelled structure is designed only to partially submerge in the cross-linking solution to ensure the continuous gelation of deposited layers along the height direction (Figure 2.17). In order to guarantee the direct contact between deposited layers before gelation and the gelled structure, a small portion of gelled structures (around 150 μm in height herein) is kept above the liquid level of the cross-linking solution during the entire printing process. The cross-linking approach, in which a newly deposited layer is stacked on the top of the gelled structure and then gelled by the calcium cations diffused upward from the cross-linking solutions below, is essential for strong interfacial bonding between two adjacent deposited layers. After the gelation of each newly deposited layer, the Z stage-based receiving platform moves downward with a certain step distance and the process is repeated until a construct is fabricated. For a better printing quality, the step size of the Z platform's downward movement should be optimized accordingly for different bioinks and cross-linking agents. In particular, 30 μm/layer and 25 μm/layer are the best downward movement step sizes for 8% alginate bioinks and cellular bioinks with a cell concentration of 5×10^6 cells ml^{-1} and 2% (w/v) alginate, respectively.

Figure 2.18 shows a vertically printed diagram of Y-shaped constructs consisting of a straight segment and two inclined overhang segments and the overhang structures including a spanning feature at the saddle location were taken into account. In this process, the calcium chloride solution can provide the dual role of liquid support and cross-linking solution, effectively avoiding the additional postprocessing procedures required by introduced solid support [31]. In addition to gel deposited materials, the cross-linking solution (calcium chloride solution herein) also serves as a support material to provide a buoyant force as shown in the inset of Figure 2.18. It can be found that the buoyant force provided by the calcium chloride solution is significant and acts as an effective support together with the adhesive force from previously deposited gel materials to hold printed structures against the gravitational force. Therefore, the printing layers were permissible to be slightly overhanging away from a previously printed layer when printing the overhang parts of Y-shaped tubes.

2.3.3.2 Bifurcated Alginate/Cellular Constructs

Moreover, Xiong et al. [49] manufactured the Y-shaped alginate tubular constructs with an inclination angle of 45° and a diameter of 5 mm under the printing

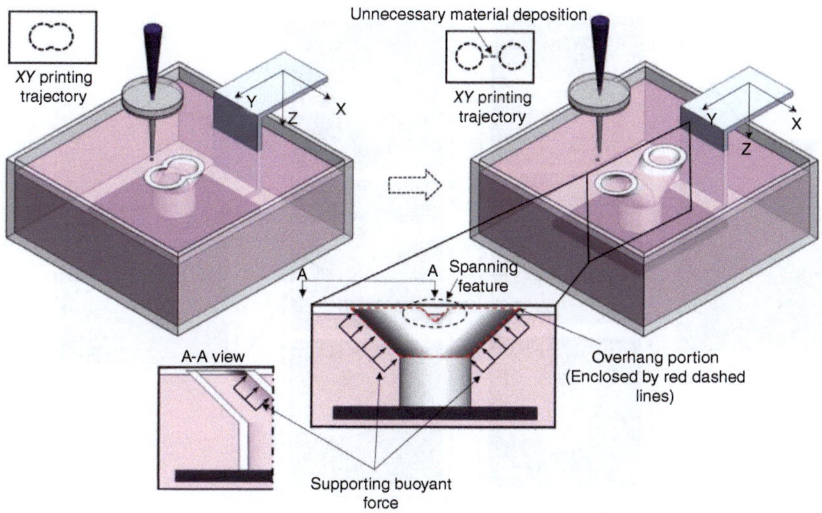

Figure 2.18 Schematics of bifurcated construct printing (inset: a typical Y-shaped bifurcated structure). Source: Xiong et al. [49]/with permission of IOP Publishing.

conditions of 2125 mJ cm^2 laser fluence, 100 mm min^{-1} substrate velocity, 10 Hz repetition rate, and 30 μm downward movement step size. It can be observed that the Y-shaped tube has an average wall thickness of 1.4 ± 0.3 mm as measured based on the bottom straight portion and a total height of around 9.5 mm (Figure 2.19a), and such a Y-shaped alginate tube takes about 120 minutes (including the ribbon change and coating time) to print. Besides, the two branch tubes with an inclination angle of 45° approximately as designed were successfully constructed with the help of the supporting force provided by the calcium chloride solution, revealing the feasibility of the proposed printing trajectory when printing the two separate branch tubes without stop, since no residual alginate material was found between the two branch tubes.

Applying the identified printing conditions (1445 mJ cm^{-2} laser fluence, 80 mm min^{-1} substrate velocity, 10 Hz repetition rate, and 25 μm downward movement step size), cell-laden Y-shaped constructs with a diameter of 5 mm, a total height of around 9.5 mm, and a wall thickness of 2.5 ± 0.3 mm were also fabricated utilizing cellular bioinks (Figure 2.19b). The resulting Y-shaped cellular tube presents a well-defined morphology, and the total printing time is around 180 minutes (including the ribbon change and coating time). It is noticed that due to the nonideal droplet formation process during the printing of cell-laden bioinks, the wall thickness of Y-shaped cellular constructs is almost doubled compared to alginate-only tubes. Based on the fluorescein staining shown in the right inset of Figure 2.19b, the post-printing cell viability of printed constructs is 68.1% immediately after printing, which increased to 70.8% after 24 hours of

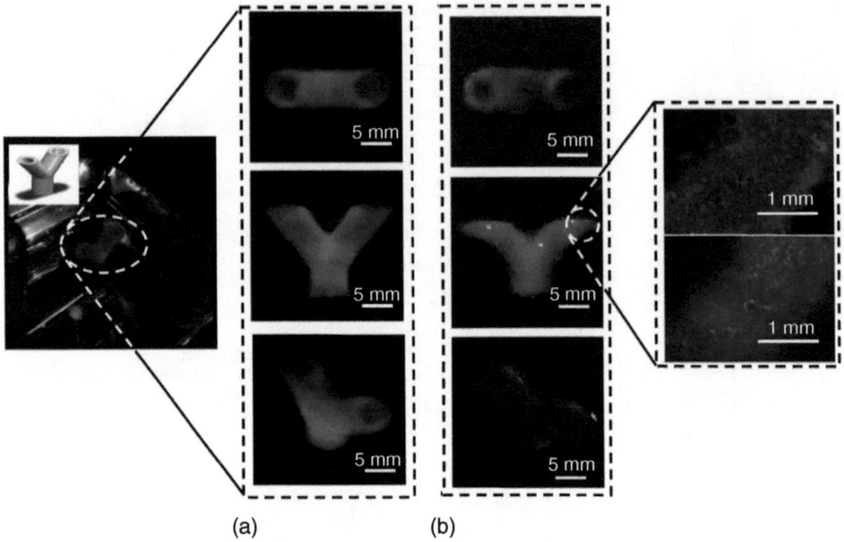

Figure 2.19 (a) Representative images of Y-shaped alginate tubes printed using 8% sodium alginate solution (inset: different views of a printed Y-shaped alginate tube). (b) Representative images of Y-shaped cellular tubes printed with 2% alginate and 5×10^6 cells ml^{-1} bioink (left inset: different views of a printed Y-shaped cellular tube, and right inset: dyed cells in blue and dyed living cells in green). Source: Xiong et al. [49]/with permission of IOP Publishing.

incubation considering the control effect. The cell viability is reasonable for effective bioprinting, and regular cell proliferation leads to an increase in cell viability after incubation. In addition, the post-printing cell viability for Y-shaped construct printing is slightly higher than that of straight construct printing, which may be attributed to less impact force damage during overhang printing.

Significant differences in print quality are observed when comparing the two print constructs obtained with cell-free (8% alginate) and cell-laden (2% alginate-based fibroblast suspension) bioinks due to the different characteristics of cellular constructs, i.e. relatively rougher surface, thicker wall, not well defined, and less uniform axially. This observation is attributed to several possible reasons. (i) An increase in alginate concentration (8% instead of 2%) leads to a higher viscosity as well as the storage and loss moduli of the bioinks, resulting in a more stable jetting and droplet formation process [26]. (ii) The random cell distribution and/or possible cell aggregation inside a formed cellular droplet may cause a nonuniform jetting trajectory, leading to variations of the droplet landing location, subsequently the deterioration of printing quality. (iii) The difference in wall thickness between alginate and cellular constructs may be attributed to the different swelling ratios of gelled alginate in different calcium cation concentration cross-linking solutions. It is expected that the improvement of jet and droplet formation processes and bioink printability may help improve the quality of printed cellular constructs.

2.4 Support Bath-Enabled Printing-then-Solidification Extrusion

2.4.1 Introduction to Support Bath-Enabled 3D Printing

Support bath-enabled 3D bioprinting, an alternative extrusion-based 3D bioprinting strategy, has emerged in recent years to create 3D constructs for various biomedical applications. Different from traditional direct ink writing, support bath-enabled 3D printing is performed by printing a 3D construct in a liquid support bath, which can stably hold the deposited construct at a liquid state for a long time. After printing, different cross-linking mechanisms (such as temperature change, UV radiation, and ionic cross-linkers) are used to cause the solidification of the construct. Finally, the solidified construct is released from the support bath and used as a functional part. Since the solidification step is separated away from the printing step, this new 3D bioprinting strategy is also called "printing-then-solidification" fashion as shown in Figure 2.20. Support bath-enabled 3D bioprinting has two main advantages: (i) bioinks are not required to possess either rapid solidification or self-supporting property [50, 51] and (ii) it is unnecessary to build up any auxiliary supporting structures because liquid support bath provides *in situ* supporting functionality during printing. As a result, this extrusion 3D bioprinting method has a wider range of printable biomaterials, higher printing efficiency, and the capacity of creating full-scale human organ and/or tissue models with complex geometries [52–55].

The selection of suitable support bath materials is of great significance during support bath-enabled 3D bioprinting. Currently, many support bath materials with different supporting mechanisms have been investigated and proposed such as unrecoverable matrix baths [56], buoyant support fluids [57], reversibly self-healing hydrogels [58], and yield-stress fluids [52–55, 59, 60]. Among these materials, yield-stress fluids are the most popular, which are usually composed of nano and/or microparticles with given inherent microstructures. Yield stress is a result of the threshold energy required to disrupt the inherent microstructures before the fluids

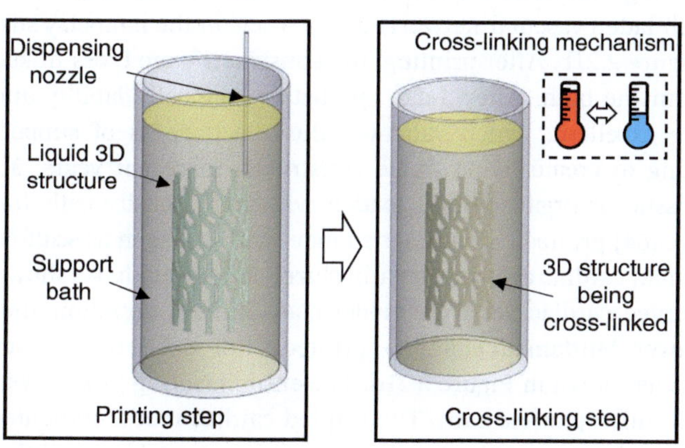

Figure 2.20 Schematic of support bath-enabled printing-then-solidification extrusion.

flow. Thus, yield-stress support bath materials have two unique properties: (i) a solid-like state when the applied stress is lower than the yield stress and (ii) a repeatable transition between liquid- and solid-like states upon different stressed conditions. For the former, except the buoyancy provided by the solvents of the yield-stress fluids, the microstructures of the fluids can provide additional supporting force to hold deposited 3D constructs stably *in situ*, which makes yield-stress fluids to have better supporting functionality than buoyant support fluids [57]. For the latter, yield-stress fluids present a liquid state when the applied stress is higher than the yield stress and recover to a solid-like state when the applied stress is removed or decreased. Thus, during printing, the higher shear stress-induced liquefied yield-stress fluid around the dispensing nozzle can fill the crevasse behind the nozzle and entrap the deposited 3D constructs in the support bath without using additional filler to facilitate this encapsulation process [56]. In addition, by designing and/or selecting suitable solvents to prepare yield-stress fluids, the chemical stability of yield-stress property can be ensured, which may prevent the dysfunctionality of supporting capability caused by some possible reactions with the reversibly self-healing hydrogels [58]. As a result, various yield-stress support bath materials have been developed and widely used in support bath-enabled 3D bioprinting.

One main purpose of cell 3D printing is to fabricate engineered organs and/or tissues, which is considered as the promising technical solution to overcome the organ shortage around the world. In organ/tissue printing, cells are mixed with hydrogel matrices as bioinks that are printed into organ-like constructs. During incubation, adjacent cellular layers undergo tissue fusion and finally complete the transformation into functional organs/tissues by self-assembly [61–63]. Support bath-enabled 3D bioprinting has been widely used to print organoids and tissue-like constructs in recent years. For example, Bhattacharjee et al. [52] successfully printed a hierarchically branched vascular network within the granular gel medium as shown in Figure 2.21a. This network with multiscale hollow channels can be used to mimic blood vessel systems in the human body for transporting nutrients and oxygen to and expelling metabolic waste from organs and tissues. In addition, Jin et al. [54] printed cell-laden vascular networked constructs in the nanoclay suspension as shown in Figure 2.21b. After printing, the constructs were cross-linked *in situ* and released from the bath. After 3-day incubation, the cell viability and metabolic activity were excellent, which validated the effectiveness of support bath-enabled 3D printing to create living tissue constructs. In recent years, 3D printing of complete tissue or organ models loaded with high-density cells has been achieved. Lee et al. [64] printed a biomimetic model with collagen as scaffold material containing human cardiac cells in a gelatin slurry support bath. As shown in Figure 2.21c, the printed cardiac ventricle model had visible contraction after 4 days of culture. Moreover, Mirdamadi et al. [55] printed a full-scale human heart model from alginate ink as shown in Figure 2.21d. The entire printing process was performed in the gelatin microparticle bath. The printed cardiac model presented mechanical properties similar to a real adult heart, making it an ideal training tool for surgical purposes. In addition, Skylar-Scot et al. [65] used embedded 3D

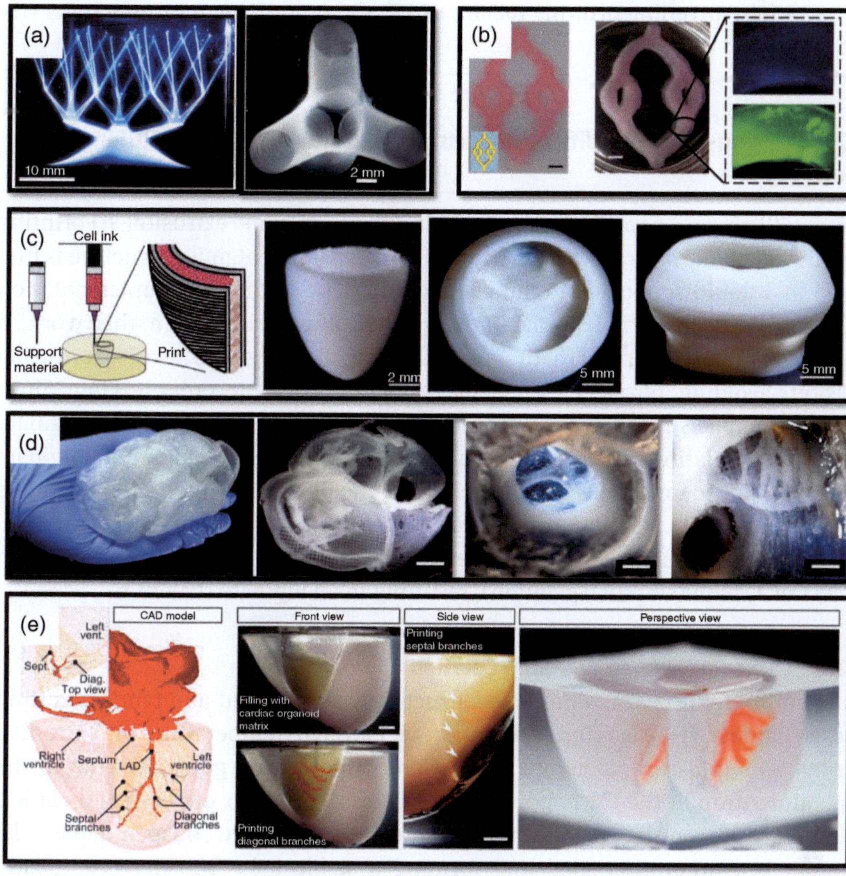

Figure 2.21 Representative applications in cell printing. (a) Printed vessel tubular networks using Carbopol ETD 2020 bath and a high-resolution image of one branch. Source: Adapted with permission from Bhattacharjee et al. [52]/American Association for the Advancement of Science – AAAS. (b) Vascular network printed in Laponite EP support bath and cross-linked gelatin/alginate construct with living 3 T3 fibroblasts. (Scale bars: 3 mm.) Source: Adapted with permission from Jin et al. [54]/American Chemical Society. (c) Human cardiac ventricle model printed in the gelatin slurry bath. Source: Reproduced with permission from Lee et al. [64]/American Association for the Advancement of Science – AAAS. (d) 3D printed heart model from alginate and its internal structure (Scale bars: 1.5 cm). Source: Adapted with permission from Mirdamadi et al. [55]/American Chemical Society. (e) Embedded 3D printing of cardiac tissue with perfusable vessel channels. Source: Adapted with permission from Skylar-Scott et al. [65]/American Association for the Advancement of Science – AAAS. (Scale bar: 1 cm.)

printing to create perfusable channels in high cell density organ building blocks (OBBs)/ECM bath, as shown in Figure 2.21e. With the help of these channels, nutrients and oxygen were able to be delivered to the cardiac tissue, enhancing the survival rate and differentiation ability of cells. Although there are many achievements in cell printing using support bath-enabled 3D printing strategy, this technique is still in its early stage. More investigations such as developing support bath materials, unveiling printing mechanisms, and optimizing posttreatments will be performed in the future.

2.4.2 Printing-then-Solidification Extrusion of Alginate and Cellular Structures

Jin et al. [59] proposed an alternative "printing-then-gelation" extrusion 3D printing method for alginate structure fabrication, in which a liquid alginate structure is first printed in a liquid support material via extrusion due to the easy implementation and high fabrication efficiency and then gelled after printing. In this work, a granular Carbopol microgel bath holds the ungelled alginate structure during extrusion, avoiding nozzle clogging caused by rapid ionic cross-linking. Since Carbopol microgels react with multivalent cations, which are needed for alginate cross-linking, gelatin is introduced as a sacrificial material to prepare an alginate and gelatin bioink for extrusion. The gelatin in the bioink gels thermally (step-one gelation) to initially stabilize the printed structure for removing from Carbopol. Then gelatin is melted and diffused away while alginate is ionically cross-linked in a 37 °C $CaCl_2$ bath (step-two gelation), resulting in an alginate structure.

2.4.2.1 Carbopol-Enabled Two-Step Gelation Approach

As shown in Figure 2.22a, a branched structure was designed and fabricated from sodium alginate (NaAlg) and gelatin at an elevated temperature by extruding layer-by-layer into a room temperature Carbopol granular gel bath. Due to the temperature drop in the Carbopol bath, the chains of gelatin underwent a conformational disorder–order transition and tended to recover the collagen triple-helix structure, becoming entangled and forming a stable gel. This step-one gelatin gelation was a slow thermal process. Fortunately, Carbopol maintained the shape of the printed structure until the gelatin was completely gelled at room temperature (Figure 2.22b). Then, the gelled alginate/gelatin structure was mechanically strong enough to be taken out of the Carbopol bath and placed in the sodium chloride (NaCl) bath to de-swell and collapse the residual granular microgels of Carbopol (Figure 2.22c). Finally, the Carbopol-free structure was immersed in an ionic cross-linking solution ($CaCl_2$) at 37 °C.

2.4.2.2 3D Bioprinting of Y-Shaped Tubular Structures

Here, a Y-shaped tubular structure was designed as a simplified basic unit of vascular constructs [66] to further demonstrate the applicability of the proposed two-step gelation fabrication approach, in which the bioink consisting of 3% (w/v) alginate and 10% (w/v) gelatin was printed in a 0.8% (w/v) Carbopol bath with neutral pH. The total height of the Y-shaped structure was 8 mm, the mean wall thickness was 1 mm, the average diameter of the base portion was 4 mm, the height was 4 mm and the horizontal plane projection diameter of the bifurcation part was 4 mm, the dip angle was 45°, and the height was 4 mm. The printing trajectories for the base, junction, and bifurcated portions, as well as the printing procedure, are illustrated

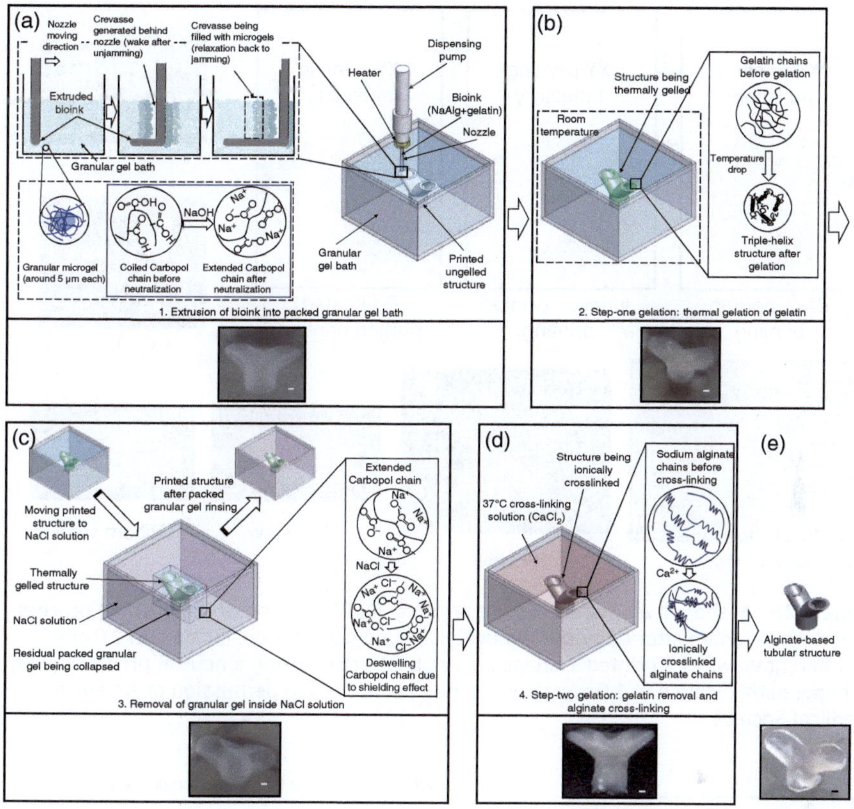

Figure 2.22 Schematic of two-step gelation approach based on reversible thermal gelation of gelatin and ionic cross-linking of sodium alginate. (a) Printing of alginate and gelatin bioink at 37 °C into a Carbopol granular gel bath to form a liquid structure. (b) Thermal gelation of gelatin in the Carbopol bath at room temperature. (c) Residual Carbopol removal at room temperature. (d) Liquefaction of gelatin and ionic cross-linking of alginate at 37 °C in a $CaCl_2$ bath. (e) Printed alginate structure. (Scale bars: 1.0 mm.) Source: Jin et al. [59]/with permission of American Chemical Society.

in Figure 2.23a, in which the connecting dashed line of the printing trajectory indicated that the pneumatic pump was turned off when the dispensing nozzle moved from one branch to the other. After immersion in the NaCl bath for 15 minutes, the printed tube was tested for Carbopol removal in the lumen by flowing a Dulbecco's Modified Eagles Medium (DMEM) solution supplemented with red dye for better visibility through the alginate/gelatin structure. As seen in Figure 2.23b, this process effectively removed Carbopol microgels from the structure. Figure 2.23c shows a printed Y-shaped alginate structure from different angles, demonstrating the effectiveness of printing 3D alginate structures using the proposed two-step gelation approach.

Using NIH 3T3 mouse fibroblast-laden bioink containing alginate and gelatin, the proposed "printing-then-gelation" approach was further applied to print straight and Y-shaped tubular constructs with living cells. The bioink contained 3% (w/v)

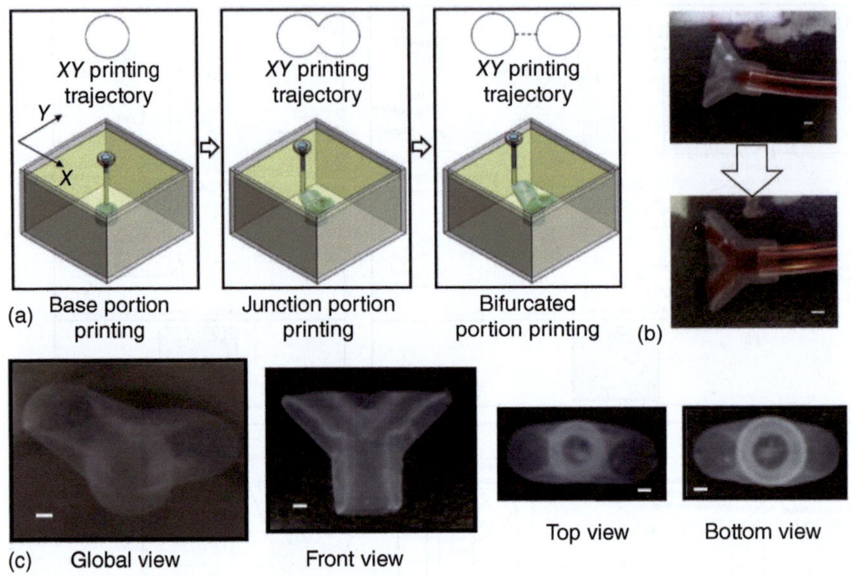

Figure 2.23 Y-shaped alginate structure printing. (a) Illustration of the printing process. (b) Testing of lumen after Carbopol removal using a dyed solution. (Scale bar: 2.0 mm.) (c) Different views of a printed Y-shaped alginate structure using a neutral pH, 0.8% (w/v) Carbopol bath. (Scale bar: 1.0 mm.) Source: Jin et al. [59]/with permission of American Chemical Society.

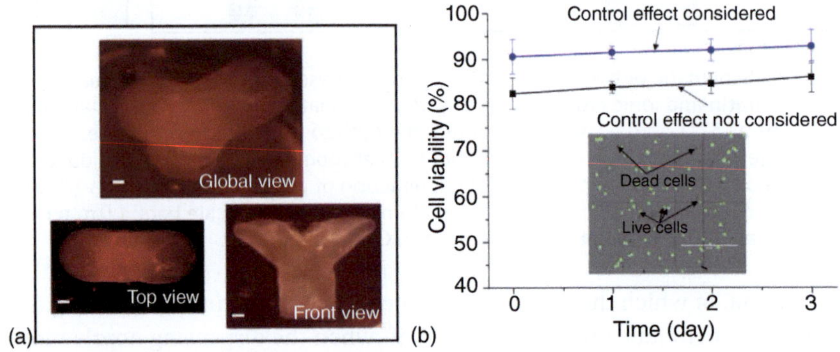

Figure 2.24 (a) Printed Y-shaped cellular construct (scale bars: 1.0 mm). (b) Post-printing cell viability test results. (Scale bar: 400 μm.) Source: Jin et al. [59]/with permission of American Chemical Society.

alginate, 10% (w/v) gelatin, and 5×10^6 cells ml^{-1} cells. Figure 2.24a illustrates a printed cellular tube with top and front views. The cell viability in the printed straight tubes was tested with unprinted bioink as the control group. It can be seen from the cell viability data as shown in Figure 2.24b that most cells can adequately survive during the proposed fabrication process with cell viability higher than 80% under representative printing conditions, demonstrating the potential of the proposed "printing-then-gelation" approach as a promising biofabrication technique.

The pressure-induced damage [66] during extrusion and alginate gelation-induced injury [67] were the main causes of cell death, while reversible cell injury during printing [52] and possible proliferation resulted in a slight increase of cell viability from 82.5% to 86.1% after printing.

2.4.3 Printing-then-Solidification of Liquid Materials in Nanoclay Suspension

The development of functional support bath materials has been of great interest for liquid material extrusion. Although Carbopol microgels [52, 59] and gelatin particles [53] are two promising yield-stress support bath materials, there are still some difficulties: ionic cross-linking material printing cannot be conducted in Carbopol because of the ionic sensitivity of Carbopol microgels, while the low melting temperature of gelatin particles (37 °C) limits the printing of thermosensitive materials with higher working temperatures in a gelatin particle bath. In Jin et al.'s work [54], Laponite nanoclay yield-stress colloid, a member of the smectite mineral family, was investigated as a functional support bath material to enable the "printing-then-solidification" fashion.

2.4.3.1 Laponite Utilized as the Support Bath Material for Extrusion Printing

Laponite is a synthetic nanoclay widely used in the fields of personal care products, coatings, and industrial applications. It consists of crystalline $Na_{0.7}Si_8Mg_{5.5}Li_{0.3}O_{20}(OH)_4$ in the form of nanoscale platelets (nanosilicates), approximately 1 nm thick and 25 nm in diameter, with very low polydispersity. Although they aggregate in the dry state, individual platelets disperse readily in aqueous solutions to form colloidal suspensions. The faces of each disc are negatively charged by the separation of sodium ions from individual platelets, while the dissociation of hydroxide ions gives the edges slight positive charges. This charge distribution drives Laponite platelets to form a stable "house-of-cards" arrangement, and the aqueous Laponite suspension exhibits a certain yield stress at equilibrium, which is the threshold energy required to break the "house-the-cards" before the suspension can flow. At a certain Laponite concentration, the charge distribution leads to the formation of the "house-of-cards" arrangement, and ionic impurities in the Laponite suspension may affect its rheological properties due to the influence of the ionic strength on electrostatic interactions. As reported, the addition of ionic solutions such as NaCl to Laponite RD dispersions, especially those with lower volume fractions ($\sim 10^{-3}$), may vary the concentration threshold. Therefore, deionized water is used as the solvent to prepare Laponite suspensions to avoid the influence of ionic impurities. Laponite is an excellent support bath material for printing different liquid building materials due to its wide range of yield stress at different grades and concentrations as well as the stable yield stress in different working conditions. The Laponite colloid is able to rapidly regain equilibrium under typical printing conditions after being disrupted by an external stimulus, as demonstrated in an imaging study. This unique yield-stress property is explored for liquid extrusion printing for the first time.

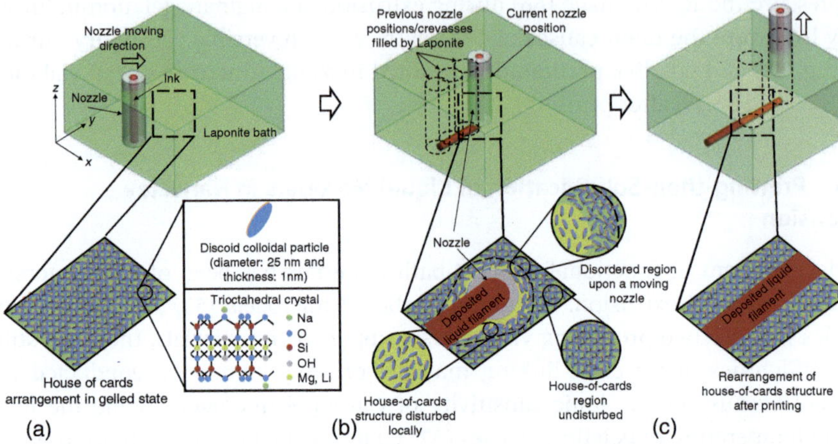

Figure 2.25 Schematic of extrusion printing in a nanoclay bath of colloidal Laponite. (a) Laponite bath with house-of-cards structure before nozzle movement. (b) When the nozzle moves, Laponite nanoclay particles around the nozzle tip are physically disturbed and fluidized to fill the crevasse due to the nozzle movement as well as cover the feature being printed. (c) After nozzle movement, Laponite particles rearrange and recover to the original house-of-cards structure and trap the deposited feature in place. Source: Jin et al. [54]/with permission of American Chemical Society.

When printing in a Laponite bath, the nanosilicates away from the nozzle retain the "house-of-cards" arrangement and maintain the gel status as shown in Figure 2.25a. Laponite nanosilicates around the nozzle tip are physically disturbed and disordered to fluidize during the movement of the nozzle (Figure 2.25b). Such gel–sol transition during shear flow under different flowing pressures has been carefully examined, and the yield-stress range of some Laponite suspensions has been further determined by standard rheological measurements. It is reported that the nanoclay suspension shows a typical gel or glass state at a low shear stress, while it transits from a gel state to a sol state at a high shear stress. Figure 2.25c shows that once the nozzle travels through the disturbed Laponite nanosilicates recover to the "house-of-cards" arrangement rapidly to trap the deposited liquid structure in place. Finally, the whole printed liquid structure is solidified in the Laponite bath in the "printing-then-solidification" fashion, which can be further solidified (second-step solidification), if needed.

2.4.3.2 Gelatin-Based Cellular Construct Fabrication

For biological applications, the ink is made from gelatin derived from collagen, a stiff helical protein with the repeating amino acid sequence glycine-X_1–X_2 in which X_1 and X_2 are often proline and hydroxyproline. Some of the fragments in the gelatin change into a triple-helical conformation when the temperature is below about 35 °C. This connection between the gelatin molecules causes the gelatin molecules to form a junction, resulting in a bulk thermal hydrogel, as shown in Figure 2.26a. Herein, in order to improve the extrudability and structural integrity of the gelatin during incubation, 2.0% (w/v) alginate is added. The bioink is printed

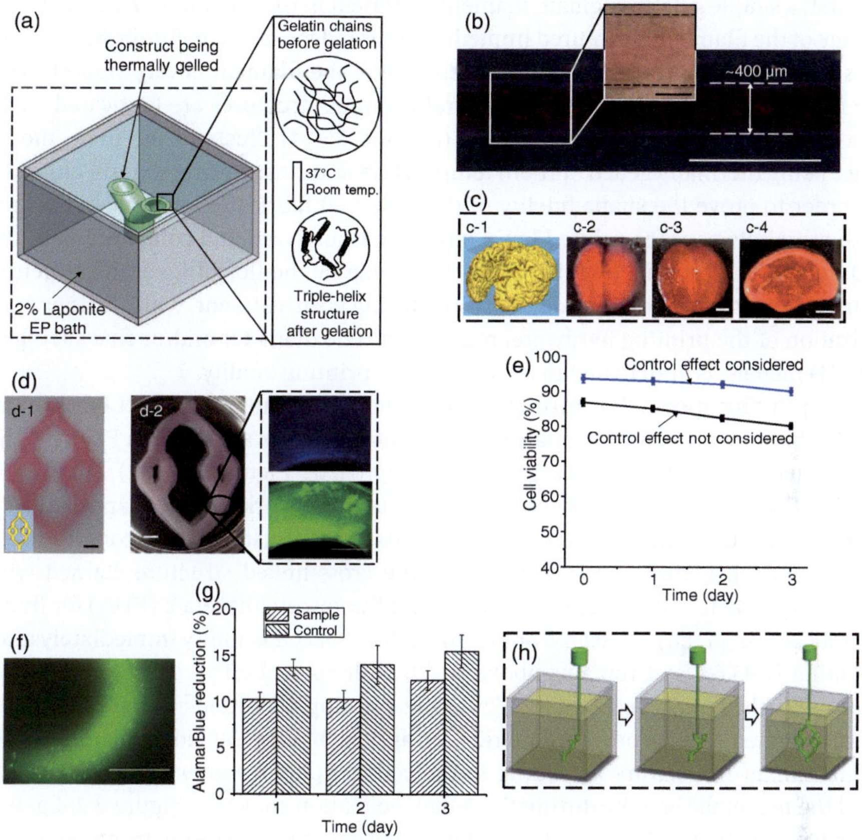

Figure 2.26 Gelatin structure printing. (a) Gelatin structure printed and then thermally cross-linked in a Laponite EP bath. (b) Morphology of a gelatin filament printed in the Laponite EP bath and a magnified view (inset). (c) Brain model as designed (c-1), thermally gelled brain structure in the Laponite EP bath (c-2), and top (c-3) and side (c-4) views of thermally gelled brain structure after Laponite rinsing. (d) Printed microvascular network in the Laponite bath (d-1) and microvascular network model (inset) and cross-linked fibroblast-based gelatin-alginate construct (d-2) (insets: segment of tube wall with all dyed cells in blue and dyed living cells in green). (e) Post-printing cell viability information (with one ± standard deviation error bars and $n = 3$). (f) Cell adhesion and proliferation after 3-day incubation. (g) Metabolic activity of cells during 3-day incubation. (h) Printing procedure of microvascular network by proposed "localized layer-by-layer" approach. (Scale bars: 1.0 mm (b), 200 μm (b, inset), 3.0 mm (c), 3.0 mm (d), 1.0 mm (d, insets), and 1.0 mm (f).) Source: Jin et al. [54]/with permission of American Chemical Society.

at an elevated temperature (37 °C) into a 37 °C Laponite EP bath and then gelled at room temperature. Accordingly, a 2.0% (w/v) Laponite EP suspension is used as the support bath material due to its suitable rheological properties. After gelatin is thermally gelled, the printed structures can be removed from the Laponite EP bath. Then residual Laponite is rinsed using an aqueous NaCl solution. Finally, the printed structures are immersed in a 37 °C $CaCl_2$ bath to cross-link sodium alginate.

First, a simple gelatin-alginate filament is printed in the Laponite EP bath, and an image of the filament is captured immediately after deposition (before cross-linking) to show the good printing quality: the surface of the filament is very smooth with no fuzzy edge phenomenon. Then several complex structures are fabricated using the gelatin-alginate bioink. Figure 2.26(c3) and 26(c4) illustrate the brain model after being thermally gelled at room temperature and rinsed using sodium chloride. In order to prove the shape fidelity of the proposed fabrication approach, the surface morphology of the printed brain structure is analyzed and compared with the 3D model. The printing quality might be affected by the flexibility of the structure during printing as well as the deformation during posttreatment, while further optimization of the printing hardware, reducing nozzle diameter and/or fine slicing of the 3D models, is anticipated to achieve higher printing quality.

The printing approach is further implemented for cellular construct fabrication. A fibroblast-laden gelatin-alginate microvascular network as shown in Figure 2.26d is printed using the proposed printing approach. Figure 2.26(d1) shows the well-defined liquid gelatin-based microvascular network in the Laponite bath with a wall thickness of ~0.8 mm, while its inset illustrates the computerized design model. Figure 2.26(d2) illustrates the cross-linked structure stained with Hoechst 33342 for all fibroblasts (in blue) and fluorescein diacetate (FDA) for living fibroblasts (in green). As seen from Figure 2.26e, the cell viability immediately after printing is 93.6% and remains above 90.0% with control effect considered, and it can be further improved by optimizing the printing conditions. It can be further demonstrated that the proposed fabrication approach has great biocompatibility and contaminant-free nature according to the cell morphology after 3-day incubation and the metabolic activity during the 3-day incubation shown in Figure 2.26f,g. It is worth noting that cells may not spread well in a gelatin-alginate environment while they are still alive. Since gelatin is selected as a thermosensitive model hydrogel to validate the proposed biofabrication approach, other applicable hydrogels and/or necessary growth cues will be used to understand the intensive biological evaluation of printed cells.

2.5 Continuous Precuring Digital Light Processing (DLP) Printing

2.5.1 Introduction to DLP Printing

Rather than operating serially as conventional inkjet or extrusion printing, DLP printing fabricates an entire plane of the object at once, which substantially decreases the building time [68]. As shown in Figure 2.27, the general setup of DLP printers involves a light source, typically UV or visible light, that illuminates a DMD chip programmed to project various digital patterns through a set of optics into a tank filled with the photocurable solution, and a platform upon the tank is used to control the height of the printing objects. Before DLP printing, an aimed CAD model is converted into STL format and sliced with preset layer thickness. During

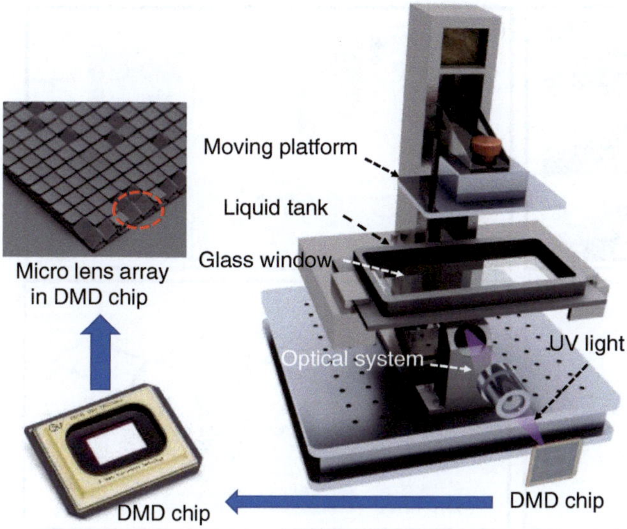

Figure 2.27 Schematic of the DLP printer.

DLP printing, one slice pattern is sent to the DMD chip, and the reflected UV light by the DMD chip is modulated to transmit the pattern information through the optical system onto the forming plane. The photocurable solution between the moving platform and the tank bottom (Figure 2.27) is photocured with a certain thickness by the modulated UV light, and then the moving platform raises a certain distance to allow the filling of the photocurable solution. After that, the next pattern information is sent, and the process will be achieved layer-by-layer until the model is built. It is worth noticing that each micromirror on the DMD chip is representative of one pixel in the digital image and thus the printing resolution can be achieved as small as 3–5 μm. As such, highly complex biomimetic structures can be readily generated with physiologically relevant topological feature sizes. Moreover, since DLP printing can be conducted by a bottom-up projection process (Figure 2.27) in a liquid environment, no support materials are needed to improve the layer artifacts [69].

In the conventional layer-by-layer DLP printing, the building regime operates in a sequential fashion where a layer is printed and then the build platform is moved upward to allow unpolymerized material to recoat the printing area prior to fabricating the next layer. Thus, a long wait time is needed during the recoating of viscous photocurable solutions [70]. Moreover, there is an intrinsic trade-off between printing efficiency and fidelity during conventional layer-by-layer DLP printing, and the printing time becomes much longer when a thinner slice is implemented for higher printing fidelity [71]. Recently, the concept of continuous DLP printing systems was developed to circumvent these challenges. Tumbleston et al. [72] developed a continuous DLP method (Figure 2.28a), named continuous liquid interface production (CLIP), which achieves continuous printing by oxygen-inhibited photopolymerization and generates a continuous liquid interface between the growing part and the

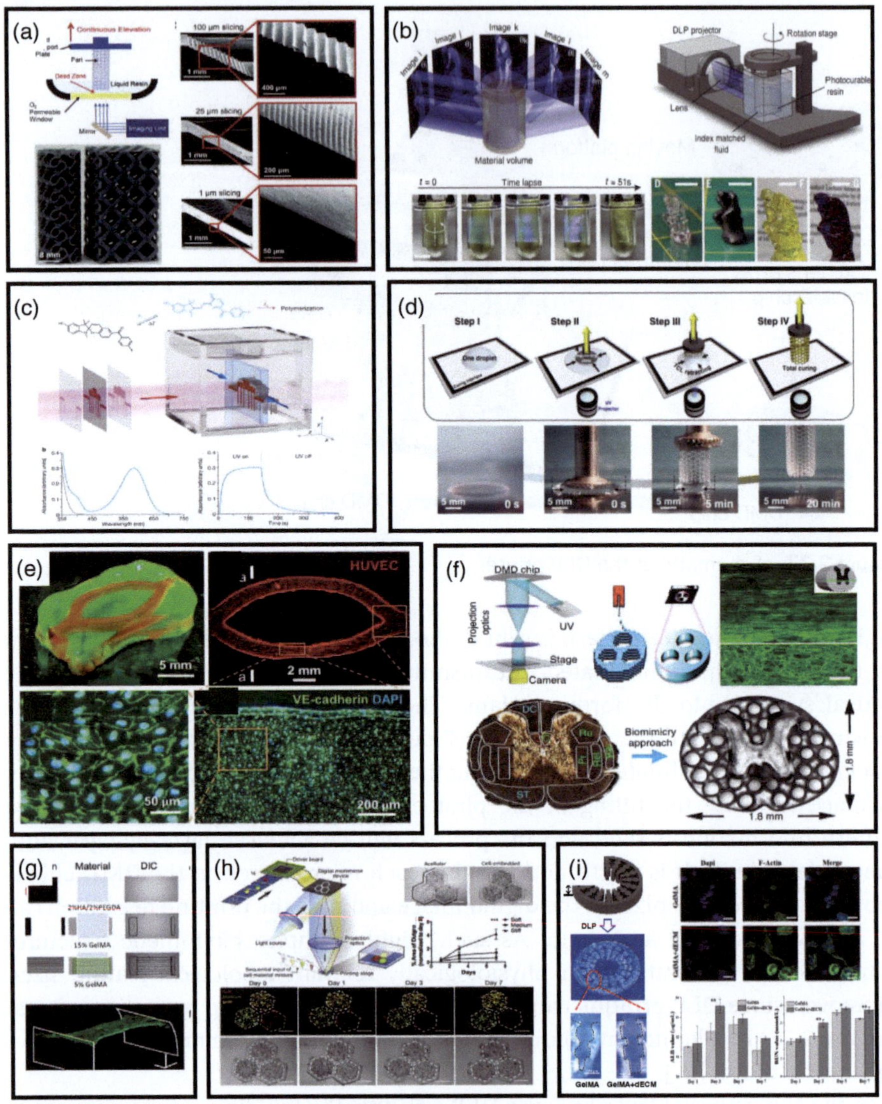

Figure 2.28 Research progress of DLP technology. (a) Continuous liquid interface production (CLIP). Source: Tumbleston [72]/with permission of American Association for the Advancement of Science – AAAS. (b) Computed axial lithography (CAL). Source: Kelly et al. [73]/with permission of American Association for the Advancement of Science – AAAS. (c) Volumetric 3D printing. Source: Regehly et al. [74]/with permission of Springer Nature. (d) Continuous 3D printing from one single droplet. Source: Zhang et al. [75]/with permission of Springer Nature; (e) Prevascularized flow channel with endothelial cells. Source: Anandakrishnan et al. [76]/with permission of John Wiley & Sons, Inc. (f) Spinal cord repair scaffold with nerve cells. Source: Koffler et al. [77]/with permission of Springer Nature. (g) DLP-printed scaffold pattern for NMVCM cells outgrowth. Source: Liu et al. [78]/with permission of Elsevier. (h) DLP-printed stent with different values of stiffness affecting liver cells function. Source: Ma et al. [79]/with permission of Elsevier. (i) DLP printing of liver microtissue with hiHep cells. Source: Mao et al. [80]/with permission of Elsevier.

exposure window. However, the construction of the dead zone (oxygen-rich zone) and the use of an entire liquid tank of the photocurable solution impede the implantation of CLIP of printing cell-laden bioinks. Kelly et al. [73] developed the computed axial lithography (CAL) (Figure 2.28b) to build models by converting vertical model construction into rotational molding depending on a new slicing method of the model. Based on the xolography method, Regehly et al. [74] first used blue light to continuously excite the forming area and then used UV light to continue forming in the orthogonal direction to achieve the volumetric 3D printing, which greatly improved the DLP forming efficiency (Figure 2.28c). Zhang et al. [75] also performed continuous DLP printing to form a droplet of superhydrophobic materials, which significantly reduced material waste (Figure 2.28d). Volumetric 3D printing and CAL technology require a large volume of printing materials in the tank, which limits their application of bioprinting with cells.

DLP printing has been used to fabricate tissue engineering scaffolds without embedded cells. Anandakrishnan et al. [76] built a large-scale model by GelMA/PEGDA bioink with an inner vascular-like flow channel, and endothelial cells were laid in the flow channel for pre-vascularization (Figure 2.28e), and Koffler et al. [77] built the DLP spinal cord scaffold for the directional attachment and growth of nerve cells to repair the spinal cord injury (Figure 2.28f). More importantly, DLP bioprinting of cell-laden bioinks has been applied for the printing of functional organs or tissues. Liu et al. [78] used DLP cell printing to construct a variety of patterns with NMVCM cell-laden GelMA bioink to explore the effect of matrix shape on myocardial beating (Figure 2.28g). Ma et al. [79] developed multimaterial DLP printing to mimic different degrees of liver cirrhosis to explore the pathological mechanism of liver cirrhosis (Figure 2.28h). Also, Mao et al. [80] constructed a liver microtissue using hiHep cell-laden GelMA/dECM bioink for repairing liver function defects (Figure 2.28i). However, previous studies on DLP cell printing were still limited by low printing efficiency and poor printing resolution and were difficult to construct complex 3D models. Thus, the novel DLP cell printing technology with high resolution and efficiency, but still keeping high cell viability, needs to be developed.

2.5.2 Theoretical Prediction of DLP Working Curve for Photocurable Materials

The accurate definition of the relation between UV light exposure time and the cured thickness of each layer is the most critical step for achieving continuous printing [81]. In previous studies, the Jacobs working curve [82] has been used to indicate the relationship between the absorbed UV light energy and the cured thickness of the photocurable materials, which can be used to precisely control the rising speed of the moving plate during DLP printing. Generally, experimental measurements are used to determine the Jacobs working curve of a certain photocurable material. However, the discrete measurements of forming structures for each photocurable material are extremely time consuming and wasteful of printing material. Moreover, it is very difficult to carry out experimental measurements of Jacobs working curves on very soft materials, since they cannot form the self-supporting construct allowing the formation of the layer measurement based on the DLP printing setup. Therefore, an analytical model to quantitatively obtain Jacobs working curves only based on material properties is of great importance for continuous DLP printing.

2.5.2.1 Analytical Model of Jacobs Working Curve

Li et al. [81] developed an analytical model of Jacobes working curve with several assumptions:

(a) The photocurable substance has two stages of absorbance during the DLP process, which are described by liquid absorbance, A_l, before materials have been cured, and solid absorbance, A_s, after completely cured.
(b) The DLP curing process can be subdivided into two phases (Figure 2.29a): in the first subphase, the material absorbs energy to polymerize small molecules, but the degree of polymerization is low and no solids are produced and this subphase ends before the solids appear. The time of this state is denoted as the threshold time, t_T. Then, in the second subphase, solids begin to appear, and their thickness increases.
(c) When the accumulated energy reaches $I_i{\cdot}t_T$ at a certain depth, where I_i is the light intensity, this depth is the curing depth (Figure 2.29b).

Here, the bottom surface is defined as $z=0$, and the UV light propagates from $-z$ to $+z$ with a light intensity I_i at $z=0$. As shown in Figure 2.30a, at an infinitesimal thickness $z=\mathrm{d}h$, the light intensity is

$$I = I_i \cdot \exp \cdot (A_l \cdot \mathrm{d}h) \tag{2.1}$$

before the solution is cured, and so after exposure to UV light for a time period of t_T, the UV light energy accumulated at $z=\mathrm{d}h$ reaches

$$E(z = \mathrm{d}h, t = t_T) = I_i \cdot \exp(A_l \cdot \mathrm{d}h) \cdot t_T \tag{2.2}$$

and the photocurable material at the bottom surface ($z=0$) is in the pre-cured state. If the photocurable material with a thickness of dh is completely cured after a further UV light exposure of time dt, the UV light energy accumulated at $z=\mathrm{d}h$ reaches the critical energy $E\ (z=\mathrm{d}h,\ t=t_T+\mathrm{d}t)=E_c=I_i \cdot t_T$. Thus, we see that the

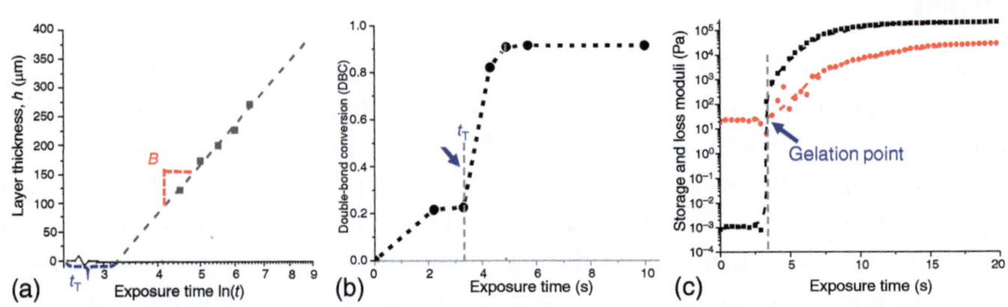

Figure 2.29 (a) Schematic diagram of working curve and meaning of each parameter. (b) Schematic diagram of the double bond conversion (DBC) curve. (c) Photorheological property photocurable materials.

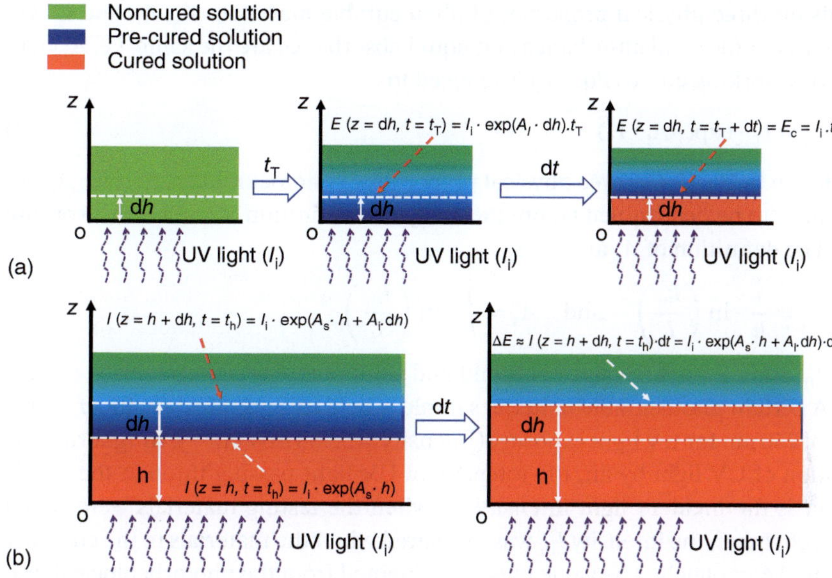

Figure 2.30 (a) Schematic of the beginning of curing with an infinitesimal thickness *dh*. (b) Schematic of curing with an infinitesimal thickness *dh* of a completely cured solid with a thickness of *h*. Source: Li et al. [81]/with permission of Elsevier.

UV light energy required for the material to cure a thickness dh after the pre-curing state is

$$E_{dh} = E_c - E(z = dh, t = t_T) = I_i[1 - \exp(A_l \cdot dh)]t_T \tag{2.3}$$

Figure 2.30b shows a schematic of the photocurable solution cured with an infinitesimal thickness $z = dh$ upon a completely cured solid with a thickness of h. The UV light energy absorbed in the process for curing the dh-thickness solution is

$$\Delta E \approx I(z = h + dh, t = t_h) \cdot dt = I_i \cdot \exp(A_s \cdot h + A_l \cdot dh) \cdot dt \tag{2.4}$$

Because in an infinitesimal thick layer, $E_{dh} = \Delta E$, and based on Eqs. (2.3, 2.4), it is shown that

$$dt = t_T \cdot \left[\frac{1 - \exp(A_l \cdot dh)}{\exp(A_s \cdot h + A_l \cdot dh)}\right] \tag{2.5}$$

the time to cure a material of thickness H from its initial state to a completely cured state (Figure 2.30b) can be obtained as

$$\begin{aligned} t_H &= t_T + \int_{t_T}^{t_H} dt = t_T - t_T \cdot \int_0^H A_l \cdot \exp(-A_s \cdot h)dh \\ &= t_T \cdot \left[\frac{A_s - A_l}{A_s} + \frac{A_l}{A_s} \cdot \exp(-A_s \cdot H)\right] \end{aligned} \tag{2.6}$$

Based on Eq. (2.6), it can be inferred that the relationship between UV light exposure time and the cured thickness, which is the Jacobs working curve, only depends on three physical properties of photocurable materials: A_s, A_l, and t_T. We observe that if the solid absorbance and liquid absorbance are the same, i.e. $A_s = A_l$, the Jacobs working curve (Eq. 2.6) is reduced to

$$t_H = t_T \cdot \exp(-A_l \cdot H) \tag{2.7}$$

As shown in Eq. (2.6), three physical properties of photocurable materials: A_s, A_l, and t_T have to be determined before the analytical definition of the Jacobs working curve. The definition of A_s and A_l is

$$A_l = \frac{1}{h} \cdot \ln\left(\frac{I_{lh}}{I_i}\right) \quad \text{and} \quad A_s = \frac{1}{h} \cdot \ln\left(\frac{I_{sh}}{I_i}\right) \tag{2.8}$$

where I_{lh} and I_{sh} are the values of liquid and solid outgoing light intensity, respectively. A certain photocurable material sample is held in a glass test chamber with a height of 250 μm for UV light exposure (intensity = 5.988 mW cm^{-2}). By ignoring the absorption of UV light by air, the intensity of UV light passing through the empty chamber is the incident light intensity, I_i, when the testing materials were added into the chamber, the intensity I_{lh} was measured when the materials in the chamber were cured completely. The value of t_T was obtained from the photorheological test, and the equalization of storage modulus and loss modulus ($G' = G''$) means that the material state changes, and the corresponding time is the curing threshold time t_T, as shown in Figure 2.29c.

2.5.2.2 Influence of UV Absorber Concentration

UV absorber is a substance that can adjust the distribution of light energy in photocurable materials. As shown in Eq. (2.6), the three physical properties of photocurable materials, A_s, A_l, and t_T, significantly influence the working curve, while the UV absorber affects the working curve by acting on these three materials physical properties. According to the Beer–Lambert's law,

$$A = -Khc \tag{2.9}$$

where A is the absorbance of materials, K is the molar absorption coefficient of the light-absorbing substance, h is the thickness of the material, and c is the density of the light-absorbing substance. Thus, the absorbance of a certain photocurable material is linearly determined by the UV absorber density. Furthermore, the energy absorbed by the thermal motion of the UV absorber molecules and the energy required for the curing reaction constitute a competitive relationship. Thus, t_T is also affected by the concentration of the UV absorber.

In addition, the UV absorber also affects the distribution of UV light on the forming surface. As shown in Figure 2.31, the UV light penetration depth or the UV light energy is not uniformly distributed on the imaging surface due to the disturbance of the optical system, and the Gaussian distribution of energy reflected by a single micromirror is generally obtained. The excess energy of the superimposed UV light will cause over-curing and the addition of a UV absorber changes the distribution of UV light, which can smooth the superimposed energy state, reduce excess energy, and avoid over-curing of UV light.

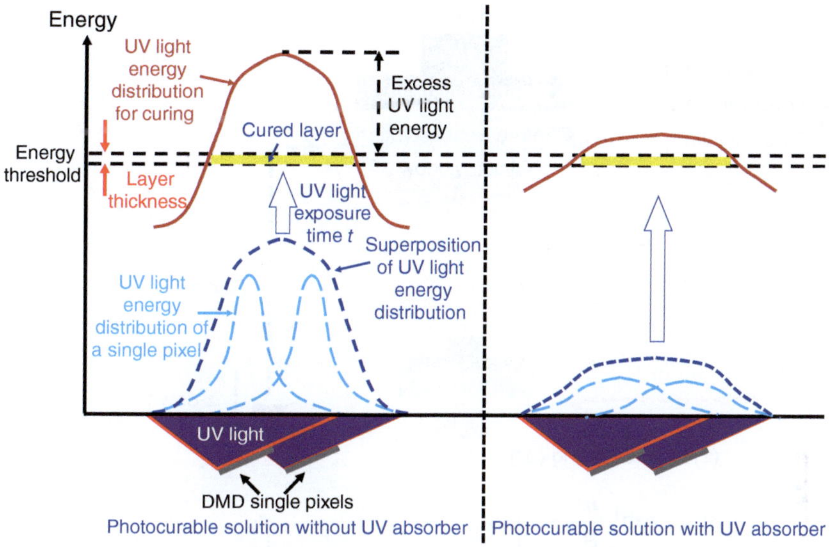

Figure 2.31 Schematic of the UV light energy distribution and accumulation in the photocurable solutions with or without the UV absorber. Source: Reproduced with permission from Li et al. [70], Elsevier.

2.5.3 Pre-curing Digital Light Processing (DLP) Printing

Taking advantage of the theoretical model of Jacobs working curve (Eq. 2.6), the curing process of the photocurable solutions can be subdivided into two phases, pre-curing and further curing, and the exposure threshold time (t_T) is used to predict the pre-curing time. After the pre-curing phase, the photocurable solution is further exposed to maintain the pre-gelled state and cured within a short "curing time" (Δt) to form a designed thickness.

2.5.3.1 The Tunable Pre-curing DLP Printing Approach

When the photocurable solution starts to be illuminated, it takes time t_T for the photocurable solution to absorb enough UV light energy to reach the pre-gelled state, which is termed pre-curing. Then, only a very short exposure time (Δt) is required for the pre-cured solution to transit from liquid to solid. When the curing process is finished, the platform moves up a distance of one layer thickness for the next printing cycle, which costs time t_R^{TP}. Moreover, the addition of the UV absorber can adjust the light intensity distribution in the photocurable solution.

As shown in Figure 2.32, the process of the tunable pre-curing DLP approach can be divided into three steps. (i) UV light modulated to carry slice image information by the DMD chip (Figure 2.31) to excite the photocurable solution in the liquid tank, and the distribution of UV light energy in the solution is shown in Figure 2.31. When the absorbed UV light energy reaches the threshold value, the photocurable solution is excited from the initial state to the pre-gelled state within the threshold time t_T and then remains in the pre-gelled state. (ii) The pre-cured photocurable solution

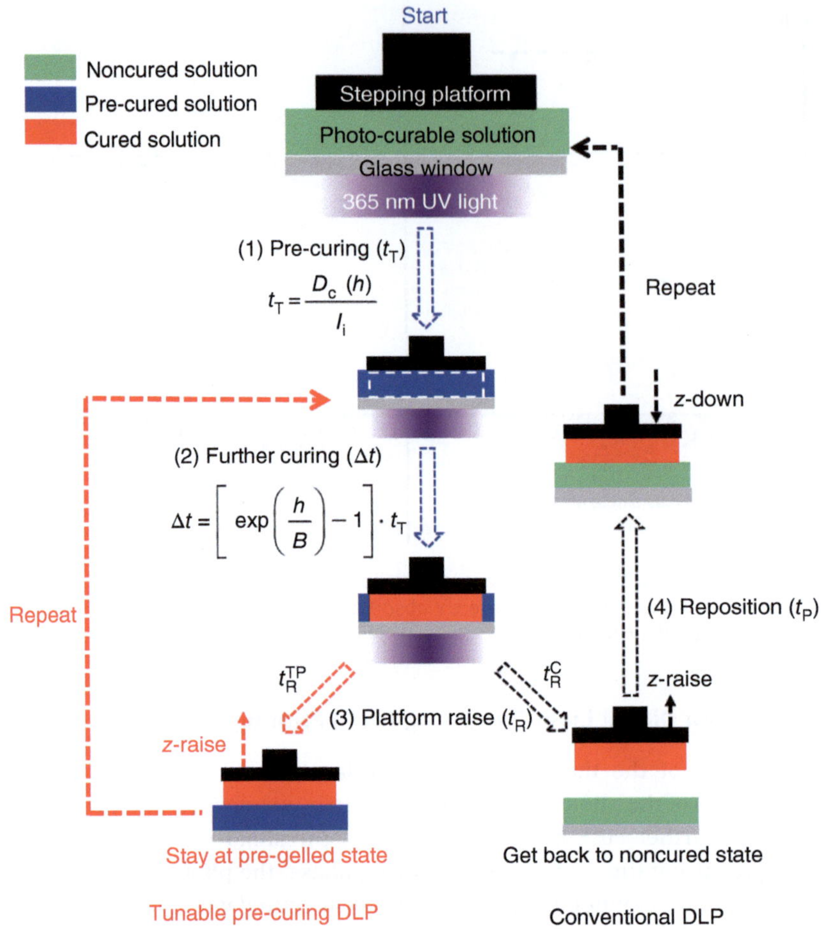

Figure 2.32 The basic steps of each printing circle of the conventional and tunable pre-curing DLP approach. Source: Reproduced with permission from Li et al. [70], Elsevier.

is cured with a slice image pattern after a further curing time Δt. In the third step, the moving platform is raised within the time t_R^{TP}. (iii) The tunable pre-curing DLP approach repeats the second and third steps of the next printing layer to form a loop of continuous DLP printing.

2.5.3.2 Improving DLP Printing Efficiency by Pre-curing DLP Printing

As shown in Figure 2.32, there are four steps in each printing cycle of the conventional DLP approach. The first three steps of both the conventional and tunable pre-curing DLP approaches are the same, including pre-curing, further curing, and recoating of the materials. In the third step of the tunable pre-curing DLP, the platform raises a height equal to the slice thickness, which spends time t_R^{TP}, and the photocurable solution always stays in a pre-gelled state. However, in the third step of conventional DLP, the platform raises a height much larger than the slice thickness, which spends time t_R^C. In addition, the whole liquid tank is filled with the photocurable solution in the conventional DLP, thus in each printing cycle, the platform needs to be repositioned to ensure the distance between the bottom of the forming structure and the bottom of the liquid tank is one slice height. There is a fourth step of conventional DLP to relocate the forming platform during each printing cycle, which spends time of t_P. Therefore, the total printing time of the conventional DLP approach (t_{total}^{C}) with n printing layers is

$$t_{total}^{C} = n \cdot \left(t_T + \Delta t + t_R^C + t_P\right) \tag{2.10}$$

The first step (pre-curing) of the tunable pre-curing DLP approach only occurs at the beginning of printing; this is because the photocurable solution is always kept in the pre-gelled state during the entire printing process (Figure 2.32). Only two steps are repeated in the printing cycles of tunable pre-curing DLP, and the total printing time with n printing layers can be estimated as

$$t_{total}^{TP} = t_T + n \cdot \left(\Delta t + t_R^{TP}\right) \tag{2.11}$$

where Δt can be estimated based on Eq. (2.6) as

$$\Delta t = t_H - t_T = t_T \cdot \frac{A_1}{A_s}[\exp(-A_s \cdot H) - 1] \tag{2.12}$$

Since t_T is related to material properties of the photocurable solution, and t_P is related to the mechanical movement, t_T and t_P are independent of the slice thickness. Furthermore, t_R^{TP} and t_R^C increase with platform movement distance, and Δt increases with slice thickness h (Eq. 2.12). Using the pyramid (Figure 2.34a) as the benchmark, the total printing times of the pyramid (5 mm) by both conventional and tunable pre-curing DLP approaches with different slice thickness are predicted by Eqs. (2.10, 2.11), respectively (Figure 2.34), where $t_T = 3.25$ seconds, $t_R^{TP} = 300$ ms in the tunable pre-curing DLP, and $t_R^C = 1800$ ms in the conventional DLP. In addition, $t_P = 1500$ ms is determined experimentally and Δt for different slice thicknesses is predicted according to Eq. (2.12). As shown in Figure 2.33, the difference of the total time between the two printing methods increases with the reduction of the printing layer thickness. For example, when the slice thickness is 100 μm, the difference of the total printing time between two DLP approaches is approximately 446.75 seconds, whereas the difference becomes 3121.75 seconds when the slice thickness is 10 μm. Besides, Figure 2.33 shows the different components of the total printing time in the two DLP approaches. The total time of Δt is the same in the tunable pre-curing and conventional DLP approaches under different values of the slice thickness (i.e. 10, 50, and 100 μm) since this period is determined by the slice thickness rather than the number of slices. However, in the conventional DLP approach, t_T, t_R^{TP} (or t_R^C) and t_P are the predominant components of total printing time, since these three times are related to the number of printing layers, which significantly increase as slice thickness decreases. Moreover, since the pre-curing only occurs at the beginning of the entire printing process and there is no need to relocate the platform in tunable pre-curing DLP (Figure 2.32), there is no t_P in the total printing time, and t_T does not increase with the number of slices. Therefore, the total printing time of the tunable pre-curing DLP approach is much shorter than that of the conventional DLP approach, especially when slice thickness is very small.

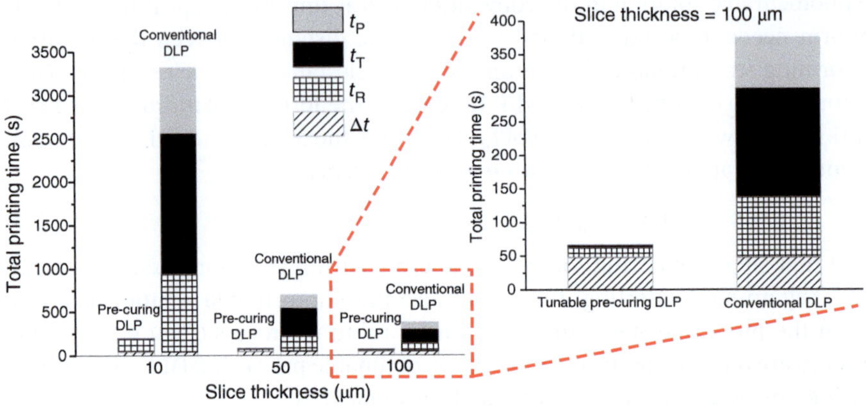

Figure 2.33 The total printing of the conventional and tunable pre-curing DLP approaches using different values of the slice thickness. Source: Reproduced with permission from Li et al. [70], Elsevier.

2.5.3.3 Validation of Pre-curing DLP Printing

As shown in Figure 2.34c,d, the staircase effect with a step height close to 100 μm is very obvious. However, the print result of the tunable pre-curing DLP approach with 10 μm slice thickness is much smoother than the result with 100 μm slice thickness by both two approaches (Figure 2.34b), which indicates a much better shape fidelity of the tunable pre-curing DLP approach. The printing time of the tunable pre-curing DLP approach with 10 μm slice thickness is shorter than that of the conventional DLP approach with 100 μm layer. There are other representative 3D constructs printed by the tunable pre-curing DLP approach shown in Figure 2.34e,f with a high printing fidelity. Therefore, the pre-curing DLP printing approach achieves a good manufacturing quality within a much shorter printing time than the conventional DLP approach.

The GelMA bioink encapsulated with PC12 cells was printed by conventional and tunable pre-curing DLP approaches into rectangular constructs with a size of $0.3 \times 5 \times 6\,mm^3$. The slice thickness for the conventional approach is 100 μm and for the tunable pre-curing DLP approach is 10 μm, respectively. The total printing time of the tunable pre-curing DLP (1.5 minutes) is much shorter than that of the conventional DLP (7.5 minutes), which greatly reduces the effect of UV light on the cells encapsulated in the photocurable hydrogels. The representative live/dead fluorescence images of PC12 cells embedded in GelMA hydrogels are shown in Figure 2.35a,b, and the values of the PC12 cell viability in the constructs printed by conventional and tunable pre-curing DLP approaches are $54.5 \pm 4.8\%$ and $90.2 \pm 6.1\%$, respectively. It can be concluded that the tunable pre-curing DLP approach is more suitable for bioprinting.

The UV light repeated exposure of the conventional DLP approach damages the cells in the bioinks and causes over-curing of the bioinks. Over-curing in the conventional DLP approach generally leads to the strong mechanical properties of materials, but the cells cannot spread well. The PC12 cells are unable to spread out in the

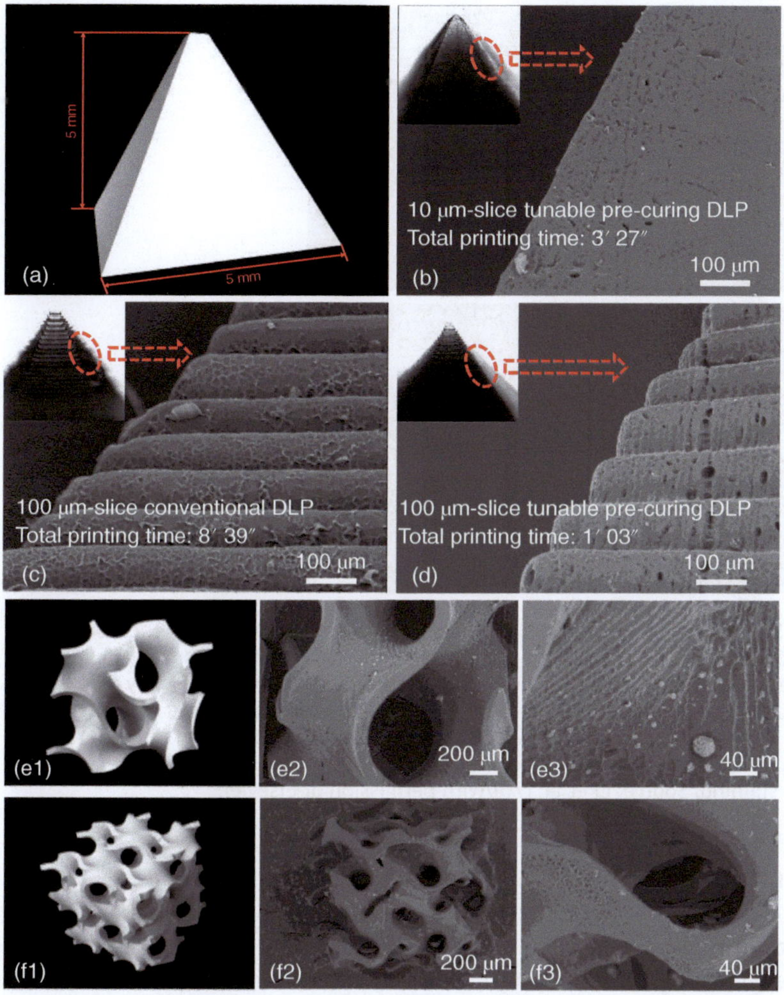

Figure 2.34 (a) The CAD model of the pyramid model. (b, c, and d) The microscope and SEM images of the pyramid models. (e, f) Triply periodic minimal surfaces model and the printed samples. Source: Li et al. [70]/with permission of Elsevier.

hydrogel printed by the conventional DLP approach (Figure 2.35c) but can grow into a fusiform in the hydrogel printed by the tunable pre-curing DLP approach (Figure 2.35d). Neurons can adaptively change the conformation and expression capacity of related proteins by sensing the stiffness and morphology of the ECM, and PC12 cells show a spreading morphology in the tunable pre-curing DLP-bioprinted hydrogels (Figure 2.35d). Therefore, the tunable pre-curing DLP bioprinting builds a more porous hydrogel construct that is similar to the natural ECM. Since the morphotypes of the neurocytes are important to neuron physiological function, the tunable pre-curing DLP approach is promising for utilization in printing thick bulk constructs with regenerating neurons, which is vital to form the biological function of 3D complex organs.

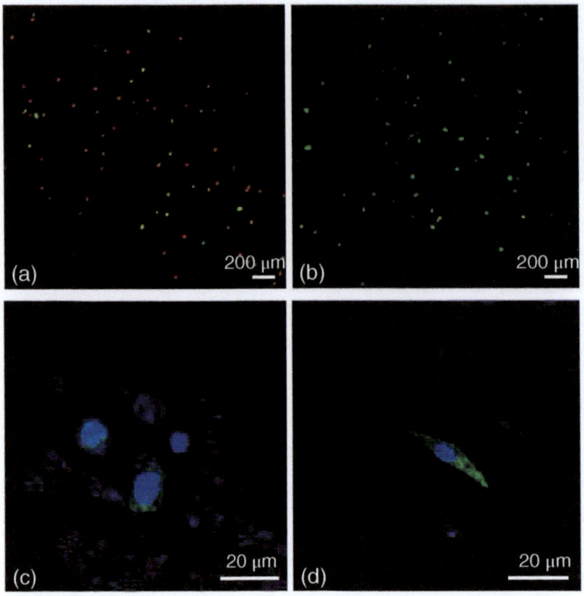

Figure 2.35 Representative live/dead fluorescence images of PC12 cells encapsulated in the printed rectangular constructs by conventional DLP with a 100 µm slice thickness (a) and the tunable pre-curing DLP with a 10 µm slice thickness (b) after 1 day of culture. Representative cellular morphology of PC12 cells encapsulated in printed rectangular constructs by the conventional DLP with a 100 µm slice thickness (c) and the tunable pre-curing DLP with a 10 µm slice thickness (d) after 3 days of culture and shown by Phalloidin staining. Source: Li et al. [70]/with permission of Elsevier.

2.6 Summary

3D cell printing has emerged as a promising solution to some of the obstacles faced by tissue engineering, and the ability to precisely control material and cell deposition as well as keep cell viability and functionality is the aim of the development of 3D cell printing. In this chapter, the recent progress of four cell printing technologies, including laser printing, inkjet printing, support-bath-enabled extrusion printing, and continuous DLP printing, has been reported. With each cell printing technology having its own working principle, the influence of each cell printing technology on the cell-laden bioinks has to be further understood to achieve the guideline for cell printing.

While remarkable achievements have been made, the field of 3D cell printing is still facing several challenges: (i) the emergence of bioinks with novel gelation mechanisms is needed to enable the development of novel cell printing technologies as well as the new biomedical applications; (ii) the quantitative analysis of the cell-laden bioink printability and shape fidelity is required to link the cell printing setup with the physical properties of cell-laden bioinks; and (iii) the potential integration of different cell printing technologies has to be achieved for building more complex tissues by combining the advantages of each technology.

References

1 Malda, J., Visser, J., Melchels, F.P. et al. (2013). 25th anniversary article: engineering hydrogels for biofabrication. *Advanced Materials* 25 (36): 5011–5028.

2 Nichol, J.W. and Khademhosseini, A. (2009). Modular tissue engineering: engineering biological tissues from the bottom up. *Soft Matter* 5 (7): 1312–1319.

3 Guillemot, F., Souquet, A., Catros, S. et al. (2010). High-throughput laser printing of cells and biomaterials for tissue engineering. *Acta Biomaterialia* 6 (7): 2494–2500.

4 Zhang, Z., Jin, Y., Yin, J. et al. (2018). Evaluation of bioink printability for bioprinting applications. *Applied Physics Reviews* 5: 041304.

5 Ferris, C.J., Gilmore, K.G., Wallace, G.G., and Panhuis, M. (2013). Biofabrication: an overview of the approaches used for printing of living cells. *Applied Microbiology and Biotechnology* 97 (10): 4243–4258.

6 Levato, R., Jungst, T., Scheuring, R.G. et al. (2020). From shape to function: the next step in bioprinting. *Advanced Materials* 32 (12): 1906423.

7 Kyle, S., Jessop, Z.M., Al-Sabah, A., and Whitaker, I.S. (2017). 'Printability' of candidate biomaterials for extrusion based 3D printing: state-of-the-art. *Advanced Healthcare Materials* 6 (16): 1700264.

8 Hölzl, K., Lin, S., Tytgat, L. et al. (2016). Bioink properties before, during and after 3D bioprinting. *Biofabrication* 8 (3): 32002.

9 Mandrycky, C., Wang, D.Z., Kim, K., and Kim, D. (2015). 3D bioprinting for engineering complex tissues. *Biotechnology Advances* 34, 34 (4): 422–434.

10 Sun, W., Starly, B., Daly, A.C. et al. (2020). The bioprinting roadmap. *Biofabrication* 12 (2): 22002.

11 Calvert, P. (2007). Printing cells. *Science* 318 (5848): 208.

12 Donderwinkel, I., Hest, J., and Cameron, N.R. (2017). Bio-inks for 3D bioprinting: recent advances and future prospects. *Polymer Chemistry* 8, 8 (31, 31): 4451–4471.

13 Slaughter, B.V., Khurshid, S.S., Fisher, O.Z. et al. (2009). Hydrogels in regenerative medicine. *Advanced Materials* 21 (32–33): 3307–3329.

14 Bingeisen, R., Pirlo, R.K., Wu, P.K. et al. (2013). Cell and organ printing turns 15: diverse research to commercial transitions. *MRS Bulletin* 38 (10): 834–843.

15 Lee, J.M. and Yeong, W.Y. (2016). Design and printing strategies in 3D bioprinting of cell-hydrogels: a review. *Advanced Healthcare Materials* 5 (22): 2856–2865.

16 Xu, H., Casillas, J., and Xu, C. (2019). Effects of printing conditions on cell distribution within microspheres during inkjet-based bioprinting. *AIP Advances* 9 (9): 95055.

17 Negro, A., Cherbuin, T., and Lutolf, M.P. (2018). 3D inkjet printing of complex, cell-laden hydrogel structures. *Scientific Reports* 8 (1): 17099.

18 Hewes, S., Wong, A.D., and Searson, P.C. (2017). Bioprinting microvessels using an inkjet printer. *Bioprinting* 7, 7: 14–18.

19 Compaan, A.M., Christensen, K., and Huang, Y. (2017). Inkjet bioprinting of 3D silk fibroin cellular constructs using sacrificial alginate. *ACS Biomaterials Science & Engineering* 3 (8): 6b, 1519–432b, 1526.

20 Takagi, D., Lin, W., Matsumoto, T. et al. (2019, 2018). High-precision three-dimensional inkjet technology for live cell bioprinting. *International Journal of Bioprinting* 5 (2): 208.

21 Ji, Y., Yang, Q., Huang, G. et al. (2019). Improved resolution and fidelity of droplet-based bioprinting by upward ejection. *ACS Biomaterials Science and Engineering* 5, 5 (8, 8): 4112–4121.

22 Teo, M.Y., Kee, S., Ravi Chandran, N. et al. (2019, 2019). Enabling free-standing 3D hydrogel microstructures with microreactive inkjet printing. *ACS Applied Materials & Interfaces* 12 (1): 1832–1839.

23 Xu, C., Zhang, M., Huang, Y. et al. (2014). Study of droplet formation process during drop-on-demand inkjetting of living cell-laden bioink. *Langmuir* 30 (30): 9130–9138.

24 Lin, Y., Huang, Y., and Chrisey, D. (2009). Droplet formation in matrix-assisted pulsed-laser evaporation direct writing of glycerol-water solution. *Journal of Applied Physics* 105: 93111.

25 Piqué, A., Chrisey, D., Auyeung, R.C.Y. et al. (1999). A novel laser transfer process for direct writing of electronic and sensor materials. *Applied Physics A* 69: S279–S284.

26 Zhang, Z., Xiong, R., Mei, R. et al. (2015). Time-resolved imaging study of jetting dynamics during laser printing of viscoelastic alginate solutions. *Langmuir* 31: 6447.

27 Zhang, Z., Xiong, R., Corr, D.T., and Huang, Y. (2016). Study of impingement types and printing quality during laser printing of viscoelastic alginate solutions. *Langmuir the ACS Journal of Surfaces & Colloids* 32: 3004.

28 Zhang, M., Krishnamoorthy, S., Song, H. et al. (2017). Ligament flow during drop-on-demand inkjet printing of bioink containing living cells. *Journal of Applied Physics* 121: 124904.

29 Zhang, Z., Xu, C., Xiong, R. et al. (2017). Effects of living cells on the bioink printability during laser printing. *Biomicrofluidics* 11: 034120.

30 Xu, C., Chai, W., Huang, Y., and Markwald, R.R. (2012). Scaffold-free inkjet printing of three-dimensional zigzag cellular tubes. *Biotechnology and Bioengineering* 109 (12): 3152–3160.

31 Christensen, K., Xu, C., Chai, W. et al. (2015). Freeform inkjet printing of cellular structures with bifurcations. *Biotechnology and Bioengineering* 112 (5): 1047–1055.

32 Yan, J., Huang, Y., and Chrisey, D. (2012). Laser-assisted printing of alginate long tubes and annular constructs. *Biofabrication* 5: 15002.

33 Dinca, V., Kasotakis, E., Catherine, J. et al. (2008). Directed three-dimensional patterning of self-assembled peptide fibrils. *Nano Letters* 8 (2): 538–543.

34 Colina, M., Serra, P., Fernández-Pradas, J.M. et al. (2005). DNA deposition through laser induced forward transfer. *Biosensors and Bioelectronics* 20 (8): 1638–1642.

35 Chrisey, D.B. (2000). The power of direct writing. *Science* 289 (5481): 879–881.

36 Barron, J., Wu, P., Ladouceur, H., and Ringeisen, B. (2004). Biological laser printing: a novel technique for creating heterogeneous 3-dimensional cell patterns. *Biomedical Microdevices* 6 (2): 139–147.

37 Ringeisen, B.R., Kim, H., Barron, J.A. et al. (2004). Laser printing of pluripotent embryonal carcinoma cells. *Tissue Engineering* 10 (3–4): 483–491.

38 Lin, Y., Yong, H., Wang, G. et al. (2009). Effect of laser fluence on yeast cell viability in laser-assisted cell transfer. *Journal of Applied Physics* 106 (4): 43106.

39 Gruene, M., Pflaum, M., Hess, C. et al. (2011). Laser printing of three-dimensional multicellular arrays for studies of cell-cell and cell-environment interactions. *Tissue Engineering Part C-Methods* 17 (10): 973–982.

40 Wei, W., Gang, L., and Huang, Y. (2019). Modeling of bubble expansion-induced cell mechanical profile in laser-assisted cell direct writing. *Journal of Manufacturing Science and Engineering* 131 (5): 51013.

41 Wang, W., Yong, H., and Chrisey, D. (2017). Numerical study of cell droplet and hydrogel coating impact process in cell direct writing. *Transactions of the North American Manufacturing Research Institution of SME* 35: 217–224.

42 Koch, L., Kuhn, S., Sorg, H. et al. (2009). Laser printing of skin cells and human stem cells. *Tissue Engineering Part C: Methods* 16 (5): 847–854.

43 Zhang, Z., Chai, W., Xiong, R. et al. (2017). Printing-induced cell injury evaluation during laser printing of 3T3 mouse fibroblasts. *Biofabrication* 9 (2): 25038.

44 Xiong, R., Zhang, Z., Chai, W. et al. (2017). Study of gelatin as an effective energy absorbing layer for laser bioprinting. *Biofabrication* 9 (2): 24103.

45 Wu, P.K. and Ringeisen, B.R. (2010). Development of human umbilical vein endothelial cell (HUVEC) and human umbilical vein smooth muscle cell (HUVSMC) branch/stem structures on hydrogel layers via biological laser printing (Bio LP). *Biofabrication* 2 (1): 14111.

46 Gaebel, R., Ma, N., Liu, J. et al. (2011). Patterning human stem cells and endothelial cells with laser printing for cardiac regeneration. *Biomaterials* 32 (35): 9218–9230.

47 Guillotin, B., Souquet, A., Catros, S. et al. (2017). Laser assisted bioprinting of engineered tissue with high cell density and microscale organization. *Biomaterials* 31 (28): 7520–7256.

48 Koch, L., Deiwick, A., Schlie, S. et al. (2012). Skin tissue generation by laser cell printing. *Biotechnology & Bioengineering* 109 (7): 1855–1863.

49 Xiong, R., Zhang, Z., Chai, W. et al. (2015). Freeform drop-on-demand laser printing of 3D alginate and cellular constructs. *Biofabrication* 7: 45011.

50 Ngo, T.D., Kashani, A., Imbalzano, G. et al. (2018). Additive manufacturing (3D printing): a review of materials, methods, applications and challenges. *Composites Part B: Engineering* 143: 172–196.

51 Mc Ormack, A., Highley, C.B., Leslie, N.R., and Melchels, F. (2020). 3D printing in suspension baths: keeping the promises of bioprinting afloat. *Trends in Biotechnology* 38 (6).

52 Bhattacharjee, T., Zehnder, S.M., Rowe, K.G. et al. (2015). Writing in the granular gel medium. *Science Advances* 1 (8): e1500655.

53 Hinton, T.J., Jallerat, Q., Palchesko, R.N. et al. (2015). Three-dimensional printing of complex biological structures by freeform reversible embedding of suspended hydrogels. *Science Advances* 1 (9): e1500758.

54 Jin, Y., Compaan, A., Chai, W., and Huang, Y. (2017). Functional nanoclay suspension for printing-then-solidification of liquid materials. *ACS Applied Materials & Interfaces* 9: 20057–20066.

55 Mirdamadi, E., Tashman, J.W., Shiwarski, D.J., and Palchesko, R.N. (2020). Fresh 3D bioprinting a full-size model of the human heart. *ACS Biomaterials Science and Engineering* 11 (6452): 6453–6459.

56 Wu, W., DeConinck, A., and Lewis, J. (2011). Omnidirectional printing of 3D microvascular networks. *Advanced Materials* 23: H178–H183.

57 Campos, D.D., Blaeser, A., Weber, M.J. et al. (2013). Three-dimensional printing of stem cell-laden hydrogels submerged in a hydrophobic high-density fluid. *Biofabrication* 5 (1): 15003.

58 Highley, C.B., Rodell, C.B., and Burdick, J.A. (2015). Direct 3D printing of shear-thinning hydrogels into self-healing hydrogels. *Advanced Materials* 27 (34): 5075–5079.

59 Jin, Y., Compaan, A., Bhattacharjee, T., and Huang, Y. (2016). Granular gel support-enabled extrusion of three-dimensional alginate and cellular structures. *Biofabrication* 8 (2): 25016.

60 Jin, Y., Song, K., Gellermann, N., and Huang, Y. (2019). Printing of hydrophobic materials in fumed silica nanoparticle suspension. *ACS Applied Materials & Interfaces* 11: 32.

61 Mironov, V., Boland, T., Trusk, T. et al. (2003). Organ printing: computer-aided jet-based 3D tissue engineering. *Trends in Biotechnology* 21 (4): 157–161.

62 Mironov, V., Visconti, R.P., Kasyanov, V. et al. (2009). Organ printing: tissue spheroids as building blocks. *Biomaterials* 30 (12): 2164–2174.

63 Chia, H.N. and Wu, B.M. (2015). Recent advances in 3D printing of biomaterials. *Journal of Jining Medical University* 9 (1): 4.

64 Lee, A., Hudson, A.R., Shiwarski, D.J. et al. (2019). 3D bioprinting of collagen to rebuild components of the human heart. *Science* 365 (6452): 482.

65 Skylar-Scott, M.A., Uzel, S., Nam, L.L. et al. (2019). Biomanufacturing of organ-specific tissues with high cellular density and embedded vascular channels. *Science Advances* 5 (9): w2459.

66 Xu, C., Christensen, K., Zhang, Z. et al. (2013). Predictive compensation-enabled horizontal inkjet printing of alginate tubular constructs. *Manufacturing Letters* 1 (1): 28–32.

67 Gudapati, H., Yan, J., Huang, Y., and Chrisey, D.B. (2014). Alginate gelation-induced cell death during laser-assisted cell printing. *Biofabrication* 6 (3): 35022.

68 Yu, C., Schimelman, J., Wang, P. et al. (2020). Photopolymerizable biomaterials and light-based 3D printing strategies for biomedical applications. *Chemical Reviews* 120: 10695–10743.

69 Shusteff, M., Browar, A.E.M., Kelly, B.E. et al. (2017). One-step volumetric additive manufacturing of complex polymer structures. *Science Advances* 3 (12): o5496.

70 Li, Y., Mao, Q., Li, X. et al. (2019). High-fidelity and high-efficiency additive manufacturing using tunable pre-curing digital light processing. *Additive Manufacturing* 30: 100889.

71 Janusziewicz, R., Tumbleston, J.R., Quintanilla, A.L. et al. (2016). Layerless fabrication with continuous liquid interface production. *Proceedings of the National academy of Sciences of the United States of America* 113 (42): 11703.

72 Tumbleston, J.R., Shirvanyants, D., Ermoshkin, N. et al. (2015). Continuous liquid interface production of 3D objects. *Science* 347 (6228): 1349.

73 Kelly, B.E., Bhattacharya, I., Heidari, H. et al. (2019). Volumetric additive manufacturing via tomographic reconstruction. *Science* 363 (6431): 1075.

74 Regehly, M., Garmshausen, Y., Reuter, M. et al. (2020). Xolography for linear volumetric 3D printing. *Nature* 588 (7839): 620–624.

75 Zhang, Y., Dong, Z., Li, C. et al. (2020). Continuous 3D printing from one single droplet. *Nature Communications* 11 (1): 1–10.

76 Anandakrishnan, N., Ye, H., Guo, Z. et al. (2021). Fast stereolithography printing of large-scale biocompatible hydrogel models. *Advanced Healthcare Materials* e2002103.

77 Koffler, J., Wei, Z., Xin, Q. et al. (2019). Biomimetic 3D-printed scaffolds for spinal cord injury repair. *Nature Medicine.*

78 Liu, J., Miller, K., Ma, X. et al. (2020). Direct 3D bioprinting of cardiac micro-tissues mimicking native myocardium. *Biomaterials* 256: 120204.

79 Ma, X., Yu, C., Wang, P. et al. (2018). Rapid 3D bioprinting of decellularized extracellular matrix with regionally varied mechanical properties and biomimetic microarchitecture. *Biomaterials* 185: 310–321.

80 Mao, Q., Wang, Y., Li, Y. et al. (2020). Fabrication of liver microtissue with liver decellularized extracellular matrix (dECM) bioink by digital light processing (DLP) bioprinting. *Materials Science and Engineering: C* 109: 110625.

81 Li, Y., Mao, Q., Yin, J. et al. (2021). Theoretical prediction and experimental validation of the digital light processing (DLP) working curve for photocurable materials. *Additive Manufacturing* 37: 101716.

82 Krishnamoorthy, S., Wadnap, S., Noorani, B. et al. (2020). Investigation of gelatin methacrylate working curves in dynamic optical projection stereolithography of vascular-like constructs. *European Polymer Journal* 124: 109487.

3

Alloy Design for Metal Additive Manufacturing

Dovgyy Bogdan and Minh-Son Pham

Imperial College London, Department of Materials, Exhibition Road, SW7 2AZ London, UK

3.1 Additive Manufacturing

One of the first developments in metal additive manufacturing (AM) dated back to 1995. Atwood et al. proposed using laser cladding as a rapid prototyping method, using fine pure aluminum powder [1]. Industrial AM of metallic components first appeared in 1998, when Optomec commercialized laser-engineered net shaping (LENS™), now known as direct energy deposition (DED). The system was developed at Sandia National Laboratories under a cooperative research and development agreement between United Technologies Pratt and Whitney [2]. By 1998, LENS was successfully used to fabricate parts from 316 stainless steel, nickel-based alloys, H13 tool steel, and titanium alloys [2]. The steady increase in equipment sales for metal AM is driving an increase in the material demand, in particular, metallic powders (Figure 3.1a). The market size for metallic powders is expected to hit the US$1 billion mark by 2022 [3]. AM now extends to over 30 different technologies covering a wide range of materials from plastic, ceramic, and composite to high-performance metallic alloys. In 2016, National Institute of Standards and Technology (NIST) estimated that total revenue from AM-related activities is worth US$44 billion [4]. The authors stated that real economic impact is difficult to calculate as a lot of companies use AM for their internal development and support for traditional manufacturing technologies. Nevertheless, these figures signify that AM technologies are maturing and are transitioning to widespread adoption across multiple industries, Figure 3.1b.

Of particular interest are metal fusion-based additive technologies. In literature, AM users often encounter confusion that multiple terms are used to refer to the same technique. For instance, laser powder bed fusion (LPBF), direct metal laser sintering (DMLS), laser cusing, direct metal printing (DMP), and selective laser melting (SLM) are all different terms for almost the same technique. Therefore, it is essential to use standard terminology. AM techniques can be categorized in several ways (e.g. by material feedstock or by melting technology). This chapter uses a terminology standard jointly developed by ASTM and ISO (ISO/ASTM 52900:2021) [5]. According to

Additive Manufacturing Technology: Design, Optimization, and Modeling, First Edition. Edited by Kun Zhou.

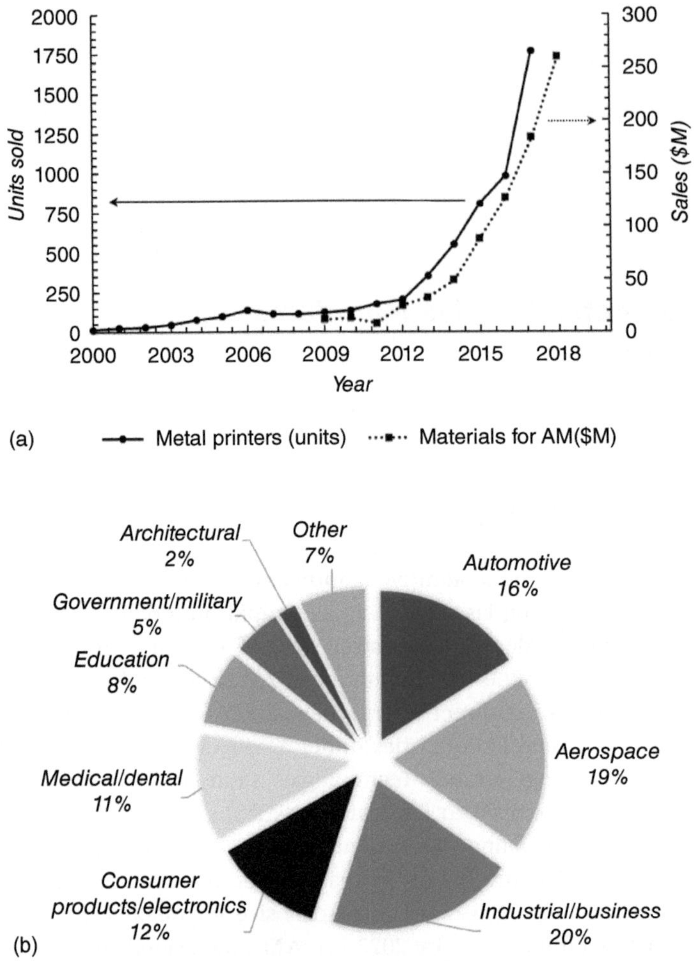

Figure 3.1 (a) Number of yearly unit sales for metal printers (solid) and market size of metal powders (dashed) AM over the 2007–2018 period. Source: Adapted from Wohelers [3]. (b) Industry segmentation of AM market in 2018. Source: Adapted from Wohelers [3].

this standard, AM processes are divided into seven categories based on the material feedstock: Vat photopolymerization, binder jetting, material jetting, material extrusion, powder bed fusion (PBF), sheet lamination, and directed energy deposition. The terminology is applicable for all materials used in AM.

3.1.1 Metal-Based Additive Manufacturing

Most metal AM research is conducted using powder bed fusion (PBF), as illustrated in Figure 3.2a. The heat source for this process is typically either a laser beam or an electron beam. Hence, PBF techniques with the two heat sources are termed laser PBF (LPBF) and electron beam powder bed fusion (EB-PBF), respectively. LPBF provides several advantages over EB-PBF in terms of broader applicability to various metals (electron bean can only be used with electrically conductive materials) and

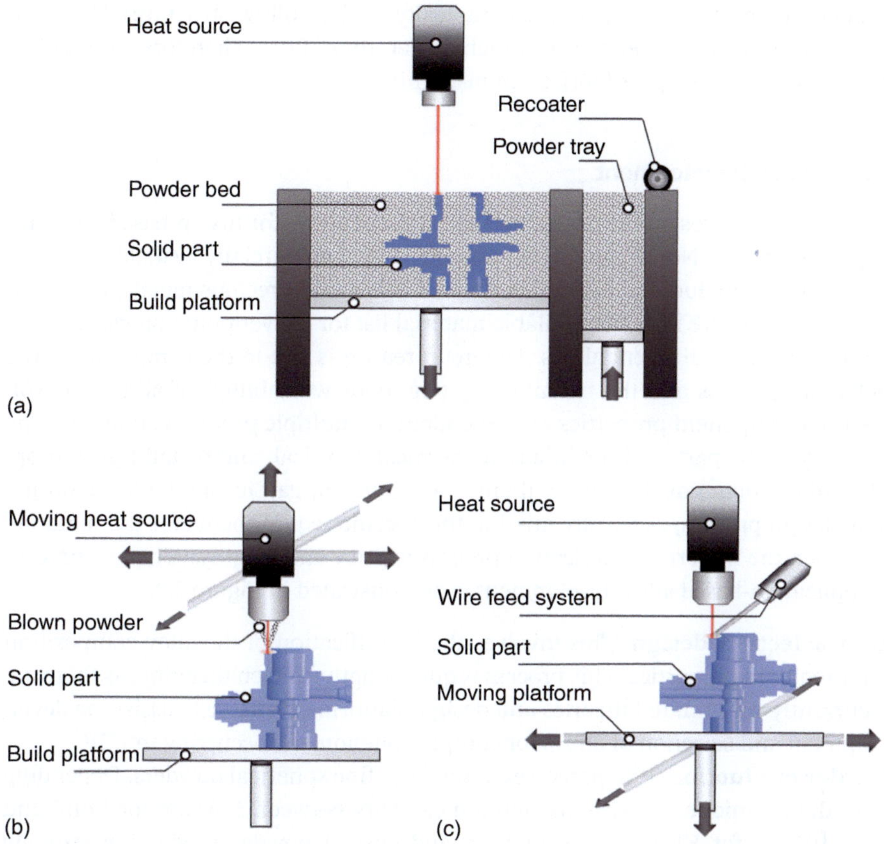

Figure 3.2 Typical fusion-based AM processes: (a) powder bed fusion (PBF), (b) and (c) direct energy deposition (DED) using powder and wire, respectively.

fine feature size (LPBF typically has a smaller melt pool, allowing for more delicate features). On the other hand, EB-PBF that provides higher deposition rates than LPBF, is carried out under a vacuum environment that helps reducing the oxidation levels during processing. The preheat of the build plate in EB-PBF also results in lower residual stresses in comparison to LPBF. Both PBF techniques are based on the same principle of operation. First, a fine powder layer is spread onto the build plate. Then, the heat source sequentially scans and melts the powder layer in a preset fashion, creating the part's first layer. Subsequently, the stage with the build platform lowers by one layer thickness, and the next layer of powder is deposited. The process repeats until the part is finished.

The DED process operates somewhat similarly to fusion welding. Both the material and energy are transferred to a location where the material is deposited. The heat source in this method can be a laser beam, an electron beam, or arc plasma. The material feed can take the form of powder, as shown in Figure 3.2b or wire Figure 3.2c. Such a method brings several advantages, such as the material deposition rate being significantly increased. The method is easier to implement for repairing parts and is capable of fabricating completely enclosed hollow structures (depending on the orientation and complexity of the hollow structure). However, the resolution of the process is much lower than PBF. Therefore, it requires significant additional post-fabrication machining.

3.1.2 Alloy Development

This chapter focuses on the design and selection of alloys for fusion-based AM with emphasis on PBF. Nowadays, there are around 30 commercially available powders for metal AM, including nonengineering materials and precious metals. However, this is still nowhere near the available material list for conventional processes, containing over 3000 different alloys. The main reason is due to the complexity of the metal AM process and the unsuitability (e.g. poor weldability) of alloys for AM. The final component properties are dependent on multiple process parameters, the geometry of the part, and the inherent chemical, physical, and metallurgical properties of the material. Therefore, there is a significant gap in the literature on the alloy design process, especially missing the specific requirements for the material. Moreover, the material qualification process is pricy and lengthy. A typical process of application-based alloy development is demonstrated in Figure 3.3.

Alloy selection/design. This involves the identification of the alloy composition for required properties. This process is quite lengthy and challenging, as there are currently very limited theories and design platforms available to assist the development and selection of alloys for complex phenomena occurring in PBF.

Powder production. PBF processes mainly use fine spherical powders. Depending on the technique, the size distribution can vary between 5–63 μm for LPBF and 45–106 μm for EB-PBF. The method and cost of powder production strongly depend on the reactivity and melting temperature of the material. The most common powder manufacturing methods that give high-quality spherical powder particles are gas and plasma atomization and plasma spheroidization, with low-volume fabrication recently enabled via ultrasonic atomization.

Fabrication. Involves searching for a window of parameters, i.e. processing window that gives an optimized quality, especially minimal porosity. For LPBF, the fundamental parameters include laser power, scanning speed (laser travel velocity), hatch spacing (lateral separation of laser tracks), layer height (nominal thickness of the layer), laser spot size (diameter of laser beam incident onto the powder surface), and scanning strategy (the pattern of laser scan for each layer).

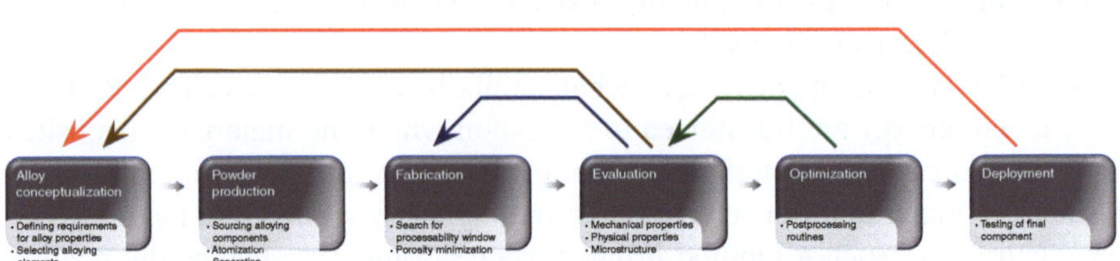

Figure 3.3 Process of alloy development for metal additive manufacturing.

Evaluation. Only after the optimal parameters are identified, one can start testing the specific required properties against initial criteria developed during alloy conceptualization. This can include monotonic and cyclic mechanical testing, creep evaluation, corrosion resistance, etc. It is rare to achieve the desired properties on the first attempt. Therefore, several iterations might be needed to achieve the desired result.

Processing optimization. This step optimizes the fabrication process and post-processes such as heat and surface treatments to enhance the material properties. Depending on the requirements, one may wish to reduce the surface roughness via electrochemical polishing, increase the homogeneity of the mechanical properties by annealing, or increase the strength via precipitation hardening. Enhanced properties are then evaluated once more, and if not satisfactory, the process can loop back to the fabrication or conceptualization steps.

Deployment. Once the lab-scale evaluation is complete, the fabricated components must undergo service-like trials. Moreover, for an alloy to be widely commercialized, standards for raw material (powder) production methods and manufacturing of the alloy to achieve required and consistent properties need to be developed.

Each of these steps requires multiple substeps, and they frequently back loop to the previous steps. The overall process is quite lengthy over several years in timeframe and can cost significant capital investment. Nevertheless, there has been massive progress in the research of evaluating existing alloys and in searching for new ones for AM.

The following sections review in greater detail the state-of-the-art alloy design process and deployment process for AM and alloy design principles specific to this application and novel methods to speed up the expansion of AM materials portfolio.

3.1.3 Available Alloys

PBF technologies show great potential in fabricating metallic components with high geometrical flexibility and reduced lead time [6], and are crucial for developing advanced materials such as architected materials [7]. Moreover, the nature of the process allows local tailoring of the microstructure (namely the solidification structure), grain morphology (shape and size), and the grain crystallographic orientation (i.e. texture) and, hence, the mechanical properties. It has been well documented that the microstructure and mechanical properties of the AM builds are generally very anisotropic and strongly influenced by the process parameters [8–12]. An alloy fabricated by AM often has higher strength and lower ductility in comparison to the same alloy fabricated by other routes. This becomes clearer if we plot a ductility–strength chart for 3000 conventionally manufactured materials and compare it to only a few alloys available to be manufactured by PBF, e.g. around 6 alloys commonly used for LPBF (Figure 3.4). Figure 3.4 also shows that LPBF not only allows tailoring the mechanical properties but also might be able to enhance them. In particular, LPBF-printed 316L steel and 718 nickel superalloy have better elongation and tensile strength than the same alloys manufactured by other processes (Figure 3.4b versus a).

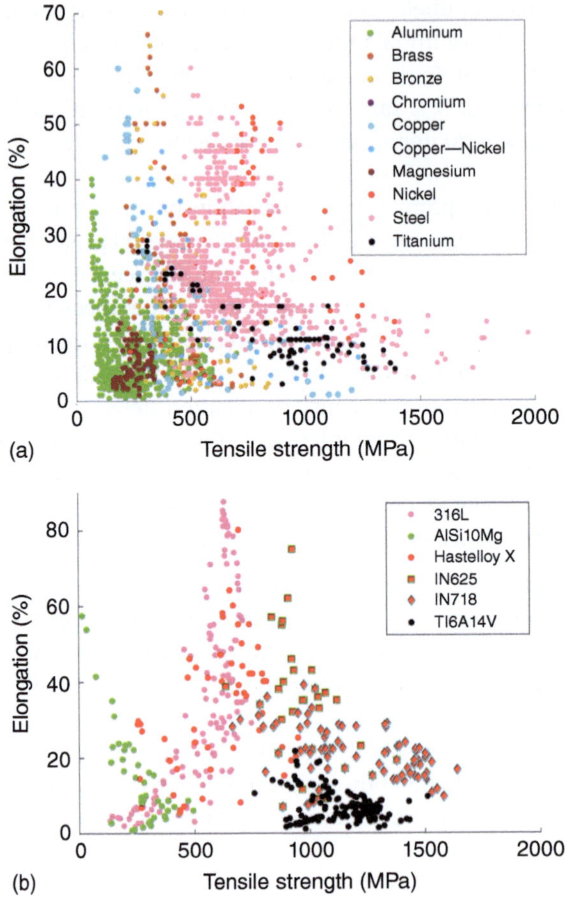

Figure 3.4 Elongation and tensile strength of alloys manufactured by conventional techniques. Source: Data from Ref. [13] (a) and via laser powder bed fusion (b). Note: the legend in (a) only states the principal element of the alloy system, e.g. aluminum refers to Al alloys. Source: Data in (b) are made available in [14].

Nevertheless, the current materials portfolio of widely commercially available alloys for fusion-based AM is relatively scarce (up to about 23 alloys). Table 3.1 gives a brief overview of mechanical requirements for AM alloys standardized by ASTM in collaboration with the International Standards Organization (ISO). Commercially, there are a few more available options (but not widely used due to either difficulty in processing or limited use cases), including precipitation-hardenable steels 17-4 [21–24] and 15-5 PH [25–27]; stainless steels 304L [28, 29], 420 [30, 31]; tool steels H11 [32, 33] and H13 [34–37]; maraging steel 300 [38–41]; Ni superalloys Hastelloy-X [42–46], Haynes 230 [47], IN738; titanium Ti6242 [48–50]; and aluminum Scalmalloy® [51–53], AlSi7Mg [54], AlSi12 [55], Al7075 [56].

However, none of these materials, except Scalmalloy, were initially designed for AM. Moreover, the defect formation mechanisms are complex and manifold, relating to both the process characteristics and inherent material properties. This is one of the main reasons why most of the materials still suffer numerous processing defects, including solid-state cracks, pores, delamination defects, warpage, extremely high internal stresses, and significant compositional changes. Such defects can significantly compromise the mechanical properties of AM alloys. This is especially important in critical load bearing and highly complex components,

Table 3.1 Minimum tensile requirements for alloys manufactured by LPBF at room temperature in different conditions.

Family	Alloy	State	Minimum tensile requirements			ASTM Standard
			Tensile strength (XY/Z MPa)	Yield stress (XY/Z MPa)	Elongation to failure (%)	
Stainless steels	316L	SR[a]	515	205	30	F3184 – 16 [15]
CoCr	Co28Cr6Mo	SR[a]	655	450	8	F3213 – 17 [16]
Superalloys	IN718	SR[a]	980/920	635/600	27	F3055 – 14a [17]
		HT[a]	1240	640/920	12	
	IN625	SR[a]	485	275	30	F3056 – 14 [18]
Titanium	Ti6Al4V	SR[a]	895	825	10	F3302 – 18 [19]
	Ti6Al4V ELI[b]	SR[a]	860	795	10	
	CP Ti	SR[a]	450	380	18	
Aluminum	AlSi10Mg	AP[a]	413	220	4	F3318 – 18 [20]
		SR[a]	241	138	10	
		T6	345	207	5	
		HIP[a]+T6	276	207	10	

a) SR – stress relieved, HT – heat treated, AP – as print, HIP – hot isostatic pressing.
b) ELI – extra low interstitial.

where safety, reliability, and performance are of the utmost paramount. In particular, under fatigue conditions and elevated temperatures, the stress cycles can assist the crack growth from numerous defects and nonequilibrium phases, substantially reducing the performance and service time. Over the past decade, the AM community has mainly tried to optimize the material performance through process parameter optimization and various post-print treatments, including hot isostatic pressing (HIP), which significantly increases the production costs. While such efforts achieve some significant improvements, challenges associated with processing defects and performance remain the main barrier to the broader industrialization of AM. Another focus currently drawing considerable attention from the community is to design new printable alloys, taking into account the unique characteristics of AM process.

The following sections review key aspects of some commonly used materials in LPBF, focusing on their microstructure, defects, and properties. The review helps to understand the characteristics of alloys fabricated by fusion-based AM and guide the design and selection of printable alloys for AM.

3.1.3.1 Ti-6Al-4V

AM brings immense opportunities for titanium alloys due to the alloys' high cost and wide use in low volume, high added value sectors [57–59]. Ti alloy ingots are over 60 times more expensive than steel. However, due to their outstanding specific strength [60] and excellent corrosion resistance [61], titanium alloys are extensively used in aerospace. Furthermore, low elastic modulus (comparable to that of a human bone) and high biocompatibility make them excellent materials for implants [62]. In both industries, there is a need for highly geometrical flexibility and reliable performance. In the aerospace industry, the lighter the components – the higher the fuel efficiency. A common term to measure the efficiency of material usage in aerospace is a buy-to-fly ratio, which compares the amount of material bought and the amount used in the final component. Typically, this value for aerospace components is around 6 : 1 but can often reach 10 : 1 or even higher for a more complex part [57]. With near net shape printing via AM, the buy-to-fly ratio can approach a 1 : 1 ratio, markedly lowering the component costs. One of the widely used titanium alloys is Ti–6Al–4V. This alloy is by far the most additively manufactured Ti alloy. It consists of dual phases: a hexagonal-close packed (HCP) α phase and a body-centered cubic (BCC) β phase, which is retained at room temperature.

Under slow cooling, low-temperature HCP α precipitates from BCC β at the transus temperature (around 1050 °C). Three typical microstructure types can be obtained for this alloy for different product purposes: equiaxed, bimodal, and lamellar. Mechanical deformation and post heat treatment can result in different α and β microstructures and, therefore, strongly affect the mechanical properties. During cooling with cooling rates exceeding 50 °Cs^{-1} [63], such as in LPBF, instead of α, martensite microstructure α′ can form via diffusionless displacive transformation. Five BCC β crystals arrange to form a body-centered tetragonal (BCT) crystal (Figure 3.5a) within, deforming it (Figure 3.5b), and displacing the central atom (Figure 3.5c) upon rapid cooling to create the unit cell of HCP α′ crystal [64]. Such transformation induces a lot of strain and internal stress in the alloy.

The typical post-fabrication martensitic microstructure of the alloy is needle-like laths (Figure 3.6a–c). These needles range from 0.02 to 3 μm depending on the order of formation (primary, secondary, or ternary) [65]. Yang et al. [65] proposed that the

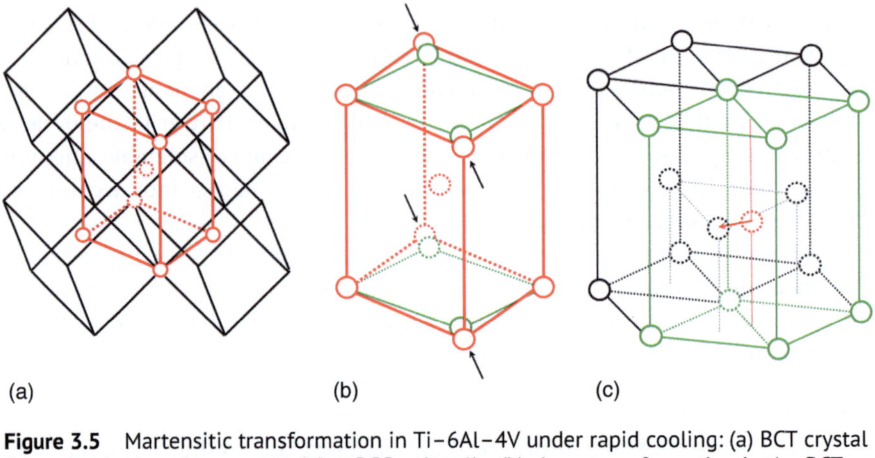

Figure 3.5 Martensitic transformation in Ti–6Al–4V under rapid cooling: (a) BCT crystal formation in an arrangement of five BCC unit cells; (b) shear transformation in the BCT crystal; (c) displacement of the central atom of BCT crystal and formation of HCP unit cell in α′ Ti martensite. Source: Adapted from Banerjee et al. [64].

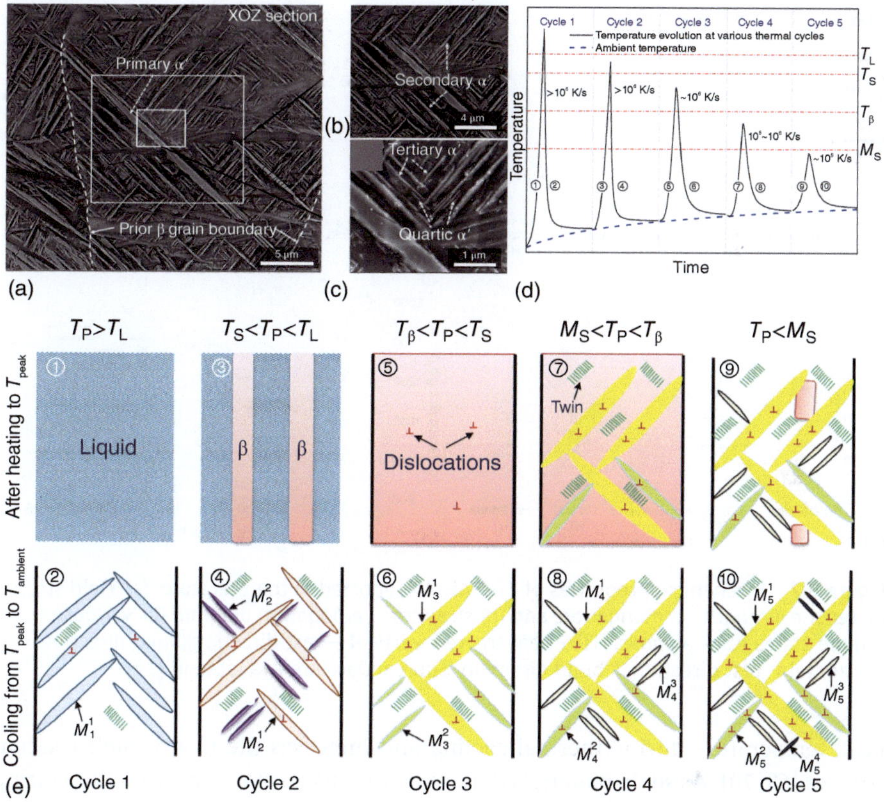

Figure 3.6 Microstructure of as-fabricated Ti–6Al–4V via LPBF (a) with multiple variants of martensite (b), (c). Schematic of the thermal cycles experienced by Ti–6Al–4V during LPBF (d) with a corresponding theory of martensite formation (e). Source: Yang et al. [65]/ With permission of Elsevier.

different order of formation is related to the inherent thermal cycling during LPBF (Figure 3.6d). Due to the repeated deposition of material in AM, the solid material experiences several cycles of heating and reheating. Such thermal cycles in the first few depositions of tracks can result in the β-martensite transformation cycle. Reheating to above the transus temperature causes a decomposition of martensite, leaving residual dislocations, which serve as nucleation sites for new martensitic needles upon cooling (Figure 3.6e).

Such martensitic transformations might seem to impose high internal stress levels, which, coupled with thermal stresses inherent to LPBF, may lead to severe deformation and crack formation. Experimental measurements report that the levels of residual stress in Ti–6Al–4V can exceed 900 MPa [66]. However, modeling efforts by Chen et al. [67] and Denlinger et al. [68] revealed that solid-state transformations have a stress-relaxation effect, reducing the residual stress levels by up to 80% [67]. Such relaxation behavior can benefit alloys with low thermal diffusivities, as they tend to accumulate the highest residual stresses [66]. Low thermal diffusivity usually correlates with a higher capacity for heat storage. This, in turn, results in increased

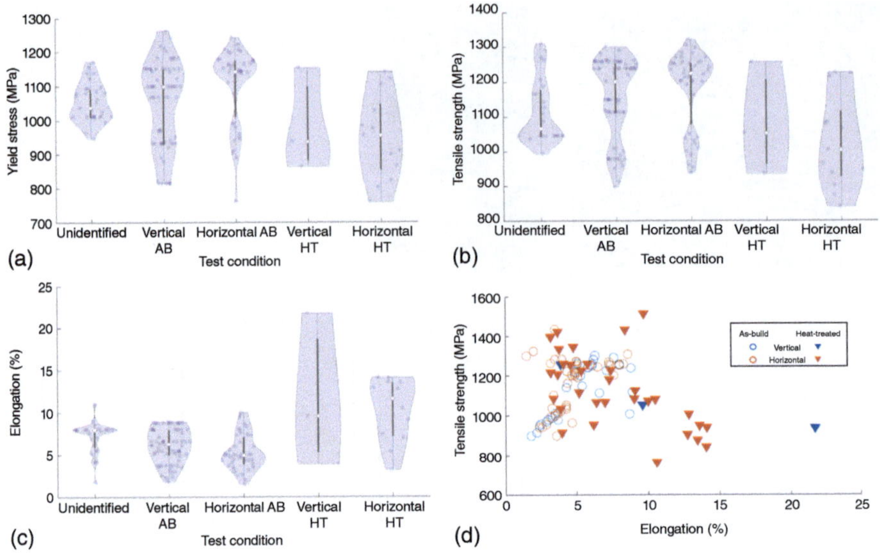

Figure 3.7 Mechanical properties of Ti–6Al–4V reported in the literature. (a) Yield stress, (b) tensile strength, (c) elongation, and (d) strength – elongation diagram for various sample states AB – as-built, HT – heat treated, vertical – test direction||build direction, horizontal – test direction ⊥ build direction. Source: Data are available in [14].

peak temperature during thermal cycling and increases the risk of high residual stresses [69, 70]. As such, materials that undergo a displacive transformation during rapid cooling can contain an inherent mechanism to reduce residual stresses via stress relaxation.

High residual stresses and metastable martensitic microstructure of Ti–6Al–4V after LPBF processing result in increased alloy strength with reduced ductility. Through heat treatment of the alloy, the microstructure can be transformed to $\alpha + \beta$. As a result, the ductility increases at the expense of strength. Overview of monotonic mechanical properties for Ti–6Al–4V is given in Figure 3.7. On average, the violin plots of mechanical properties suggest that mechanical properties are more or less consistent across different loading directions. Data for the as-build condition also show separation into two domains: one with good mechanical properties and one with depreciated. Such behavior is an indication that quite a few tested samples had poor quality. This domain is visible on the lower left corner of Figure 3.7d, where samples with low ductility and low strength are located.

Another issue observed in the PBF of Ti–6Al–4V is the compositional change [71–74]. Some alloying elements preferentially evaporate during processing. This depends on the volatility of the elements. In the case of Ti–6Al–4V, Al is the most volatile. The higher the energy density used to manufacture the sample – the more significant the loss of Al. Figure 3.8a shows the dependence of Al content on the volumetric energy density used during LPBF. This issue in the LPBF of Ti–6Al–4V is not very severe. However, under vacuum conditions of EB-PBF, Al losses are much more pronounced (Figure 3.8b). Losses are reported to be up to 25% from

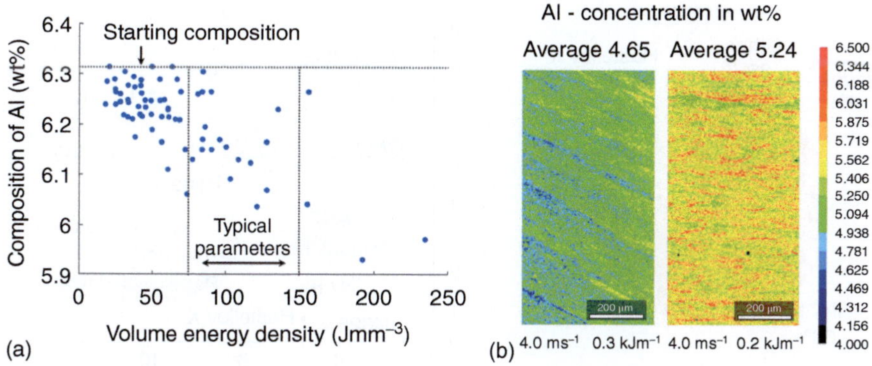

Figure 3.8 Compositional variations in Ti–6Al–4V: (a) Aluminum content in LPBF manufactured samples as a function of the energy density employed. Source: Zhang et al. [71]/With permission of Elsevier; (b) compositional distribution of Al in EB-PBF manufactured Ti–6Al–4V at two different beam energy densities. Source: Juechter et al. [72]/with permission of Elsevier.

the original composition [72]. Moreover, the aluminum distribution becomes inhomogeneous due to remelting of some regions when the subsequent track is being deposited.

3.1.3.2 Superalloys

Another family of high-performance engineering alloys is superalloys. These alloys have excellent heat resistance, retaining their strength at elevated temperatures. The most widespread superalloys are based on the Ni system. Nickel-based alloys typically are divided into solid solution and precipitation strengthened. The latter are used in the most demanding conditions. The basis of their superior mechanical properties at elevated temperatures is the formation of a coherent-ordered faced-centred cubic (FCC) $L1_2$ γ' precipitate in an FCC γ matrix. Ordered phases show much higher resistance to dislocation movements. The chemically ordered arrangement of atoms in γ' induces an energy penalty when dislocations travel through them. When a dislocation shears through a γ' precipitate, it disrupts the order arrangement. Restoring the arrangement requires another dislocation on the same slip system to shear the precipitate, creating an antiphase boundary (APB) between the two dislocations. Such dislocations are termed superdislocations. The primary slip systems in FCC materials are $\{111\}\langle\bar{1}10\rangle$, with an example of superdislocation dissociation reaction: $a[110] \rightarrow \frac{a}{2}[110] + \text{APB} + \frac{a}{2}[110]$ on traveling on the $(\bar{1}11)$ plane. At elevated temperatures, some of the $\frac{a}{2}[110]$ screw dislocations can cross-slip onto a non-close packed (001) plane. This is a thermodynamically driven process enabled to lower energy for APB on {100} family of planes. However, as {100} planes are not close-packed, dislocation glide is significantly impaired, making the dislocation sessile. This is termed Kear–Wilsdorf lock [75] and is responsible for anomalies in the yield stress of Ni superalloys, exhibiting yield stress increases with increasing temperature. When temperature increases to be above a critical temperature, the unlocking mechanism becomes active, and

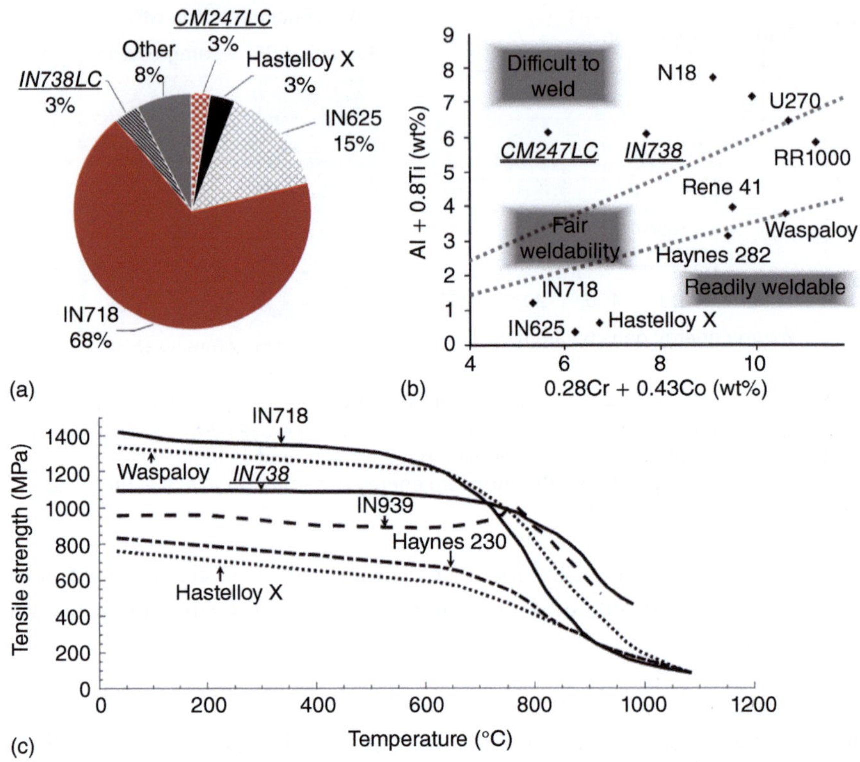

Figure 3.9 (a) Fraction of research on Ni-based superalloys in powder bed fusion segmented by individual alloys. Source: Sanchez et al. [76]/With permission of Elsevier. (b) Weldability diagram of Ni-based superalloys based on Al, Ti, Cr, and Co contents. Source: Haakens and Matthey [77]/U.S Department of Energy/Public Domain. (c) Strength of several superalloys with respect to the temperature. Source: Adapted from Eriksson [78].

dislocations start cross-slipping back to {111} planes resulting in the reduction of yield stress. Numerous superalloys have been developed for specific end-uses with a varying fraction of γ'. However, not all are processable via AM. The most studied Ni superalloy in LPBF (as well as in general AM) is IN718. Around 70% of studied superalloys in AM research [76] were dedicated to IN718 (Figure 3.9a). Such dominance is thanks to a combination of high weldability (Figure 3.9b) and excellent alloy mechanical properties (Figure 3.9c). Two other well-studied superalloys in AM are IN625 and Hastelloy X. The three alloys are among the "weldable" category [77].

3.1.3.2.1 Hastelloy X It is known in welding that γ' content directly correlates with the weldability of the alloy and the tendency to form cracks during processing. Hastelloy X, a solid solution strengthened superalloy without γ', is widely reported to be weldable and highly resistant to cracking [79–81], see Figure 3.9b. However, the alloy is susceptible to solidification cracking during LPBF processing [43, 46, 82, 83]. The mechanisms for solidification crack formation in Hastelloy-X are still being disputed. Tomus et al. suggested that the amount of minor alloying elements such as Mn and Si plays a significant role in crack formation [84]. The author suggests that the increased concentration of these elements decreases the material's solidification temperature, causing an increase in microsegregation toward the grain boundaries. Segregation weakens the grain boundaries and enables the shrinkage strains to initiate cracks. Therefore, a reduction in Mn and Si contents was shown to benefit by almost eliminating solidification cracks. Harrison et al. suggested an alternative to tackle the solidification cracking in Hastelloy X via improving the intrinsic strength of the matrix. By slightly increasing the contents of solid solution strengtheners (W and Mo), the authors predicted an 8 MPa increase in matrix strength. Additionally, a reduction of the content of tramp elements (O, N, P, Cu, Pd) was performed. However, no details on the extent of reduction were provided. As a result, the modified alloy showed an 80% reduction in crack density. An investigation by Tomus et al. employed a hot tearing sensitivity analysis to demonstrate the importance of Si and C contents on crack formation [83], suggesting the highest risk for crack formation for $0.4 < \text{wt\% Si} < 0.6$ and $0.03 < \text{wt\% C} < 0.05$. Sanchez-Mata et al. did not observe any solidification cracks in as-print Hastelloy-X [85], whereas Wang et al. found that solidification cracks were quite pronounced [46]. All the studies mentioned above reported a slightly different alloy composition, suggesting that minor compositional changes can lead to solidification crack formation in Hastelloy X. Table 3.2 summarizes the measured alloy composition and the crack density from various studies.

Such vastly different cracking behaviors inevitably result in inconsistent mechanical properties (Figure 3.10). This is especially evident when comparing

Table 3.2 Compositions of Hastelloy X manufactured by LPBF with corresponding reported crack density.

References	Ni	Cr	Fe	Mo	Co	W	Mn	Si	C	Cracking
Tomus et al. [83]	Bal.	21.38	18.43	8.82	1.83	0.86	<0.01	0.11	0.01	–
	Bal.	21.1	17.9	8.8	2	0.9	0.14	0.36	0.06	+
Sanchez-Mata et al. [85]	Bal.	21.2	17.6	8.8	2	N/A	<0.1	0.2	0.06	–
Marchese et al. [86]	Bal.	21.4	18.7	9.35	1.85	0.92	N/A	0.37	0.051	$3\,\text{mm mm}^{-2}$[a]
Harrison et al. [43]	Bal.	21.8	18.59	9.4	1.77	1.05	0.22	0.31	0.054	$15\,\text{mm}^{-2}$[b]
	Bal.	21.3	19.5	9	1.04	0.56	0.48	0.32	0.057	$35\,\text{mm}^{-2}$[b]
Wang et al. [46]	Bal.	22.6	19.6	8.9	1.5	0.89	0.39	0.65	0.05	$4.5\,\text{mm}^{-2}$[b]
Han et al. [87]	Bal.	21.2	17.6	8.8	2	0.6	0.1	0.2	0.06	0.65%[c]

a) Crack density measured as mean crack length per unit area.
b) Crack density measured as the number of cracks per unit area.
c) Crack density measured as a volume fraction using 3D CT scanning.
"+" and "–" represent the reported presence and absence of cracks in the reference, respectively.

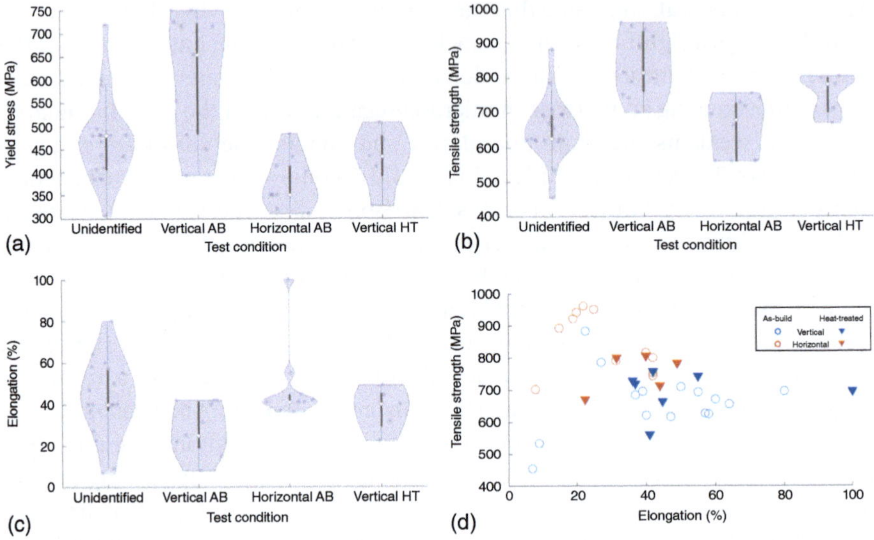

Figure 3.10 Mechanical properties of Hastelloy-X reported in the literature. (a) Yield stress, (b) tensile strength, (c) elongation, and (d) strength – elongation diagram for various sample states AB – as-built, HT – heat treated, vertical – test direction build direction, horizontal – test direction ⊥ build direction. Source: Data are available in [14].

yield stress and ultimate tensile stress along the build direction and perpendicular to the build direction. Solidification cracks typically form along high angle grain boundaries, which in fusion AM (including LPBF) are commonly aligned parallel to the build direction. Therefore, applying stress perpendicular to the crack direction imposes a higher stress concentration factor, hence the significant anisotropy in monotonic tensile behavior in Figure 3.10a,b. The presence of cracks in combination with process-induced porosity is even more detrimental in fatigue loading conditions. The fatigue life of Hastelloy X was reduced up to 1 order of magnitude compared to that of HIPed samples [45]. It was reported that most of the cracks were initiated from surface (or sub-surface) pores [45], whereas the fatigue samples in the study were polished to reduce the effects of surface roughness and sub-surface porosity. In addition, given the anisotropy often seen in AM alloys, the fatigue tests in [45] were tested parallel to the build direction, so it might not show the complete picture of the fatigue behavior.

3.1.3.2.2 Inconel 718 IN718 is one of the most printed alloys in PBF and the most studied AM Ni superalloys. It is a precipitation-hardening Ni-based superalloy with good weldability and capable of bearing high loads up to 700 °C (Figure 3.9b,c). Under rapid cooling conditions of PBF processes, this alloy solidifies as a primarily solid solution of chemically disordered FCC. Reports have shown that at the final stages of solidification, a small fraction of Laves phase forms [88]. Phases such as hexagonal Laves and orthorhombic δ have low ductility and cause a negative impact on mechanical properties. Through solution heat treatment above 1010 °C, one can dissolve the unwanted Laves and δ [89]. Subsequent aging forms

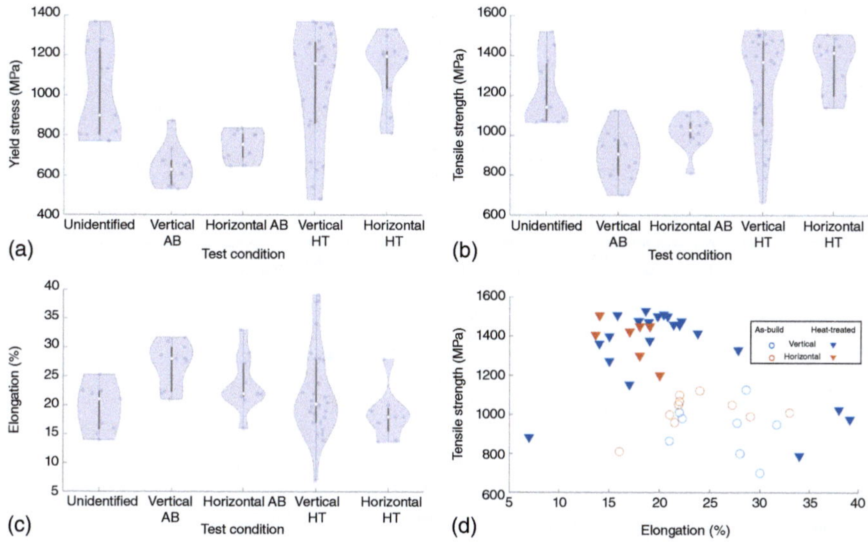

Figure 3.11 Mechanical properties of IN718 reported in the literature. (a) Yield stress, (b) tensile strength, (c) elongation, and (d) strength elongation diagram for various sample states. Note that AB is the as-built, HT represents heat treated, vertical denotes the test direction || build direction, and horizontal denotes the test direction ⊥ build direction. Source: Data are available in [14].

strengthening precipitates of $L1_2$ $Ni_3(Al, Ti)$ γ' and ordered tetragonal Ni_3Nb γ'', both of which substantially improve the strength of the alloy. From the materials science standpoint, IN718 is an example where AM provides many benefits. For example, LPBF-printed IN718 could have lower creep rates as compared to the wrought variant under high temperature and stress creep conditions [88]. From the standpoint of defects, if adequately processed, little to no cracks are observed. This allows to achieve a stable mechanical performance of the material (Figure 3.11). Strength in horizontal and vertical directions usually differs by around 100 MPa because of the defect distribution and crystallographic texture. Meanwhile, different heat treatments allow to achieve a wide range of desirable mechanical properties (Figure 3.11d).

3.1.3.3 316L Stainless Steel

Fe-based alloys are the most versatile and used family of alloys, ranging from hard tool steels with strength exceeding 1600 MPa, to ductile and corrosion-resistant stainless steels. Among this family, AISI 316L stainless steel is to this day the most widely used material in AM. It is an austenitic single-phase FCC alloy with high resistance to corrosion, excellent ductility, and damage tolerance. The alloy has wide industrial applications in power generation, construction, oil and gas, biomedical applications, food processing, and pharmaceuticals, thanks to its outstanding corrosion resistance over a wide range of temperatures, high oxidation resistance, and superior creep-fatigue resistance at high temperatures. However, the main characteristic of this steel for AM is its excellent printability [21, 90–93]. It is well

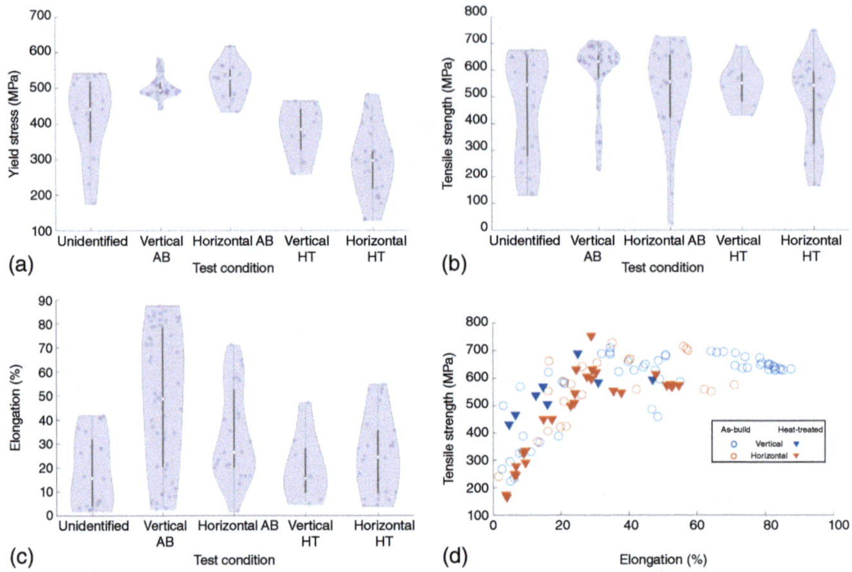

Figure 3.12 Mechanical properties corresponding to different sample orientations and material conditions of 316L Stainless Steel reported in the literature. (a) Yield stress, (b) tensile strength (c) elongation, and (d) strength – elongation relation for various sample states. Note that AB is the as-built, HT represents heat treated, vertical denotes the test direction || build direction, and horizontal denotes the test direction ⊥ build direction. Source: Data are available in [14].

documented that non-precipitation hardened alloys possess high strength after PBF processing. However, in most cases, this comes with a reduction of ductility. Most interestingly, 316L steel (and similar grades of stainless steel such as 304) fabricated by AM has strength and ductility that exceed the conventional material properties [92–94] (Figure 3.12). PBF processing promotes the twinning-induced plasticity (TWIP) in the material during deformation [92, 93]. This mechanism typically is not observed in low Mn steels processed by other processes; however, the TWIP in AM 316L predominantly occurs to achieve excellent ductility (up to 80% [95]), offering a unique way to mitigate the negative effect induced by porosity. Apart from the typical processing defects such as lack of fusion, keyhole porosity and residual stress, the material is rather insusceptible to the crack formation in AM. Mechanical properties along and perpendicular to the build direction reported for 316L are pretty similar (Figure 3.12). Interestingly, Figure 3.12d shows the variation of strength and ductility, reflecting the change in print quality of the material across studies in literature.

3.1.3.4 AlSi10Mg

AlSi10Mg is an alloy initially developed for casting processes. It is commonly used in industrial fields such as automotive and aerospace because of its high specific strength (strength normalized by density) coupled with relatively low thermal expansion coefficient (20 μm mK^{-1}) compared to other Al-based alloys [96, 97]. Low thermal expansion is achieved through the addition of Si in the structure.

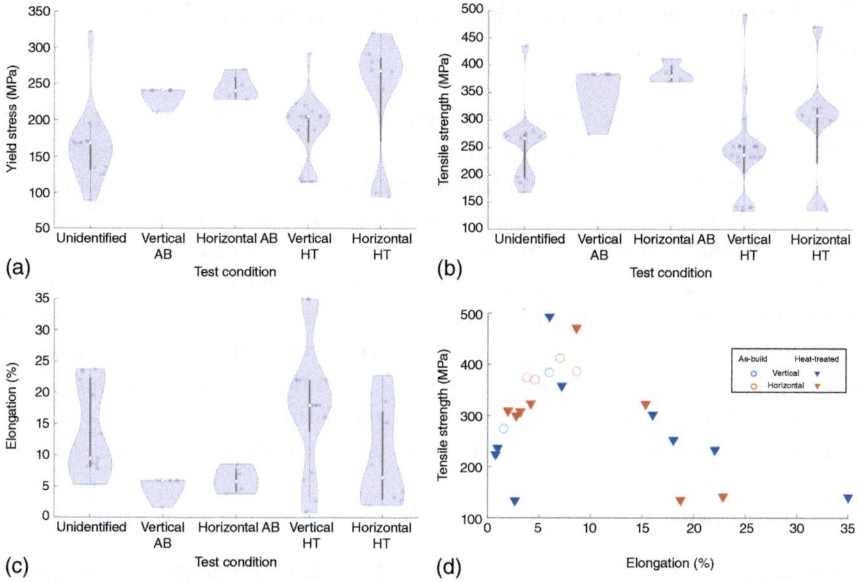

Figure 3.13 Mechanical properties corresponding to different sample orientations and material conditions of AlSi10Mg reported in the literature. (a) Yield stress, (b) tensile strength, (c) elongation, and (d) strength – elongation relation for various sample states. Note that AB is the as-built, HT represents heat treated, vertical denotes the test direction||build direction, and horizontal denotes the test direction ⊥ build direction. Source: Data are available in [14].

This property is beneficial when combatting dimensional inaccuracies during processing [98]. Alloying Al with Si to near-eutectic compositions (12.6 wt% Si) also reduces the freezing range (i.e. the difference between solidus and liquidus temperature). A small freezing range is beneficial as it reduces the chemical segregation, increasing the microstructure homogeneity and giving a lower susceptibility to hot tearing [99, 100]. The addition of Mg introduces precipitation hardening capabilities in the alloy [101, 102]. However, the traditional heat treatment procedure T6 was often used to reduce the strength of the alloy [103] (Figure 3.13a,b). Moreover, Zhang et al. have shown that the fatigue resistance of relieved AlSi10Mg manufactured by LPBF has dropped significantly after T6 heat treatment, solutionizing and stress relieving as compared to the as-print version [104]. Solutionizing step during the heat treatment dissolves the eutectic Al—Si shell surrounding the solidification cells and homogenizes the solute distribution. However, if the alloy is subjected to direct aging, the eutectic shell is retained, and strengthening precipitates are nucleated in the core of the solidification cells. Such heat treatment achieves over 40% strength increase compared to the traditional sequence while retaining a decent level of ductility [105]. Overall, the material processed via LPBF is reported to exhibit better strength than that of the traditional cast counterpart [103]. Mechanical properties of the LPBF-printed parts in as-built condition are similar in vertical and horizontal directions (Figure 3.13). The alloy is generally reported to have good printability, resistance to hot tearing, and significant dimensional distortions [103].

3.2 Melting and Cooling Processes and Associated Defects

The previous section presented a brief overview of the commonly printed materials, their defects, and their properties in comparison to traditional processes. This section aims to provide a deeper understanding of the defect formation mechanisms and establish design criteria to guide the design of new alloys for LPBF. It is, therefore, essential to understand the characteristics of fusion-based metal AM processes. Here, a general overview of the principles, current understanding of underlying mechanisms, and influence of critical processing parameters will be provided. This understanding will be essential when considering how the material's physical properties interact with the processing conditions. The chapter starts with the mechanisms and phenomena occurring during fusion-based AM. Subsequently, the major types of defects mentioned in the previous section will be discussed.

3.2.1 The Process

Fusion-based powder bed processes utilize high-density energy sources (laser, electron beam, or plasma) to melt and subsequently fuse a thin layer of powder to the solid metal (build plate for the first layer or previously deposited tracks/layers). This is achieved by forming a melt pool from fine metallic powder. Once the melt pool solidifies, it forms a physical and metallurgical bonding with the previously deposited material (or build plate). The selection of process parameters such as the power of the energy source, the speed of the moving energy source (i.e. scan speed), hatch spacing (lateral distance between deposition tracks), and layer thickness (vertical distance between deposition tracks) need to be carefully defined to achieve melt pool dimensions that can ensure the bonding. Sufficient overlap between melt pools is required to avoid regions with unfused powder (i.e. lack of fusion). The melt pool formation in LPBF involves the following processes [42, 106–108]:

1. The metal powder absorbs the high-density energy, beginning to melt (Figure 3.14a).
2. The liquid surface temperature reaches boiling point, resulting in localized vapor depression due to recoil pressure [112] (Figure 3.14b,c).
3. Some of the incident energy propagates through the vapor depression and molten metal to melt more powder (Figure 3.14b).
4. Marangoni forces (surface-tension-driven convection) induce convection currents, which redistribute the temperature and expand the melt pool [112, 113] by conducting the heat to the surrounding area at the solid–liquid interface [114] (Figure 3.14b).
5. The vapor depression grows and fluctuates, depending on the power density and the scan speed [115]. If the energy density is high, the vapor depression will transition from semi-circular shape to conical and further to keyhole shape, causing a substantial increase in the melt pool depth as the heat from the energy source travels through the vapor depression toward the bottom of the melt pool (Figure 3.14d).

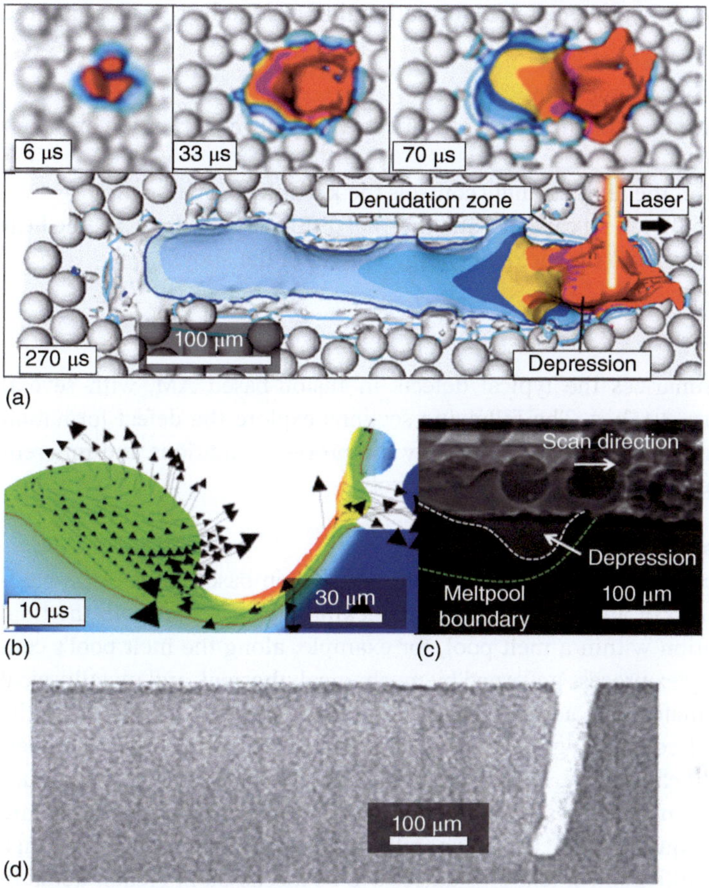

Figure 3.14 Time sequence of melt pool formation in 316L at laser powder of 200 W and scan speed is 1.5 m s^{-1} from the top view (a) and side cross-section view (b). Source: Khairallah et al. [109]/With permission of Elsevier. (c) In situ X-ray imaging of LPBF of Ti–6Al–4V at 210 W and 0.6 m s^{-1}. Source: Hojjatzadeh et al. [110]/from Springer Nature/CC BY 4.0. (d) In situ X-ray imaging of LPBF of AlSi10Mg at 265 W and 0.5 m s^{-1}. Source: Parab et al. [111]/with permission of John Wiley & Sons, Inc.

6. As the laser beam travels along the scanning direction, the surrounding powder is dragged into the melt pool under the vapor-driven drag force in combination with gravity, then subsequently melts [106, 116].
7. Energy from the melt pool is mostly dissipated via conduction through previously deposited solid material (or build plate) and to a lesser extent through conduction to loose powder, convection, and radiation [117].

Melt pool formation in the EB-PBF process is somewhat similar. The differences are in the energy source–matter interaction. Electron interaction with matter typically occurs in the surface and subsurface regions [118] with the interaction size of up to several tens of micrometers, whereas laser irradiation is typically absorbed in the surface region within tens of nanometers [119, 120].

Key features governing the melt pool formation that need to be considered in the alloy design are

- Energy absorptivity of the alloy
- The surface tension of the molten metal affects the heat redistribution and melt pool shape.
- Vapor pressure of the alloy strongly controls the keyhole formation.
- Melting temperature, heat capacity, and thermal diffusivity (i.e. how fast the heat is conducted away from the incident energy source) of the alloy.

3.2.2 Defects

Figure 3.15a summarizes the typical defects in fusion-based AM, with several examples in Figure 3.15b–g. The following sections explore the defect formation mechanisms in more detail and explain how the process conditions and inherent material properties affect the formation of defects.

3.2.2.1 Solidification Cracks

Solidification cracks (Figure 3.15b) are common in all fusion-based processes, including welding and AM of alloys. Solidification cracking typically occurs in the final stage of solidification within a melt pool, for example, along the melt pool's centerline. It is a complex process governed by mechanical, thermal, and metallurgical properties of the material in addition to the solidification parameters, such as the thermal gradient, freezing range (difference between liquidus and solidus temperatures), and solidification rate. Solidification cracks are observed to form because of the excessive thermal stresses exerted on the mushy zone, which causes fracture of the remaining liquidus film [121]. Generally, alloys with a wide freezing range tend to form a solidification crack. This defect will be discussed in greater detail in Section 3.4.

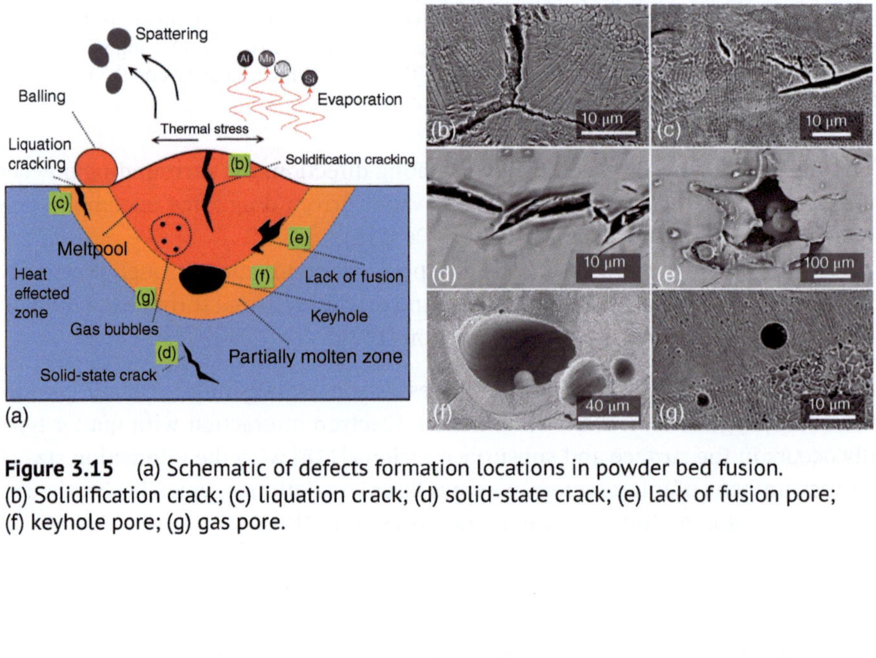

Figure 3.15 (a) Schematic of defects formation locations in powder bed fusion. (b) Solidification crack; (c) liquation crack; (d) solid-state crack; (e) lack of fusion pore; (f) keyhole pore; (g) gas pore.

3.2.2.2 Liquation Cracks

Liquation cracks (Figure 3.15c) can occur in the heat-affected zone (HAZ). As the alloy solidifies, the first solid to form is often depleted of the solute (below nominal concentration), while the final solid to form is enriched in solute (above nominal concentration). The enrichment of solute atoms lowers the local melting temperature. As the successive tracks are deposited, previously deposited regions experience multiple reheating cycles (i.e. each region experiences multiple times as a HAZ). Once reheated, solute-enriched regions in HAZ can melt while the surroundings are still in a solid state. Upon cooling, thermal strain induced by the shrinkage of surrounding solids can tear the thin films apart to form liquation cracks (i.e. liquation cracking) locally along the solute-enriched regions. Such solute-enriched regions are typically located at the solidification grain boundaries. Factors like hot ductility (plasticity limit at elevated temperature) and grain boundary segregation are thought to be the critical parameters in the material's susceptibility to liquation cracking [114]. The liquation cracking is less severe in the AM process (compared to conventional welding), as the base material and the filler material in AM are identical, and the segregation is minimal due to high cooling rates in AM. However, the presence of thermal cycles due to the deposition of successive layers can melt solute-rich regions of previous tracks and form liquation cracks for some alloys with a high tendency of chemical segregation. Reducing interstitial alloying elements can minimize the grain boundary segregation, reducing the liquation cracking susceptibility.

3.2.2.3 Solid-State Cracking and Residual Stress

Solid-State cracks (Figure 3.15d) generally refer to cracks that occur in the solid state after the solidification of the alloy. The underlying mechanisms are complex and often related to loss in ductility over a specific temperature range because of the temporal and spatial changes in microstructure. Multiple solid-state cracking phenomena have been reported. Strain aging cracking and ductility dip cracking (DDC) are two of the most common solid-state cracking phenomena. While strain aging cracking is commonly observed in precipitation-hardened alloys because of the increase in strength with the compromise of ductility during the aging post-heat-treatment process, DDC occurs from solidification associated with a sudden drop in material ductility at certain temperatures between T_m and $0.5T_m$ during cooling. Such loss in ductility is associated with several factors, including chemical segregation, phase transformations, and precipitation of secondary phases occurring upon cooling. Due to the rapid cooling, high internal thermal strain can exceed the ductility limit at elevated temperatures, forming cracks along weakened regions which are typically observed at high angle grain boundaries or on the interphase between two phases [114, 122]. Materials with a significant loss in ductility at elevated temperatures can be very susceptible to the DDC as the part cools [123, 124]. An intrinsic loss in ductility of an alloy with reducing temperature (such as ductile-brittle transition) can also be responsible for the cracking in the solid state. This is particularly dangerous in BCC materials if the ductile to brittle transition (DBT) temperature is higher than room temperature. The formation of porosity is unavoidable in PBF. Irregular pores with sharp corners can result in high stress concentration, which can initiate cracks under the thermal strain. Hence, materials with low fracture toughness are also susceptible to solid-state cracking, particularly when porosity is irregular. Furthermore, due to repeated heating and cooling cycles, internal stresses start to build up in the deposited material. The remaining stress after processing is the residual stress. The level of residual stress strongly depends on the component's geometry, processing parameters, and material properties, in particular, the thermal expansion coefficient and elastic modulus. Residual stresses also tend to distort the geometry and cause cracks in printed components in a few days and weeks after the fabrication.

3.2.2.4 Lack-of-Fusion Porosity

Lack-of-fusion porosity (Figure 3.15e) is typically associated with insufficient molten metal (such as excessively small melt pool, incomplete melting or powder denudation from the previous scan). Such can occur because of insufficient energy density input to the melt pool when the distance between the neighboring melt tracks or layers is larger than the melt pool dimensions. In such a case, the pore contains partially or fully unmelted powder particles. Lack of fusion can be minimized by fine-tuning the process parameters to achieve an optimal overlap. One thing to note is that melt pool size directly depends on the thermal distribution of the part. On the first few layers, the thermal distribution is controlled by the preheating of the build-plate. However, as the build continues, heat can accumulate in the component, and the temperature can reach up to 550 °C in some cases [125], which inevitably impacts the melt pool shape. Hence, the optimized overlap between the melt tracks for a tall build with a constrained heat flow path is not suitable for short parts with a large cross-sectional area conducting the heat away. The extent of the heat build-up depends on the material's thermal diffusivity. Materials with low thermal diffusivities accumulate more heat and reach higher temperatures. Therefore, particular care must be taken when working with such materials. Melt pool instabilities can increase the risk of forming the lack-of-fusion porosity. Such instabilities can arise due to several reasons, including the variations in energy delivered to the melt pool as a result of hardware limitations, uneven surface caused by the balling, presence of high melting temperature particles (e.g. oxides), and inhomogeneous powders. The rough surface directly influences the flow characteristics of the molten metal [126]. The presence of oxides on the surface can cause a reduction in wettability and consequently disturbances in the flow characteristics and, finally, a lack-of-fusion pore [127, 128].

3.2.2.5 Gas Pores

Gas pores (Figure 3.15f) are another common defect in the LBPF process. These defects can form in several ways. First, entrapped gas already present in the raw powder is released as bubbles in melting. However, due to the rapid solidification and turbulent convection flow in LPBF, the gas bubbles cannot escape the molten pool and are frozen in solid after solidification. Second, gas bubbles can be generated because of the evaporation of volatile elements. As the solubility of gasses in molten metal is high, when the liquid cools down, the solubility drops and gases precipitate out in the form of a gas bubble [106].

3.2.2.6 Keyhole Porosity

Keyhole porosity, Figure 3.15g forms due to the collapse of an unstable keyhole during the deposition process. For quite some time, the mechanisms of keyhole formation were not well understood. However, recent advances in X-Ray in situ imaging and multiphysics simulations shed light on the process [107, 115, 129–131]. In simple terms, once a keyhole reaches a particular aspect ratio, it begins to fluctuate, and metal vapor fumes get entrapped by solidified material. The keyhole formation mechanism will be discussed in greater detail in Section 3.1.

3.2.2.7 Compositional Changes

Preferential evaporation of alloying elements is common in fusion processes. Particularly when low pressures and prolonged exposure to high temperatures are involved, such as vacuum smelting. For example, high manganese steels are known to show high losses of manganese during processing [132]. Such behavior is directly related to the temperature and temperature-dependent vapor pressures of the alloying elements. In PBF, particularly in EB-PBF, where a vacuum environment is required, this can cause substantial evaporation of the volatile alloying elements, as demonstrated for Ti–6Al–4V in Section 3.1.3.1 and Figure 3.8. Significant evaporation of main alloying elements will inevitably cause changes in an alloy's physical and mechanical properties. Section 3.2 explores methods to quantify the evaporation losses during LPBF.

3.2.2.8 Balling

Balling occurs at high scanning speed because of the increased capillary instabilities associated with uneven heating of the powder and increased solidification rates. Gu and Shen [133] developed a process map for balling phenomena in AlSi10Mg alloy, while Zhou et al. [134] explained the effects of material properties on the balling tendency. Balling phenomenon is a mechanism through which a liquid metal droplet minimizes its surface area when in contact with a solid material. The extent of balling, i.e. how close the shape is to a sphere, depends on the surface tension of the liquid and temperature [134, 135]. At high scanning speed, the interaction time between the energy source and the powder is very short, making the heating and melting less uniform. This also increases the melt pool fluctuations, oscillations and Plateau–Rayleigh instabilities [136]. As the remaining portion of liquid on the top of the melt pool solidifies, instead of spreading evenly, they may bead up and form sphere-like balls, i.e. a balling defect. Such uneven top surface can cause significant issues when the subsequent layer of powder is being deposited. The powder cannot spread evenly near a balling defect, resulting in a lack of fusion. Furthermore, such a rough surface can damage the powder spreading component (e.g. recoater blade). This, in turn, may cause the whole print job to fail. The material properties aspect in the formation of balling defects will be explored in greater depth in Section 3.3. Processing effects on the formation of Plateau–Rayleigh instabilities are investigated in-depth in previous works [126, 134, 137, 138]

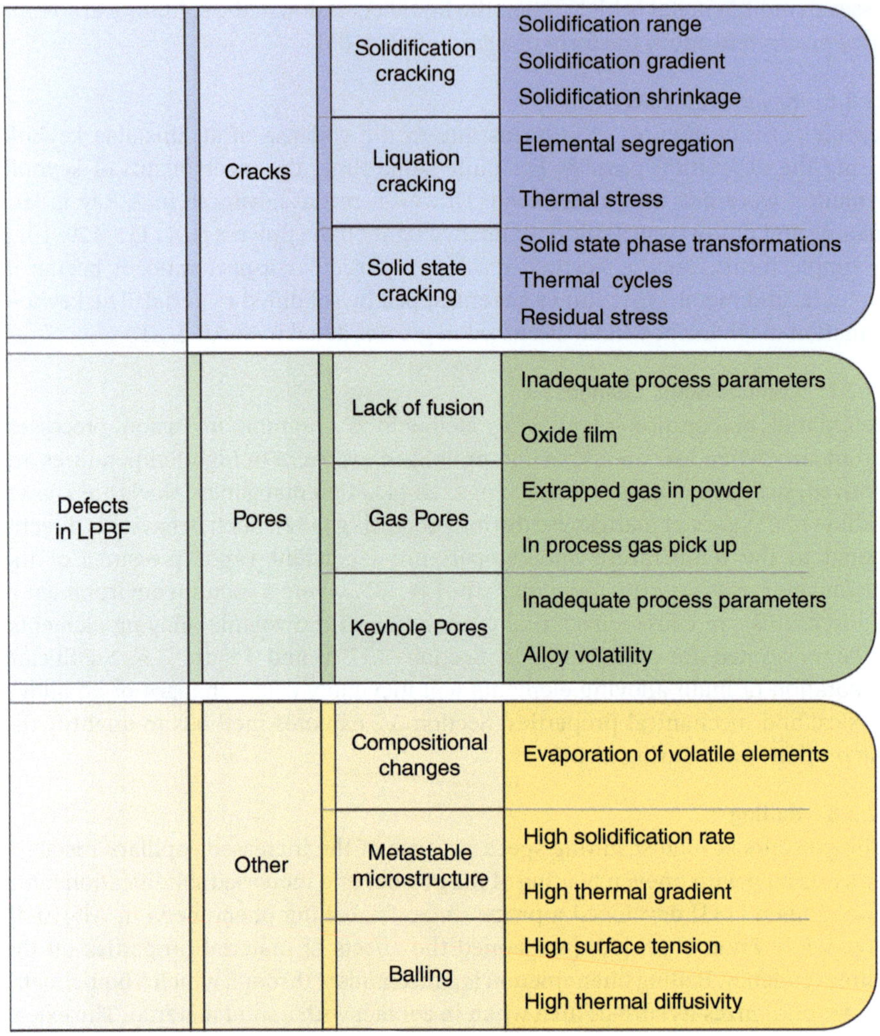

Figure 3.16 Classification of defects in LPBF and factors that influence the formation of respective defects.

3.2.2.9 Summary

Following the discussion of the commonly observed defects that can form in PBF, we need to identify the relevant material properties that affect the defect formation and develop the criteria to assess the risk of forming a given type of defect. Figure 3.16 summarizes the knowledge presented in the previous sections concerning the mechanisms responsible for defect formation.

3.2.3 Roles of Material Chemical–Physical Properties

Considering the defects formation mechanisms outlined in Sections 3.2.2.1–3.2.2.8, several essential material properties have to be considered when assessing the susceptibility to the formation of given defects. The following sections provide some insights into these properties and how their effects on the defect formation differ between conventional manufacturing and PBF.

3.2.3.1 Absorptivity/Backscattering Coefficient

The amount of energy absorbed by metal powder and the melt pool depends on several factors. Depending on the energy source used, different mechanisms take place. For the laser melting, photon energy is transferred through the free electrons via inverse Bremsstrahlung mechanisms [139] and subsequently to lattice phonons via collisions [120]. The interaction type between photons and surface electrons is strongly correlated to the electron density of the electronic band structure of the metal/alloy and photon frequency [120]. Hence, the absorptivity of the laser is strongly dependent on the chemical element. For example, at the typical wavelength of lasers employed in LPBF (~ 1060 nm for Nd: YAG lasers), Fe and Ni have much higher absorptivity than Cu and Al [140] (Figure 3.17a).

For the electron beam melting, incident electrons collide with outer shell electrons of atoms, transferring the kinetic energy, therefore heating the material. A fraction of the incident beam can be scattered back, depending on the atomic number, crystallographic orientation, and the incident energy of the electrons [141]. Materials with higher atomic numbers backscatter more electrons, reducing the process efficiency [142] (Figure 3.17b).

In both cases presented earlier, the amount of power absorbed also strongly depends on the incident angle with respect to the powder surface. Furthermore, due to the uneven surface of the powder, multiple reflections/scattering also happen. As such, the absorptivity of energy is also dependent on the surface roughness, and energy absorptivity of the powder-covered surface is usually higher than that of polished surfaces [143–146] (Figure 3.18).

3.2.3.2 Heat Capacity and Enthalpy of Melting

As new powder enters the melt pool, it undergoes rapid heating and subsequently melting process. The energy loss associated with these processes depends on the

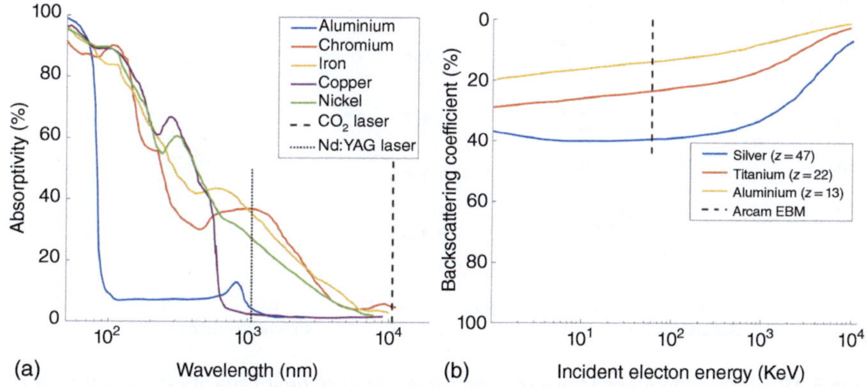

Figure 3.17 (a) Absorptivity of incident photons onto a metallic surface as a function of photon wavelength. Source: Adapted from Bass et al. [140]. (b) Backscattering coefficient of incident electrons onto a metallic surface as a function of electron energy. Source: Adapted from Kanaya and Okayama [141].

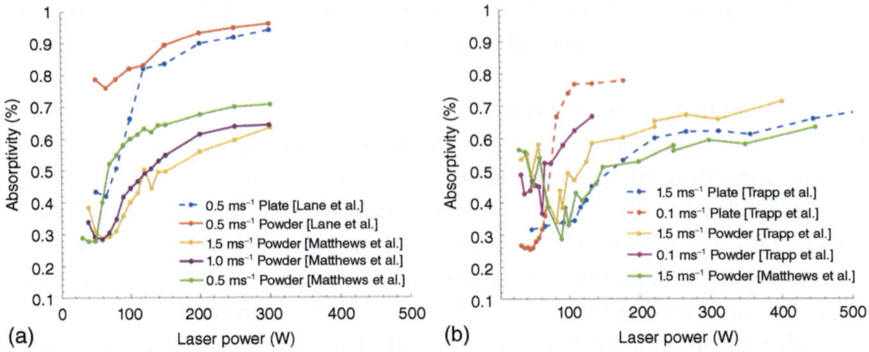

Figure 3.18 Absorptivity of metal plates and powders under incident laser radiation at a wavelength of 1020 nm with varying laser power and scan speed in IN625 (a) and 316L (b). Source: Data were taken from Lane et al. [145], Matthews et al. [146], and Trapp et al. [147].

material's specific heat capacity in solid and liquid states and the latent enthalpy of melting. Additionally, heat capacity indicates the amount of thermal energy stored in the deposited material. Hence, two other important physical properties are the specific heat capacity and enthalpy. As discussed in the previous paragraph, temperature build-up can cause significant difficulties during processing. The heat capacity of the material should be evaluated alongside the thermal conductivity. These factors combined give an estimation for the heat build-up during the manufacturing process. Figure 3.19 gives a general overview of the heat capacity (Figure 3.19a and enthalpy of melting (Figure 3.19b for different alloy families. For instance, magnesium alloys have a high heat capacity and relatively low thermal conductivity, increasing the risk of associated formation of defects due to heat build-up during the manufacturing process. In contrast, copper alloys have low heat capacity, very high conductivity, and relatively high melting temperature. Such a combination of properties would suggest that the heat transfers away from the melt pool at a very high rate, requiring a high heat input from the energy source.

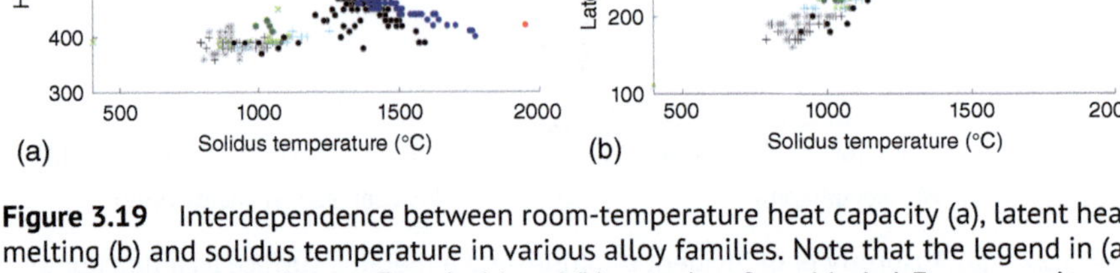

Figure 3.19 Interdependence between room-temperature heat capacity (a), latent heat of melting (b) and solidus temperature in various alloy families. Note that the legend in (a) and (b) are the same. Source: Data in (a) and (b) are taken from MaskeltFrom repository [13].

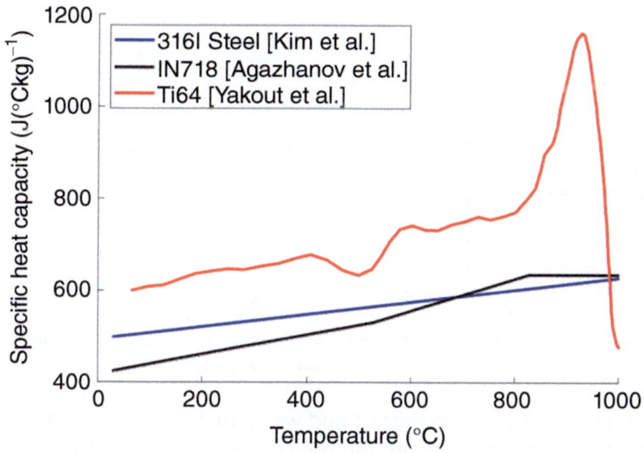

Figure 3.20 Specific heat capacity of wrought 316L, IN718, and Ti–6Al–4V alloys with respect to temperature. Source: Data are taken from Kim et al. [148], Agazhanov et al. [149], and Yakout et al. [150].

Note that the heat capacity is a function of the material's temperature and microstructure (in particular, the crystal structure). Figure 3.20 shows the specific heat capacity of several commonly used AM alloys. 316 L austenitic stainless steel has a monotonic and rather linear dependence between heat capacity and temperature. IN718 is a precipitation-hardening alloy with several secondary phases forming at elevated temperatures, resulting in the complex dependence of heat capacity on the temperature. Ti6Al4V is an $(\alpha + \beta)$ dual-phase alloy, transitioning to pure beta at ~1000 °C. Such multiple phase transformations are causing strong nonlinear behavior of the material's heat capacity.

3.2.3.3 Thermal Conductivity

The input power required to melt the powder and maintain a stable melt pool depends on how fast the heat within the melt pool is carried away. As conduction is the dominant heat transfer mechanism in PBF [117], the material's thermal conductivity plays a significant role. Several works demonstrated that the thermal conductivity of AM metals slightly (up to 15%) differs from the conventional counterpart [148, 151, 152] (Figure 3.21a). However, these differences are not substantial and are mainly attributed to the microstructure of the printed parts [151]. Additionally, it is essential to consider the residual heat accumulated in many previous tracks [125, 153]. Literature shows that heat accumulation can have several effects [154–157]. First of all, when a subsequent track is just deposited next to the recently solidified scan, the region remains at elevated temperatures, increasing the risk of overheating the melt pool and forming a keyhole pore [153]. This effect is especially apparent at the beginning and end of scan vectors. Second, heat build-up in the bulk of printed parts has been found to relate to an increase in porosity and instabilities of the melt pool in 316l stainless steel [125]. Third, the accumulated heat can stimulate the precipitation of undesirable and brittle phases [154]. The uneven distribution of the heat build-up due to variations in

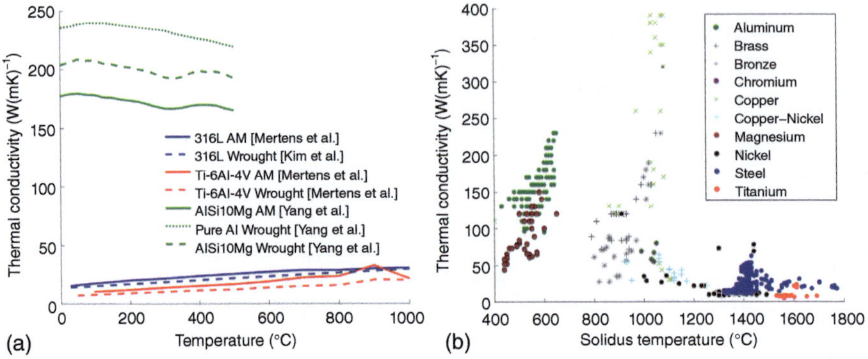

Figure 3.21 (a) Thermal conductivity of LPBF processed and wrought 316L, Ti–6Al–4V, and Al–Si10Mg alloys with respect to temperature. (b) Interdependence between room temperature thermal conductivity and solidus temperature in various alloy families. Source: Data in (a) are taken from Mertens et al. [152], Kim et al. [148], and Yang et al. [151]. Data in (b) are taken from MakeItFrom repository [13].

part geometry and thermal pathways can result in nonuniform microstructure and phase distribution in the component. Therefore, great care must be taken when processing materials with low thermal conductivity and high melting temperatures, such as steels, titanium, and nickel alloys (Figure 3.21b).

3.2.3.4 Surface Tension

Melt pool dynamics are mainly driven by drag forces and thermo-capillary convection (i.e. Marangoni effect) [110]. Liquid tends to flow from regions with low surface tension (γ) to regions with high surface tension (Figure 3.22). As surface tension is strongly dependent on temperature, different regions in the melt pool have different surface tension due to the temperature distribution. In most pure metals, surface tension has a negative slope with respect to temperature, i.e. as temperature rises

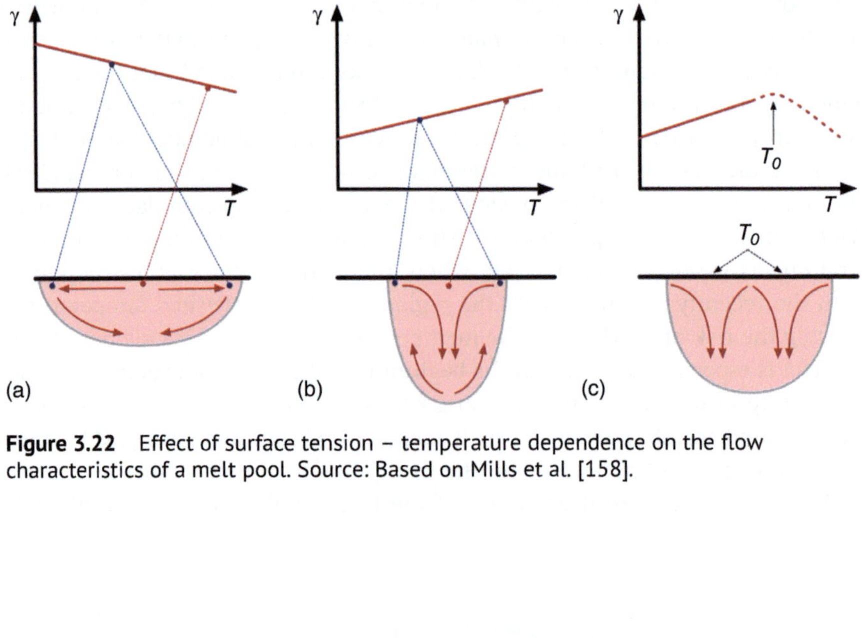

Figure 3.22 Effect of surface tension – temperature dependence on the flow characteristics of a melt pool. Source: Based on Mills et al. [158].

and surface tension decreases [159] (Figure 3.22a). As a result, these forces trigger a circulation flow of liquid metal. The intensity and the direction of the flow are directly related to the surface tension–temperature dependence for given materials [158, 160–162].

Welding literature has extensively investigated the effects of surface-active elements like S and O on surface tension. S is a highly effective micro-alloying element in steels, capable of reversing the negative slope of $\partial\gamma/\partial T$ at concentrations above 40 ppm [158]. For instance, SO_2 gas is sometimes used during gas tungsten arc (GTA) welding to increase the weld penetration depth [163]. Similarly, one can utilize these methods in AM to manipulate the melt pool flow characteristics.

The intensity of the flow characteristics depends on the magnitude of the surface tension–temperature slope. Table 3.3 shows the values for the $\partial\gamma/\partial T$ for some commercial alloys. Some of the alloy compositions are proprietary, hence were not stated. However, the authors indicate that the large differences in surface tension gradients between the three types of carbon steel investigated in Table 3.3 are due to different concentrations of tramp elements (e.g. S, P, and O).

Table 3.3 Surface tension measurements on parabolic flights by the Thermolab project.

Alloy	Liquidus temperature (K)	Surface tension at liquidus temperature (Nm^{-1})	$\partial\gamma/\partial T$ (Nm^{-1} K^{-1})
CMSX-4	1657	1.85	$-9.01 \cdot 10^{-4}$
IN738LC	1608	1.85	$-1.48 \cdot 10^{-3}$
MM247LC	1641	1.86	$-1.36 \cdot 10^{-3}$
C263	1641	1.74	$-6.85 \cdot 10^{-4}$
C-Steel1	1753	1.61	$-4.47 \cdot 10^{-4}$
C-Steel2	1816	1.38	$-1.65 \cdot 10^{-4}$
C-Steel3	1816	1.53	$-12.3 \cdot 10^{-4}$
Tooling-Steel	1808	1.73	$-3.38 \cdot 10^{-4}$
Cr-Steel	1750	1.76	$-9.04 \cdot 10^{-4}$
Cu-alloy	1340	1.15	$-7.28 \cdot 10^{-4}$
Al–75 at%Ni	1683	1.63	$-5.88 \cdot 10^{-4}$
Al–65 at%Ni	1823	1.65	$-1.01 \cdot 10^{-3}$
Al–31.5 at%Ni	1523	1.16	$-1.14 \cdot 10^{-4}$
CuSnP	1301	1.08	$-4.77 \cdot 10^{-4}$
CuSn8	1305	1.05	$-3.74 \cdot 10^{-4}$
Rene90	1633	1.71	$-4.30 \cdot 10^{-4}$
ReneN5	1645	1.70	$-4.91 \cdot 10^{-4}$
M-SX	1648	1.58	$-4.95 \cdot 10^{-4}$

Source: Egry et al. [164]/With permission of Elsevier.

3.2.3.5 Boiling Temperature and Volatility

Upon interaction of the energy source with the powder or liquid, the heat is not distributed uniformly [165, 166]. Local temperature variations are inevitable, and there is local vaporization of chemical elements. The intensity of the vaporization strongly depends on the difference between the melting and boiling temperature of the material and the volatility of constituent elements. Upon interaction with liquid metal, the laser beam creates a vapor depression due to the recoil pressure of the metal vapor jet [167]. The surface pressure on the liquid–gas interface is strongly dependent on the atmospheric pressure and temperature [168]. A combination of ambient pressure and liquid metal temperature governs the intensity of evaporation. Depending on the thermodynamic properties, some metals evaporate more readily than others [169]. Vapor pressure describes the equilibrium pressure of vaporized material at a particular temperature. Higher vapor pressure implies that a substance (metal in this case) has a thermodynamically higher tendency to vaporize at a given temperature. Metals with high vapor pressures are referred to as having high volatility. Figure 3.23a shows the vapor pressure dependence on temperature for pure metals commonly used as principal elements (solid line) and alloying elements (dashed line). Vapor pressure of the alloy typically follows the trend governed by the base metal. From Figure 3.23a, it is evident that Fe-, Ni-, Co-, Ti-, Cu-, and Al-based alloys would have similar volatilities. However, Mg-based alloys have several orders of magnitude higher vapor pressure. This would make them highly prone to keyhole instability and excessive evaporation [172]. Figure 3.23b shows the vapor pressures of some typical contaminants in metallic alloys. Figure 3.23b suggests that the alloy should lose some sulfur and phosphorus over alloying metals at typical processing temperatures, providing some benefit of refinement.

During AM, high material volatility results in more evaporation and, therefore, larger vapor depression. As the dependence of the vapor pressure on temperature is exponential, the vapor depression rapidly transitions into a keyhole after a threshold temperature is passed. In the keyhole mode, the absorptivity of the incident energy sharply increases due to the increased number of reflections/scattering (increased surface to volume ratio). Alongside the excessive evaporation, this can result in melt pool instability and, subsequently, keyhole pores.

Another issue is related to differences in the volatility of alloying elements that result in preferential evaporation and changes in chemical composition. Calculations of individual partial vapor pressures of elements in an alloy are complex, as it depends on the pure metal vapor pressure and the activity coefficient for the elements in the alloy. Ivanchenko et al. have outlined a method to calculate the activity coefficients for a ternary system through the excess Gibbs-free energy [173]. Zhang et al. used this methodology and overlaid it with evaporation kinetics to study the effect of processing parameters on Al loss in Ti–6Al–4V alloy [71]. Figure 3.24c–f show the calculation of the thermodynamic phase diagram of the equilibrium concentrations of alloying elements in the gas phase at various temperatures for 316L steel, Ti–6Al–4V, IN718, and Al-Si10-Mg alloys. It can be seen that the vapor pressures of pure metals are closely related to the equilibrium gas-phase composition. For instance, Al is much more volatile than Ti and V

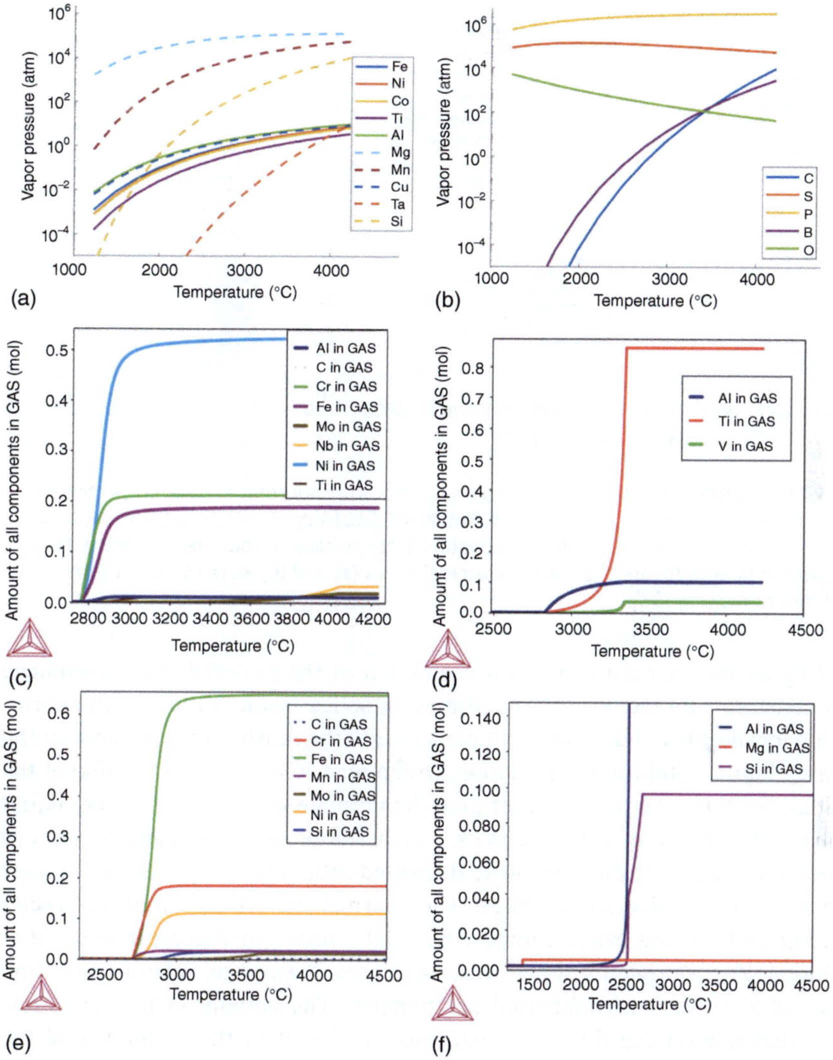

Figure 3.23 (a) Vapor pressure of metallic elements at various temperatures. Source: Adapted from Alcock et al. [170] and (b) vapor pressures of typical light elements in metallic alloys at various temperatures. Source: Adapted from Lide [171]. (c–f) Amount of alloying elements in the gaseous phase as a function of temperature in IN718, Ti–6Al–4V, 316L, and AlSi10Mg, respectively, calculated using ThermoCalc.

(Figure 3.23d). Hence, Al preferentially evaporates from the Ti–6Al–4V alloy. A similar effect can be seen in Al-Si10-Mg alloy. As one of the most volatile elements, magnesium vigorously evaporates even before any Al or Si even starts to evaporate.

3.2.3.6 Thermal Expansion and Contraction

The molten track solidifies from the melt pool boundary inwards. While the center of the melt pool is still liquid, the outer boundaries shrink due to cooling,

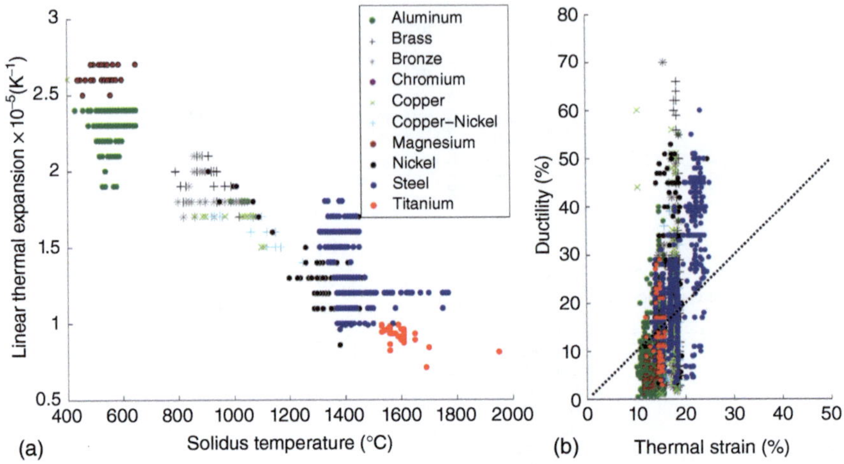

Figure 3.24 Interdependence between room temperature thermal expansion coefficient and solidus temperature in various alloy families. (b) Ductility of various alloy families with respect to estimated thermal strain upon cooling from solidus to room temperature. Note that legend in (a) and (b) are the same. Source: Data in (a) and (b) were obtained from MakeItFrom repository [13].

depending on the thermal expansion coefficient of the material. Such shrinkage with a constraint imposed by the surrounding solids results in strain and corresponding tensile stress. Such stress propagates into the mushy zone and acts on the remaining liquid–solid interface regions, which tend to be at the centerline of the deposition bead. If the stress is higher than the ultimate strength of the solid–liquid interphase, the shrinkage causes a crack. Furthermore, as the deposition of a new bead heats and expands the previously deposited solid, the part develops complex tensile and compressive stress states. The thermal history and multiple cycles of heating and cooling cause complex residual stresses in the component after deposition [174–177]. Such residual stresses can cause warpage, delamination, and cracking of additively manufactured components. The amount of residual stress strongly depends on the thermal expansion coefficient of the material and the temperature gradient. At high temperatures, metals have very high ductility and can accommodate the strain plastically. As the temperature decreases, the yield stress typically increases with loss in ductility. Therefore, it is vital to consider the shrinkage strain of the material in relation to the ductility of the material. Figure 3.24a shows the room temperature coefficient of thermal expansion (CTE) of materials with respect to the solidus temperature of the material. Assuming that the CTE does not change significantly with temperature, a simple estimation of thermal strain upon cooling can be made by multiplying the CTE by the temperature change from solidus to room temperature. If one compares this estimate with the ductility at a given temperature of the alloy, one could assess the risk of defects associated with shrinkage. In Figure 3.24b, alloys with ductility above the dashed line (1 : 1 line) would have good resistance to shrinkage-related defects. This estimation is very oversimplified, as ductility is strongly related to the material's microstructure,

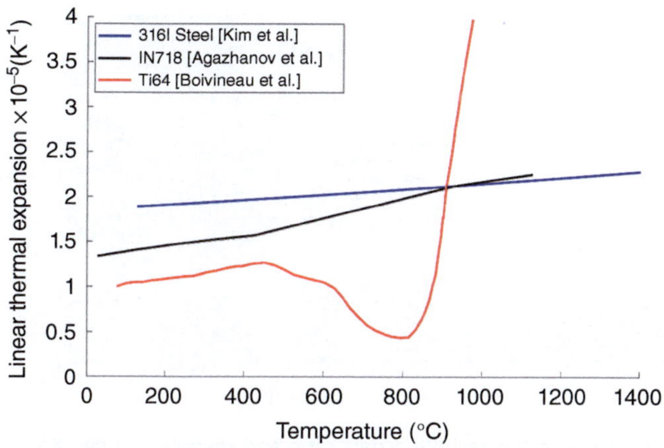

Figure 3.25 Linear thermal expansion of wrought 316L, IN718, and Ti–6Al–4V alloys with respect to temperature. Source: Data are taken from Kim et al. [148], Agazhanov et al. [149], and Boivineau et al. [178].

whereas the CTE of materials is strongly dependent on temperature and solid-state transformations. Figure 3.25 shows the temperature dependence of CTE for some common additively manufactured materials.

3.3 Alloy Design Methodology

In the past few years, several works focused on the design of alloys for rapid solidification processes and AM in particular. These efforts were focused on establishing mathematical equations representing the links between material chemical/physical properties and processing to the quality of printed builds. Such relationships were used to guide the tailoring of the alloy composition to reduce the defect density and improve mechanical properties, with some studies taking into account application-specific property requirements and processing conditions.

3.3.1 Keyhole Formation

Gan et al. recently developed several dimensionless parameters that fully explain the keyhole formation for multiple materials [129]. One of such parameters, keyhole number *Ke*, Eq. (3.1), was directly correlated to the keyhole aspect ratio (e^*) (Figure 3.26a). *Ke* is defined as [129]:

$$Ke = \frac{\eta P}{(T_l - T_{\text{ambient}})\pi \rho C_p \sqrt{\alpha v r_0^3}} \tag{3.1}$$

where η is the absorptivity, P is the power, T_l is the liquidus temperature, T_{ambient} is the build plate temperature, C_p is the specific heat capacity, ρ is the mass density, α is the thermal diffusivity, v is the scan speed, and r_0 is the beam diameter. The keyhole

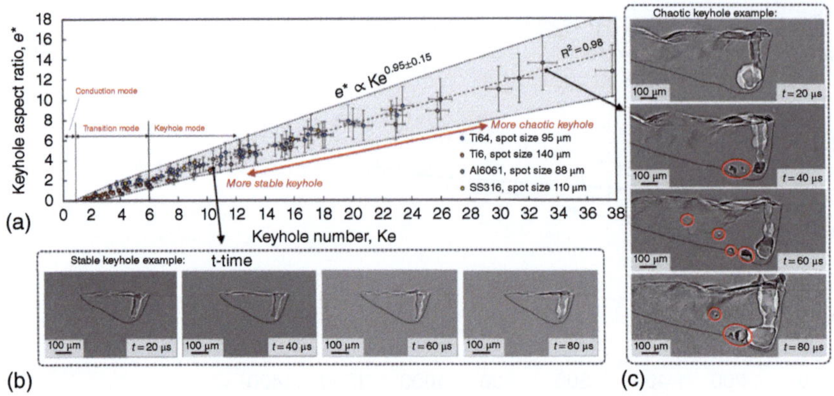

Figure 3.26 (a) Correlation between the keyhole aspect ratio and keyhole number (Ke defined in the text). (b), (c) Examples of stable and unstable keyholes, respectively. Source: Gan et al. [129]/with permission of Springer Nature.

aspect ratio is calculated from the *Ke* as: $e^* = 0.4(Ke - 1.4)$, while the variation of e^*, $\Delta e^* = 0.36Ke^{0.86}$. As the keyhole aspect ratio increases, the keyhole transitions from being stable (Figure 3.26b) to unstable (Figure 3.26c). The research group then developed a scaling law for the porosity based on a dimensionless variable equal to the keyhole number normalized energy density. For more details, the reader is referred to the original article [129]. The critical takeaway from this research is the dependence of keyhole number on the material properties C_p, ρ, α, and η.

However, absorptivity should be taken with caution. It is dependent not only on the inherent material absorptivity for a particular wavelength of radiation but also on external factors, such as reflection from the powder and vapor plume attenuation. The reflection of the powder is, in particular, significant. The extent of the vapor plume effect depends on the quality of inert gas shielding [179], and the intensity of material evaporation which is governed by the volatility of the material.

3.3.2 Evaporation of Alloying Elements

The extent of evaporation depends on the temperature of the melt pool, its surface area, and the duration of time over which that area is exposed to high temperature and partial pressures of the alloying elements. Ivanchenko et al. have outlined a method to calculate the activity coefficients for a ternary system through the excess Gibbs-free energy [173]. Zhang et al. used this methodology and overlaid it with evaporation kinetics to study the effect of processing parameters on Al loss in Ti–6Al–4V alloy [71]. Mukherjee et al. [180] presented a simplified method estimating the level of vaporization for each element during PBF processing. For simplicity, it was assumed that liquid alloys display ideal behavior near the boiling temperature and, therefore, the equilibrium vapor pressure of an element (P_i) in an alloy can be estimated using Eq. (3.2):

$$P_i = X_i P_i^0 \tag{3.2}$$

where X_i is the volume fraction of element i and P_i^0 is the equilibrium vapor pressure of the element over its pure liquid (as shown in Section 3.2.3.5 in Figure 3.24a). The intensity of the vaporization of element i (vaporization flux J_i) can then be estimated through Langmuir Eq. (3.3):

$$J_i = \frac{\lambda P_i}{\sqrt{2\pi M_i T}} \tag{3.3}$$

where λ is condensation fraction of vaporized elements, M_i is the molecular weight, and T is the temperature. Mass of the vaporized element (Δm_i) can be estimated by considering the surface area of the melt pool (A_s) and deposition time of each track (i.e., track length L to scan speed v ratio) resulting in Eq. (3.4):

$$\Delta m_i = \frac{L A_s J_i}{v} \tag{3.4}$$

This equation allows calculating the new composition of the alloy. The weight percentage of the alloying elements that remained in the alloy after the deposition is defined as $W_i^{\,d}$. One can approximate the volume of the deposited track as the cross-sectional area of the melt pool at the deepest point perpendicular to the scanning direction (A_t) multiplied by the track length, resulting in Eq. (3.5):

$$W_i^{\,d} = \frac{L A_t \rho W_i - \Delta m_i}{L A_t \rho W_i - \sum_i \Delta m_i} \tag{3.5}$$

where W_i is the weight percentage of element i in powder and ρ is the mass density of the alloy.

Even though these calculations require some knowledge on the melt pool dimensions, they provide valuable information about the alloy processability. The authors of the work calculated the melt pool dimensions through modeling, which requires knowledge of the thermophysical properties of the alloy.

3.3.3 Balling Defects

Zhou et al. [134] investigated balling phenomena during LPBF of tungsten. To quantify the extent of the balling, the authors simplified the process to a liquid metal droplet wetting its own solid. Under the conditions of LPBF, the spreading of the droplet is driven by the capillary force and counteracted by inertia. The time (τ) required for a droplet of diameter d to spread can be estimated via Eq. (3.6):

$$\tau = \sqrt{\frac{\rho d^3}{\sigma}} \tag{3.6}$$

where ρ is the mass density of the molten droplet and σ is the surface tension. The authors then estimated the solidification time (τ_{sol}) of the droplet using Eq. (3.7) developed by Gao and Sonin [181]:

$$\tau_{\text{sol}} = 2\frac{d^2}{3\alpha} \ln\left(\frac{T_0 - T_{\text{ambient}}}{T_l - T_{\text{ambient}}}\right) \tag{3.7}$$

where α is the thermal diffusivity, T_0 is the temperature of the droplet, T_{ambient} is the ambient temperature, and T_l is the liquidus temperature of the material. By comparing the two terms, one can estimate whether a droplet of a given size would have enough time to spread before it solidifies. In this manner, the authors demonstrated that W droplets of 100 μm in diameter would need ~86 μs to spread while taking 43 μs to solidify. Hence, the droplet would solidify before fully spreading, resulting in a balling defect.

3.3.4 Solidification Cracking Models

Several concepts attempt to explain the solidification cracking phenomena. The shrinkage-feeding model assumes that the cracking occurs at the terminal stage of solidification when the liquid feed into the mushy zone (rate of feeding, ROF) becomes insufficient to compensate the solidification shrinkage (rate of shrinkage, ROS). Feurer et al. [182] showed that cracking occurs when ROF < ROS. ROS is directly proportional to the thermal contraction of the material and the cooling rate, while ROF depends on accessibility to the dendritic network by the liquid. In the LPBF process, thermal gradients and solidification rates are higher than those in welding, which hinders the formation of the secondary dendritic arms. Therefore, the hindrance to the flow of liquid induced by the dendritic network is lower. There were numerous further developments of this theory. The accessibility to the interdendritic regions is related to the rate of crystal growth, which, in turn, is proportional to the rate of the solid formation (f_s) with respect to the temperature (T) at the final solidification stage; df_s/dT is called the solidification gradient. The role of solidification gradient has been emphasized in a number of studies. Benoit et al. compared the cracking behavior of two IN625 powders during LPBF with slightly different Si contents [183]. Powder with higher silicon content had numerous solidification cracks, whereas powder with low Si content had no cracks. By analyzing the solidification behavior using the Scheil model, the authors found that the solidification gradient of the Si-rich powder was much higher. Kou [184, 185] did extensive work on Al–Si alloys and concluded that the difficulty in liquid feeding into the interdendritic (or inter-cellular in LPBF) spacing during the final stage of solidification has a strong influence on the crack formation in Al alloys. As the alloy solidifies, the strain develops in the direction opposite to the solidification direction due to shrinkage and the constrained solid. The amount of strain is proportional to the coefficient of thermal contraction of the material and the temperature change in the region. This strain is opposed by the growing crystal and feeding of the liquid in between the solidifying structures. If strain exceeds the contributions of crystal growth and liquid feeding, the solid–liquid interface breaks down, and a solidification crack can occur. This rationale was formulated and described in Eq. (3.8) [185]

$$\underbrace{\frac{d\varepsilon_{\text{local}}}{dT}}_{(1)} > \underbrace{\sqrt{1-\beta}\frac{d\sqrt{f_s}}{dT}}_{(2)} + \frac{1}{dT/dt}\frac{d}{dz}\left[\underbrace{\left(1-\sqrt{1-\beta}\sqrt{f_s}\right)v_z}_{(3)}\right], \quad \text{when } \sqrt{f_s} \to 1 \tag{3.8}$$

where $\varepsilon_{\text{local}}$ is the local thermal strain due to cooling, T is the temperature, β is the solidification shrinkage, f_s is the fraction of solid, t is the time, z is the position along the growth direction, and v_z is the intergranular liquid flow velocity. Part (1) of

Eq. (3.8) denotes the tensile deformation between the grooving crystals, which leads to crack formation; part (2) denotes the contribution of the crystal growth from neighboring dendrites toward one another prior to bonding; and part (3) of Eq. (3.8) quantifies the feeding of the liquid into the interdendritic region, supporting the growth of the crystals. Kou then correlates the length of the liquid channel with the solidification gradient at a terminal stage, defined by $\frac{d(T)}{d(f_s^{0.5})}$, when $0.87 < f_s < 0.94$.

3.3.5 Solid-State Defects

Hunt and coworkers [186] published one of the first works evaluating the suitability of alloy for AM. They considered the thermal shock resistance arising from the thermal stresses (σ_{thermal}) during the rapid cooling and dimensional stability in their design. The thermal stresses are related to the temperature difference experienced (ΔT), elastic modulus (E), and thermal expansion coefficient, α (Section 3.2.3.6), as shown in Eq. (3.9)

$$\sigma_{\text{thermal}} = E \cdot \Delta T \cdot \alpha \tag{3.9}$$

E and α are material-dependent constants, while ΔT is process dependent and evaluated in the region surrounding the melt pool. ΔT is a function of laser power (P), k is the thermal conductivity (Section 3.2.3.3), d is the melt pool depth, and F_{shape} is the melt pool shape, Eq. (3.10).

$$\Delta T = \frac{P}{2\pi \cdot k \cdot d} F_{\text{shape}} \tag{3.10}$$

The extended Rosenthal model developed for welding can be used to estimate the d and F_{shape} [187]. The shock resistance index was defined according to Eq. (3.11) after combining Eqs. (3.9) and (3.10) and normalizing by the ultimate tensile strength (σ_{UTS}) of the material

$$\text{Shock Resistance} = \frac{\sigma_{\text{UTS}} \cdot 2\pi \cdot k \cdot d}{P \cdot E \cdot \alpha} \tag{3.11}$$

Mukherjee with coworkers considered thermal distortions from the perspective of strains [180]. By applying the Buckingham pi-theorem, the authors derived a dimensionless strain equation for the LPBF process, Eq. (3.12):

$$\varepsilon^* = \frac{\alpha \Delta T}{EI} \frac{t}{F\sqrt{\rho}} H^{3/2} \tag{3.12}$$

where H is the heat input per unit length (absorbed power/scanning speed), ΔT is the temperature difference between liquidus and ambient temperatures, I is the second moment of inertia, and t is the characteristic timescale (deposition time). While F is a Fourier number characterizing the thermal transfer during AM defined by Eq. (3.13)

$$F = \frac{\beta}{vw} \tag{3.13}$$

where β is the thermal diffusivity, v is the scanning speed, and w is the track length. By fitting the ε^* against the experimentally measured maximum thermal strains of

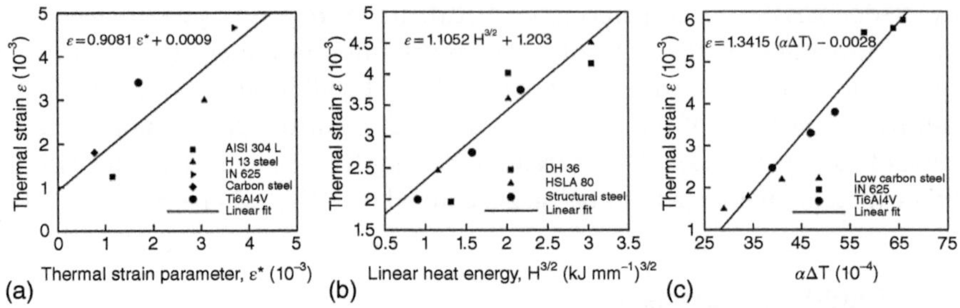

Figure 3.27 Relation between experimentally measured thermal strain in various alloys and (a) thermal strain parameter from Eq. (3.12), (b) linear heat energy, and (c) thermal expansion multiplied by temperature difference. Source: Mukherjee et al. [180]/Springer nature/CC BY.

various alloys, the authors identified an expression for the maximum thermal strain, Eq. (3.14), during LPBF:

$$\varepsilon = \frac{0.9081\alpha\Delta T}{EI}\frac{t}{F\sqrt{\rho}}H^{3/2} + 0.0009 \tag{3.14}$$

From Figure 3.27, one can see that Eq. (3.14) is quite effective at evaluating thermal strain during AM. Figure 3.27b,c shows the influence of the individual factors in Eq. (3.14) on the thermal strain. While linear heat energy is mostly process-dependent, volumetric expansion strongly depends on the material, in particular, the thermal expansion coefficient and melting temperature.

Harrison et al. [43] modified Eq. (3.11), considering the reported inverse dependency of E and α [188] and that the absorptivity decreases with increase in the thermal conductivity k. They proposed a new crack susceptibility parameter (ς) defined in Eq. (3.15).

$$\varsigma = \frac{\sigma_{\text{UTS}}}{\sigma_T} \tag{3.15}$$

where σ_T is the thermal stress. Eq. (3.15) might be helpful to rank alloys according to the risk factor of crack formation.

Hunt et al. [186] related a dimensional stability to the maximising of the ratio of thermal conductivity to thermal expansion [186].

$$\text{Dimentinal stability} = \frac{k}{\alpha} \tag{3.16}$$

The researchers suggested that materials with high shock resistance, Eq. (3.11) and dimensional stability are better suited for AM [186]. Although the approach successfully predicted the printability of many common alloys, it could not explain the printability of stainless steel 316L. In particular, the shock resistance and dimensional stability of 316L are much lower than maraging or tool steels. However, the latter has very poor printability, while 316L is generally considered as highly printable [92].

3.3.6 Modifications to Solidification Behavior

Martin *et al.* [189] proposed to modify the metal powder, Figure 3.28a with nucleation seeds (i.e. inoculants), and Figure 3.28b to promote the nucleation of crystals and equiaxed grain microstructure to minimize the solidification cracking. As discussed in Section 3.4, thermal contraction from the solidifying solid exerts tension onto the liquid film. If the liquid film is broken, a solidification crack is formed. It is widely known that high thermal gradient and rapid solidification result in the epitaxial cellular growth of crystal, worsening the susceptibility to the solidification cracking (Figure 3.28c). However, this can be avoided by the use of seeds to promote nucleation in solidification. High melting-point particles act as nucleation sites, forming equiaxed grains in the solidifying liquid, unconstrained by the nearby solid. The addition of these high melting point nanoparticles also resulted in grain refinement (Figure 3.28e,f). Moreover, Scheil solidification analysis reveals that the addition of Zr reduces the solidification gradient at the final stage of solidification (refer to Section 3.4). Such change in the solidification mode and microstructure substantially reduced the number of cracks, Figure 3.28e,f. The enhancement in solidification and microstructure led to improvement in mechanical properties of the alloy (Figure 3.28h). Crack-free Zr-modified Al7075 showed a one order of magnitude increase in strength and regained ductility. This approach can also be expected to improve fatigue performance. Zr and Sc tend to form $Al_3(Zr, Sc)$ $L1_2$ ordered coherent precipitates. During conventional processing, agglomeration of such precipitates can act as crack initiation sites during fatigue loading [190]. However, if adequately distributed, such intermetallics can significantly improve the fatigue properties of Al–Si–Cu–Mg alloys [191]. The turbulent dynamics in melting and rapid cooling in AM process minimize the agglomeration. Meanwhile, an equiaxed microstructure can increase the isotropy of mechanical properties.

There has been a substantial number of studies investigating Zr and Sc additions to 2XXX, 5XXX, and 7XXX series aluminum alloys. Since the development of Scalmalloy [192], numerous investigations have been conducted by adding Zr and/or Sc to conventional Al alloys [51, 193–196]. Even though most of the research used a prealloyed powder, a significant reduction and even elimination of solidification cracking were achieved along with substantial grain refinement.

Within the Al-based alloys, a study carried out by Thapliyal et al. is of particular interest [196]. The authors performed a compositional optimization before fabrication. Their focus was on the Al—Cu system with additions of Sc and Zr. CALPHAD was used to identify the solubility limits for alloying elements. They subsequently calculated the Scheil solidification curves to provide insights into the solidification crack susceptibility, Figure 3.29a, as previously discussed in Section 3.4. The authors considered the solidification cracking index formulated by Kou and the critical temperature range (ΔT_{CTR}), defined as the temperature range between solidus and 0.95 fraction of solid. Composition deemed

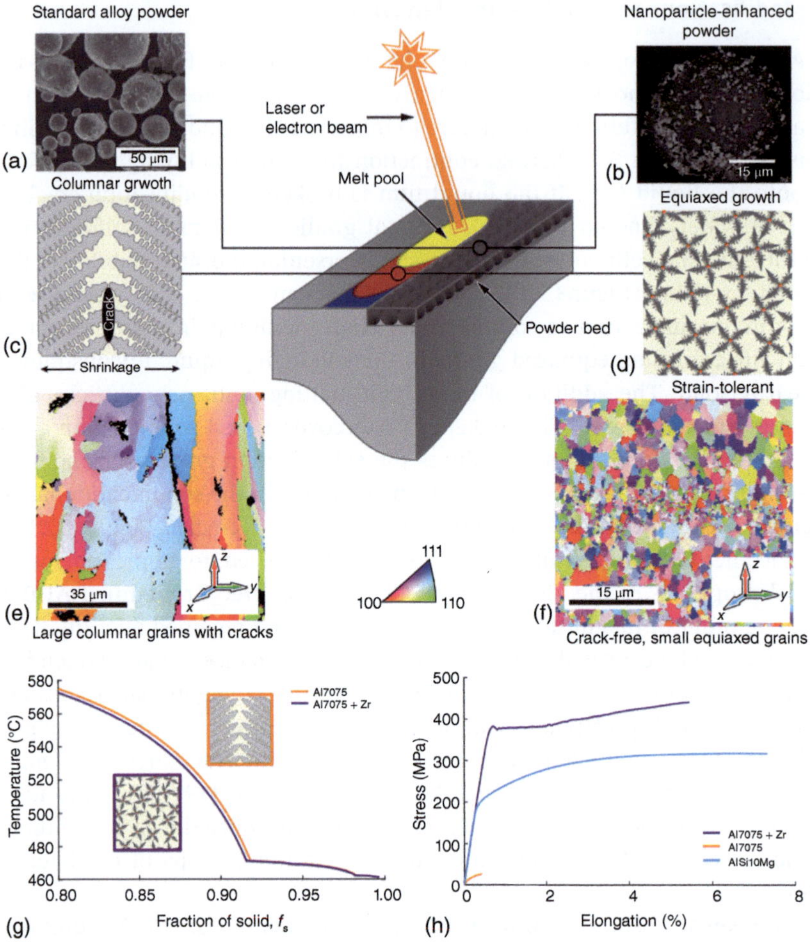

Figure 3.28 Schematic diagram of the modifying conventional powder for LPBF (a) with seeding nanoparticles (b). The effect of seeds on solidification mode from dendritic (c) to equiaxed (d). Effect of the solidification mode on grain structure and cracking susceptibility (e) and (f). (g) The effect of Zr on the Scheil solidification behavior of Al7075 alloy at the terminal stage of solidification and (h) resulting mechanical properties of modified alloy compared to unmodified version and AlSi10Mg. Source: Martin et al. [189]/with permission of Springer Nature.

optimal at Al–1.5Cu–0.8Sc–0.4–Zr. The researchers also predicted the strength of the alloy after heat treatment by evaluating the Peierls–Nabarro stress, solid solution strengthening, precipitation strengthening, and dislocation strengthening contributions [197–199] (Figure 3.29b). It was expected that the alloy could reach strength levels of 420–450 MPa. The measured strengths of the material in both as-fabricated and after heat treatment were lower than predicted (Figure 3.29c). Nevertheless, the material was reported to be highly printable with no evidence of solidification cracks.

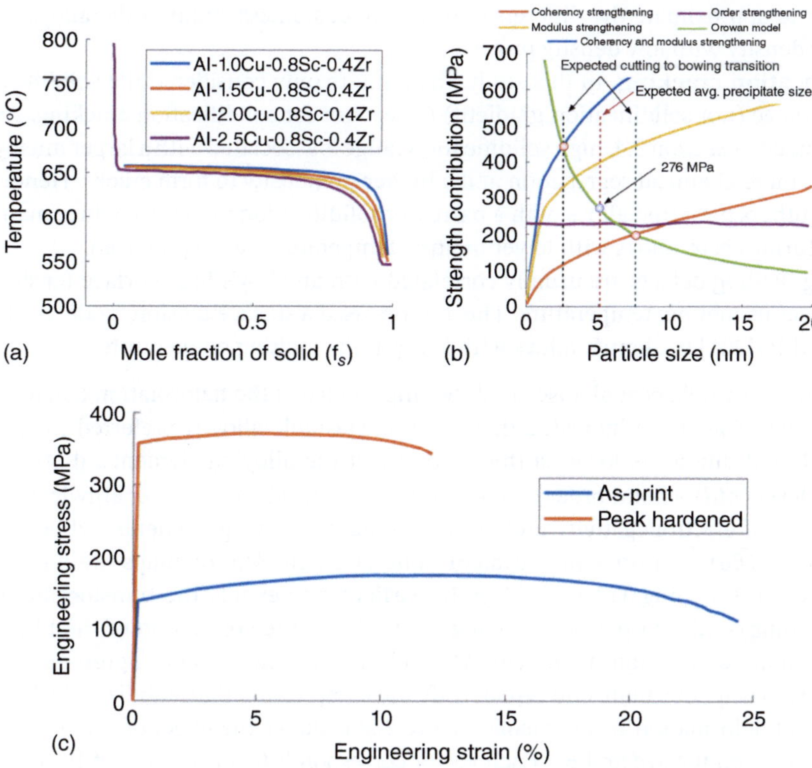

Figure 3.29 (a) Scheil solidification curves for different Cu contents in Al−xCu−0.8Sc−0.4Zr alloy. (b) Estimated strength contributions of various strengthening mechanisms in Al−1.5Cu−0.8Sc−0.4Zr alloy as a function of precipitate size. (c) Mechanical properties of as-print and precipitation hardened alloy. Source: Thapliyal et al. [196]/With permission of Elsevier.

3.3.7 Examples of Alloy Design for Additive Manufacturing

This section explores several research papers in a greater depth that focus on designing an alloy for a particular alloy system, taking the processing conditions and application into account.

3.3.7.1 Titanium Alloy for Medical Applications

Ackers et al. [200] used several factors to design Ti-based alloys for medical applications. The authors used the dimensional strain parameter, Eq. (3.12), developed by Mukherjee et al., solidification range of the alloy, mean d-orbital energy level, mean bond order, and surface tension [200]. The authors identified several defect types and methods to minimize them.

Thermal strain. Calculation of ε^* in Eq. (3.12) requires thermophysical data for an alloy. Instead, the authors assumed that the thermophysical properties of the alloys with similar compositions did not vary much. This assumption allowed them to approximate the difference in the dimensionless strain of the alloys by their density with low density preferred.

Solidification cracking. In this study, the authors only considered the solidification range (not solidification gradient) to account for solidification cracking. As discussed in Section 3.4, high solidification range is associated with a larger mushy zone, more chemical segregation and a higher propensity to form cracks. Hence, the authors preferred alloys with a narrower solidification range. To reduce manufacturing costs, alloys with lower melting temperatures were preferred.

Balling. Balling defects are usually correlated with an alloy's high surface tension (γ) near its melting temperature. The authors used a surface tension model by Su and Mills [201] to identify alloys with low γ at the melting temperature.

Additionally, Ackers et al. discussed the importance of the deformation behavior of the alloy. They have indicated that a beta-metastable alloy is preferred for the medical field due to its lower stiffness. To select the alloying elements, d-orbital energy level ($\overline{Md}$) – bond order ($\overline{Bo}$) diagrams were used. D-orbital energy level is a function of electronegativity and atomic radius of alloying elements, while the bond order ($\overline{Bo}$) quantifies the covalent bond strength. $\overline{Md}$–$\overline{Bo}$ maps are a common tool in designing Ti-based alloys in particular to enable the transformation (or twinning)–induced plasticity. Figure 3.30a shows the correlation of the phase stability and deformation mode with $\overline{Md}$ and $\overline{Bo}$, with known alloys presented as dots [202]. Using this map, one can identify the composition that stabilizes the beta phase and deformation mechanisms. Figure 3.30b shows the effect of some alloying elements on the $\overline{Md}$ and $\overline{Bo}$. Each vector corresponds to an addition of 10 at% of the alloying element. To evaluate the effects of multiple elements, one can calculate the sum of the individual contributions. This works used a rule of mixtures based on individual orbital energies and bond orders of alloying elements (Figure 3.30c). Similar diagrams for Ni-based ad Fe-based alloys can be found in a recent overview by Morinaga [203].

Alloys for biomedical applications require high compliance to match as close as possible to the compliance of the bone. In Ti alloys, niobium and tantalum act as a beta-phase stabilizer, which aids the reduction of elastic modulus. On the other hand, moving joints of the implant are subjected to wear conditions. Therefore,

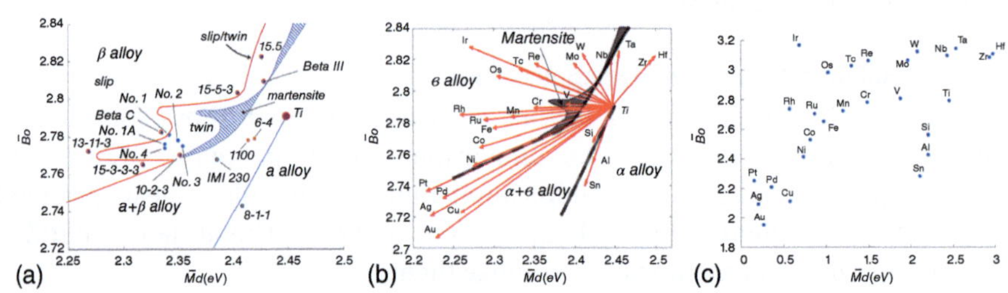

Figure 3.30 (a) Influence of $\overline{Md}$ and $\overline{Bo}$ on the phase stability of Ti alloys; (b) influence of alloying elements on phase stability; vector length corresponds to 10 at% of an alloying element; (c) $\overline{Md}$ and $\overline{Bo}$ for several metals alloyed with BCC Ti. Source: Adapted from Morinaga et al. [202].

the authors considered additions of Zr, Sn, Ta, and Si to enable precipitation of secondary phases. Thermodynamic properties of the alloys (phases, solidification range) along with alloy density were predicted using commercial CALPHAD software – ThermoCalc under equilibrium conditions.

Through this approach around 70 000 different alloys were generated in a compositional range Nb:0–40, Mo:0–12, Fe:0–10, Ta:0–10, Zr:0–10, and Sn:0–10. A ranking system was applied to guide the selection process, sorting the alloys according to the combination of properties of interest. The researcher subsequently used an effective approach of blending elemental powders to achieve the required nominal composition. The team has evaluated 22 different alloys. Printability was evaluated via the surface roughness and internal structure of the printed disk. Alloys with a smooth surface and lack of visible cracks were deemed printable. The approach employed in this study seems very effective at rapidly evaluating numerous possible alloy compositions via CLPHAD and empirical relations. It allows to substantially narrow down the candidate list according to the specific requirements of the given application and printability criteria. The use of blended elemental powders mitigates the lengthy and costly process of manufacturing the prealloyed powders [204, 205]. At the same time, the slotted platform with repeated remelting of multiple alloys at once provides means to rapidly screen the behavior of several alloys during LPBF conditions. Even though such a method does not fully replicate the thermal cycles and cooling rates because of the reduced conductivity of the powder and the absence of a thermal passway to the build plate, it provides a glance at material microstructure and properties after LPBF.

3.3.7.2 Creep-Resistant Ni-Based Superalloy

Tang et al. performed a very comprehensive study on the alloy design of precipitation-hardened Ni-based superalloys [206]. They considered the compositional and microstructural effects on the formation of solidification cracks and solid-state cracks together with the yield stress and creep performance for high-temperature applications. The team used CALPHAD-based approach to estimate the flow stress, following the work of Crudden et al. [207] along with design rules developed by Reed et al. for the creep strength [208]. The Scheil module in ThermoCalc was used to evaluate the solidification range to indicate solidification crack tendency. The researchers also considered the susceptibility of forming a topological close-packed phase using the design rules by Reed et al. [208]. A database of over 10^7 potential alloys was generated in their study (Figure 3.31a,b). These data were then analyzed by Pareto frontier to identify the optimal compositions with respect to multiple requirements. The requirements were set based on existing alloys and their performance during PBF. In particular, the group used IN718 as an example of a highly printable alloy, which is rather insusceptible neither to solidification cracking nor solid-state cracking. IN738LC has been frequently reported to suffer from solidification cracking; therefore, the solidification range limit was set at 285K (IN738LC) and 265K (IN718). As for the solid-state cracking, the authors used a composition based index of $M_{SAC} = [Al] + 0.5[Ti] + 0.3[Nb] + 0.15[Ta]$, where SAC stands for strain age cracking. In this case, IN939 was used as a reference, which was reported to form solid-state cracks with its index of 4.3 wt%; hence, the limit for M_{SAC} was set to be 4 wt%. Using these criteria, the authors developed and printed two new alloys named ABD-850AM and ABD-900AM alongside several known alloys (CM247LC and IN939) for comparison. Neither ABD-850AM nor ABD-900AM showed solidification or solid-state cracks, unlike the known alloys (Figure 3.31c,d). In-depth crack characterization and identification were performed. The authors applied Kou's solidification gradient approximation [185] to evaluate the solidification cracking susceptibility (see Section 3.4). IN718 and ABD-900AM alloys showed better solidification characteristics compared to IN738LC and CM247LC. As the primary objective was to obtain good creep resistance on pair with high printability, Figure 3.31f shows the Larson Miller Parameter (LMP) for the creep resistance in the developed ABD-900AM and known alloys produced via LPBF. Creep properties of the ABD-900AM were predicted to approach that of CM247LC and higher than that of the printable IN718. Nevertheless, further research is needed to establish more quantitatively defined design criteria to achieve desired alloy properties.

3.3.7.3 High Strength Co-Based Superalloy for High-Temperature Applications

Murray and coworkers strived to develop a Co-based superalloy with good printability, aiming for high-temperature strength and high oxidation resistance [209]. The goal to achieve good high-temperature strength was through a high volume fraction of $L1_2$ γ' stable to at least 1100 °C and a liquidus temperature exceeding 1300 °C. To achieve a good oxidation resistance, the authors considered the formation of protective coherent α-Al_2O_3. A narrow freezing range was preferred to minimize the risks of forming a solidification crack during PBF processes. Authors acknowledge that alloys containing a high volume fraction of γ' have a significant risk of strain-age cracking. Therefore, they suggested that a temperature gap between solidus and γ' formation temperature should be larger than 50 °C. A computational combinatorial approach was used to optimize the process, starting from a ternary system of Co–Al–W and moving to a multicomponent alloy. Density functional theory (DFT) was utilized to find optimal alloying elements to increase the γ' fraction and APB energy (hence improving the strength). Analysis suggested the use of Ti, Ta, Nb, and Ni. CALPHAD was employed to tune the composition to achieve a required range for the γ' solvus temperature. Due to the difficulties in modeling the oxide coherency, a combinatorial library approach was developed. Ion plasma deposition with five electrodes containing Co, Ni, Al, W, Cr, and Ta was used to create 234 different alloy variants rapidly. These variants were screened for the oxide with α-Al_2O_3 scales and verified with CALPHAD to contain a suitable amount of γ'. Finally, an alloy with a composition of Co–36.4Ni–13.2Al–6Cr–3.5Ta–1W (at%) was created and termed SB-CoNi-10. Minor additions of 0.08B–0.08C–0.018Hf–0.002Y (at%) were incorporated to increase the grain boundary strength, promote carbide formation, and enhance oxide scale adhesion. Alloy had a liquidus temperature of 1381 °C, the solidus of 1329 °C and γ' solvus of 1204 °C under equilibrium conditions. Therefore, the freezing range was 52 °C, and the solidus-γ' solvus gap was 125 °C to minimize the solidification and

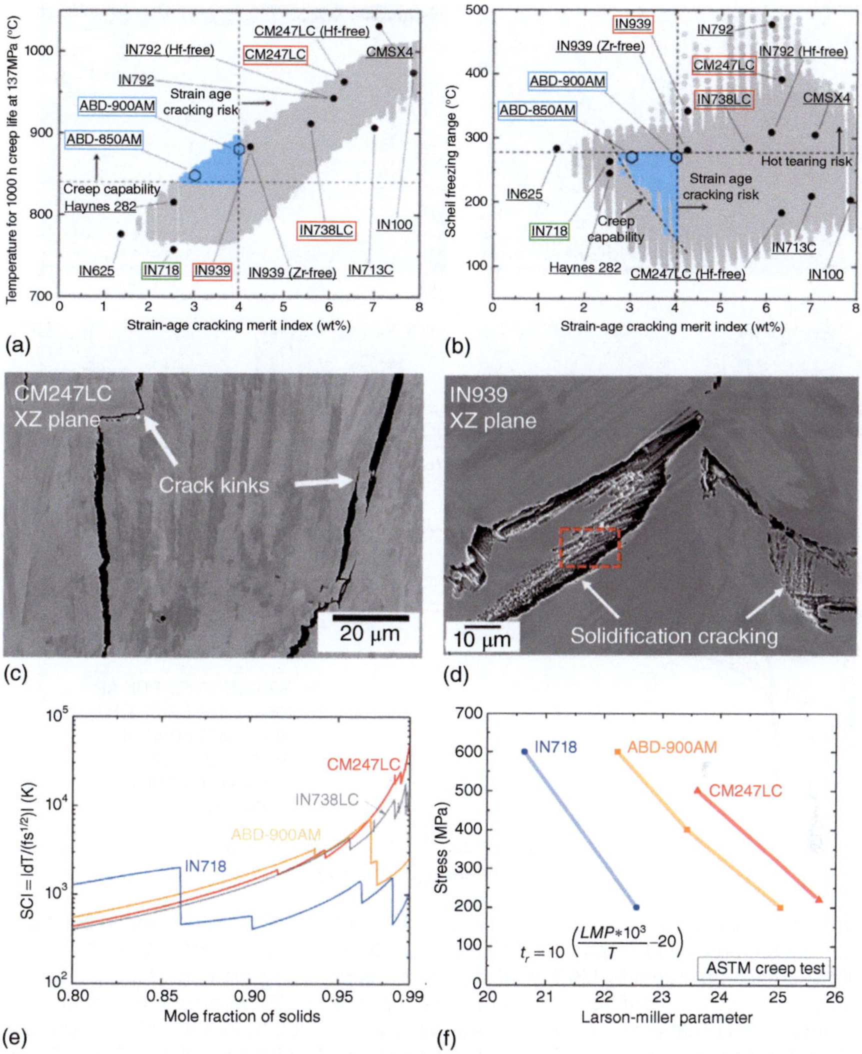

Figure 3.31 (a, b) Computational alloy design spaces evaluated by Tang et al. [206] for printable Ni-base superalloys. Heritage alloys are presented for reference as dots. Alloys developed in the study are marked with blue boxes, alloys used as a reference for development are outlined in green (printable) and red (with defects) boxes. Examples of defects forming in heritage alloys are presented in (c) solid-state cracks and (d) solidification cracks. Solidification crack index calculated for developed and heritage alloys are shown in (e). Comparison of creep resistance between developed and heritage alloys shown in (f) using Larson Miller parameter (LMP). For the reader's convenience, the equation relating the LMP and creep rupture time (t_r [h]) at a temperature T [K] is provided in the bottom left of the (f). Source: Tang et al. [206]/Elsevier/CC BY.

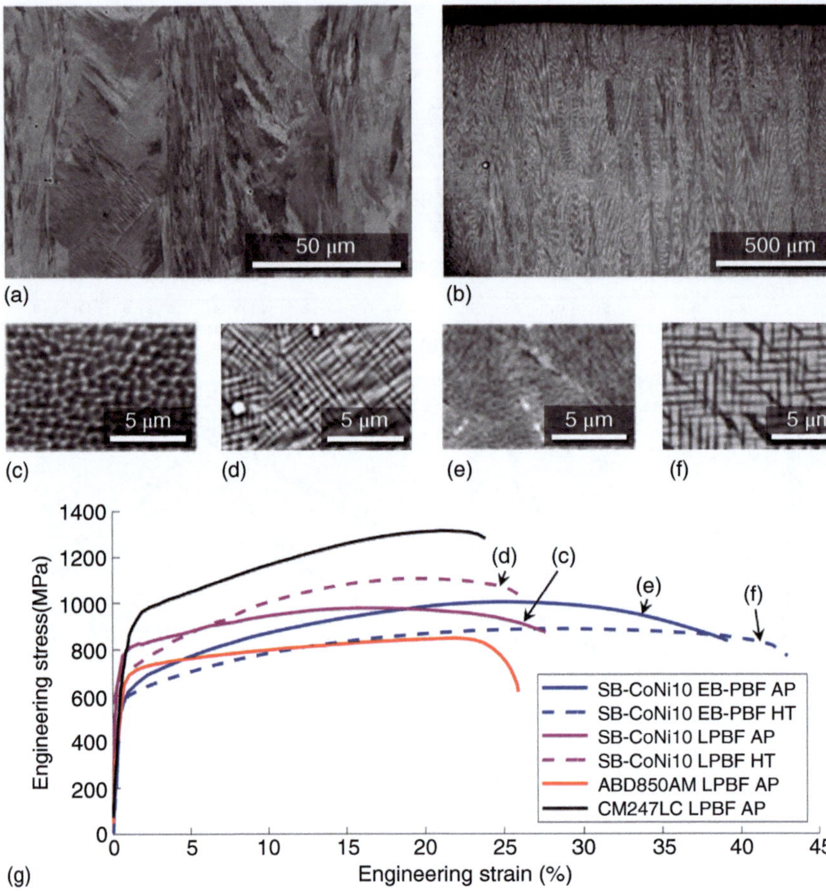

Figure 3.32 (a, b) Microstructure of as-printed (AP) SB-CoNi-10 alloy developed by Murray et al. Manufactured by LPBF and EB-PBF, respectively. (c, d) Fine microstructure of SB-CoNi-10 manufactured by LBPF in AP and heat treated (HT) condition, respectively. (e, f) Fine microstructure of SB-CoNi-10 manufactured by EB-PBF in AP and HT conditions, respectively. Note that preheated build plate was used at 1000 °C for EB-PBF and 200 °C for LPBF. (g) Mechanical properties of superalloys developed by Murray et al. (SB-CoNi-10*) and Tang et al. (ABD850AM**) compared to heritage CM247LC**. Note the references to the microstructure of the tested material in (c–f). Source: *Murray et al. [209]/with permission of Springer Nature and **Tang et al. [206]/from Elsevier/CC BY 4.0.

solid-state cracking. Experiments showed that γ′ fraction could reach up 70% after aging at 1000 °C. The alloy had been successfully printed via both LPBF and EB-PBF methods, producing an almost crack-free material (Figure 3.32a,b). As anticipated, the LPBF samples showed a solid solution appearance, with no evidence of γ′ formation (Figure 3.31c). Further heat treatment (HIP+SH + aging) achieved the desired γ–γ′ microstructure, Figure 3.31d, achieving the alloy design goals. Interestingly, the EB-PBF samples presented a γ–γ′ microstructure in as-printed condition, as shown in Figure 3.32e. The authors attributed it to the high preheat temperature of the build-plate (1000 °C), which caused an inhomogeneous size distribution of the γ′ along the build height. Heat treatment (HIP + SH + aging) significantly coarsened the γ'. The microstructure characteristics explain the differences in the mechanical testing of the specimens, Figure 3.32g, where HT reduced the strength of the EB-PBF sample while increasing the strength of LPBF specimens.

3.4 Summary

Fusion-based additive manufacturing of metallic components is a promising technology offering plenty of benefits for a range of applications. Each application has specific working conditions requiring unique material properties. Engineering alloys have been developed for decades to obtain desired properties suitable to be manufactured by specific manufacturing techniques for given working conditions. Because of the availability of materials suitable for fusion-based AM processes, it is still in infancy and metallurgical issues remain as significant challenges to the wide industrialization for the fusion-based AM. In particular, high cooling rates and thermal gradients in fusion-based AM often lead to unique microstructures such as fine solidification cells containing dense dislocations at cellular boundaries. Directional cooling gives rise to high texture and anisotropy of mechanical properties. Multiple cycles of remelting with a high-energy density source can lead to excessive evaporation of alloying elements. At the same time, the build-up of residual stresses can warp and distort the printed components. Keyhole and lack-of-fusion pores are often reported to be detrimental to printed components, significantly limiting the reliability of AM-fabricated parts. Despite substantial efforts in minimizing the defect formation via optimizing the process parameters to identify a processability window with optimal consolidation, the challenges remain persistent because of the unsuitability of alloys for AM. This prompts another direction of research and developments addressing the metallurgical and mechanical issues by selecting (and designing) alloys that are more suitable for fusion AM. Prior research in welding literature provides a significant understanding of underlying mechanisms responsible for the defect formation and microstructure in fusion-based AM. Several theories have been developed to evaluate the tendency to form a specific type of defects. Recent investigations demonstrated that such theories are useful and effective in selecting and developing alloys that have better resistance to the formation of defects commonly found in fusion-based AM, particularly for solidification cracking, to improve the mechanical properties. This chapter has provided a state-of-the-art understanding of mechanisms responsible for the defect formation in PBF processes, particularly the effect of inherent material properties. The chapter also presents methodology formulations and necessary considerations to select and design the alloy composition to achieve high resistance to the formation of porosity and cracks. Key examples of recent developments in alloys for AM are reviewed to highlight the effectiveness and limitations in the current alloy design efforts. Nevertheless, the authors acknowledge that controlling and tuning the processing conditions during fusion-based AM are equally essential to achieving a defect-free component. Consequently, holistic considerations of both material characteristics and processing conditions should be developed to achieve high-performing materials using fusion-based AM for safety-critical applications.

Abbreviations

AP	As-printed
AM	additive manufacturing
ASTM	American Society for Testing and Materials
CALPHAD	calculation of phase diagrams
DED	direct energy deposition
EB-PBF	electron beam powder bed fusion
HT	heat treated
ISO	International Standards Organization
LENS	laser-engineered net shaping
LPBF	laser powder bed fusion
PBF	powder bed fusion

References

1 Koch, J.L. and Mazumder, J. (1995). Rapid prototyping by laser cladding. *Laser Cladding Journal of Laser Applications* 11: 556–565.

2 Clint, A., Griffith, M., Harwell, L. et al. (1998). Laser engineered net shaping (LENS): a tool for direct fabrication of metal parts. *International Congress on Applications of Lasers & Electro-Optics (ICALEO)* 1998: E1.

3 Wohelers, T. (2018). Wohlers Report 2018: Additive Manufacturing and 3D Printing State of the Industry, 344 pages; FYI, link of report: https://wohlersassociates.com/product/wohlers-report-2018/.

4 Scott, T. J., Beaulieu, T. J., Rothrock, G. D., and O'connor, A. C. (2016). Economic Analysis of Technology Infrastructure Needs for Advanced Manufacturing: Additive Manufacturing, NIST GCR 16-006. http://dx.doi.org/10.6028/NIST.GCR.16-006.

5 A. 52900:2015 (2015). Standard Terminology for Additive Manufacturing – General Principles – Terminology. *ASTM International*, vol. i, pp. 1–9.

6 Royal Academy of Engineering (2013). Additive manufacturing : opportunities and constraints. *Royal Academy of Engineering*, no. May 2013 21.

7 Pham, M.-S.S., Liu, C., Todd, I., and Lertthanasarn, J. (2019). Damage-tolerant architected materials inspired by crystal microstructure. *Nature* 565 (7739): 305–311.

8 Kudzal, A., McWilliams, B., Hofmeister, C. et al. (2017). Effect of scan pattern on the microstructure and mechanical properties of Powder Bed Fusion additive manufactured 17-4 stainless steel. *Materials and Design* 133: 205–215.

9 Geiger, F., Kunze, K., and Etter, T. (2016). Tailoring the texture of IN738LC processed by selective laser melting (SLM) by specific scanning strategies. *Materials Science and Engineering A* 661: 240–246.

10 Ishimoto, T., Hagihara, K., Hisamoto, K. et al. (2017). Crystallographic texture control of beta-type Ti–15Mo–5Zr–3Al alloy by selective laser melting for the

development of novel implants with a biocompatible low Young's modulus. *Scripta Materialia* 132: 34–38.

11 Thijs, L., Montero Sistiaga, M.L., Wauthle, R. et al. (2013). Strong morphological and crystallographic texture and resulting yield strength anisotropy in selective laser melted tantalum. *Acta Materialia* 61 (12): 4657–4668.

12 Livescu, V., Knapp, C.M., Gray, G.T. et al. (2018). Additively manufactured tantalum microstructures. *Materialia* 1: 15–24.

13 Iron Boar Labs Ltd. (2021). MakeItFrom.com: Material Properties Database. https://www.makeitfrom.com. (Accessed: 29 Jul 2021).

14 Dovgyy, B., Santos Maldonado, C.-T., and Pham, M.-S. (2021). Process parameters and properties of laser powder bed fusion alloys (Version 1) [Data set]. *Zenodo*, November 2021.

15 ASTM International (2016). ASTM F3184 (2016) Standard Specification for Additive Manufacturing Stainless Steel Alloy (UNS S31603) with Powder Bed Fusion. *Annu. B. ASTM Stand.*, vol. i, pp. 1–9, 2016.

16 ASTM (2017). ASTM F3213-17 Standard for Additive Manufacturing – Finished Part Properties – Standard Specification for cobalt-28 chromium-6 Molybdenum via Powder Bed Fusion. *ASTM Standard*, no. January, pp. 1–9.

17 A. International (2014). Standard Specification for Additive Manufacturing Nickel Alloy (UNS N07718) with Powder Bed Fusion F3055. *ASTM Standard*, no. March, pp. 1–8.

18 ISO/ASTM (2021). F3056-14. Standard Specification for Additive Manufacturing Nickel Alloy (UNS N06625) with Powder Bed Fusion *ASTM Standard*, vol. 14, no. March 2014, pp. 1–8.

19 ASTM F3302. (2018). Standard for Additive Manufacturing – Finished Part Properties – Standard Specification for Titanium Alloys via Powder Bed Fusion *ASTM Standard*, vol. i, no. February, pp. 1–8.

20 Standard for Additive Manufacturing-Finished Part Properties-Specification for AlSi10Mg with Powder Bed Fusion-Laser Beam, ASTM F3318-18, Vol. 10.04, pages 8. https://doi.org/10.1520/F3318-18.

21 Mower, T.M. and Long, M.J. (2016). Mechanical behavior of additive manufactured, powder-bed laser-fused materials. *Materials Science and Engineering A* 651: 198–213.

22 Lin, X., Cao, Y., Wu, X. et al. (2012). Microstructure and mechanical properties of laser forming repaired 17-4PH stainless steel. *Materials Science and Engineering A* 553: 80–88.

23 Hu, Z., Zhu, H., Zhang, H., and Zeng, X. (2017). Experimental investigation on selective laser melting of 17-4PH stainless steel. *Optics and Laser Technology* 87: 17–25.

24 Rashid, R., Masood, S.H., Ruan, D. et al. (2017). Effect of scan strategy on density and metallurgical properties of 17-4PH parts printed by Selective Laser Melting (SLM). *Journal of Materials Processing Technology* 249: 502–511.

25 López-Castro, J.D., Marchal, A., González, L., and Botana, J. (2017). Topological optimization and manufacturing by Direct Metal Laser Sintering of an aeronautical part in 15-5PH stainless steel. *Procedia Manufacturing* 13: 818–824.

26 Coffy, K.M. (1989). *Microstructure and Chemistry Evaluation of Direct Metal Laser Sintered 15–5 PH Stainless Steel*. University of Central Florida.
27 Rafi, H.K., Starr, T.L., and Stucker, B.E. (2013). A comparison of the tensile, fatigue, and fracture behavior of Ti-6Al-4V and 15-5 PH stainless steel parts made by selective laser melting. *International Journal of Advanced Manufacturing Technology* 69 (5–8): 1299–1309.
28 Qu, S., Huang, C.X., Gao, Y.L. et al. (2008). Tensile and compressive properties of AISI 304L stainless steel subjected to equal channel angular pressing. *Materials Science and Engineering A* 475 (1–2): 207–216.
29 Abd-Elghany, K. and Bourell, D.L. (2012). Property evaluation of 304L stainless steel fabricated by selective laser melting. *Rapid Prototyping Journal* 18 (5): 420–428.
30 Song, M., Lin, X., Liu, F. et al. (2016). The purification of AISI 420 stainless steel in laser solid forming. *Materials and Design* 89: 1035–1040.
31 Zhao, X., Wei, Q., Song, B. et al. (2015). Materials and manufacturing processes fabrication and characterization of AISI 420 stainless steel using selective laser melting fabrication and characterization of AISI 420 stainless steel using selective laser melting. *Materials and Manufacturing Processes* 3011: 1283–1289.
32 Casati, R., Coduri, M., Lecis, N. et al. (2018). Microstructure and mechanical behavior of hot-work tool steels processed by Selective Laser Melting. *Materials Characterization* 137: 50–57.
33 Huber, F., Bischof, C., Hentschel, O. et al. (2019). Laser beam melting and heat-treatment of 1.2343 (AISI H11) tool steel – microstructure and mechanical properties. *Materials Science and Engineering A* 742: 109–115.
34 Krell, J., Röttger, A., Geenen, K., and Theisen, W. (2018). General investigations on processing tool steel X40CrMoV5-1 with selective laser melting. *Journal of Materials Processing Technology* 255: 679–688.
35 Cormier, D., Harrysson, O., and West, H. (2004). Characterization of H13 steel produced via electron beam melting. *Rapid Prototyping Journal* 10 (1): 35–41.
36 Breidenstein, B., Brenne, F., Wu, L. et al. (2018). Effect of post-process machining on surface properties of additively manufactured H13 tool steel. *HTM Journal of Heat Treatment and Materials* 73 (4): 173–186.
37 Mazur, M., Leary, M., McMillan, M. et al. (2016). SLM additive manufacture of H13 tool steel with conformal cooling and structural lattices. *Rapid Prototyping Journal* 22 (3): 504–518.
38 Yasa, E., Kempen, K., Kruth, J.P. et al. (2010). Microstructure and mechanical properties of 18Ni300 maraging steel after Selective Laser Melting. *Proceedings Solid Freeform Fabrication Symposium* 383–396.
39 Mooney, B., Kourousis, K.I., and Raghavendra, R. (2019). Plastic anisotropy of additively manufactured maraging steel: influence of the build orientation and heat treatments. *Additive Manufacturing* 25: 19–31.
40 Shakerin, S., Hadadzadeh, A., Amirkhiz, B.S. et al. (2019). Additive manufacturing of maraging steel-H13 bimetals using laser powder bed fusion technique. *Additive Manufacturing* 29: 100797.

41 Bai, Y., Yang, Y., Wang, D., and Zhang, M. (2017). Influence mechanism of parameters process and mechanical properties evolution mechanism of maraging steel 300 by selective laser melting. *Materials Science and Engineering A* 703: 116–123.

42 Keshavarzkermani, A., Marzbanrad, E., Esmaeilizadeh, R. et al. (2019). An investigation into the effect of process parameters on melt pool geometry, cell spacing, and grain refinement during laser powder bed fusion. *Optics and Laser Technology* 116: 83–91.

43 Harrison, N.J., Todd, I., and Mumtaz, K. (2015). Reduction of micro-cracking in nickel superalloys processed by Selective Laser Melting: a fundamental alloy design approach. *Acta Materialia* 94: 59–68.

44 Montero-Sistiaga, M.L., Liu, Z., Bautmans, L. et al. (2020). Effect of temperature on the microstructure and tensile properties of micro-crack free hastelloy X produced by selective laser melting. *Additive Manufacturing* 31: 100995.

45 Han, Q., Vanmeensel, K., Fan, H. et al. (2018). Laser powder bed fusion of Hastelloy X: effects of hot isostatic pressing and the hot cracking mechanism. *Materials Science and Engineering A* 732: 228–239.

46 Wang, H., Chen, L., Dovgyy, B. et al. (2021). Micro-cracking, microstructure and mechanical properties of Hastelloy-X alloy printed by laser powder bed fusion: as-built, annealed and hot-isostatic pressed. *Additive Manufacturing* 39: 101853.

47 Bauer, T., Dawson, K., Spierings, A.B., Wegener, K.. (2015). Microstructure and mechanical characterisation of SLM processed Haynes® 230. Proceedings Solid Freeform Fabrication Symposium. https://doi.org/10.3929/ethz-a-010584903.

48 Lopez, M., Pickett, C., Arrieta, E. et al. (2020). Effects of postprocess hot isostatic pressing treatments on the mechanical performance of EBM fabricated TI-6Al-2Sn-4Zr-2Mo. *Materials (Basel)* 13 (11): 2604.

49 Fan, H. and Yang, S. (2020). Effects of direct aging on near-alpha Ti–6Al–2Sn–4Zr–2Mo (Ti-6242) titanium alloy fabricated by selective laser melting (SLM). *Materials Science and Engineering A* 788: 139533.

50 Carrozza, A., Aversa, A., Fino, P., and Lombardi, M. (2021). A study on the microstructure and mechanical properties of the Ti-6Al-2Sn-4Zr-6Mo alloy produced via Laser Powder Bed Fusion. *Journal of Alloys and Compounds* 870: 159329.

51 Awd, M., Tenkamp, J., Hirtler, M. et al. (2018). Comparison of microstructure and mechanical properties of Scalmalloy® produced by selective laser melting and laser metal deposition. *Materials* 11 (1): 17. https://doi.org/10.3390/ma11010017.

52 Knoop, D., Lutz, A., Mais, B., and von Hehl, A. (2020). A tailored AlSiMg alloy for laser powder bed fusion. *Metals (Basel)* 10 (4): 514.

53 Cordova, L., Bor, T., de Smit, M. et al. (2020). Effects of powder reuse on the microstructure and mechanical behaviour of Al–Mg–Sc–Zr alloy processed by laser powder bed fusion (LPBF). *Additive Manufacturing* 36: 101625.

54 Kimura, T. and Nakamoto, T. (2016). Microstructures and mechanical properties of A356 ($AlSi_7Mg_{0.3}$) aluminum alloy fabricated by selective laser melting. *Materials and Design* 89: 1294–1301.

55 Dai, D., Gu, D., Zhang, H. et al. (2018). Influence of scan strategy and molten pool configuration on microstructures and tensile properties of selective laser melting additive manufactured aluminum based parts. *Optics and Laser Technology* 99: 91–100.

56 Montero-Sistiaga, M.L., Mertens, R., Vrancken, B. et al. (2016). Changing the alloy composition of Al7075 for better processability by selective laser melting. *Journal of Materials Processing Technology* 238: 437–445.

57 J. Allen, Rolls-Royce plc (2006). "An investigation into the comparative costs of additive manufacture vs. machine from solid for aero engine parts," Defense technical information center.

58 Froes, F.H. and Imam, M.A. (2010). Cost affordable developments in titanium technology and applications. *Key Engineering Materials* 436: 1–11.

59 Froes, F.H.S., Gungor, M.N., and Ashraf Imam, M. (2007). Cost-affordable titanium: the component fabrication perspective. *JOM* 59 (6): 28–31.

60 Welsch, G., Boyer, R., and Collings, E.W. (1998). *Materials Properties Handbook: Titanium Alloys*, 2nde. Materials Park, OH: ASM International.

61 Cahn, R.W., Haasen, P., and Kramer, E.J. (2008). *Materials Science and Technology: A Comprehensive Treatment*, vol. 1–2, VCH.

62 Sidambe, A.T. (2014). Biocompatibility of advanced manufactured titanium implants – a review. *Materials (Basel, Switzerland)* 7 (12): 8168–8188.

63 Sieniawski, J., Ziaja, W., Kubiak, K., and Motyk, M. (2013). Microstructure and mechanical properties of high strength two-phase titanium alloys. In: *Titanium Alloys – Advances in Properties Control* (ed. J. Sieniawski and W. Ziaja). London: InTech https://www.intechopen.com/chapters/44359.

64 Banerjee, D., Muraleedharan, K., and Strudel, J.L. (1998). Substructure in titanium alloy martensite. *Philosophical Magazine A: Physics of Condensed Matter, Structure, Defects and Mechanical Properties* 77 (2): 299–323.

65 Yang, J., Yu, H., Yin, J. et al. (2016). Formation and control of martensite in Ti-6Al-4V alloy produced by selective laser melting. *Materials and Design* 108: 308–318.

66 Bartlett, J.L. and Li, X. (2019). An overview of residual stresses in metal powder bed fusion. *Additive Manufacturing* 27. Elsevier: 131–149.

67 Chen, S. guang, Zhang, Y. du, Wu, Q. et al. (2021). Effect of solid-state phase transformation on residual stress of selective laser melting Ti6Al4V. *Materials Science and Engineering A* 819: 141299.

68 Denlinger, E.R. and Michaleris, P. (2016). Effect of stress relaxation on distortion in additive manufacturing process modeling. *Additive Manufacturing* 12: 51–59.

69 Tseng, K.H. and Chou, C.P. (2002). The effect of pulsed GTA welding on the residual stress of a stainless steel weldment. *Journal of Materials Processing Technology* 123 (3): 346–353.

70 Dieter, G.E. and Bacon, D. (1976). *Mechanical Metallurgy*. New York: McGraw-Hill.

71 Zhang, G., Chen, J., Zheng, M. et al. (2020). Element vaporization of Ti-6Al-4V alloy during selective laser melting. *Metals (Basel)* 10 (4): 435.

72 Juechter, V., Scharowsky, T., Singer, R.F., and Körner, C. (2014). Processing window and evaporation phenomena for Ti-6Al-4V produced by selective electron beam melting. *Acta Materialia* 76: 252–258.

73 Klassen, A., Scharowsky, T., and Körner, C. (2014). Evaporation model for beam based additive manufacturing using free surface lattice Boltzmann methods. *Journal of Physics D: Applied Physics* 47 (27): 275303.

74 Semiatin, S.L., Ivanchenko, V.G., Akhonin, S.V., and Ivasishin, O.M. (2004). Diffusion models for evaporation losses during electron-beam melting of alpha/beta-titanium alloys. *Metallurgical and Materials Transactions B: Process Metallurgy and Materials Processing Science* 35 (2): 235–245.

75 Hirsch, P.B. (1996). Kear-wilsdorf locks, jogs and the formation of antiphase- boundary tubes in Ni3Al. *Philosophical Magazine A: Physics of Condensed Matter, Structure, Defects and Mechanical Properties* 74 (4): 1019–1040.

76 Sanchez, S., Smith, P., Xu, Z. et al. (2021). Powder bed fusion of nickel-based superalloys: a review. *International Journal of Machine Tools and Manufacture* 165: 103729.

77 Haakens, M.H. and Matthey, J.H.G. (1982). New approach to the weldability of nickel-base as-cast and powder metallurgy superalloys. *Welding Journal* V 61 (N 11): 25–30.

78 Eriksson, R. (2013). *Thermal Barrier Coatings: Durability Assessment and Life Prediction*. Linköping University.

79 McCoy, H. E., King, J. F., and Rittenhouse, P. I. (1981). Weldability evaluations and weldment properties of Hastelloy X.

80 Shimizu, S. and Mutoh, Y. (1984). Weldability and weld performance of a special grade hastelloy-X modified for high-temperature gas-cooled reactors. *Nuclear Technology* 66 (1): 44–53. https://doi.org/10.13182/NT84-A33453.

81 Graneix, J., Beguin, J.-D., Pardheillan, F. et al. (2014). Weldability of the superalloys Haynes 188 and Hastelloy X by Nd:YAG. *MATEC Web of Conferences* 14: 13006.

82 Banoth, S., Palleda, T.N., Shimazu, S., and Kakehi, K. (2021). Yttrium's effect on the hot cracking and creep properties of a Ni-based superalloy built up by additive manufacturing. *Materials (Basel)* 14 (5): 1143.

83 Tomus, D., Rometsch, P.A., Heilmaier, M., and Wu, X. (2017). Effect of minor alloying elements on crack-formation characteristics of Hastelloy-X manufactured by selective laser melting. *Additive Manufacturing* 16: 65–72.

84 Tomus, D., Jarvis, T., Wu, X. et al. (2013). Controlling the microstructure of Hastelloy-X components manufactured by selective laser melting. *Physics Procedia* 41: 823–827.

85 Sanchez-Mata, O., Wang, X., Muñiz-Lerma, J.A. et al. (2018). Fabrication of crack-free nickel-based superalloy considered non-weldable during laser powder bed fusion. *Materials (Basel)* 11 (8): 1–9.

86 Marchese, G., Basile, G., Bassini, E. et al. (2018). Study of the microstructure and cracking mechanisms of Hastelloy X produced by laser powder bed fusion. *Materials (Basel)* 11 (1): 106.

87 Han, Q., Gu, Y., Setchi, R. et al. (2019). Additive manufacturing of high-strength crack-free Ni-based Hastelloy X superalloy. *Additive Manufacturing* 30: 100919.

88 Pröbstle, M., Neumeier, S., Hopfenmüller, J. et al. (2016). Superior creep strength of a nickel-based superalloy produced by selective laser melting. *Materials Science and Engineering A* 674: 299–307.

89 Gallmeyer, T.G., Moorthy, S., Kappes, B.B. et al. (2020). Knowledge of process-structure-property relationships to engineer better heat treatments for laser powder bed fusion additive manufactured Inconel 718. *Additive Manufacturing* 31 (October 2019): 100977.

90 Tolosa, I., Garciandía, F., Zubiri, F. et al. (2010). Study of mechanical properties of AISI 316 stainless steel processed by 'selective laser melting', following different manufacturing strategies. *International Journal of Advanced Manufacturing Technology* 51 (5–8): 639–647.

91 Kuznetsov, P.A., Zisman, A.A., Petrov, S.N., and Goncharov, I.S. (2016). Structure and mechanical properties of austenitic 316L steel produced by selective laser melting. *Metally* 2016 (4): 930–934.

92 Pham, M.S., Dovgyy, B., and Hooper, P.A. (2017). Twinning induced plasticity in austenitic stainless steel 316L made by additive manufacturing. *Materials Science and Engineering A* 704: 102–111.

93 Wang, Y.M., Voisin, T., McKeown, J.T. et al. (2017). Additively manufactured hierarchical stainless steels with high strength and ductility. *Nature Materials* 17 (October): 63–70.

94 Ghayoor, M., Lee, K., He, Y. et al. (2020). Selective laser melting of 304L stainless steel: role of volumetric energy density on the microstructure, texture and mechanical properties. *Additive Manufacturing* 32: 1–26.

95 Yakout, M., Elbestawi, M.A., and Veldhuis, S.C. (2019). Density and mechanical properties in selective laser melting of Invar 36 and stainless steel 316L. *Journal of Materials Processing Technology* 266: 397–420.

96 Hidnert, P. and Krider, H.S. (1952). Thermal expansion of aluminum and some aluminum alloys. *Journal of Research of the National Bureau of Standards* 48 (3): 209–220. https://nvlpubs.nist.gov/nistpubs/jres/048/jresv48n3p209_a1b.pdf.

97 Monteiro, W.A. (ed.) (2014). *Light Metal Alloys Applications*. InTech.

98 Fulcher, B. A., Leigh, D. K., and Watt, T. J. (2014). Comparison of ALSI10MG and AL 6061 processed through DMLS. Solid Freeform Fabrication Symposium.

99 Uludağ, M., Çetin, R., and Dispinar, D. (2018). Freezing range, melt quality, and hot tearing in Al-Si alloys. *Metallurgical and Materials Transactions A* 49 (5): 1948–1961.

100 Dahle, A. K. (1996). Mushy zone properties and castability of aluminium foundry alloys. p. 212.

101 Yildirim, M. and Özyürek, D. (2013). The effects of Mg amount on the microstructure and mechanical properties of Al–Si–Mg alloys. *Materials and Design* 51: 767–774.

102 Caceres, C.H., Davidson, C.J., Griffiths, J.R., and Wang, Q.G. (1999). The effect of Mg on the microstructure and mechanical behavior of Al-Si-Mg casting alloys. *Metallurgical and Materials Transactions A, Physical Metallurgy and Materials Science* 30 (10): 2611–2618.

103 Trevisan, F., Calignano, F., Lorusso, M. et al. (2017). On the selective laser melting (SLM) of the AlSi10Mg alloy: process, microstructure, and mechanical properties. *Materials* 10 (1): 76.

104 Zhang, C., Zhu, H., Liao, H. et al. (2018). Effect of heat treatments on fatigue property of selective laser melting AlSi10Mg. *International Journal of Fatigue* 116: 513–522.

105 Fite, J., Eswarappa Prameela, S., Slotwinski, J.A., and Weihs, T.P. (2020). Evolution of the microstructure and mechanical properties of additively manufactured AlSi10Mg during room temperature holds and low temperature aging. *Additive Manufacturing* 36: 101429.

106 Leung, C.L.A., Marussi, S., Atwood, R.C. et al. (2018). In situ X-ray imaging of defect and molten pool dynamics in laser additive manufacturing. *Nature Communications* 9 (1): 1355.

107 Martin, A.A., Calta, N.P., Khairallah, S.A. et al. (2019). Dynamics of pore formation during laser powder bed fusion additive manufacturing. *Nature Communications* 10 (1): 1–10.

108 Leung, C.L.A., Marussi, S., Towrie, M. et al. (2019). The effect of powder oxidation on defect formation in laser additive manufacturing. *Acta Materialia* 166: 294–305.

109 Khairallah, S.A., Anderson, A.T., Rubenchik, A., and King, W.E. (2016). Laser powder-bed fusion additive manufacturing: physics of complex melt flow and formation mechanisms of pores, spatter, and denudation zones. *Acta Materialia* 108: 36–45.

110 Hojjatzadeh, S.M.H., Parab, N.D., Yan, W. et al. (2019). Pore elimination mechanisms during 3D printing of metals. *Nature Communications* 10 (1): 3088.

111 Parab, N.D., Zhao, C., Cunningham, R. et al. (2018). Ultrafast X-ray imaging of laser-metal additive manufacturing processes. *Journal of Synchrotron Radiation* 25 (5): 25.

112 Basu, S. and DebRoy, T. (1992). Liquid metal expulsion during laser irradiation. *Journal of Applied Physics* 72 (8).

113 DebRoy, T., Basu, S., and Mundra, K. (1991). Probing laser induced metal vaporization by gas dynamics and liquid pool transport phenomena. *Journal of Applied Physics* 70 (3): 1313–1319.

114 Kou, S. (2002). *Welding Metallurgy*, vol. 822, no. 2. Hoboken, NJ, USA: Wiley.

115 Cunningham, R., Zhao, C., Parab, N. et al. (2019). Keyhole threshold and morphology in laser melting revealed by ultrahigh-speed X-ray imaging. *Science (80-.)*. 363 (6429): 849–852.

116 Ly, S., Rubenchik, A.M., Khairallah, S.A. et al. (2017). Metal vapor micro-jet controls material redistribution in laser powder bed fusion additive manufacturing. *Scientific Reports* 2017 71 7 (1): 1–12.

117 Schoinochoritis, B., Chantzis, D., and Salonitis, K. (2017). Simulation of metallic powder bed additive manufacturing processes with the finite element method: a critical review. *Proceedings of the Institution of Mechanical Engineers, Part B* 231 (1): 96–117.

118 Everhart, T.E. and Hoff, P.H. (1971). Determination of kilovolt Electron energy dissipation vs penetration distance in solid materials. *Journal of Applied Physics* 42 (13): 5837–5846.

119 Swinehart, D.F. (1962). The Beer-Lambert law. *Journal of Chemical Education* 39 (7): 333.

120 Brown, M.S. and Arnold, C.B. (2010). Fundamentals of laser-material interaction and application to multiscale surface modification. In: *Laser Precision Microfabrication*, Springer Series in Materials Science, vol. 135 (ed. K. Sugioka, M. Meunier and A. Piqué), 91–120. Berlin, Heidelberg: Springer https://doi.org/10.1007/978-3-642-10523-4_4.

121 Campbell, J. (2003). OBE FREng, "Castings second edition," p. 354.

122 DuPont, J.N., Lippold, J.C., and Kiser, S.D. (2009). *Welding Metallurgy and Weldability of Nickel Based Alloys*. Wiley.

123 Shi, X., Wang, H., Feng, W. et al. (2020). The crack and pore formation mechanism of Ti–47Al–2Cr–2Nb alloy fabricated by selective laser melting. *International Journal of Refractory Metals and Hard Materials* 91: 105247.

124 Braun, J., Kaserer, L., Stajkovic, J. et al. (2019). Molybdenum and tungsten manufactured by selective laser melting: analysis of defect structure and solidification mechanisms. *International Journal of Refractory Metals and Hard Materials* 84: 104999.

125 Williams, R.J., Piglione, A., Rønneberg, T. et al. (2019). In situ thermography for laser powder bed fusion: effects of layer temperature on porosity, microstructure and mechanical properties. *Additive Manufacturing* 30: 100880.

126 Zhou, X., Wang, D., Liu, X. et al. (2015). 3D-imaging of selective laser melting defects in a Co–Cr–Mo alloy by synchrotron radiation micro-CT. *Acta Materialia* 98: 1–16.

127 Kempen, K., Thijs, L., Van Humbeeck, J., and Kruth, J.-P. (2015). Processing AlSi10Mg by selective laser melting: parameter optimisation and material characterisation. *Materials Science and Technology* 31 (8): 917–923.

128 Read, N., Wang, W., Essa, K., and Attallah, M.M. (2015). Selective laser melting of AlSi10Mg alloy: process optimisation and mechanical properties development. *Materials and Design* 65: 417–424.

129 Gan, Z., Kafka, O.L., Parab, N. *et al.*, "Universal scaling laws of keyhole stability and porosity in 3D printing of metals." *Nature Communications* 12, 2379 (2021). https://doi.org/10.1038/s41467-021-22704-0

130 DePond, P.J., Fuller, J.C., Khairallah, S.A. et al. (2020). Laser-metal interaction dynamics during additive manufacturing resolved by detection of thermally-induced electron emission. *Communications Materials* 1: 92. https://doi.org/10.1038/s43246-020-00094-y.

131 Wang, L., Zhang, Y., and Yan, W. (2020). Evaporation model for keyhole dynamics during additive manufacturing of metal. *Physical Review Applied* 10: 64039.

132 Chu, J., Bao, Y., Li, X. et al. (2021). Kinetic study of Mn vacuum evaporation from Mn steel melts. *Separation and Purification Technology* 255: 117698.

133 Gu, D. and Shen, Y. (2009). Balling phenomena in direct laser sintering of stainless steel powder: metallurgical mechanisms and control methods. 30 (8): 2903–2910.

134 Zhou, X., Liu, X., Zhang, D. et al. (2015). Balling phenomena in selective laser melted tungsten. *Journal of Materials Processing Technology* 222: 33–42.

135 X. Zhao, S. Xu, and J. Liu, "Surface tension of liquid metal: role, mechanism and application." *Frontiers in Energy* 11, 535–567 (2017). https://doi.org/10.1007/s11708-017-0463-9

136 Gunenthiram, V., Peyre, P., Schneider, M. et al. (2017). Analysis of laser–melt pool–powder bed interaction during the selective laser melting of a stainless steel. *Journal of Laser Applications* 29 (2): 022303.

137 Gusarov, A. V. and Smurov, I. (2010). Modeling the interaction of laser radiation with powder bed at selective laser melting. *Physics Procedia* 5: 381–394.

138 Yadroitsev, I., Gusarov, A., Yadroitsava, I., and Smurov, I. (2010). Single track formation in selective laser melting of metal powders. *Journal of Materials Processing Technology* 210: 1624–1631.

139 B. N. Chicbkov, C. Momma, S. Nolte, F. Yon Alvensleben, and A. Tiinnermann, "*Femtosecond, Picosecond and Nanosecond Laser Ablation of Solids*," 1996.

140 M. Bass, *Handbook of Optics*. Volume II: Devices, Measurements, and Properties, 1995.

141 Kanaya, K. and Okayama, S. (1972). Penetration and energy-loss theory of electrons in solid targets. *Journal of Physics D: Applied Physics* 5 (1): 308.

142 Klassen, A., Bauereiß, A., and Körner, C. (2014). Modelling of electron beam absorption in complex geometries. *Journal of Physics D: Applied Physics* 47 (6): 065307.

143 C. D. Boley, S. A. Khairallah, and A. M. Rubenchik, "*Calculation of Laser Absorption by Metal Powders in Additive Manufacturing*," 2015.

144 Deisenroth, D., Mekhontsev, S., and Lane, B. (2020). Measurement of mass loss, absorbed energy, and time-resolved reflected power for laser powder bed fusion. *Proc. SPIE Photonics West, San Francisco*, CA, [online], https://doi.org/10.1117/12.2547491.

145 Lane, B., Zhirnov, I., Mekhontsev, S. et al. (2020). Transient laser energy absorption, co-axial melt pool monitoring, and relationship to melt pool morphology. *Additive Manufacturing* 36: 101504.

146 Matthews, M., Trapp, J., Guss, G., and Rubenchik, A. (2018). Direct measurements of laser absorptivity during metal melt pool formation associated with powder bed fusion additive manufacturing processes. *Journal of Laser Applications* 30 (3): 032302.

147 Trapp, J., Rubenchik, A.M., Guss, G., and Matthews, M.J. (2017). In situ absorptivity measurements of metallic powders during laser powder-bed fusion additive manufacturing. *Applied Materials Today* 9: 341–349.

148 Kim, C.S. (1975). *Thermophysical Properties of Stainless Steels*. IL, USA: Argonne.

149 Agazhanov, A.S., Samoshkin, D.A., and Kozlovskii, Y.M. (2019). Thermophysical properties of Inconel 718 alloy. *XXXV Siberian Thermophysical Seminar. Journal of Physics: Conference Series* 1382: 12175.

150 Yakout, M., Elbestawi, M.A., and Veldhuis, S.C. (2020). A study of the relationship between thermal expansion and residual stresses in selective laser melting of Ti-6Al-4V. *Journal of Manufacturing Processes* 52: 181–192.

151 Yang, P., Deibler, L.A., Bradley, D.R. et al. (2018). Microstructure evolution and thermal properties of an additively manufactured, solution treatable AlSi10Mg part. *Journal of Materials Research* 33 (23): 4040–4052.

152 Mertens, A., Reginster, S., Paydas, H. et al. (2014). Mechanical properties of alloy Ti–6Al–4V and of stainless steel 316L processed by selective laser melting: influence of out-of-equilibrium microstructures. *Powder Metallurgy* 57 (3): 184–189.

153 Hooper, P.A. (2018). Melt pool temperature and cooling rates in laser powder bed fusion. *Additive Manufacturing* 22: 548–559.

154 Huang, S., Kumar, P., Yeong, W.Y. et al. (2022). Fracture behavior of laser powder bed fusion fabricated Ti41Nb via in-situ alloying. *Acta Materialia* 225: 117593.

155 Huang, S., Narayan, R.L., Tan, J.H.K. et al. (2021). Resolving the porosity-unmelted inclusion dilemma during in-situ alloying of Ti34Nb via laser powder bed fusion. *Acta Materialia* 204: 116522.

156 Zhong, Y., Rännar, L.-E., Liu, L. et al. (2017). Additive manufacturing of 316L stainless steel by electron beam melting for nuclear fusion applications. *Journal of Nuclear Materials* 486: 234–245.

157 Song, B., Dong, S., Liu, Q. et al. (2014). Vacuum heat treatment of iron parts produced by selective laser melting: microstructure, residual stress and tensile behavior. *Materials and Design* 54: 727–733.

158 Mills, K.C., Keene, B.J., Brooks, R.F., and Shirali, A. (1998). Marangoni effects in welding. *Philosophical Transactions of the Royal Society A - Mathematical Physical and Engineering Sciences* 356 (1739): 911–925.

159 Pearson, J.R.A. (1958). On convection cells induced by surface tension. *Journal of Fluid Mechanics* 4 (5): 489–500.

160 Kou, S. and Wang, Y.H. (1986). Weld pool convection and its effect. *Welding Journal* 65 (3).

161 Y. Wang, Q. Shi, and H. L. Tsai, "Modeling of the effects of surface-active elements on flow patterns and weld penetration." *Metallurgical and Materials Transactions B* 32, 145–161 (2001). https://doi.org/10.1007/s11663-001-0017-7

162 Heiple, C. R. and Roper, J. R. (1982). Mechanism for Minor Element Effect on GTA Fusion Zone Geometry Surface active impurities alter surface tension gradients in the molten weld pool thereby changing fluid flow patterns and fusion zone geometry, Welding Research Supplement, pp. 97-s–102-s; https://app.aws.org/wj/supplement/WJ_1982_04_s97.pdf.

163 Heiple, C. and Burgardt, P. (1985). Effects of SO_2 torch gas additions on GTA weld shape. *Welding Journal* 64 (6): 35–36.

164 Egry, I., Ricci, E., Novakovic, R., and Ozawa, S. (2010). Surface tension of liquid metals and alloys – recent developments. *Advances in Colloid and Interface Science* 159 (2). Elsevier B.V.: 198–212.

165 Kaplan, A.F.H. (2012). Absorptivity modulation on wavy molten steel surfaces: the influence of laser wavelength and angle of incidence. *Applied Physics Letters* 101 (15): 151605.

166 Mościcki, T., Hoffman, J., and Szymański, Z. (2006). Modelling of plasma plume induced during laser welding. *Journal of Physics D: Applied Physics* 39 (4): 685–692.

167 Semak, V. and Matsunawa, A. (1997). The role of recoil pressure in energy balance during laser materials processing. *Journal of Physics D: Applied Physics* 30 (18): 2541–2552.

168 Pang, S., Hirano, K., Fabbro, R., and Jiang, T. (2015). Explanation of penetration depth variation during laser welding under variable ambient pressure. *Journal of Laser Applications* 27 (2): 022007.

169 Waseda, Y., Jacob, K.T., Tsuchiya, Y., and Tamaki, S. (1978). Vapour pressure of liquid metals and alloys. *Zeitschrift fur Naturforsch. - Section A: Journal of Physical Sciences* 33 (8): 940–945.

170 Alcock, C.B., Itkin, V.P., and Horrigan, M.K. (1984). Vapour pressure equations for the metallic elements: 298–2500K. *Canadian Metallurgical Quarterly* 23 (3): 309–313.

171 Lide, D.R. (2004). *CRC Handbook of Chemistry and Physics*, 84e, 2003-2004. CRC Press.

172 Zhang, B., Liao, H., and Coddet, C. (2012). Effects of processing parameters on properties of selective laser melting Mg–9%Al powder mixture. *Materials and Design* 34: 753–758.

173 Ivanchenko, V.G., Ivasishin, O.M., and Semiatin, S.L. (2003). Evaluation of evaporation losses during electron-beam melting of Ti-Al-V alloys. *Metallurgical and Materials Transactions B* 34: 911–915. https://doi.org/10.1007/s11663-003-0097-7.

174 Mercelis, P. and Kruth, J.P. (2006). Residual stresses in selective laser sintering and selective laser melting. *Rapid Prototyping Journal* 12 (5): 254–265.

175 Simson, T., Emmel, A., Dwars, A., and Böhm, J. (2017). Residual stress measurements on AISI 316L samples manufactured by selective laser melting. *Additive Manufacturing* 17: 183–189.

176 Deng, D., Peng, R.L., Brodin, H., and Moverare, J. (2018). Microstructure and mechanical properties of Inconel 718 produced by selective laser melting: sample orientation dependence and effects of post heat treatments. *Materials Science and Engineering A* 713: 294–306.

177 Yadroitsev, I. and Yadroitsava, I. (2015). Evaluation of residual stress in stainless steel 316L and Ti6Al4V samples produced by selective laser melting. *Virtual and Physical Prototyping* 10 (2): 67–76.

178 Boivineau, M., Cagran, C., Doytier, D. et al. (2006). Thermophysical properties of solid and liquid Ti-6Al-4V (TA6V) alloy. *International Journal of Thermophysics* 27 (2): 507–529.

179 Reijonen, J., Revuelta, A., Riipinen, T. et al. (2020). On the effect of shielding gas flow on porosity and melt pool geometry in laser powder bed fusion additive manufacturing. *Additive Manufacturing* 32: 101030.

180 Mukherjee, T., Zuback, J.S., De, A., and DebRoy, T. (2016). Printability of alloys for additive manufacturing. *Scientific Reports* 6.

181 Gao, F. and Sonin, A.A. (1994). Precise deposition of molten microdrops: the physics of digital microfabrication. *Proceedings of the Royal Society of London. Series A, Mathematical and Physical Sciences* 444 (1922): 533–554.

182 Feurer, U. (1977). Influence of alloy composition and solidification conditions on dendrite arm spacing, feeding and hot tearing properties of aluminium alloys. *Proceedings of the International Symposium, Delft* 131–145.

183 Benoit, M.J., Mazur, M., Easton, M.A., and Brandt, M. (2021). Effect of alloy composition and laser powder bed fusion parameters on the defect formation and mechanical properties of Inconel 625. *International Journal of Advanced Manufacturing Technology* 114 (3–4): 915–927.

184 Kou, S. (2015). A simple index for predicting the susceptibility to solidification cracking. *Welding Journal* 94 (12): 374s–388s.

185 Kou, S. (2015). A criterion for cracking during solidification. *Acta Materialia* 88: 366–374.

186 Hunt, J., Derguti, F., and Todd, I. (2014). Selection of steels suitable for additive layer manufacturing. *Ironmaking and Steelmaking* 41 (4): 254–256.

187 Nunes, A.C. (1983). An Extended Rosenthal Weld Model A moving heat source weld model can be extended to include effects of phase changes and circulations in the weld pool. *American Welding Society* 62: 165–170.

188 Toda-Caraballo, I., Galindo-Nava, E.I., and Rivera-Díaz-del-Castillo, P.E.J. (2013). Unravelling the materials genome: symmetry relationships in alloy properties. *Journal of Alloys and Compounds* 566: 217–228.

189 Martin, J.H., Yahata, B.D., Hundley, J.M. et al. (2017). 3D printing of high-strength aluminium alloys. *Nature* 549 (7672): 365–369.

190 Kosturek, R., Śniezek, L., Torzewski, J., and Wachowski, M. (2020). Low cycle fatigue properties of sc-modified AA2519-T62 extrusion. *Materials (Basel)* 13 (1).

191 Liu, G., Blake, P., and Ji, S. (2019). Effect of Zr on the high cycle fatigue and mechanical properties of Al–Si–Cu–Mg alloys at elevated temperatures. *Journal of Alloys and Compounds* 809: 151795.

192 Schmidtke, K., Palm, F., Hawkins, A., and Emmelmann, C. (2011). Process and mechanical properties: applicability of a scandium modified Al-alloy for laser additive manufacturing. *Physics Procedia* 12: 369–374.

193 Zhou, L., Pan, H., Hyer, H. et al. (2019). Microstructure and tensile property of a novel AlZnMgScZr alloy additively manufactured by gas atomization and laser powder bed fusion. *Scripta Materialia* 158: 24–28.

194 Agrawal, P., Gupta, S., Thapliyal, S. et al. (2021). Additively manufactured novel Al-Cu-Sc-Zr alloy: microstructure and mechanical properties. *Additive Manufacturing* 37: 101623.

195 Zhou, L., Hyer, H., Park, S. et al. (2019). Microstructure and mechanical properties of Zr-modified aluminum alloy 5083 manufactured by laser powder bed fusion. *Additive Manufacturing* 28: 485–496.

196 Thapliyal, S., Komarasamy, M., Shukla, S. et al. (2020). An integrated computational materials engineering-anchored closed-loop method for design of aluminum alloys for additive manufacturing. *Materialia* 9: 100574.

197 Kumar, N. and Mishra, R.S. (2013). Additivity of strengthening mechanisms in ultrafine grained Al-Mg-Sc alloy. *Materials Science and Engineering A* 580: 175–183.

198 Chen, Y., Hu, Q., Pan, S. et al. (2019). Influences of Cu content on the microstructure and strengthening mechanisms of Al-Mg-Si-xCu alloys. *Metals* 9 (5): 524.

199 Seidman, D.N., Marquis, E.A., and Dunand, D.C. (2002). Precipitation strengthening at ambient and elevated temperatures of heat-treatable Al(Sc) alloys. *Acta Materialia* 50 (16): 4021–4035.

200 Ackers, M.A., Messé, O.M.D.M., and Hecht, U. (2021). Novel approach of alloy design and selection for additive manufacturing towards targeted applications. *Journal of Alloys and Compounds* 866: 158965.

201 Su, Y., Mills, K.C., and Dinsdale, A. (2005). A model to calculate surface tension of commercial alloys. *Journal of Materials Science* 40 (9–10): 2185–2190.

202 M. Morinaga, N. Yukawa, T. Maya, K. Sone, and H. Adachi. (1988). Theoretical design of titanium alloys.

203 Morinaga, M. (2016). Alloy design based on molecular orbital method. *Materials Transactions* 57 (3): 213–226.

204 Sing, S.L., Huang, S., Goh, G.D. et al. (2021). Emerging metallic systems for additive manufacturing: in-situ alloying and multi-metal processing in laser powder bed fusion. *Progress in Materials Science* 119 (November 2020): 100795.

205 Mosallanejad, M.H., Niroumand, B., Aversa, A., and Saboori, A. (2021). In-situ alloying in laser-based additive manufacturing processes: a critical review. *Journal of Alloys and Compounds* 872: 159567.

206 Tang, Y.T., Panwisawas, C., Ghoussoub, J.N. et al. (2021). Alloys-by-design: application to new superalloys for additive manufacturing. *Acta Materialia* 202: 417–436.

207 Crudden, D.J.J., Mottura, A., Warnken, N. et al. (2014). Modelling of the influence of alloy composition on flow stress in high-strength nickel-based superalloys. *Acta Materialia* 75: 356–370.

208 Reed, R.C., Tao, T., and Warnken, N. (2009). Alloys-by-design: application to nickel-based single crystal superalloys. *Acta Materialia* 57 (19): 5898–5913.

209 Murray, S.P., Pusch, K.M., Polonsky, A.T. *et al.* A defect-resistant Co–Ni superalloy for 3D printing. *Nature Communications* 11, 4975 (2020). https://doi.org/10.1038/s41467-020-18775-0

4

Laser and Arc-Based Methods for Additive Manufacturing of Multiple Material Components – From Design to Manufacture

Chao Wei[1,2], *Heng Gu*[1], *Zhizhou Zhang*[1], *Yuan-Hui Chueh*[1], *and Lin Li*[1]

[1] *The University of Manchester, Laser Processing Research Centre, Department of Mechanical, Aerospace and Civil Engineering, Faculty of Science and Engineering, Oxford Road, Manchester, M13 9PL, UK*
[2] *Ningbo Institute of Materials Technology and Engineering, Chinese Academy of Sciences, 1219 West Zhongguan Road, Zhenhai District, Ningbo, Zhejiang, Manchester 315201, P.R. China*

List of Abbreviations

3D	three dimension
AM	additive manufacturing
BCC	body-centered cubic
CAD	computer-aided design
CALPHAD	calculation of phase diagrams
CFD	computational fluid dynamics
CNC	computer numerical control
CS	cold spraying
DED	directed energy deposition
DEM	discrete element modeling
EDS	energy-dispersive spectroscopy
EHLA	extreme high-speed laser material deposition
FCC	face-centered cubic
FFF	fused filament fabrication
FGM	functionally graded material
GMAW	gas metal arc welding
GTAW	gas tungsten arc welding
HAZ	heat-affected zone
HDPE	high-density polyethylene
HEA	high-entropy alloy
IR	infrared
L-DED	laser-based directed energy deposition
LME	liquid metal embrittlement
L-PBF	laser-based powder bed fusion
MD	molecular dynamics

Additive Manufacturing Technology: Design, Optimization, and Modeling, First Edition. Edited by Kun Zhou.

MIG	metal inert gas
MMAM	multi-material additive manufacturing
NAB	nickel aluminum bronze
PAW	plasma arc welding
PBF	powder bed fusion
PC	polycarbonate
PEEK	polyetheretherketone
PET	polyethylene terephthalate
QR	quick response
SLA	stereolithography
SLM	selective laser melting
SS	stainless steel
TIG	tungsten inert gas
TMC	titanium matrix composite
VC	vanadium carbide
VOF	volume of fluid
WAAM	wire arc additive manufacturing
WPDL	wire and powder deposition by laser

4.1 Background

Over the last three decades, additive manufacturing (AM) technologies have progressed rapidly. A variety of AM technologies, such as powder bed fusion (PBF) and directed energy deposition (DED) have been successfully commercialized for the manufacture of metallic components and have been widely used in the aerospace, automotive, medical, and defense industries. Table 4.1 summarizes the laser and arc-based metal component AM methods. The application of metal AM is not just for rapid prototyping, but more and more for batch production of high value-added products. The user's demand is no longer just to quickly fabricate complex parts but also to integrate more functions with the structure of the component by locally tailoring its properties. There may be required properties of higher surface hardness, better corrosion resistance, and lower thermal expansion coefficient at predesigned zones in AM-processed parts [2]. This purpose can be achieved by integrating dissimilar materials with different properties into one part through AM. This is commonly called multi-material AM (MMAM). The definition of MMAM is an AM process in which no fewer than two types of materials are physically deposited to form a single component under computer control according to preprogrammed codes [3].

All the seven types of AM methods listed in the ASTM F2792 standard (Standard Terminology for Additive Manufacturing Technologies) have been used for MMAM. As shown in Table 4.2, except for DED, PBF, and sheet lamination, the energy input in the rest four groups of the AM methods is insufficient to melt the metallic compositions directly. These AM methods can only bond micro-/nanoscale metallic particles together through adhesives to form three-dimensional (3D) "green" parts. The metallic particles in the green parts must undergo subsequent heat treatment to

Table 4.1 Classification of PBF and DED technologies involving either a laser beam or an electric arc source.

AM method classification (ASTM F2792)	Specific AM process	Registered commercial names
Directed energy deposition (DED) – excluding E-beam-based methods	Laser powder-fed DED (L-DED-powder)	Laser engineered net shaping, direct metal deposition, laser metal deposition, laser freeform fabrication, etc.
	Laser wire-fed DED (L-DED-wire)	—
	Wire and arc AM (WAAM)	Shaped metal deposition.
Laser powder bed fusion (L-PBF)	Laser-PBF (L-PBF)	Selective laser sintering, selective laser melting, and direct metal laser sintering.

Source: Adapted from Yusuf et al. [1].

Table 4.2 Materials processed by AM technologies.

AM Method classification (ASTM F2792)	Materials				Remark
	Polymer	**Metal**	**Ceramic**	**Compositions**	
Directed energy deposition	—				Metal-based material
Powder bed fusion					Metal-based material
Sheet lamination				—	
Vat photopolymer-ization					Polymer-based material
Material extrusion					Polymer-based material
Binder jetting				—	
Material jetting				—	

The build material is fused or cured during an AM process.
The build material is unmelted and only temporally bonded during an AM process, which is then sintered in the post-process.

remove the binder and then be fused or sintered together. Dissimilar metallic materials may significantly differ in their thermal physical properties (e.g., melting point and thermal expansion coefficient). In such a condition, using a post heat treatment process to sinter or fuse different metal materials with different melting points on the same part can easily lead to defects such as deformation, cracking, and lack of fusion. Therefore, in theory, these methods are not suitable for metal-based MMAM.

Also, the sheet/foil feeding mechanism restricts the sheet lamination method's flexibility in fabricating multi-material parts with complex structures. Hence, most of the current MMAM technologies are based on DED and PBF approaches.

This chapter introduces DED-based (i.e., L-DED and WAAM) and PBF-based (i.e., L-PBF) MMAM technologies, including the fundamentals of MMAM-processed component design, working principles, materials and component characteristics, modeling and simulation, potential applications, and future challenges.

4.2 MMAM components design

MMAM components design procedure can be divided into four steps: user needs identification, material selection, MMAM method selection, and material composition and part geometry design, as illustrated in Figure 4.1 [4].

In Step 1, the designer needs to fully understand the user's functional requirements and the corresponding technical specifications, confirm whether the MMAM methods can be used to produce the required functionally integrated multi-material parts, and analyze the potential technical difficulties in the process. In addition, designers and manufacturing engineers need to estimate the lead time and manufacturing costs of designing and manufacturing multi-material integrated parts.

Each function of a multi-metal materials part may be realized with one type of materials in the material selection step. In the fusion MMAM process, this material needs to be combined with other materials and merged into a single component. Therefore, the selected material combination must have appropriate metallurgical compatibility. Otherwise, defects such as pores, cracks, and deformation and generation of intermetallic phases may occur. The selection of material combination fundamentally determines whether MMAM can successfully manufacture multi-material parts. The quality function deployment-based method has been applied to a single material product's material selection decision-making process [5]. However, this method is not entirely suitable for the material selection process for multi-material MMAM, because it usually does not include compatibility data of dissimilar materials. Unlike the large variety of appropriate materials for traditional processing methods such as casting, each metal AM technology only has a limited type of commercial metal materials for selection. For example, the casting method can produce all eight series of aluminum alloys (from AA1XXX to AA8XXX), and each series has different grades (e.g., the A6 series of aluminum alloys have 16 grades) [6]. On the other hand, the commonly used aluminum alloy grades suitable for laser AM have only nine grades (e.g., $AlSi_{10}Mg$, 6061) [7]. So, it can be based on

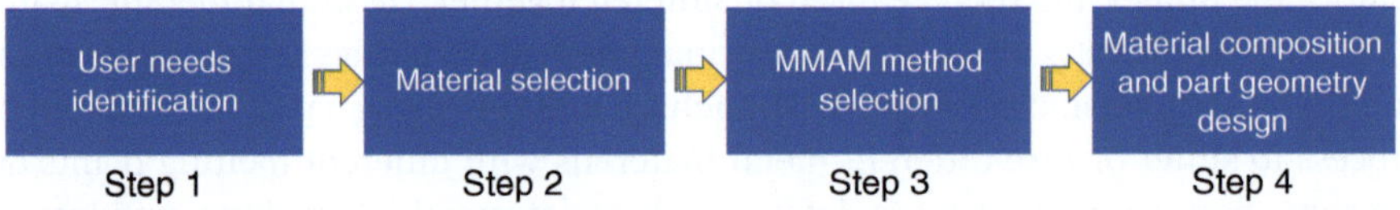

Figure 4.1 MMAM components design procedure.

the compatibility data of dissimilar materials accumulated in traditional processing such as welding, casting, and powder metallurgy, to determine whether the selected multi-material combination is suitable for MMAM.

L-DED, L-PBF, and WAAM have all been used in MMAM. The primary consideration for selecting a suitable candidate method from the three is whether the method can process the material combination determined in step 2. This will be discussed in detail in Sections 4.3–4.5 of this chapter. The size, structural complexity, and surface roughness requirements of multi-material parts are also essential factors in selecting processing methods. In addition, processing efficiency and processing cost constraints should also be considered. The comparison of the three MMAM methods in these aspects will be explained in Section 4.6 of this chapter.

Design of multi-material components to be fabricated through an MMAM method can be divided into two categories: material composition design and geometric design. The physical properties of dissimilar materials such as melting point and thermal expansion eco-efficiency may be different. In addition, material wettability and possible precipitation of intermetallic phases are factors to be considered for the suitability of their combinations and resulting properties. The face-to-face material bonding method can easily cause defects such as cracks and debonding at the material interface. Japanese researchers first proposed functionally graded materials (FGMs) [6] to add a transition zone composed of materials mixture between two pure materials to change the material composition gradually. It can effectively improve the two materials' metallurgical bonding to avoid various defects caused by material mutations in sharp transition. Besides, the composition and structure of FGM can be precisely designed to create tailor-made multifunctional performance [2]. As a result, it is a trend to design FGM structures in multi-material parts. However, the current computer-aided design (CAD) software usually does not support FGM structural design function, because it needs to define the geometric dimensional attributes of the part and simultaneously establish dissimilar materials' composition attributes. Therefore, the concept of CAD design based on voxels has been proposed [8]. The voxel-based CAD design method defines a 3D component as an entity composed of many small volume units (i.e., voxels). Therefore, a unique material parameter attribute can be added to each voxel. The binder-jetting-based polymer 3D printer (PolyJet™) developed by Stratasys has used voxel-based CAD design software for the manufacturing of 3D polymer components having different colors. In addition, there are other voxel-based CAD design tools, such as Monolith, nTopology, and ParaMatters, which could be explored and modified for AM of multi-metallic materials, although these have not yet been used for this purpose.

4.3 Multi-material L-DED

4.3.1 Introduction of L-DED

A typical L-DED system (see Figure 4.2a) consists of a laser module delivering a focused/defocused laser beam, a raw material feeding module dispensing powder

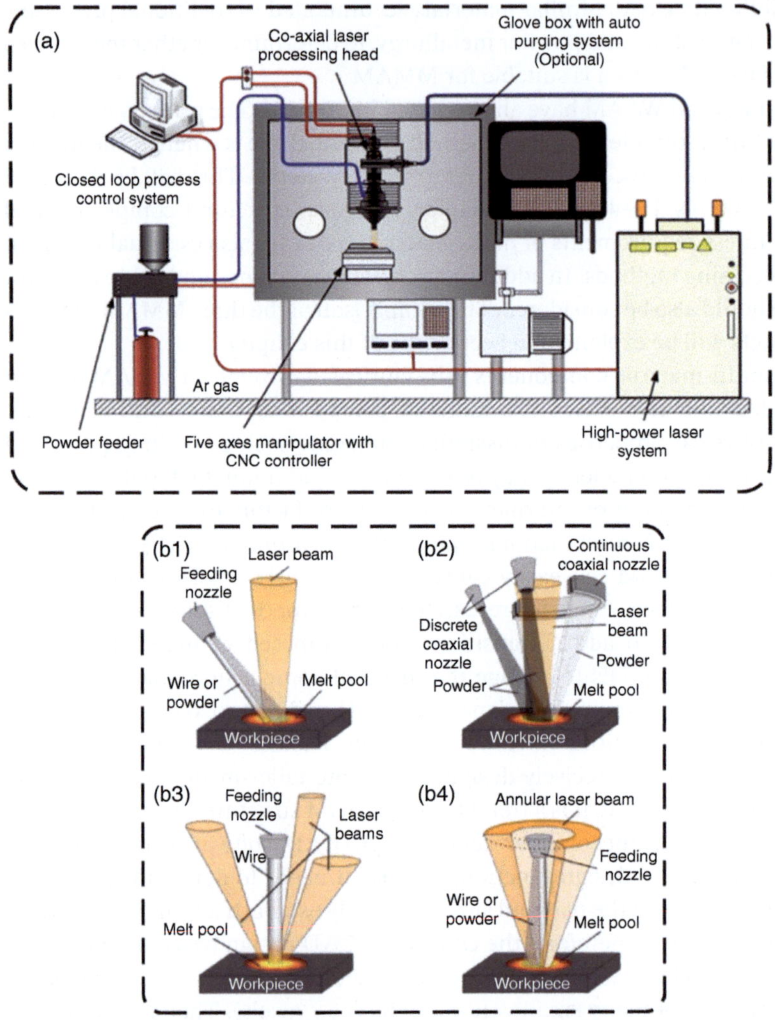

Figure 4.2 (a) Typical DED system setup. Source: Mishra et al. [9]/with permission of Elsevier, deposition head configurations with (b1) a material feeding nozzle lateral with an axial laser beam, (b2) coaxial with an axial laser beam, (b3) powder feeding axially with three laser beams, and (b4) powder feeding axially with an annular laser beam. Source: Govekar et al. [10]/with permission of Elsevier.

or wire into the melt pool, and a motion and control module driving the workpiece translation in at least x, y, and z directions.

The deposition head is a critical component of an L-DED system, which integrates laser optics, material feeding nozzles, inert gas pipelines, and sensors. The material feeding nozzles can be lateral with or coaxial with an axial laser beam (see Figure 4.2b1,b2, respectively). Coaxial feeding improves omnidirectionality of material deposition, and the inert gas surrounding the molten pool prevents the liquid metal material from being oxidized. Although the off-axis single-nozzle powder feeding head is relatively cheap and straightforward, the molten pool is directional, and the size of the molten pool is asymmetrical. It is not suitable for the manufacture of parts with complex shapes. Therefore, a commercial L-DED system normally uses a coaxial feeding deposition head [11]. Also, there is a deposition head design using the annular laser beam and axial powder feeding [10, 12], as shown in Figure 4.2b3 and b4. The advantage of such a design is that it can obtain higher powder collection efficiency (>80%) and modify the deposited layer properties by adjusting the laser beam intensity distribution [12], although this technology has not yet been used in multi-material L-DED. Researchers from the University of Manchester have demonstrated an ultrasonic-assisted coaxial gas-free powder feeding method in L-DED to achieve close to 100% powder collection efficiency and rapid material switching [13].

The laser beam creates a small molten pool (usually 0.25–1 mm in diameter and 0.1–0.5 mm in depth) on the substrate. Powder particles sprayed from nozzles are melted after entering the molten pool and solidify as the laser beam moves away. In the recent extreme high-speed laser material deposition (EHLA) technology developed by Fraunhofer Institute for Laser Technology (ILT), powders are melted before entering the molten pool [14]. The solidified material forms a narrow track through the relative movement of the substrate and the laser beam. Multiple consecutive overlapping tracks constitute a build layer. The deposition head moves a layer of a thickness (typically 0.25–0.5 mm) relative to the substrate, and a new build layer is produced. An L-DED-processed 3D component is made through repeating the aforementioned steps [15].

The small molten pool and high laser beam scanning speed in the L-DED process result in a high molten pool cooling rate (10^3–10^5 °C s^{-1}) and a large thermal gradient. A high cooling rate contributes to microstructural benefits which are hard to achieve via conventional processes, including less diffusion-controlled solid-state phase transformations, more supersaturated solutions and nonequilibrium phases, ultrafine grains dramatically reduced elemental segregation, and fine secondary phase particles [11].

4.3.2 Material Feeding Mechanism in Multi-Material L-DED

Powder is usually used in L-DED-based MMAM. In an L-DED system, the powder feeding system comprises three modules: a powder metering unit, a powder conveyance unit, and one or more deposition heads, as presented in Figure 4.3a. Pneumatic powder feeding using an inert gas, e.g., argon, as the powder carrier, is the most popular method for transporting the powder from a powder metering unit to an L-DED deposition head. However, other conveyance methods, including the screw/auger method [17], vibration-assisted method [18], gravity-driven method [19], and the electrostatic method [20], have been investigated. The deposition head configuration in the L-DED system determines the powder mixing mechanism during MMAM.

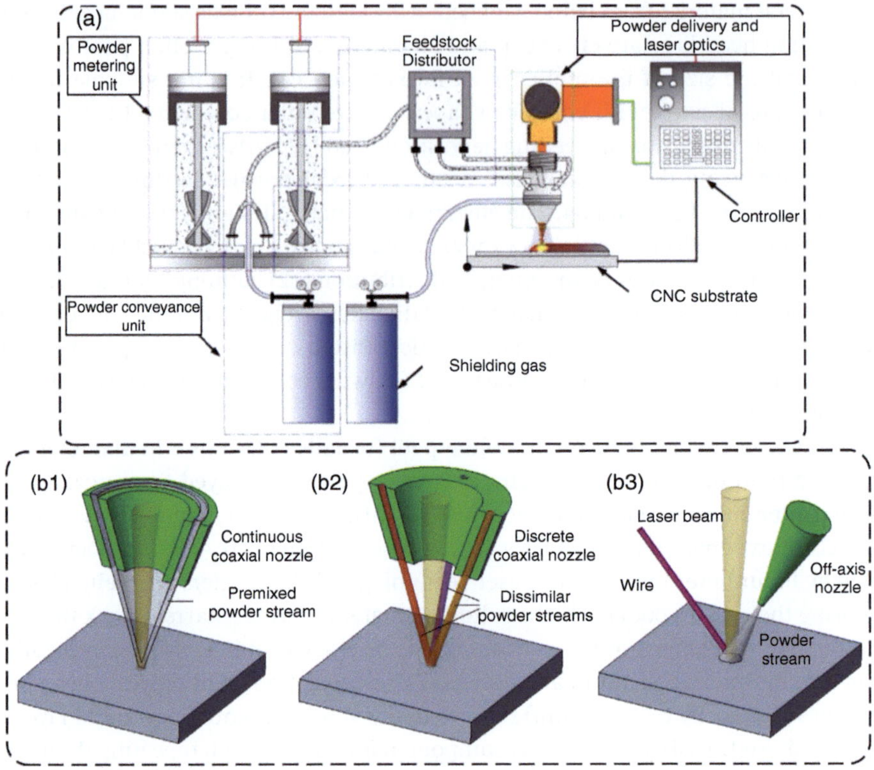

Figure 4.3 (a) Schematic diagram of the powder feeding system in an L-DED setup. Source: Singh et al. [16]/with permission of Elsevier, (b1) a continuous coaxial nozzle to feed premixed powder composition, (b2) a discrete coaxial nozzle to feed dissimilar powders which will be mixed and melted after entering the melt pool, and (b3) simultaneous wire and off-axis powder feeding for multi-material L-DED.

4.3.2.1 Continuous Coaxial Powder Feeding

FGM components can be fabricated with any L-DED system equipped with at least two powder metering units and a continuous coaxial powder deposition head. Dissimilar powders are stocked in different powder metering units separately and are premixed homogeneously in a feedstock distributor before being transported to the constant coaxial powder deposition head by the carrier gas, as shown in Figure 4.3b1.

The drawback of this setup is that, if the powder mixture is composed of powders with a significantly different density such as copper alloys and aluminum alloys, it is difficult to maintain the powder mixture's homogeneity during its transportation from the powder feedstock distributor to the deposition head. The flow velocity of lighter and smaller particles is higher than that of heavier and bigger particles, resulting in the composition segregation in the melt pool [21, 22]. A "powder holding space" design feature is proposed to be added into the continuous coaxial powder deposition head, acting as a mixing chamber for multiple feed materials (either uniformly or creating a gradient) before being dispensed into the melt pool [23, 24].

The other drawback of this approach is that it cannot fabricate small FGM/multi-material features on the same build layer during multi-material L-DED. The powder mixture in the powder feed line has to be thoroughly exhausted. The other powder mixture with different composition can be transported into the continuous coaxial powder deposition head [25].

4.3.2.2 Discrete Coaxial Powder Feeding

A discrete coaxial powder deposition head is more suitable for dispensing dissimilar powders to form material property gradients along and across the build direction. Different powder streams jetted from different nozzles in the discrete coaxial deposition head are blended in the melt pool to form a predesigned composition, as illustrated in Figure 4.3b2. The powder composition volume fraction can be adjusted by controlling the carrier gas's flow velocities for different powder materials. In addition, because different powders are injected into the molten pool at different angles, this causes strong Marangoni convection, strengthening the mixing of different materials and reducing the segregation of material components [16].

The flow velocity of the particles deposited by a discrete coaxial powder deposition head is much lower than that by a continuous coaxial powder deposition head (3.76 and 11.28 $m\,s^{-1}$ [26]). Some of the particles could bounce off the surface of the molten pool, which results in the powder capture efficiency of the discrete coaxial powder deposition head being lower than that of a continuous coaxial powder deposition head [26]. Under the typical operating powder feed rate, the powder capture efficiency ranges from 20% to 75% for a four-nozzle discrete coaxial powder deposition head [27], and a single powder feeding nozzle of 70–80% [28], respectively.

4.3.2.3 Simultaneous Wire and Powder Feeding

Notably, the powder mixture's unmelted dissimilar powder compositions used in the aforementioned two deposition methods are challenging to be separated for recycling, which may cause environmental pollution [29]. Researchers from the University of Manchester demonstrated simultaneous wire and powder feeding approach for multi-material L-DED [30] and named this new process as wire and powder deposition by laser (WPDL). Due to the different forms of materials, and the fed wire is often 100% melted, unused powdered metal can be easily recycled for reuse.

Figure 4.3b3 shows that the wire material from a wire feeder and the powder material from an off-axis powder feeding nozzle (which can be replaced by a coaxial powder deposition head [31]) are fed into the melt pool simultaneously during WPDL. An FGM structure can be fabricated by adjusting the wire's feed rate and that of the powder separately. The deposition efficiency of multi-material WPDL is higher than that of pure powder-based multi-material L-DED [32], because the molten pool absorbs more energy from laser irradiation during WPDL [31].

4.3.3 Materials and Characteristics in Multi-Material L-DED

Almost all metal materials that can be welded are suitable for L-DED processing, including aluminum (Al)-based alloys, nickel (Ni)-based alloys, copper (Cu)-based

Table 4.3 Summary of multi-material L-DED investigations.

Material	Ni	SS	Ti	Fe	Fe_3Al
Al			[34, 35]		
Ni			[36–39]		
Cu	[40–42]			[43–46]	
SS	[47–49]	[50]	[51–55]	[56]	[57]
Ti			[58]		
Mo			[59]		
Ceramic		[60]	[61–63]		

Bimetal; bimetal FGM with diffusion barrier layers.

alloys, niobium, stainless steel (SS), tantalum, titanium (Ti)-based alloys, tungsten, and iron (Fe)-based alloys [33]. A wide range of material combinations has been applied in multi-material L-DED as summarized in Table 4.3. L-DED is used to process bimetallic FGM, especially the combination of Ti and SS with other materials, as highlighted in Table 4.3. The L-DED-fabricated components have material gradients with the composition changing from one component to another by controlling different powders' feed rates during the L-DED process. Brittle intermetallic phases are formed during laser melting of Ti–SS or Ti–Ni. Addition of a third metallic element in the transition zone between Ti and SS/Ni during multi-material L-DED has been used to avoid the harmful second phases.

4.3.3.1 L-DED of Ni-Cu Bimetal

Ni-based superalloys are commonly used materials for aeroengines and liquid rocket engine combustion chambers because of their high mechanical strength at high temperatures. Cu-based alloys have excellent thermal conductivity but are not resistant to high temperatures. Combining these two materials can reduce creep fatigue damage and extend the life of the rocket engine combustion chambers.

Copper has good compatibility with many other metal materials during the melting process. Therefore, copper has been widely used as a buffer material between incompatible metal materials to prevent harmful intermetallic compounds from being produced during the melting and solidification of metals [64]. Cu–Ni FGM manufactured by L-DED has a reliable metallurgical bond [40]. It shows a 150% higher thermal diffusivity than the pure Inconel 718 alloy, which is attributed to copper's high thermal conductivity [65]. It is worth noting that during the L-DED manufacturing of Cu–Ni FGM, the nickel alloy powder cannot be fully melted [41]. It is because the high melting point Ni alloy powder does not transfer enough heat to the molten pool [37]. The WPDL method could solve this problem, in which the material with a lower melting point (Cu alloys) should be in powder form, and the material with a higher melting point (Ni alloys) should be in the form of wire [41]. Because of the higher molten pool temperature and larger molten pool size obtained in WPDL, the high melting point wire material can be fully melted, thereby obtaining good element mixing and smooth material gradient [36].

4.3.3.2 L-DED of Ni–SS Bimetal

Ni alloys and SS are commonly used materials in light-water nuclear reactors. According to the temperature, pressure, nuclear radiation dose, and stress level under working conditions of the nuclear reactor, different materials or material combinations are selected. Ni-based alloys have high strength, good corrosion resistance and oxidation resistance, and good creep resistance and sound fatigue resistance at moderate temperatures, but they are expensive. SS has excellent corrosion resistance and low cost, but its strength is lower than that of nickel alloys. Therefore, L-DED technology was employed to study the feasibility of joining the Ni alloy and SS to combine the advantages of the two.

The microhardness of L-DED-fabricated 316L-Inconel 718 FGM increases linearly with the increase of Inconel 718 content [49]. With the increase of Inconel 718 content, a high-hardness carbide (NbC) appears, which leads to an increase in the microhardness of FGM [48]. Using calculation of phase diagrams (CALPHAD) phase stability calculation, one can predict that the metal monocarbides in the form of (Mo and Nb) C will appear in the 304L-Inconel 625 FGM (Figure 4.4a) processed by L-DED [47]. Therefore, L-DED of SS–Ni is a promising processing method for selectively controlling FGM structures' hardness and wear resistance.

The metallurgical bonding of SS and nickel alloys is good, with fewer micro-defects. It is because the two types of alloys' crystal structures in the

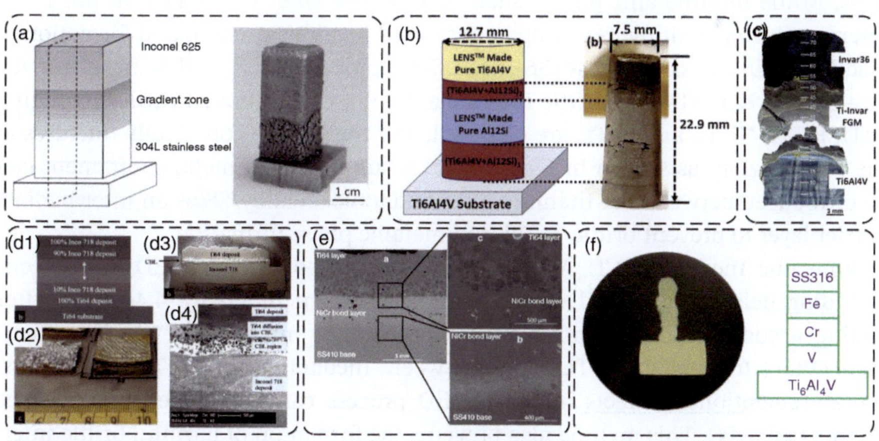

Figure 4.4 L-DED-processed (a) 304L–Inconel 625 FGM sample. Source: Carroll et al. [47]/Elsevier, (b) Ti_6Al_4V–$Al_{12}Si$ FGM sample. Source: Zhang et al. [35]/with permission of Elsevier, and (c) Ti_6Al_4V–Invar36 FGM sample. Source: Bobbio et al. [36]/with permission of Elsevier, (d1) schematic of Inconel 718–Ti_6Al_4V FGM and (d2) de-bonded Inconel 718–Ti6Al4V FGM, (d3) polished cross section of the bimetallic Inconel 718 and Ti_6Al_4V sample with a compositional bond layer and (d4) its magnified microstructure image. Source: Onuike et al. [37]/with permission of Elsevier, (e) microstructure of the cross section of the Ti_6Al_4V–SS410 with a NiCr bond layer. Source: Sahasrabudhe et al. [51]/with permission of Elsevier, and (f) microstructure schematic of the cross section of the Ti_6Al_4V–SS316 with a V-Cr-Fe bond layer. Source: Li et al. [54]/with permission of Elsevier.

solid and molten state are the same single-sided center cubic crystal structures, and there is no allotropic phase transition [66]. The thermal expansion coefficients of the two are also similar [66]. The main components of the two alloys, including Fe, Cr, and Ni, have good solubility. These three factors make L-DED suitable for processing Ni–SS FGM [47].

4.3.3.3 L-DED of Ti–Al Bimetal

Aluminum alloys have low absorptivity and high reflectivity to the laser beam at near infrared (IR) wavelengths [67]. Therefore, it is challenging to process aluminum alloys with L-DED. It has been found that [35] during L-DED processing Ti_6Al_4V–$Al_{12}Si$ (see Figure 4.4b for the sample), the low laser absorptivity of aluminum alloy causes the high melting point titanium alloy powder to be unable to melt completely. Also, if the laser power is low, the intermetallic phase $TiSi_2$ will appear at the bimetallic interface, and if the laser power is high, the Ti_3Al phase will occur [30]. These intermetallic phases can cause interface cracks, as shown in Figure 4.4b.

4.3.3.4 L-DED of Ti-Ni FGM and Ti–SS FGM with Diffusion Barrier Layers

It is difficult to directly fabricate defect-free titanium alloy–nickel alloy structure through L-DED because the crystallographic difference between the two materials leads to the formation of brittle Ti–Ni phases after the two materials are melted together, which will reduce the bonding strength and lead to macro defects such as cracking and delamination. L-DED processing of Ti–SS FGM has similar problems. Brittle intermetallic phases such as TiFe and $TiFe_2$ will appear at the Ti–SS interface during melting, resulting in decreased bonding strength and formation of cracks [67]. In an extreme case [36], i.e., L-DED processing of Ti_6Al_4V-Invar (64 wt% Fe + 36 wt% Ni) FGM, all brittle intermetallic phases mentioned earlier including FeTi, Fe_2Ti, Ni_3Ti, and $NiTi_2$ are observed, and the sample completely fractures at the FGM region, as shown in Figure 4.4c. A suitable third metal component can be introduced between the titanium alloy and a nickel alloy/SS as an intermediate barrier layer to prevent brittle Ti–Ni intermetallic phase formation.

Bimetallic Inconel 718-Ti_6Al_4V FGM manufactured using L-DED has defects, including delamination and cracking [37], as shown in Figure 4.4d1–d2. After adding a compositional bond layer composed of a mixture of "vanadium carbide (VC) + Inconel 718 + Ti_6Al_4V" between Inconel 718 and Ti_6Al_4V layers, the aforementioned defects in the L-DED process can be avoided entirely (see Figure 4.4d3–d4). This method also prevents the formation of harmful brittle intermetallic phases, including Ti_2Ni and $TiNi_3$. The newly formed Cr_3C_2 additional phase improves the Ti–Ni bimetal's metallurgical bonding strength and reduces the thermal stress at the bimetal interface. In the study of L-DED of TA15-Inconel 718 bimetal structure [38], Nb/Cu interlayers are used to achieve the same purpose, effectively avoiding the generation of brittle Ti–Ni and Ti–Cu phases. It is worth noting that pure copper should not be used as an intermediate transition layer material for L-DED processing of Ti–Ni bimetallic structures [38], because it will produce a brittle Ti–Cu phase [39].

Adding a NiCr intermediate layer at the Ti_6Al_4V–SS410 interface can avoid delamination caused by brittle intermetallic compounds and residual stress during L-DED processing Ti_6Al_4V–SS410 bimetal structure [51], as shown in Figure 4.4e. Using an intermediate layer composed of niobium (Nb) in the L-DED-fabricated Ti_6Al_4V–SS410 bimetal structure can also achieve the same purpose and avoid the formation of harmful second phases [55]. Because, Nb element is a β-stabilizer to Ti and forms ε-Fe_2Nb stable phase. Vanadium (V) element is not suitable as an intermediate layer material of Ti_6Al_4V-304L FGM processed by L-DED [52], because it will generate brittle intermetallic compound (FeTi) and a brittle Fe-V-Cr sigma phase. A brittle σ-FeV phase may also appear at the interface of V-304L [53], which will cause the component to crack. Thermodynamic modeling can predict the formation of the σ-FeV phase mentioned earlier [53]. Besides, a more complex composition transition path ($Ti_6Al_4V \rightarrow V \rightarrow Cr \rightarrow Fe \rightarrow$ SS316) could avoid brittle intermetallic phases and cracks during L-DED of Ti_6Al_4V–SS316 FGM [54], as shown in Figure 4.4f.

4.3.3.5 L-DED of Fe–Cu Bimetal

Articek et al. [43] showed for the first time the L-DED-produced H13-Cu FGM structure with a 10% gradient of composition (Figure 4.5a), proving that the graded composition change avoided the low-ductility martensite in high carbon steel (H13) from causing cracks. This study also found that the copper-iron powder presented element segregation and a copper-rich phase (Figure 4.5b). It might be due to spinodal decomposition in the supercooled melt pool before crystallization, separating phases with different crystal structures, including body-centered cubic (BCC) structure Fe phase (αFe) and face-centered cubic (FCC) structure Cu (γCu) [68].

The other studies on direct bonding of Cu to the H13 substrate reported interfacial cracks [44, 45], as shown in Figure 4.5c, which might be caused by the liquid metal embrittlement (LME) phenomenon [46]. The researchers avoided the crack defect by adding a nickel alloy (D22) transition layer between Cu and H13. The thermal conductivity of the Cu-D22-SS system was increased by about 100%, and the thermal diffusivity was increased by about 370% [44].

Li et al. [46] studied the mechanical properties and high corrosion resistance of a nickel aluminum bronze (NAB)/15-5 PH bimetal structure processed by L-DED and realized a crack-free NAB/15-5 PH interface. It was found that an intermediate layer composed of Fe_xAl-phase dendrites appeared at the interface between the two materials, as shown in Figure 4.5d. The brittle Fe_xAl dendrites limited the deformation of NAB and 15-5 PH, causing stress concentration at the interface. When stretched along the building direction, the composite material tended to crack at the interface of the Fe_xAl and Cu phases resulting from their crystal incoherence.

4.3.3.6 L-DED of Ti-ceramic Material System

$Ti_6Al_4V + Al_2O_3$ mixture manufactured by L-DED shows severe thermal expansion and equiaxed Ti grains and some unmelted Al_2O_3 particles appearing at the material interfaces [61]. α-Ti and β-Ti phases appear in Ti_6Al_4V area. The corundum phase appears in the pure Al_2O_3 area, and a combination of titanium and corundum

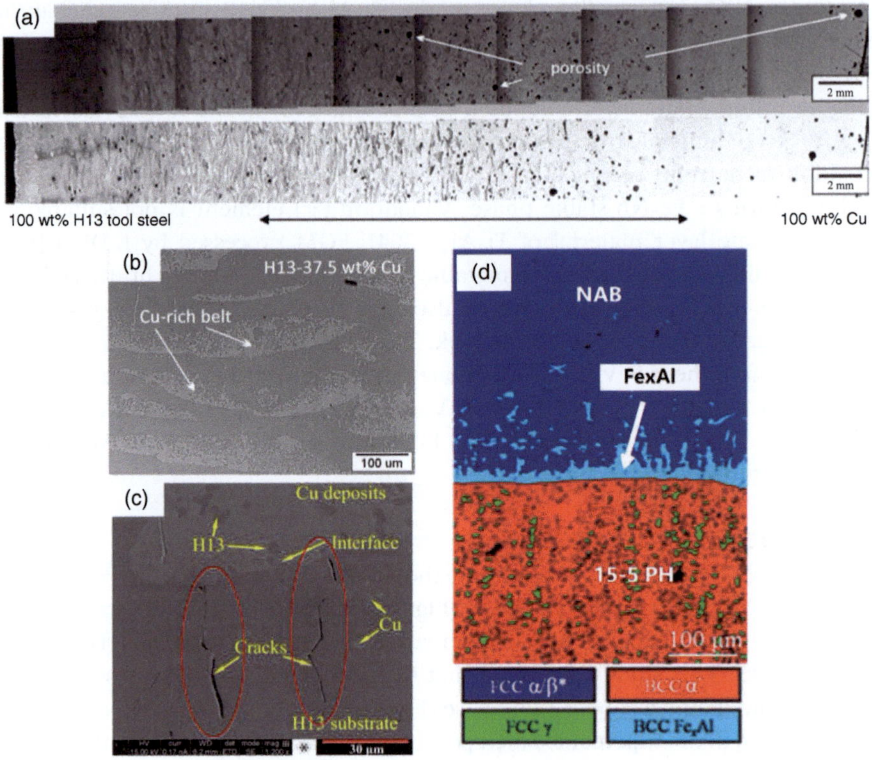

Figure 4.5 (a) The H13-Cu FGM structure with the composition varying in 10% increments. Source: Articek et al. [43]/with permission of the University of Maribor (b) and its the microstructure micrograph showing element segregation. Source: Articek et al. [43]/with permission of the University of Maribor; (c) the cracks at the Cu-H13 interface. Source: Zhang et al. [45]/with permission of Elsevier; (d) the Fe_xAl intermetallic phase at the NAB/15-5PH material interface. Source: Li et al. [46]/with permission of Elsevier.

phases appears in the $Ti_6Al_4V + Al_2O_3$ area [61]. The TiCp content changes the tensile strength and elongation of L-DED-processed TiCp-Ti_6Al_4V FGM. If the TiC content is 5%, the tensile strength of the TiCp-Ti_6Al_4V sample is increased by nearly 12.3%. If the volume fraction of TiC exceeds 5%, the tensile strength and elongation of the composition decrease [62]. It is because the number and size of unmelted TiCp and dendritic TiC phases in the original state increase with the increase in TiCp volume fraction. L-DED can produce defect-free Ti-TiCp FGM samples with TiCp content up to 95 vol% [63].

4.4 Multi-material L-PBF

4.4.1 Introduction of L-PBF

Selective laser sintering was invented at the University of Texas at Austin, United States, in the 1980s, using CO_2 laser to sinter polymer materials selectively. Later,

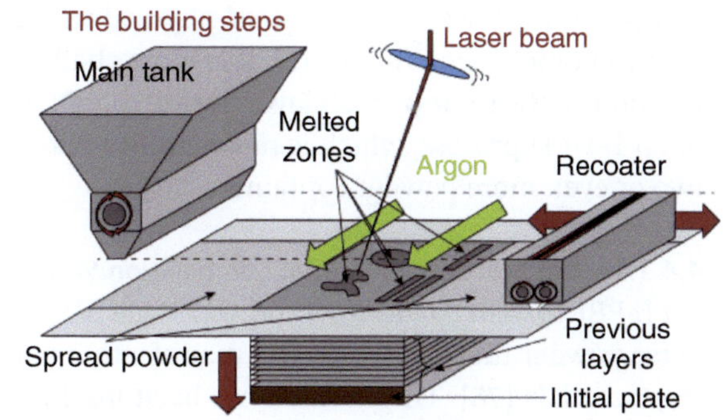

Figure 4.6 A schematic diagram of a standard L-PBF system. Source: Sun et al. [71]/with permission of Elsevier.

by using other high-energy beams such as a fiber laser at a 1070-nm wavelength, the technology was extended to process metallic and ceramic materials [69]. Fraunhofer ILT introduced selective laser melting (SLM) technology by direct laser melting of metal powders to produce fully dense metallic objects [70]. A typical L-PBF system is shown in Figure 4.6. A powder spreading device such as a blade first spreads the metal powder on a build substrate (i.e., the initial plate in Figure 4.6) in the processing chamber. Then the laser beam guided by an X–Y galvo scanner selectively melts the powder on the substrate according to the predesigned pattern of the current layer. Then the substrate moves down for a thickness of one layer, and the powder spreading device spreads a new layer of metal powder on the processed layer for laser processing. The aforementioned steps are repeated until the entire part is produced. The entire process is carried out in a gas-protected processing chamber to prevent the metal from reacting with other gases at high temperatures and removes all splashes and fumes that may block the laser transmission path [72].

The instantaneous heating and rapid movement of the laser in the L-PBF process causes the powder to melt and solidify quickly. This leads to less diffusion-controlled solid phase transitions, more supersaturated solutions and nonequilibrium phases, ultrafine grains with greatly reduced elemental segregation, and fine second phase particles. These advantages lead to the excellent mechanical properties of parts processed using L-PBF.

Compared with other metal AM technologies such as L-DED and WAAM, the advantages of L-PBF are higher geometrical resolution of processed components, higher material utilization, and a wider range of material options (including metals, ceramics, and polymers). Therefore, L-PFB is a promising MMAM method.

4.4.2 Material Deposition Mechanism in Multi-Material L-PBF

The composition distribution in a component composed of multiple materials can be homogeneous, changed layer by layer, or spatially varied [73]. Multi-material objects can be fabricated by one of the following methods: using special raw materials including premixed discrete multiple materials, composite materials, and porous material adapted to be infiltrated by a second material [74], or employing modified AM manufacturing processes capable of carrying out multiple-material AM. Currently, standard L-PBF machines are suitable to fabricating parts with a homogenous composition. However, the objects' physical variations cannot be flexibly managed using a homogeneous alloy [75]. It is more challenging to change the material composition on the same powder layer during LBPF [76], compared with powder feeding in an L-DED process, which is more easily achieved by using a powder feeder with two or even more powder containers [77].

4.4.2.1 Unidirectional Material Composition Variation

An L-PBF system having an improved blade recoating unit, which can store and disperse powder independently, can manufacture parts composed of vertically changing materials [78]. This method has been used to produce copper/SS samples [79] and silver/copper samples [80], both of which have sharp material interfaces. The sudden transformation of the material is prone to high-stress concentration at the interface and even causes delamination under complex loading conditions [81]. This problem is solved by using a gradient interface and a mixture of two materials, which is called FGM [82]. Therefore, a special powder mixer has been developed to produce a bimetallic Ti_6Al_4V/In718 sample whose material composition gradually changes in the vertical direction [83] and other similar approaches [84–87]. All these powder dispersing mechanisms are only suitable for producing parts with vertical material gradients. The use of premixed powders has been shown to produce H13/Cu FGM samples with gradual material composition [88].

4.4.2.2 Spatial material composition variation

There are three general ways for discrete dry powder deposition over the same layer: (i) depositing over the whole layer with the first material followed by laser melting and selective removal of unwanted material and then spreading the second powder material to cover an entire powder layer; (ii) accurately depositing each material on the designed position; and (iii) a combination of aforementioned two methods [89].

4.4.2.2.1 Spreading and Removal Powders on an Entire Powder Layer The process of completely recoating the powder layer with a powder blade and then selectively removing the powder can achieve the purpose of depositing two powder materials on the same build layer. Still, the aforementioned process causes dissimilar powders' cross-contamination on the same layer, and the dimensional accuracy of the parts along the build direction is low [89]. It is also possible to use a soft blade with comb teeth to remove all unmelted powder [90], but this will result in uneven material interfaces and poor surface roughness. Because the soft teeth experience irreversible deformation after repeatedly scraping the solidified part's top surface, the surface quality of the solidified powder layer is reduced [85].

4.4.2.2.2 Selective Powder Deposition Ultrasound is a mechanical wave transmitted at a frequency higher than 20 kHz or a frequency beyond the normal range of human hearing [91]. Piezoelectric oscillators or magnetostriction oscillators can generate ultrasonic waves [92]. They carry long-distance transmission through solid or liquid media without any significant energy loss [93]. Ultrasonic vibration is ideal for dry powder deposition because it can break agglomerated powder clusters and increase the powder's fluidity [94]. This method has been used to draw 2D powder patterns with fine features [95–98]. Researchers from the UK show a circuit pattern composed of dissimilar metal powders with a characteristic resolution higher than 100 μm [98]. Researchers at the University of Manchester [99] integrated this method with laser-based AM. The low processing speed is an obstacle to integrating ultrasonic powder deposition and AM [100]. Making the thickness of the ultrasonically deposited powder layer uniform is also a challenge [101].

Electrostatic powder deposition is based on an electrographic process. It uses electrostatics to print toner particles onto a paper/substrate [102]. This method is suitable for spreading powders over a large area rapidly. The working principle of the electrographic process integrated into an L-PBF system is similar to that of the office laser printer, which includes six steps: charging, imaging, powder development, powder transfer, fusing, and cleaning [103]. In such an AM system, powder patterns can be transferred via a belt [104], a drum [105], or an automated build platform [106]. Aerosint SA developed a unique dual-material patterning drum device for L-PBF to deposit two types of powders on the same powder layer successfully [107]. It is essential to investigate the effects of powder types, deposition parameters, and temperature on the powder transfer [108]. Otherwise, powder particles may randomly drop on the powder bed, causing cross-contamination of powders [109].

Pneumatic pressure-assisted flow has been investigated for dry powder deposition [95, 110–112]. This method uses inert gas to blow the dry powder stored in the hopper/pipette out of a tiny nozzle. A unstable powder flow rate is the most significant disadvantage of this method [110], so there is no report that it has been used for L-PBF.

4.4.2.3 Hybrid Methods for Discrete Powder Deposition

Multi-material L-PBF must improve processing efficiency and powder utilization, reducing cross-contamination of different powders [113]. An alternative is to integrate it with the conventional PBF system, that is, the main powder is spread over a large area through the powder blade, and then the second powder or more types of powders are selectively deposited through an ultrasonic powder distributor. The unmelted main powder is selectively removed through a micro-vacuum device [25]. The whole process is shown in Figure 4.7a1. The specific experimental setup is shown in Figure 4.7a2. The working principle of the selective vacuum-based powder removal device is to use a vacuum producer to generate a negative pressure so that the powder particles near the exhausting tool are sucked away and collected in the powder separator such as a micro-vortex system [115]. Researchers from Germany [113, 116] designed a similar hybrid L-PFB system. Researchers from the University of Manchester produced a set of 3D and 2D samples (Figure 4.7a3–a5) to demonstrate the feasibility of this processing route. After adding a radial vibration motor to the ultrasonic powder dispenser to loosen the agglomerated powder block close to the powder feeding nozzle, the powder's fluidity can be significantly improved [117]. Thanks to this dual vibrating powder dispenser's compact structure,

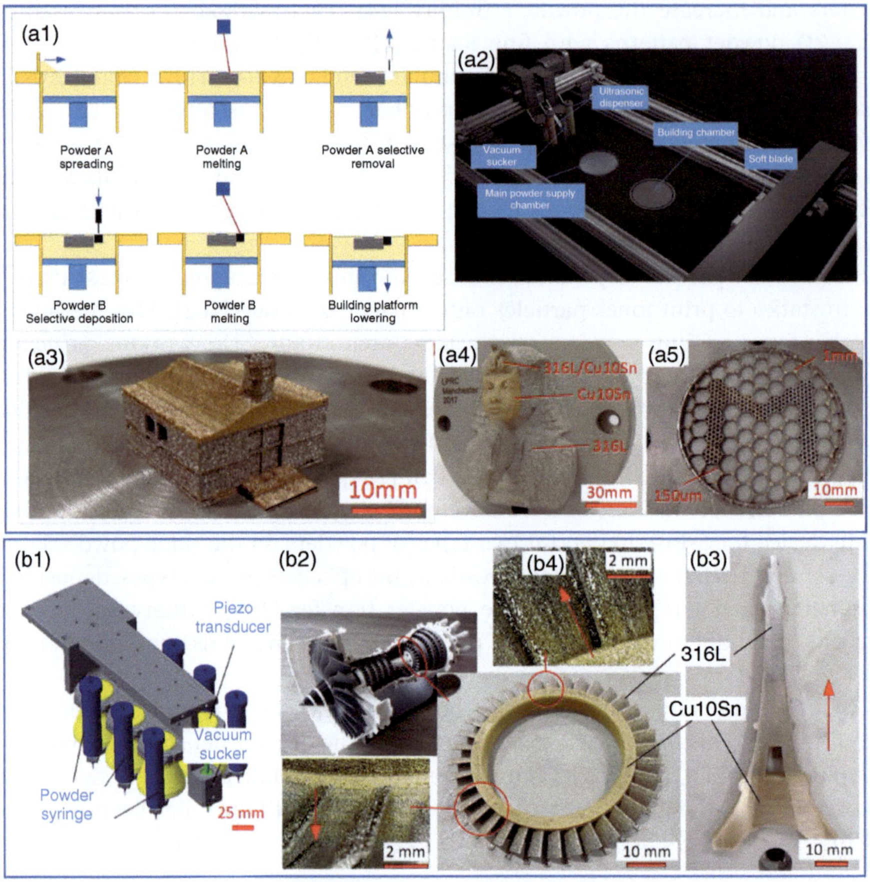

Figure 4.7 (a1) Flow chart of ultrasonic-assisted L-PBF. Source: Wei et al. [25]/with permission of Elsevier, (a2) a modified L-PBF experimental system integrating ultrasonic powder dispensing and powder vacuum suction functions. Source: Chao Wei et al. [100]/IOP Publishing/CC BY 3.0, (a3)–(a5) a set of 3D samples fabricated via multi-material L-PBF. Source: Wei et al. [25]/with permission of Elsevier, (b1) an ultrasonic powder dispensers array design, (b2) a turbine disk sample, and (b3) an Eiffel Tower sample composed of Cu/SS FGM. Source: Wei et al. [114]/with permission of American Society of Mechanical Engineers (ASME).

multiple powder dispensers can be easily integrated to form a powder feeding array [114]. This device is ideal for the manufacturing of FGM. It has been successfully used in manufacturing a 3D turbine disk sample (Figure 4.7b2) and a 2D Eiffel Tower FGM sample (Figure 4.7b3) composed of Cu/SS.

In addition to the modified L-PFB systems mentioned earlier, combining L-PBF with another AM method is an alternative solution: processing one material with L-PBF and processing another material with another AM method [101, 118–120]. Besides, a standard L-PBF process has been applied to melt a powder material on a dissimilar material substrate [121, 122]. The bimetallic components are comprised of the fused powder layers and the substrate.

4.4.3 Materials and Characteristics in Multi-Material L-PBF

4.4.3.1 L-PBF of Multiple Metallic Materials

Three routes for joining dissimilar metals include direct joining, gradient path, and intermediate section [75], as shown in Figure 4.8. Using the direct jointing method, one can bond one material directly with another. Using the gradient path method, one can gradually transfer the material composition from one material to the other. An intermediate section composed of a third material can be used as a buffer layer between two incompatible materials. All of these methods have been used for L-PBF of multiple metallic materials.

The state of the art of multi-material L-PBF based on different processing routes for different material combinations is introduced in the following sections.

4.4.3.1.1 Direct Joining Method

Fe–Cu Bimetal L-PBF has been used to join SS and copper alloys, which are difficult to bond with conventional approaches (e.g., welding). During welding, significant differences in the two materials' thermal conductivity and melting points can lead to hot cracking in HAZs of the steel [123].

At the material interface of the 316L-C18400 sample produced with L-PBF, there is obvious interdiffusion of Fe and Cu elements to each other, forming an excellent metallurgical bond [124]. However, due to the infiltration of Cu element into the austenite grain boundaries in the SS, some dendritic cracks are observed in the mixed region near the 316L side [125]. These dendritic cracks propagate rapidly in the tensile test and accelerate the fracture failure of the specimen. The refined grains due to repeated remelting and recrystallization in L-PBF improve the bonding strength of the 316L-Cu10Sn dissimilar material interface [126]. Remelting the 18Ni300-CuSn

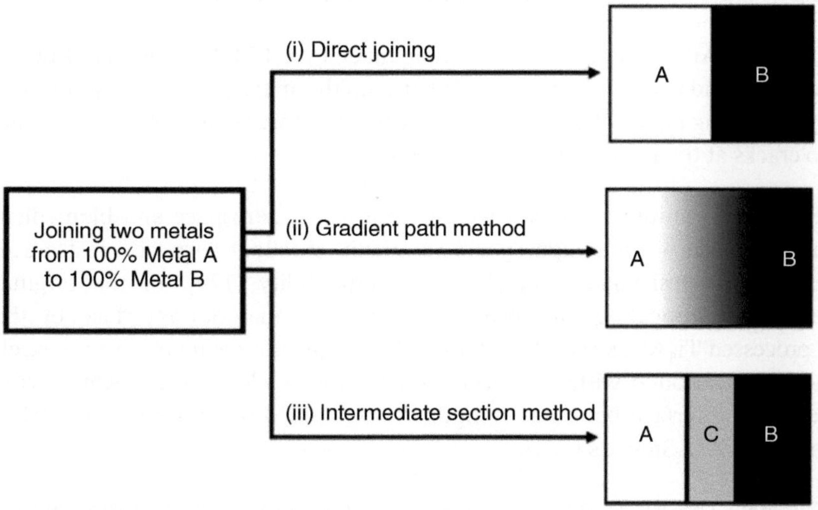

Figure 4.8 Processing routes for joining bimetallic materials. Source: Yan et al. [75]/with permission of Elsevier.

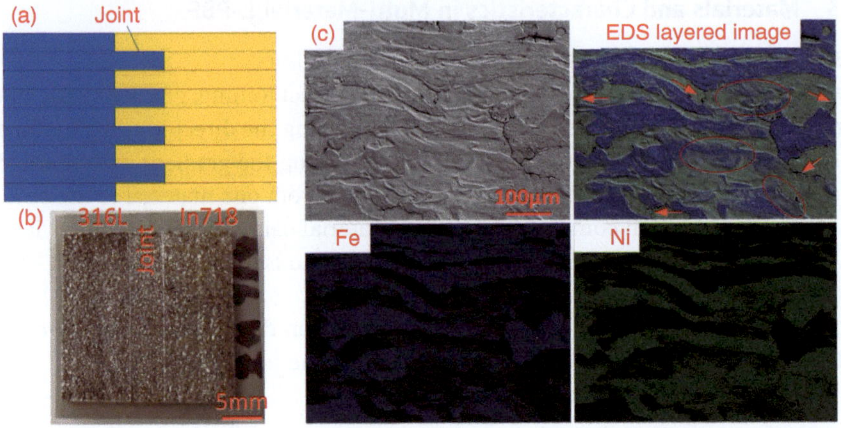

Figure 4.9 (a) A schematic diagram of the cross-finger structure at the materials' interface Source: Wei et al. [25]/with permission of Elsevier. (b) a square sample with the composition changing from 316L to In718 [25], and (c) the SEM image and EDS maps of the microstructure at the 316L/In718 interface Source: Wei et al. [25]/IOP Publishing/CC BY 3.0.

bimetal interface cannot improve the tensile strength, making the interface more brittle and reducing the elongation [127].

Different material distribution strategies have been developed to improve the metallurgical bonding strength at the interface of dissimilar materials. During L-PBF fabrication of tool steel–copper alloys, a 0.2-mm-wide powder overlap is set horizontally and then remelted during processing [116]. A cross-finger structure is designed at the material interface of the L-PBF printed 316L/In718 sample (Figure 4.9a,b), which can increase the contact area between the two materials, thereby enhancing the bonding strength. Its energy-dispersive spectroscopy (EDS) mapping ((Figure 4.9c) clearly distinguishes the material distribution.

Al–Cu Bimetal An $AlSi_{10}Mg$-C18400 sample made by L-PBF has apparent mutual diffusion of Al and Cu at the interface. The uneven thermal expansion of aluminum and copper during rapid solidification and the brittle phase (Al_2Cu) between metals leads to cracks at the material's interface [128].

Ti–Ti Bimetal Titanium alloys are widely used in the aerospace and biomedical industries because of their high specific strength, excellent corrosion resistance, excellent fatigue resistance, and good biocompatibility [129]. In the original state and after annealing and heat treatment, the material interface of the L-PBF-processed $Ti_5Al_{2.5}Sn/Ti_6Al_4V$ bimetallic sample always maintains an excellent metallurgical bond without defects. A 70-μm-wide element diffusion area is obtained. The interface bonding strength of dissimilar titanium alloys is higher than that of $Ti_5Al_{2.5}Sn$ parts [130].

Al–Al Bimetal The $Al_{12}Si$–$Al_{3.5}Cu_{1.5}MgSi$ dissimilar aluminum alloy sample produced by L-PBF consists of four areas: an $Al_{12}Si$ area, an interface, a texture strengthening area, and an AlCuMgSi area. The cell size (eutectic Al–Si) in the interfacial region is larger than that in the $Al_{12}Si$ region. The growth of these eutectic Al–Si cells is interrupted by the intermetallic phase Al_2Cu. The interface's metallurgical bonding is good, and the fracture occurs in the $Al_{12}Si$ zone because of the unmelted $Al_{12}Si$ powder [131].

Fe–Fe Bimetal There are γ phase and α phase in the H13 zone and 316L zone at the interface of the 316L-H13 bimetallic sample treated with L-PBF, and the epitaxial grains grow at the entire interface. The Marangoni convection effect determines the solidification of the molten 316L/H13 mixture's microstructure during the process [132].

4.4.3.1.2 Gradient Path Method The direct joining of two metals with mismatched thermal properties is prone to causing microcrack defects. Therefore, an FGM zone is added between the two base metals to alleviate the aforementioned problem [133].

The L-PBF system, with an in situ powder mixing device, can be used to produce vertical 316L–MS1 FGM samples. Due to the gradual change in composition, the material boundary between 316L and MS1 disappears, and no macro cracks are observed [87]. Using ultrasonic-assisted L-PBF, one can fabricate 316L–$Cu_{10}Sn$ FGM samples with gradual material changes in the horizontal direction (Figure 4.10b1,b2). Although there are no large cracks, many unmelted 316L powder particles are observed (Figure 4.10 b3–b6), which are embedded in the solidified $Cu_{10}Sn$ alloy. With the increase in the copper alloy content, this phenomenon becomes more obvious [114]. Similar results have been reported [121, 127, 134, 135].

During L-PBF, most of the energy of the incident laser beam is absorbed by the powder on the powder layer's surface. In contrast, the powder at the bottom of the powder layer only absorbs very little energy [136]. Studies on L-PBF of the bimetallic Invar36/$Cu_{10}Sn$ have confirmed this phenomenon [137]. The energy density that can initially melt the high melting point pure metal (Invar36) cannot melt the Invar36-$Cu_{10}Sn$ mixture. The solidified powder layer shows obvious delamination (Figure 4.11a). The copper alloy is completely melted, and the high

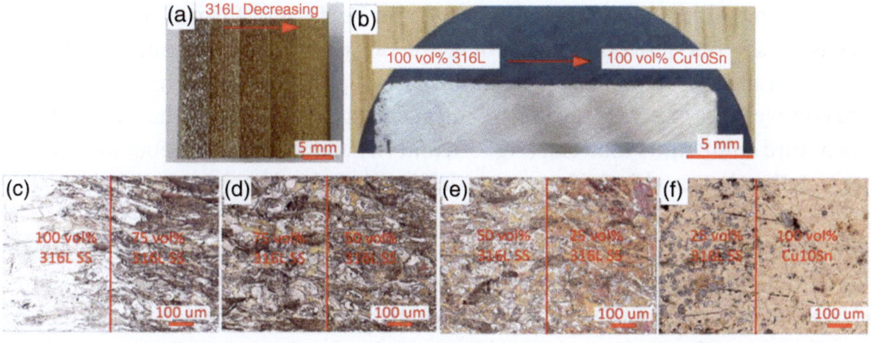

Figure 4.10 (a) Top view and (b) cross section of a horizontal 316L/Cu10Sn FGM square sample, and (c)–(f) the microstructure at the material interface. Source: Wei et al. [114]/with permission of American Society of Mechanical Engineers (ASME).

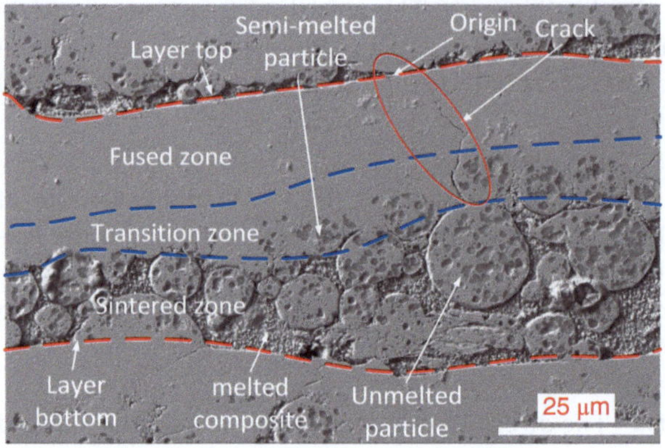

Figure 4.11 Morphological characteristics of a solidified Invar36/Cu10Sn powder layer. Source: Wei et al. [137]/with permission of Elsevier.

melting point Invar36 particles are melted when they are located on the top of the powder layer, and the unmelted Invar36 particles shielded by $Cu_{10}Sn$ powder are embedded in melted $Cu_{10}Sn$. It is due to the low laser absorptivity and high thermal conductivity of the copper alloy. When the copper powder partially covers the powder layer's upper surface, the copper powder reflects a significant amount of the laser beam. The absorbed energy dissipates quickly, because copper has a high thermal conductivity. Therefore, the Invar36 powder located under the copper powder cannot obtain sufficient energy to be melted. Increasing the energy density can overcome this. Reducing hatch distance or layer thickness can increase the laser beam penetration depth and suppress defects caused by a lack of fusion [138].

Due to the low compatibility and mutual solubility of the two materials Fe and $Al_{12}Si$, there are many visible cracks [83] in the Fe–$Al_{12}Si$ FGM fabricated by L-PBF. L-PBF manufactured Ti_6Al_4V–In718 FGM contains the intermetallic phase Ti_2Ni, increasing the In718 content, which will eventually cause the sample to crack [84].

4.4.3.1.3 Intermediate Section Method Bimetallic FGM comprising a titanium alloy and steel is attractive for industrial applications. It can combine the excellent mechanical properties of titanium alloys and the low cost of the steel. However, brittle intermetallic phases, including FeTi and Fe_2Ti, will form if the Fe content exceeds 0.1 at% [139]. Accumulation of brittle phases at the grain boundaries of the fused material leads to cracks [140]. In the case of joining a titanium alloy with SS, more complex intermetallic phases are formed [141]. An intermediate section made up of a third compatible material (Ag, Cu, and Ni) which does not generate harmful intermetallic phases with the base materials can be added between a titanium alloy and steel to avoid the brittle phases [142].

An L-PBF-fabricated Ti_6Al_4V/316L bimetallic sample with a 0.5-mm copper alloy (CuA) interlayer [134] shows that the Ti_6Al_4V/CuA interface containing three detrimental phases is the critical interface to affect the component mechanical strength. Using the melt pool non-homogeneity results in tough reinforcement phase into a relatively brittle matrix, and controlling the laser energy input helps increase the proportion of the relatively tougher interfacial α'-Ti phase and decrease that of other brittle phases. This α′-Ti phase deflects the cracks from the somewhat brittle TiA/CuA interface toward the ductile CuA interlayer.

In summary, the bimetallic material interface manufactured using L-PBF usually has a good metallurgical bond. If the two materials' thermal properties are very different, defects such as cracks, pores, and unmelted powders are likely to appear at the interface. Brittle intermetallic compounds are also widely observed if two materials have low compatibility. The ductility and the bimetallic part's fatigue properties become worse due to the aforementioned defects [143].

4.4.3.2 L-PBF of Hybrid Metal/Ceramic Materials

Metal/ceramic composite materials can significantly improve the surface hardness, wear resistance, chemical inertness, and temperature stability of parts [144]. Earlier investigations use ceramic as reinforcement particles which are usually lower than 5 wt% of a composite [145, 146].

Figure 4.12a1–a3,b1–b3 show three types of metal/ceramic powder combinations for L-PBF and corresponding microstructures [147]. It confirms that L-PBF can be used for molten metal/ceramic mixtures.

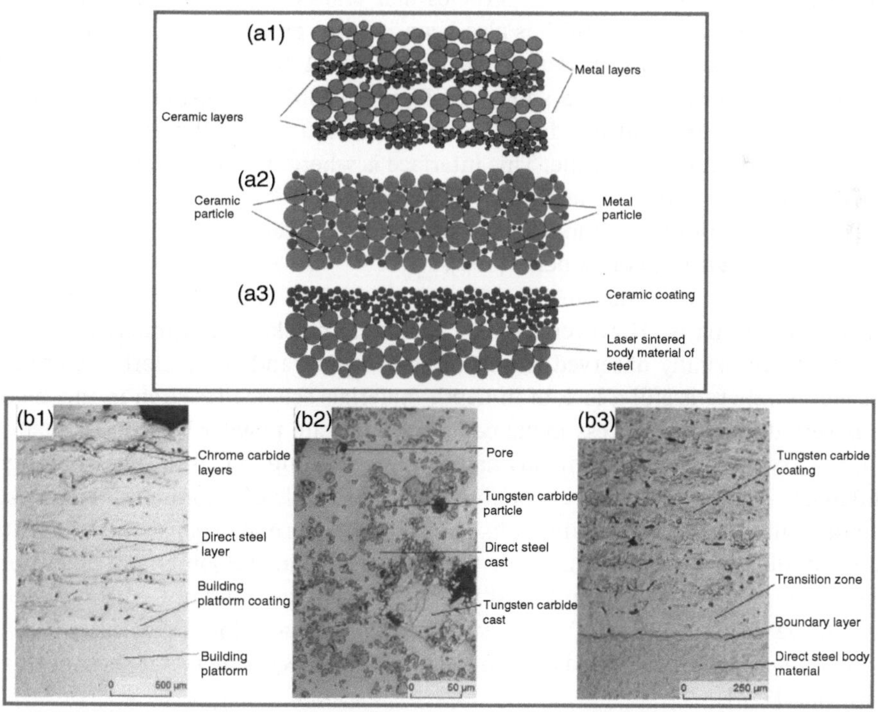

Figure 4.12 Schematic diagrams of (a1) metal/ceramic building in layers, (a2) metal/ceramic powder composite, (a3) ceramic coating on fused metal layers; (b1)–(b3) microstructure of L-PBF-processed samples with earlier-related metal/ceramic combinations. Source: Trenke et al. [147]/with permission of Elsevier.

In the study of using L-PBF to melt the ceramic powder mixture ($ZrO_2 + Al_2O_3$) on a tool steel substrate, the remelting strategy significantly improves the bonding strength of the ceramic and metal interface, because this method leads to the formation of a jagged, interlocking microstructure at the metal/ceramic interface [148, 149].

Boride-reinforced titanium matrix composite (TMC) is a promising material for aerospace applications. An excessively high scanning speed during L-PBF of TiB_2–Ti_6Al_4V bimetal causes the TiB_2 track to exhibit an unstable surface, which presents excessive shrinkage, splash, and transverse cracks. The laser energy density range of 0.67–1.50 J mm^{-1} (scanning speed 300–600 mm s^{-1} and laser power 400–450 W) is a suitable process parameter range for producing stable TiB_2 tracks with smooth surfaces [150]. The material interface in L-PBF-processed TiB_2–Ti_6Al_4V bimetal shows a nano-hardness gradient of TiB_2/TiB/Ti_6Al_4V [151, 152]. TiB phase formation is due to the in situ reaction between Ti and TiB_2 caused by laser irradiation [153].

Researchers from the University of Manchester [89] have demonstrated a metal-soda lime glass dual-material pendant printed with ultrasonic-assisted L-PBF (Figure 4.13a1,a2). There is good mechanical bonding at the interface of the metal/soda-lime glass. Still, there are also apparent defects, including insufficient melting of the glass particles and cracks in the molten glass layer. The ultrasonic-assisted L-PBF has also been used to process metal/ceramic FGM samples [86], the cross section is shown in Figure 4.13b1, and the logo composed of horizontal cermet FGM is shown in Figure 4.13b2. No element diffusion is observed at the material interface. The fracture occurs near the interface of the ceramic matrix composite zone. This interface is where the material changes from the ductile phase to the brittle phase.

The aforementioned technology has been utilized to fabricate ceramic/metal 3D circuits and electronic components [155].

In summary, the current multi-material L-PBF technology can process metal/ceramic parts. However, defects, including cracks and unmelted ceramic powders, are usually observed in the ceramic region and the material interface. At present, there is still a lack of in-depth materials science research in this field. It is very difficult to process metal-ceramic composite powder by L-PBF. First of all, the crystal structure of metals and ceramics is different, resulting in a large difference in melting points. Secondly, the strong bonds of ceramic crystals make element diffusion extremely difficult. Thirdly, their thermal expansion coefficients are very different, resulting in large thermal stresses at the joints and cracks on the ceramic side. The use of ultrafast lasers as the energy input of L-PBF may help to achieve direct printing of metal-ceramic parts. In addition, combining L-PBF with other processing methods is also a solution for processing metal-ceramic dual materials [156].

4.4.3.3 L-PBF of Hybrid Metal/Polymer Materials

Components combining both metal and polymer are widely used in industries. Due to highly diverse material properties between metal and polymer, AM of such a

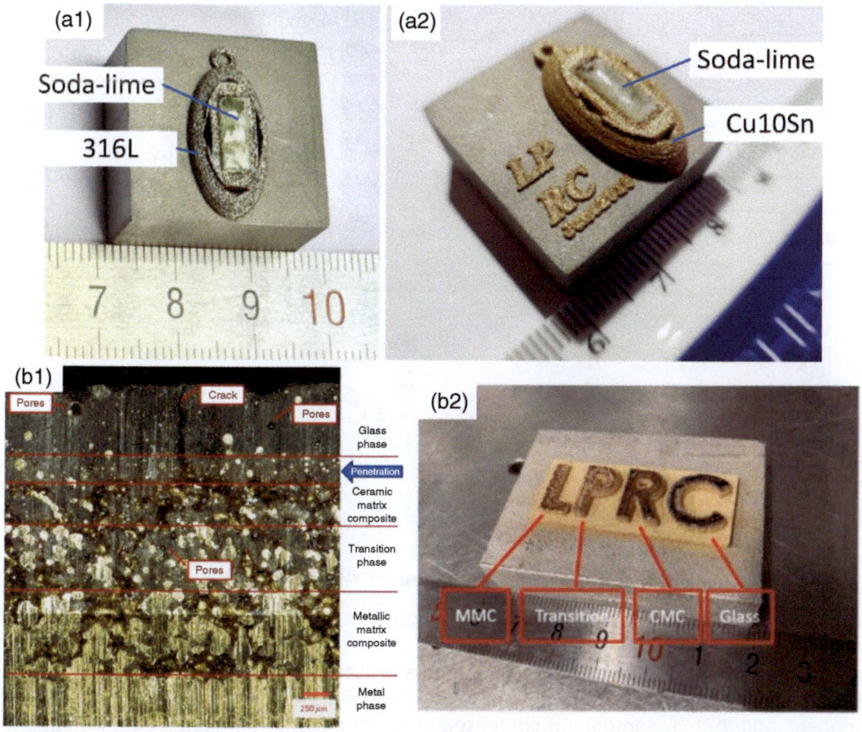

Figure 4.13 (a1) A 316 L-glass pendant, (a2) a Cu10Sn-glass pendant. Source: Zhang et al. [154]/with permission of American Society of Mechanical Engineers (ASME), (b1) the cross section of the vertical gradient metal/glass sample, and (b2) a laboratory logo composed of horizontal metal/glass FGM. Source: Zhang et al. [86]/with permission of Elsevier.

material combination in a single component and process is still challenging. L-PBF has a unique capability of melting both metallic and polymeric materials directly [157, 158].

Combining L-PBF with welding techniques (such as ultrasonic welding [159] and laser welding [160]) can manufacture metal/plastic components. However, the earlier-mentioned gradual hybrid AM method's disadvantage is that it is not suitable for producing complex geometrical parts and increases processing time and cost. Different from the earlier-mentioned stepwise method, researchers at the University of Manchester [118] have shown an inline integrated MMAM technology that uses L-PBF for melting metal powder and uses the fused filament fabrication (FFF) method to melt polymer filaments to produce mixed metal/polymer components (Figure 4.14a1,a2). A macro-mechanical interlocking structure is designed at the material interface. External pressure is applied to cause the molten polymer to flow into the interlocking structure during the process to increase the vertical bond strength. The disadvantage is that it cannot combine metals and polymers in the horizontal direction. The ultrasonic-assisted L-PBF mentioned earlier has been used to print the powder mixture composed of metal (316L, $Cu_{10}Sn$) and polymer (PA11) [161], as shown in Figure 4.14b1 and b2 for samples. It has been found

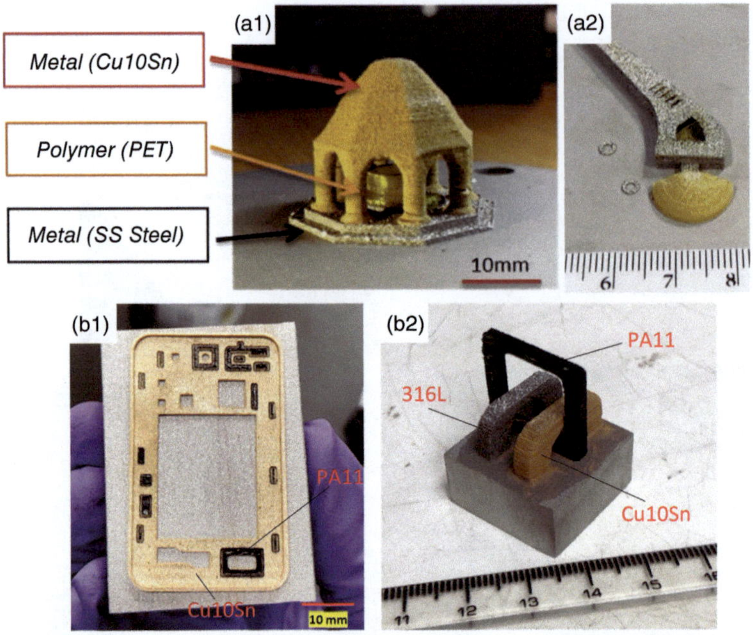

Figure 4.14 (a1) and (a2) Cu-PET-SS components processed by inline integrated L-PBF-FFF approach. Source: Chueh et al. [118]/with permission of Elsevier, (b1) 2D component, and (b2) 3D component composed of metal/polymer fabricated by L-PBF in one process. Source: Chueh et al. [161]/with permission of Elsevier.

that adding a moderate amount of polymer powder into the metallic powder could improve the surface roughness of fused metallic components [161].

Based on the aforementioned discussion, it is seen that unitizing AM to directly joining metal/polymer is challenging. An alternative solution is the AM of polymer matrix composites [162]. Because of reinforcements, various functional properties (e.g., enhanced mechanical strength, thermal conductivity, or electrical conductivity) can be incorporated into the polymer matrix while maintaining their properties [163, 164]. L-PBF demonstrates excellent potential for fabricating various polymeric composites [165]. L-PBF has been utilized to produce porous FGM structures consisting of polyetheretherketone (PEEK) and polycarbonate (PC) matrix with alternating layers of magnetic nano oxides of Ni (NiO) and non-magnetically of Cu (CuO) [166, 167]. Hysteresis phenomena are found in alternating ferromagnetic Ni-PEEK (Ni-PC) and nonmagnetic Cu-PEEK (Cu–PC) layers. It is found that both tensile and compressive modulus increase, while the strain at break and yield decreases as the increase of glass bead volume fraction in the L-PBF-processed FGM structures composed of Nylon 11-glass powder mixtures [168, 169].

Researchers [170] from the University of Manchester have demonstrated both vertical and horizontal PA11/Cu10Sn FGM samples (Figure 4.15a,b), PA11/Al_2O_3/Cu_{10}Sn sample (Figure 4.15c), and PA11/Cu_{10}Sn sample (Figure 4.15d).

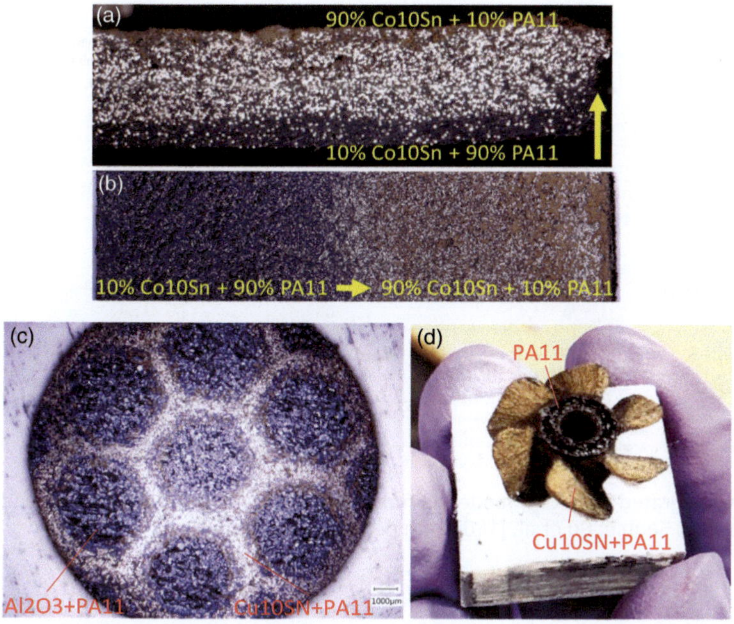

Figure 4.15 Printed $Cu_{10}Sn$/PA11 (a) vertical FGM and (b) horizontal FGM. Printed multiple functional polymeric composite (c) a shoe sole structure for tribological applications, and (d) multiple functional 3D functionally graded motor blades [170].

These samples are fabricated via an ultrasonic-assisted L-PBF process with an ultrasonic dispenser to achieve stable feeding of polymer/metal/glass powder mixtures that are otherwise easily agglomerated.

The challenge for L-PBF of hybrid metal/polymer mixture is how to avoid decomposition and gasification of low melting point polymer powders. The resulting fume also adversely affects the molten pool of the metal. Solutions include using metal powder and polymer powder with melting points as close as possible [171]. The metal part and the polymer part are kept at a certain distance in the component design to avoid the polymer's thermal decomposition due to the high temperature of the molten metal's heat-affected zone (HAZ) [170].

4.4.3.4 Modeling and Simulation of Multi-Material L-PBF Processes

Simulations for the L-PBF processes can be categorized into three groups: macroscopic, mesoscopic, and microscopic [172]. A limited number of studies are based on the macroscopic method [133] and the microscopic method [173] for multi-material L-PBF modeling. Most of the investigations in this field have been carried out on the mesoscopic scale. These simulations generally include two steps, i.e., discrete element modeling (DEM) and computational fluid dynamics (CFD) modeling.

A DEM tool has been used to evaluate the influence of the volume fraction and diameter ratio of the SiC/316L powder mixture. It is found that the remaining volume ratio of the powder mixture has a significant impact on its fluidity [117]. Also, the DEM tool is used to analyze the heat transfer process between 316L

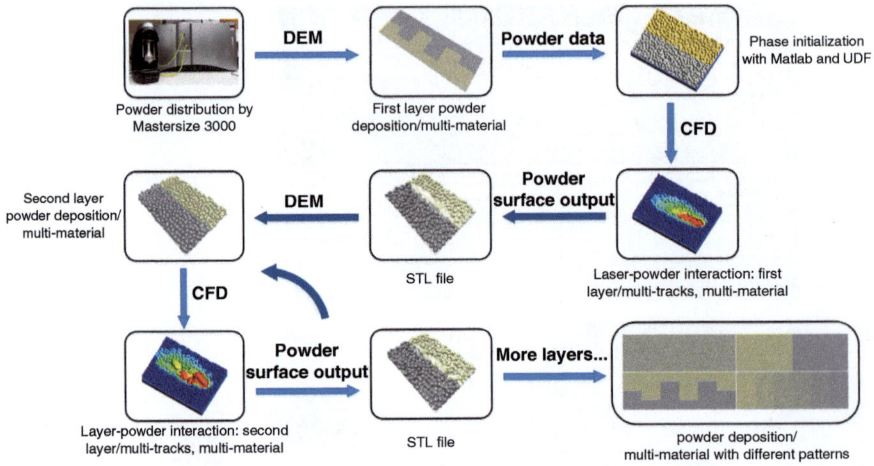

Figure 4.16 An integrated DEM-CFD modeling framework for multi-track, multi-layer, and multi-material L-PBF. Source: Gu et al. [180]/with permission of Elsevier.

powder and Ti_6Al_4V powder without considering the melting and solidification process [174].

For molten pool simulation when treating the powder bed as a continuum, the volume of fluid (VOF) model is used to study the effect of laser energy input on the dynamics of the molten pool of the TiC-$AlSi_{10}Mg$ composite during the L-PBF process [175]. Similar methods have also been used to study the effect of thermodynamics on the migration and distribution of composite powders, including AlN–$AlSi_{10}Mg$ [176], TiB_2-Ti_6Al_4V [177], and TiC–$AlSi_{10}Mg$ [178]. For powder/melt pool simulations containing discrete particles, molecular dynamics (MD) methods can be used to simulate the melting and solidification process of Fe–Al multi-material powders in a multilayer L-PBF system [179].

An integrated DEM–CFD modeling framework (see Figure 4.16) has been built to simulate the multi-track, multi-layer, and multi-material L-PBF processes of $Cu_{10}Sn$–316L [180]. It is found that the convection inside the molten pool will cause phase migration and increase the element diffusion at the interface of the two materials. Investigators from the University of Manchester [181] have developed a similar coupled DEM–CFD model to examine the effect of mixed powder composition of In718–$Cu_{10}Sn$ on the solidified track morphology. The simulation results indicate that hybrid melting-sintering regions appear on both sides of the solidified track. It ultimately leads to fluctuations in the track, and reducing the hatch distance can suppress this defect. The aforementioned modeling framework was also used to study the molten pool dynamics of the $Cu_{10}Sn$-Invar36 material system. The study [182] found that an increase in the copper content in the powder mixture caused the temperature of the powder bed to decrease, resulting in a shallower molten pool depth of the Invar36 powder and forming defects such as spheroidization and lack of fusion of Invar particles.

Due to the lack of experimental data, it is challenging to obtain material properties for mixed powders, especially in the high-temperature range [183, 184], such as surface tension coefficient [185] and laser absorptivity [186]. It brings difficulty to predict the thermal history for multi-material L-PBF accurately.

4.5 Multi-Material WAAM

4.5.1 Introduction of Multi-Material WAAM

WAAM is an AM technology based on an automated arc welding process that uses the arc as a heat source to continuously melt welding wires to form 3D solid parts [187]. A schematic diagram of the WAAM system is shown in Figure 4.17a. A WAAM system usually consists of a computer numerical control (CNC) system, a power supply, a welding torch, an automatic wire feeding system, a workbench or robot system, and some accessories (such as shielding gas, preheating, or cooling system) [192]. A typical WAAM robot system uses a computer to program the parts' geometry and process parameters to be manufactured and collect sensor signals. The robot controller is used to control the movement of the welding torch for metal deposition. The programmable welding power source is used to control the current, frequency, and other parameters in the welding process. The arc is formed between a consumable wire electrode and a conductive substrate [193]. The base metal/solidified metal and filler wire are melted by the arc heat generated between the electrode and the base metal/solidified metal. It is then solidified to form a solid component layer

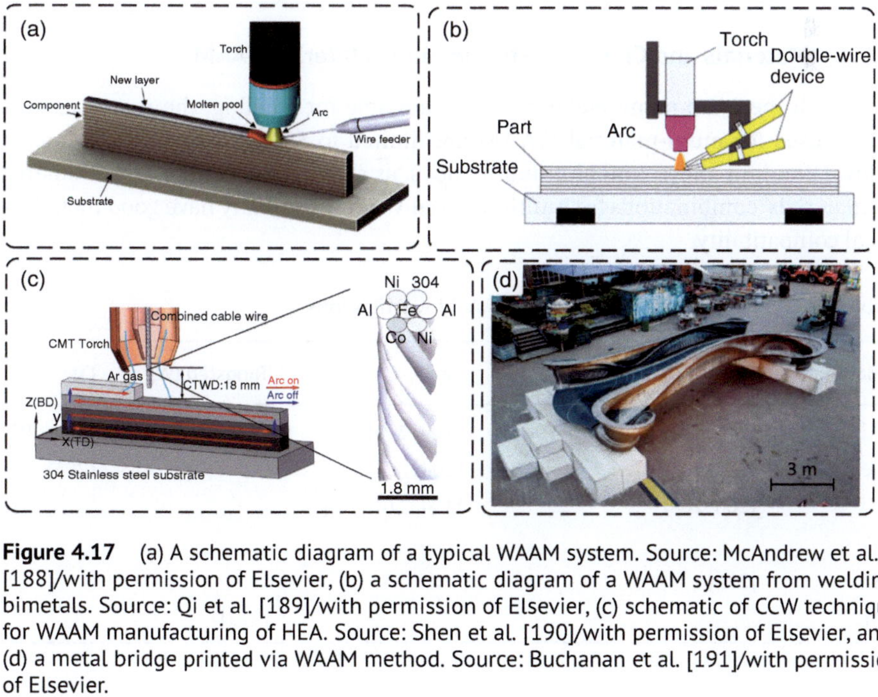

Figure 4.17 (a) A schematic diagram of a typical WAAM system. Source: McAndrew et al. [188]/with permission of Elsevier, (b) a schematic diagram of a WAAM system from welding bimetals. Source: Qi et al. [189]/with permission of Elsevier, (c) schematic of CCW technique for WAAM manufacturing of HEA. Source: Shen et al. [190]/with permission of Elsevier, and (d) a metal bridge printed via WAAM method. Source: Buchanan et al. [191]/with permission of Elsevier.

by layer. Fusion welding technologies including gas metal arc welding (GMAW) or metal inert gas (MIG) welding, gas tungsten arc welding (GTAW) or tungsten inert gas (TIG) welding, and plasma arc welding (PAW) have been used for WAAM [187].

4.5.2 Material Feeding Mechanism of Multi-Material WAAM

WAAM has been used to manufacture large multi-metal parts and FGM parts. Compared with single-material WAAM, the difference of the process setup of multi-material WAAM is to set two wire feeding devices close to the torch (see Figure 4.17b), which simultaneously feed dissimilar metal wires into the molten pool during welding [194]. The wire feeding speed can be controlled to realize FGMs with varying composition fractions [195].

The use of pretwisted combined cable wire (CCW) with multi-element composition for WAAM is the other route for WAAM of multi-material components, as shown in Figure 4.17c. This method is suitable for the development of new materials such as high-entropy alloys (HEAs).

Compared with energy-beam-based metal AM methods such as L-DED and L-PBF, WAAM has the advantage that it can manufacture super-large metal structural parts. For example, researchers from the UK [191] demonstrated a WAAM-processed metal bridge with a width of 2.5 m and a span of 10 m, as shown in Figure 4.17d. However, the big melt pool in WAAM leads to the low geometrical resolution of WAAM-processed metallic components, and these components usually need to be subjected to subsequent machining to achieve the designed dimensions [188].

4.5.3 Materials and Characteristics in Multi-Material WAAM

Although there are many metal materials suitable for welding, the materials currently used for multi-material WAAM are limited to SS, iron-based alloys, nickel alloys, titanium alloys, and aluminum alloys, as shown in Table 4.4. The elements in materials' combinations for multi-material WAAM generally have good metallurgical compatibility.

Table 4.4 Summary of multi-material WAAM investigations.

Material	SS	Ti	Fe	Tungsten (W)	Cu
Al		[196–198]	[199, 200]		[189, 201]
Fe	[202, 203]		[204]		[205]
Ni	[206–209]		[194, 210, 211]		
Cu					
SS	[212]				
Mo				[213]	

4.5.3.1 WAAM of SS–Fe/SS Bimetals

The bimetal parts made of low carbon steel and SS produced by WAAM have poor ductility. After heat treatment in the temperature range of 800–1100 °C for 30 minutes to 2 hours, the ultimate tensile strength and elongation of the earlier-mentioned bimetallic components increased by 35% and 250%, respectively [202]. The WAAM-processed SS-Fe (22Cr-ER70S6) FGM has defect-free fine grains, so its tensile strength is 6% higher than the base material. There is no chromium enrichment on the bimetallic interface's carbon-manganese steel side, which ensures the corrosion resistance of the duplex SS [203]. By changing the wire feeding speed in the metal deposition process, the molten pool's chemical composition can be modified in the process. Besides, changing the welding path, the low ductility phase's positions and the high-ductility phase in the cured bimetallic material can be controlled as needed [212].

4.5.3.2 WAAM of SS–Ni Bimetals

The super austenitic SS– Hastelloy (SS904L–C276) FGM manufactured based on GMAW's WAAM process shows higher tensile strength than the forged SS904L alloy. The existence of the second phase and the microstructure changing along the structural direction lead to the anisotropy of the aforementioned FGM's mechanical properties. The dilution of the two materials' elements leads to a sharp increase in the microhardness at the bimetal interface, which ultimately results in the fatigue life of the FGM mentioned earlier being 28–35% lower than that of the SS904L forging [206, 207].

No welding defects are found in the material interface of WAAM-processed YS308L–Ni6082 FGM. The bimetal interface's metallurgical bonding strength is equal to the welding strength of the corresponding welding, single base metal [80]. Intermediate layers inserted between the YS308L and Ni6082 regions gradually change the FGM sample's chemical composition, thereby avoiding residual thermal stress at the interface of the bimetals processed by WAAM [209].

4.5.3.3 WAAM of Ti–Al Bimetals

The WAAM method has been used to synthesize a Ti–Al alloy in situ [196, 197], and the chemical composition of the alloy is changed by adjusting the feeding speed of pure titanium wire and pure aluminum wire, respectively. The volume fraction and morphology of different Ti–Al alloy phases are closely related to the Al content in the alloy. As shown in Figure 4.18a, from bottom to top, with the increase of Al content, the phases can be characterized in the following order: $\alpha+\beta\rightarrow\alpha+\alpha2\rightarrow\alpha2\rightarrow\alpha2+\gamma\rightarrow\gamma$ [196]. Also, during WAAM, the addition of Nb element in the form of foil to the Ti–Al molten pool generates the Ti_6Al_7Nb alloy. Compared with the Ti_6Al alloy, the ultimate tensile strength and elongation of the Ti_6Al_7Nb alloy are increased by 35%–47%, respectively [197]. The improvement of mechanical properties is attributed to the nanoscale lath β phase formed after the grain refinement and Nb element addition. When processing Ti_6Al_4 V/$Al_{6.25}$Cu bimetal using WAAM, the materials' deposition sequence also has a significant effect on the phase formation [198].

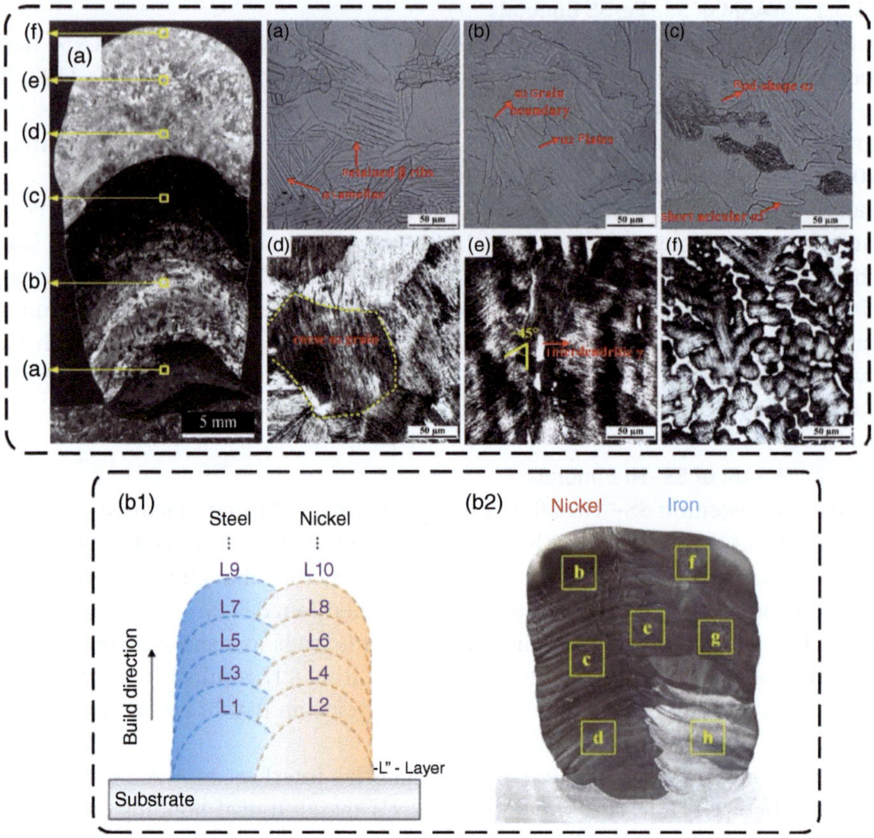

Figure 4.18 (a) The microstructure of the cross section of a Ti–Al alloy and related phases at different regions along the build direction. Source: Wang et al. [196]/with permission of the University of Wollongong, (b1) the schematic diagram of the interweaving deposition strategy, and (b2) the microstructure of the Ni-Fe bimetal cross section. Source: Wu et al. [210]/with permission of Elsevier.

4.5.3.4 WAAM of Fe–Al Bimetals

The Fe–FeAl FGM with a designed composition gradient has been successfully manufactured using the WAAM process [199, 200]. At room temperature, the tensile strength and ductility values of the FGM samples mentioned earlier are similar to previous studies.

4.5.3.5 WAAM of Fe–Ni Bimetals

An interweaving deposition strategy (Figure 4.18b1) used in the WAAM processing of SS-nickel bimetallic samples can improve the material interface's metallurgical bonding strength, making the tensile strength at the interface greater than ER70S-6 steel wire and ERNi-1 nickel wire [210]. Because the interweaving strategy produces interlocking and mutually different microstructures (Figure 4.18b2), it eliminates the potential weak fusion boundary at the interface. Simultaneously, the mixed solid solution of iron and nickel that appears at the interface further strengthens the interface strength.

The fully dense Fe–Fe_3Ni FGM sample with a smooth composition gradient prepared by WAAM has a sharp hardness gradient. The post-heat treatment significantly improves this problem. Before the heat treatment, Fe–Fe_3Ni FGM contains only fcc-Fe_3Ni and bcc-α-Fe phases, but not FCC-γ-Fe. The heat treatment effectively dissolves the bcc-α-Fe phases in the FGM mentioned earlier [194, 211]. Neutron diffraction is suitable for characterizing the aforementioned thermal lattice evolution process.

4.5.3.6 WAAM of Cu-involved Multi-Metals

The phases in Al-Cu-Mg FGM processed by the WAAM process are mainly α-Al and S phases. The increase of the copper/magnesium ratio increases the θ phase in the microstructure and also leads to the rising of the microhardness and tensile strength [189]. The homogeneous annealing process can be used to reduce the content of intermetallic compounds (including $CuAl_2$ and Cu_9Al_4) in the Cu–Al bimetallic samples treated by WAAM, thereby improving the mechanical properties including hardness, tensile strength, and elongation of the Cu–Al alloy. As the annealing temperature increases, the aforementioned bimetallic samples' mechanical properties are significantly improved [201].

The low-carbon steel–silicon bronze bimetal prepared by the WAAM method has no cracks and pores at the material interface, achieving a good metallurgical bond. In addition, there is no Cu element on the steel side of the material interface, and Fe element penetrates the bronze side in the form of particles and chunks [205].

4.6 Comparison of Multi-Material AM Technologies

Table 4.5 gives a comparison of multi-material L-DED, L-PBF, and WAAM.

The surface roughness and dimensional accuracy of parts processed by WAAM using wire materials are far worse than those in L-DED and L-PBF using powder materials. Since the heat input from the non-focused heat source (i.e., an arc) in the WAAM process is much greater than that of the small focus laser spot in the L-DED and L-PBF processes, the melt pool and HAZ of WAAM are larger than those of L-DED and L-PBF, resulting in poor surface roughness and inferior dimensional accuracy. The residual thermal stress in the WAAM process is much greater than that in the L-DED and L-PBF processes, which will cause greater thermal deformation in WAAM parts [13]. Therefore, parts manufactured with WAAM usually require subsequent machining to meet the design dimensions. In the L-DED and L-PBF process, thermal stress and thermal deformation may also occur, but they can be overcome by optimizing process parameters, optimizing the scanning/deposition path, and setting up auxiliary support structures [215]. Compared with powder materials, wire materials have obvious advantages, including higher deposition efficiency, less material waste, and cheaper materials. The processing accuracy of L-PBF is better than that of L-DED, because the powder size, laser spot, and layer thickness used by the former are smaller than those used by the latter (powder: 10–50 μm for L-PBF and 50–150 μm for L-DED [16], laser spot: 50–80 μm for L-PBF and 1.0–4.0 mm for

Table 4.5 Comparison of multiple metallic material AM technologies.

Feature		L-DED	L-PBF	WAAM	References
Material combinations		Metal-metal, metal-ceramic	Metal-metal, metal-ceramic, and metal-polymer	Metal-metal	—
Surface roughness		Ra 15–60 μm	Ra 5–18 μm	Ra 45–200+μm	[214]
Resolution	XY	0.1–1 mm	0.02–0.1 mm	2–50 mm	[214]
	Z	0.1–1 mm	0.02–0.1 mm	1–10 mm	[214]
Maximum component dimension		>1 m	>1 m	>1 m	—

L-DED [216], layer thickness: <100 μm for L-PBF and 0.25–0.5 mm for L-DED [11]), so the molten pool and HAZ are smaller.

A single galvo scanner can only process parts with a cross section less than 250 × 250 mm. At present, many commercial L-PBF systems have integrated multiple galvo scanners and multiple laser beams to achieve large-size part fabrication [217]. However, the maximum component dimension is still less than 1 m [218]. On the other hand, L-DED and WAAM can handle huge parts, as long as the deposition head's travel range is large enough.

Compared with powder and laser-based DED and PBF processes, WAAM using wire and arc discharge as energy sources has some advantages. WAAM has a very high deposition rate. The deposition efficiency is a parameter used to describe the actual deposited material's ratio to the total used material. The deposition efficiency of L-DED is only 14%, while the material utilization rate of WAAM is as high as 100%. The energy efficiency of WAAM is also higher than that of L-DED and L-PBF. The energy efficiency of WAAM can be as high as 90%, and L-DED is 30–50%. In addition, the cost of WAAM systems is much lower than those using metal powders [208, 209].

4.7 Potential Applications of Multi-Material AM

Applications of multiple metallic materials AM are usually based on materials' differences in thermal conductivity, thermal expansion coefficient, and corrosion resistance. Applications for metal/ceramic components usually utilize metal/ceramic combinations' differences in heat resistance, wear resistance, biocompatibility, and surface hardness. Applications for metal/polymer usually utilize the differences in biocompatibility, electrical conductivity, and wear resistance.

L-DED is used to fabricate aerospace shaft sleeve parts (Figure 4.19a) whose radial material is gradually transitioned from A286 SS to Invar36 [56]. The thermal expansion of this part is much smaller than that of pure SS. The National

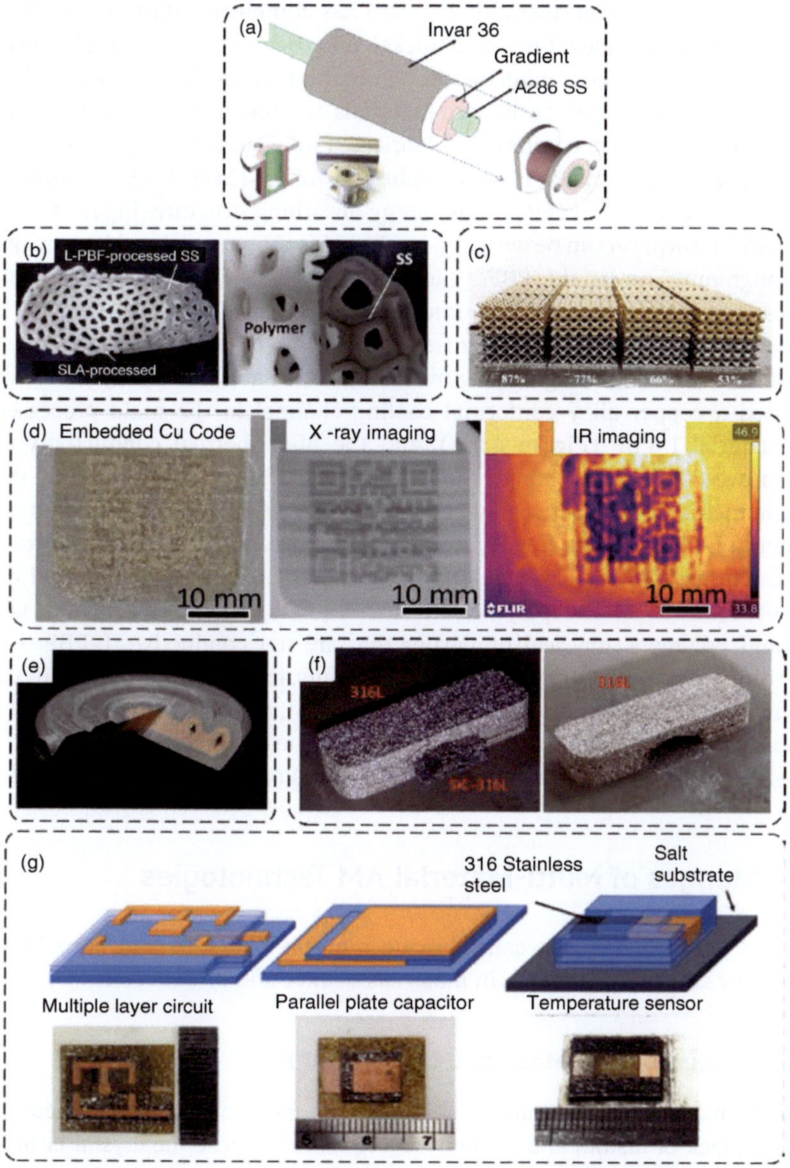

Figure 4.19 (a) L-DED-manufactured aerospace shaft sleeve parts composed of A286 SS and Invar36. Source: Hofmann et al. [56]/with permission of Springer Nature, (b) the bioinspired metal-polymer arm orthosis fabricated through L-PBF and SLA. Source: Silva et al. [219]/with permission of Institution of Mechanical Engineers, (c) L-PBF-printed copper-steel bimetal porous energy-absorbing structure. Source: Zhang et al. [127]/with permission of Elsevier/CC BY-NC-ND 4.0, (d) the copper QR code embedded in the L-PBF processed stainless steel components, and its X-ray image and IR image. Source: Wei et al. [220]/with permission of American Society of Mechanical Engineers (ASME), (e) the copper alloy conformal cooling channel in the steel mold with a copper conformal cooling channel processed by L-PBF. Source: Aerosint [221]/with permission of Aerosint Société Anonyme, (f) a 316L bridge sample with 316L-SiC composite support structure after removing the support structure. Source: Wei et al. [222]/with permission of American Society of Mechanical Engineers (ASME), and (g) L-PBF-printed 3D circuit, a parallel plate capacitor, and a temperature sensor. Source: Cheng et al. [155]/with permission of Elsevier.

Aeronautics and Space Administration (NASA) has combined L-PBF and L-DED to produce the copper alloy liner and nickel alloy jacket of the liquid rocket combustor, respectively. The combustor has higher thermal conductivity and higher working temperature because of the two materials combination [101]. L-PBF and stereolithography (SLA) are combined to fabricate bioinspired metal-polymer arm orthosis (Figure 4.19b) to make it lightweight and robust [219]. L-PBF is used to manufacture copper-steel bimetal porous energy-absorbing structure (Figure 4.19c), and the energy absorption can be designed by changing the material and its porosity [127]. Through multi-material L-PBF, a high-density copper alloy is embedded into SS parts made by AM, used for anti-counterfeiting of high value-added metal parts [220]. Differences in material density caused embedded security code to be detected by X-ray and infrared (IR) imaging (Figure 4.19d). Aerosint SA company [221] has demonstrated a copper alloy conformal cooling channel in the mold processed by multi-material L-PBF (Figure 4.19e), and the high thermal conductivity of copper improves the heat exchange speed of the mold. Taking advantage of the difference in melting points between metal and ceramic, researchers also have used multi-material L-PBF to make fragile and easy-to-remove metal-ceramic composite support structures for AM (Figure 4.19f). Copper alloy circuits are deposited on ceramic substrates [222] and polymer substrates [223] using L-PBF, which use ceramics and polymers' physical properties that are not conductive. L-PBF was utilized to directly manufacture electronic components, including multilayer circuit boards, chip capacitors, and temperature sensors [155], as shown in Figure 4.19g. These electronic components could work and realize the design function.

Potential applications for metal-based MMAM are listed in Table 4.6.

4.8 Challenges of Multi-Material AM Technologies

Although the MMAM methods can fabricate multiple materials and even FGM structures, they still face challenges in materials science and process control.

4.8.1 Challenges in Multi-Material L-DED and L-PBF

During multi-material L-DED and L-PBF, defects include brittle intermetallic phases, LME, lack of fusion, and element segregation. Suppose the crystal lattices of the two materials do not match. In that case, the elements lack solubility, and the melting points and densities of the two materials differ significantly, and these defects are likely to occur. Tables 4.7 and 4.8 summarize these defects in multi-material L-DED and L-PBF, respectively. These MMAM-specific defects will further lead to typical laser-based metallic AM defects such as porosity, balling, residual stresses-induced cracks and delamination [242], and eventually cause a decrease in the mechanical properties of the material.

Before conducting multi-material L-PBF studies, the compatibility of different materials must be carefully reviewed. One should avoid the formation of harmful intermetallic phases and cracks caused by LME. The cooling rate has an important

Table 4.6 Potential applications for metal-based multi-material AM.

Category	Materials	Application	Industry	Processing technologies	References
Metal/metal	Copper-stainless steel	The bimetal base plate used on automatic cooling control instrument	Aerospace	L-DED, L-PBF, WAAM	[224]
	Invar-stainless steel	Radially graded FGM inserts used on carbon fiber composite components	Aerospace	L-DED, L-PBF, WAAM	[225]
	Nickel alloy-stainless steel	Nuclear reactor components used in the nuclear power plant or the miniature nuclear-electricity energy conversion system for space exploration	Nuclear	L-DED, L-PBF, WAAM	[66]
			Aerospace	L-DED, L-PBF, WAAM	[226]
	Tungsten-copper	The tungsten/copper FGM diverter plate directly exposed to the plasma in steady state magnetic fusion reactors, i.e., nuclear reactors	Nuclear	L-DED, L-PBF, WAAM	[227]
	Copper-aluminum	Aluminum-copper joint used on solar energy collectors or lightweight electric motors and actuators	Electrics	L-DED, L-PBF, WAAM	[228]
	Copper-titanium	Heat exchanger	Aerospace	L-DED, L-PBF, WAAM	[229]
	Copper-aluminum	Heat exchanger	Automotive	L-DED, L-PBF, WAAM	[230]
Metal/ceramic	Titanium-hydroxy-apatite	A graded interface for bone-in orthopedic implants	Biomedical	L-DED, L-PBF	[231]
	Stainless steel-hydroxyapatite			L-DED, L-PBF	[231]
	Metal-ceramic	Metal/ceramic FGM heterogeneous turbine blade used in aircraft engines	Aerospace	L-DED, L-PBF	[232]
	Aluminum-alumina	New generation ceramic/metal FGM armor for combat vehicles	Defense	L-DED, L-PBF	[233]
	NiAl-alumina	Heat-resistant valve piston for internal combustion engines	Automotive	L-DED, L-PBF	[234]

Table 4.6 (Continued)

Category	Materials	Application	Industry	Processing technologies	References
Metal/polymer	CoCrMo alloy-PLA	Metal/polymer implant with a controllable drug delivery profile	Biomedical	L-PBF	[235, 236]
	Stainless steel-PLA			L-PBF	
	Aluminum-PA11	Metal/polymer electronic devices	Electrics	L-PBF	[237]
	Nickel-PEEK	Micro/nano electro-mechanical systems	Electrics	L-PBF	[166]
	Copper-PEEK			L-PBF	
	Copper-HDPE (high-density polyethylene)	3D circuit with imbedded electrical components	Electrics	L-PBF	[223]
	PA11-Glass	Functional ball/cylinder bearing support	Biomedical	L-PBF	[170]
	PA11-Al_2O_3			L-PBF	

Table 4.7 Summary of defects reported in multi-material L-DED investigations.

Material combination		Intermetallic phases	Lack of fusion	References
Ni	Ni–Cu	——	●x025CF;	[40–42]
	Ni–SS	——	——	[47–49]
SS	SS–SS	——	——	[50]
Ti	Ti–Al	●x025CF;	●x025CF;	[34, 35]
	Ti–Ni	●x025CF;	——	[36–39]
	Ti–SS	●x025CF;	——	[51–55]
	Ti–Ti	——	——	[58]
	Ti–Mo	——	●x025CF;	[59]
Fe	Fe–Cu	——	●x025CF;	[43]
	Fe–SS	——	——	[56]
Metal	SS304–VC	——	——	[60]
	SS316L–Fe_3Al	——	——	[57]
	Ti–ceramic	——	●x025CF;	[61–63]

Note: symbol "●x025CF;" means including; the symbol "—— " means not including.

Table 4.8 Summary of defects reported in multi-material L-PBF investigations.

Material combination		Intermetallic phases	LME	Lack of fusion	References
Al	Al–Al	●x025CF;	——	——	[131]
	Al–Cu	●x025CF;	——	——	[128]
	Al–Ti	——	——	——	[67]
	Al–Fe	●x025CF;	——	——	[121]
Cu	Cu–SS	——	●x025CF;	●x025CF;	[25, 114, 124–126, 180, 238]
	Cu–Ni	——	——	●x025CF;	[181]
	Cu–Fe	——	●x025CF;	●x025CF;	[88, 99, 116, 127, 135, 137]
	Cu–W	——	——	●x025CF;	[122]
SS	SS–Ti	●x025CF;	——	——	[134]
	SS–Ni	——	——	——	[239]
	SS–Fe	——	——	——	[132]
Ti	Ti–Ti	——	——	——	[130]
	Ti–Ni	●x025CF;	——	——	[83]
Fe	Fe–Fe	——	——	——	[240]
Metal	Cu–polymer	——	——	●x025CF;	[118, 161, 241]
Metal	SS–glass	——	——	●x025CF;	[117, 154]
	Ti–ceramic	●x025CF;	——	——	[150–152], 178]
	Fe–ceramic	——	——	●x025CF;	[147, 148]

Note: symbol "●x025CF;" means including; the symbol "——" means not including.

influence on the phase transition and grain growth during AM of FGM [243], so the CALPHAD method can be used to predict the phase formation during processing based on the phase diagram. The use of targeted methods (such as controlling the processing temperature) avoids the formation of second phases [244]. Another practical solution to these problems is to add an intermediate layer composed of elements with good metallurgical compatibility related to the two base materials [245, 246]. Combining an AM method with other processing methods is also a possible solution to fabricating nonmiscible materials. For example, combining AM and cold spraying (CS) to join W-Cu [247] and combining L-PBF and SLA to bind SS-polymers.

The large difference in the two materials' melting points will cause the low melting point material to evaporate during the melting process. The high melting point material will not melt sufficiently, causing defects such as element segregation, pores, and powder infusibility, which will eventually lead to a decrease in the processed parts' mechanical properties. Homogenization heat treatment can reduce element segregation [248]. The simultaneous use of laser power in the medium range and high scanning speed can also suppress low melting point metal elements' evaporation and reduce segregation [249].

Suppose there is a significant difference in the density of various elements in the powder mixture and the liquid's viscosity. In that case, it will lead to element segregation, including element-rich areas and insufficient diffusion [250]. Elements with higher material mass density sink to the bottom of the molten pool and gather near the molten pool boundary. High-density materials should be placed on the bottom and low-density materials on the top in the FGM design. Also, the smooth material gradient transition will suppress component segregation. Heat treatment can fine-tune the phase distribution and phase transition and release internal stress, thereby improving mechanical properties [251].

How to recycle the unmelted powder mixture produced by MMAM is a technical challenge. Separating different powder materials from the mixture for reuse can significantly reduce material costs and reduce environmental pollution [29]. Possible methods for the separation of dissimilar powders mixtures is to utilize the different physical properties such as magnetism [252], distinct particle size distribution [253], different material wettability [254], and distinct particle inertia caused by different powder density [255].

4.8.2 Challenges in Multi-Material WAAM

Table 4.9 summarizes the possible defects of WAAM processing of different materials [256]. Defects in WAAM may occur due to many reasons, such as wrong programming strategy, dynamic instability of molten pool due to improper processing parameters, and residual thermal deformation [257].

In the WAAM process, deformation and residual stress will inevitably occur. Deformation is caused by the thermal expansion and contraction of parts during repeated melting–cooling cycles and is especially noticeable when WAAM processes large thin-walled structures. When using WAAM to manufacture bimetal parts, the

Table 4.9 Typical defects in WAAM.

Material	Process	Defect or feature					
		Porosity	Cracking	Delamination	Oxidation	Substrate adherence	Surface finish
Ti_6Al_4V	TIG	No	No	No	Light	Good	Smooth
	Plasma	No	No	No	No	Good	Smooth
	CMT	No	No	No	Light	Good	Smooth
	DCEP-GMAW	No	No	No	Light	Medium	Poor
H08Mn2Si steel	DE-GMAW	low	No	No	No	Good	Waviness
Copper–coated steel	GMAW	No	No	No	Light	Good	Medium-rough
ER4043 Al alloy	CMT	High	No	No	Light	Good	Smooth
	VP-GTAW	No	No	No	No	Good	Medium-rough
AA2319 Al alloy	CMT	High	No	No	No	Good	Smooth
	CMT-PADV	No	No	No	No	Good	Smooth
5356 Al alloy	VP-GTAW	No	Yes	No	No	Good	Smooth
Inconel 625	PPAD	High	Yes	No	No	Good	Smooth
	GTAW	No	No	No	No	Good	Smooth
Inconel 718	GMAW	Medium	Yes	Yes	No	Good	Smooth
AZ31 Mg alloy	PMIG	No	No	No	Light	Medium	Medium-rough
Nickel–Al–Cu	CMT	No	No	No	No	Good	Smooth
Steel–bronze bimetal	GMAW	No	No	No	No	Good	Smooth
Steel–nickel bimetal	GMAW	No	No	No	No	Good	Medium-rough
Intermetallic Fe/Al	GTAW	High	Yes	No	Serious	Medium	Medium-poor
Intermetallic Al/Ti	GTAW	Low	Yes	No	No	Good	Rough
Intermetallic Al/Cu	GTAW	No	No	Yes	Light	Poor	Rough

Source: Adapted from Wu et al. [256].

difference in the thermal expansion coefficient and the material's melting point will cause more residual stress and deformation. Therefore, when using WAAM to process bimetallic materials, it is necessary to control the interpass temperature precisely. Solutions include optimization of WWAM deposition path, post-heat treatment, interlayer cold rolling, interpass cooling, shot peening, and ultrasonic impact treatment [256].

During WAAM of bimetals (such as Al–Cu, Al–Ti, and Al–Fe), defects such as cracking and delamination may occur. The mutual solubility and chemical activity of dissimilar metals are very different, so the phase balance between the metals is broken freely, which causes crack propagation along the grain boundaries. Inconel nickel alloys are prone to solidification cracks due to a liquid film's existence when the ends are solidified. The measures below can be used to reduce crack defects, including optimizing the mixed wire and its composition, reducing the cooling rate during the deposition process, and avoiding solid solution treatment to improve strength [256].

4.9 Summary and Outlook

4.9.1 Summary

This chapter has introduced the background of MMAM and multi-material components design tools. This chapter focuses on the research status of three commonly used metal AM methods (i.e., L-DED, L-PBF, and WAAM) for MMAM, including their respective multi-material feeding/deposition mechanisms and materials science research progress. The available dissimilar material combinations, surface roughness, and processing resolution that could be processed by the aforementioned three methods are compared, and their potential applications are described. The challenges to be faced by these three methods for MMAM are discussed.

4.9.2 Outlook

MMAM technologies have many potential applications in the aerospace, automotive, medical, nuclear, and defense industries, usually based on materials' differences in thermal conductivity, thermal expansion coefficient, and corrosion resistance.

However, the differences in materials' physical properties and the compatibility of elements in the materials make it very challenging to process high-quality multi-metallic parts through MMAM. Defects prone to occur in multi-material L-PBF and multi-material L-DED include defects caused by thermal stress, cracks caused by LME, brittle intermetallic phases formed in the process, uneven chemical composition of materials, exothermic redox reactions, and cross-used different powders pollution. The defects to be created easily in multi-material WAAM include thermal deformation, cracking, and delamination. All the aforementioned defects can be classified into two key material science challenges, including how to achieve sufficient melting of different elements inside the material mixture and how to achieve reliable bonding of dissimilar materials at the material interfaces.

Current studies on MMAM have demonstrated the feasibility of using a certain AM technology to produce metal-based multi-material systems. The research on defect formation mechanism and associated suppression methods is still at the early stage, compared with general single metallic material AM technologies. The present technological maturity of metallic MMAM is far from reaching the industrial application standard. Therefore, it is necessary to conduct more in-depth research on these technologies, especially theoretical research on phase transition prediction, thermodynamic calculation, modeling, and numerical simulation. These tasks are essential to reduce processing defects and reveal mechanisms behind the phenomena. The use of artificial intelligence techniques, such as machine learning, should help increase the speed to archive the aforementioned investigation aims.

References

1 Yusuf, S.M., Cutler, S., and Gao, N. (2019). Review: the impact of metal additive manufacturing on the aerospace industry. *Metals (Basel)* 9: 4. https://doi.org/10.3390/met9121286.

2 Li, Y., Feng, Z., Hao, L. et al. (2020). A review on functionally graded materials and structures via additive manufacturing: from multi-scale design to versatile functional properties. *Advanced Materials Technologies* 5: 23. https://doi.org/10.1002/admt.201900981.

3 Vaezi, M., Chianrabutra, S., Mellor, B., and Shoufeng, Y. (2013). Multiple material additive manufacturing – Part 1 : a review. *Virtual and Physical Prototyping* 8: 19–50. https://doi.org/10.1080/17452759.2013.778175.

4 Yao, X., Moon, S.K., Bi, G., and Wei, J. (2018). A multi-material part design framework in additive manufacturing. *International Journal of Advanced Manufacturing Technology* 99: 2111–2119. https://doi.org/10.1007/s00170-018-2025-7.

5 Prasad, K. and Chakraborty, S. (2013). A quality function deployment-based model for materials selection. *Materials and Design* 49: 525–535. https://doi.org/10.1016/j.matdes.2013.01.035.

6 Mukhopadhyay, P. (2012). Alloy designation, processing, and use of AA6XXX series aluminium alloys. *International Scholarly Research Notices Metallurgy* 2012: 1–15. https://doi.org/10.5402/2012/165082.

7 Mauduit, A., Pillot, S., Gransac, H. et al. (2017). Study of the suitability of aluminum alloys for additive manufacturing by laser powder-bed fusion. *UPB Scientific Bulletin, Series B: Chemistry and Materials Science* 4: 219–238.

8 Chenyan, L. and Shikai, J. Multi-attribute voxelization technology of AMF for 3D printing. In: *2019 IEEE 6th International Conference on Industrial Engineering and Applications*, vol. 2019, 81–85. https://doi.org/10.1109/IEA.2019.8714990.

9 Mishra, G.K., Paul, C.P., Rai, A.K. et al. (2021). Experimental investigation on laser directed energy deposition based additive manufacturing of Al_2O_3 bulk structures. *Ceramics International* 47: 5708–5720. https://doi.org/10.1016/j.ceramint.2020.10.157.

10 Govekar, E., Jeromen, A., Kuznetsov, A. et al. (2018). Annular laser beam based direct metal deposition. *Procedia CIRP* 74: 222–227. https://doi.org/10.1016/j.procir.2018.08.099.

11 Gibson, I., Rosen, D., and Stucker, B. (2015). Directed energy deposition processes. In: *Additive Manufacturing Technology*, 19–39.

12 Kuznetsov, A., Jeromen, A., Levy, G. et al. (2016). Annular laser beam cladding process feasibility study. *Physics Procedia* 83: 647–656. https://doi.org/10.1016/j.phpro.2016.08.067.

13 Wang, W., Pinkerton, A.J., and Li, L. (2008). A gas-free powder delivery system for 100% deposition efficiency in direct laser deposition. In: *ICALEO 2008 – 27th International Congress on Applications of Lasers and Electro-optics (ICALEO)*, vol. 801, 415–423. https://doi.org/10.2351/1.5061341.

14 Li, T., Zhang, L., Bultel, G.G. et al. (2019). Extreme high-speed laser material deposition (EHLA) of AISI 4340 steel. *Coatings* 9: 1. https://doi.org/10.3390/coatings9120778.

15 Thompson, S.M., Bian, L., Shamsaei, N., and Yadollahi, A. (2015). An overview of direct laser deposition for additive manufacturing; part I: transport phenomena, modeling and diagnostics. *Additive Manufacturing* 8: 36–62. https://doi.org/10.1016/j.addma.2015.07.001.

16 Singh, A., Kapil, S., and Das, M. (2020). A comprehensive review of the methods and mechanisms for powder feedstock handling in directed energy deposition. *Additive Manufacturing* 35: 101388. https://doi.org/10.1016/j.addma.2020.101388.

17 Jinping, Q., Baoshan, S., Yanhong, F., and Hezhi, H. (2006). Dependence of solids conveying on screw axial vibration in single screw extruders. *Journal of Applied Polymer Science* 102: 2998–3007. https://doi.org/10.1002/app.24658.

18 Dunst, P., Bornmann, P., Hemsel, T., and Sextro, W. (2018). Vibration-assisted handling of dry fine powders. *Actuators* 7: 1–11. https://doi.org/10.3390/act7020018.

19 Thayalan, V. and Landers, R.G. (2006). Regulation of powder mass flow rate in gravity-fed powder feeder systems. *Journal of Manufacturing Processes* 8: 121–132. https://doi.org/10.1016/S1526-6125(06)80007-1.

20 Balachandran, W., Hu, D., Ghadiri, M. et al. (1997). The study of the performance of an electrostatic valve used for bulk transport of particulate materials. *IEEE Transactions on Industry Applications* 33: 871–878. https://doi.org/10.1109/28.605726.

21 Li, W., Karnati, S., Zhang, Y., and Liou, F. (2018). Investigating and eliminating powder separation in pre-mixed powder supply for laser metal deposition process. *Journal of Materials Processing Technology* 254: 294–301. https://doi.org/10.1016/j.jmatprotec.2017.11.045.

22 Balu, P., Leggett, P., and Kovacevic, R. (2012). Parametric study on a coaxial multi-material powder flow in laser-based powder deposition process. *Journal of Materials Processing Technology* 212: 1598–1610. https://doi.org/10.1016/j.jmatprotec.2012.02.020.

23 Jeantette, F.P., Keicher, D.M., Romero, J.A., and Schanwald, L.P. (1996). Method and system for producing complex-shape objects, US006046426A.

24 Sato, A. and Ishikawa, Y. (2006). Powder metal cladding nozzle, US20060266740A1.

25 Wei, C., Li, L., Zhang, X., and Chueh, Y.H. (2018). 3D printing of multiple metallic materials via modified selective laser melting. *CIRP Annals* 67: 245–248. https://doi.org/10.1016/j.cirp.2018.04.096.

26 Zhong, C., Pirch, N., Gasser, A. et al. (2017). The influence of the powder stream on high-deposition-rate laser metal deposition with inconel 718. *Metals (Basel)* 7: 9. https://doi.org/10.3390/met7100443.

27 Haley, J.C., Schoenung, J.M., and Lavernia, E.J. (2018). Observations of particle-melt pool impact events in directed energy deposition. *Additive Manufacturing* 22: 368–374. https://doi.org/10.1016/j.addma.2018.04.028.

28 Koike, R., Takemura, S., Kakinuma, Y., and Kondo, M. (2018). Enhancement of powder supply efficiency in directed energy deposition based on gas-solid multiphase-flow simulation. *Procedia CIRP* 78: 133–137. https://doi.org/10.1016/j.procir.2018.09.061.

29 Horn, M., Prestel, L., Schmitt, M. et al. (2020). Multi-material additive manufacturing – Recycling of binary metal powder mixtures by screening. *Procedia CIRP* 93: 550–555. https://doi.org/10.1016/j.procir.2020.04.098.

30 Pinkerton, A.J., Syed, W.U.H., and Li, L. (2007). Theoretical analysis of the coincident wire-powder laser deposition process. *Journal of Manufacturing Science and Engineering, Transactions of the ASME* 129: 1019–1027. https://doi.org/10.1115/1.2752828.

31 Syed, W.U.H., Pinkerton, A.J., and Li, L. (2006). Combining wire and coaxial powder feeding in laser direct metal deposition for rapid prototyping. *Applied Surface Science* 252: 4803–4808. https://doi.org/10.1016/j.apsusc.2005.08.118.

32 Syed, W.U.H., Pinkerton, A.J., and Li, L. (2006). Simultaneous wire- and powder-feed direct metal deposition: an investigation of the process characteristics and comparison with single-feed methods. *Journal of Laser Applications* 18: 65–72. https://doi.org/10.2351/1.2164485.

33 Mahamood, R.M. (2018). Laser metal deposition process of metals, alloys, and composite materials. *Spring* https://doi.org/10.1007/978-3-319-64985-6_5.

34 Liu, Y., Liu, C., Liu, W. et al. (2018). Microstructure and properties of Ti/Al lightweight graded material by direct laser deposition. *Materials Science and Technology (United Kingdom).* 34: 945–951. https://doi.org/10.1080/02670836.2017.1412042.

35 Zhang, Y. and Bandyopadhyay, A. (2019). Direct fabrication of bimetallic Ti6Al4V+Al12Si structures via additive manufacturing. *Additive Manufacturing* 29: 100783. https://doi.org/10.1016/j.addma.2019.100783.

36 Bobbio, L.D., Otis, R.A., Borgonia, J.P. et al. (2017). Additive manufacturing of a functionally graded material from Ti-6Al-4V to Invar: Experimental characterization and thermodynamic calculations. *Acta Materialia* 127: 133–142. https://doi.org/10.1016/j.actamat.2016.12.070.

37 Onuike, B. and Bandyopadhyay, A. (2018). Additive manufacturing of Inconel 718 – Ti6Al4V bimetallic structures. *Additive Manufacturing* 22: 844–851. https://doi.org/10.1016/j.addma.2018.06.025.

38 Shang, C., Wang, C., Xu, G. et al. (2019). Laser additive manufacturing of TA15 – Inconel 718 bimetallic structure via Nb/Cu multi-interlayer. *Vacuum* 169: 108888. https://doi.org/10.1016/j.vacuum.2019.108888.

39 Shang, C., Xu, G., Wang, C. et al. (2019). Laser deposition manufacturing of bimetallic structure from TA15 to inconel 718 via copper interlayer. *Materials Letters* 252: 342–344. https://doi.org/10.1016/j.matlet.2019.06.030.

40 Onuike, B., Heer, B., and Bandyopadhyay, A. (2018). Additive manufacturing of Inconel 718 — Copper alloy bimetallic structure using laser engineered net shaping (LENS™). *Additive Manufacturing* 21: 133–140. https://doi.org/10.1016/j.addma.2018.02.007.

41 Li, L., Syed, W.U.H., and Pinkerton, A.J. (2006). Rapid additive manufacturing of functionally graded structures using simultaneous wire and powder laser deposition. *Virtual and Physical Prototyping* 1: 217–225. https://doi.org/10.1080/17452750601141523.

42 Syed, W.U.H., Pinkerton, A.J., Liu, Z., and Li, L. (2007). Single-step laser deposition of functionally graded coating by dual 'wire–powder' or 'powder–powder' feeding—A comparative study. *Applied Surface Science* 253: 7926–7931. https://doi.org/10.1016/j.apsusc.2007.02.174.

43 Articek, U., Milfelner, M., and Anzel, I. (2013). Synthesis of functionally graded material H13/Cu by LENS technology. *Advances in Production Engineering & Management* 8: 169–176.

44 Zhang, X., Pan, T., Flood, A. et al. (2021). Investigation of copper/stainless steel multi-metallic materials fabricated by laser metal deposition. *Materials Science and Engineering A* 811: 141071. https://doi.org/10.1016/j.msea.2021.141071.

45 Zhang, X., Sun, C., Pan, T. et al. (2020). Additive manufacturing of copper – H13 tool steel bi-metallic structures via Ni-based multi-interlayer. *Additive Manufacturing* 36: 101474. https://doi.org/10.1016/j.addma.2020.101474.

46 Li, B., Han, C., Lim, C.W.J., and Zhou, K. (2022). Interface formation and deformation behaviors of an additively manufactured nickel-aluminum-bronze/15-5 PH multimaterial via laser-powder directed energy deposition. *Materials Science and Engineering A* 829: 142101. https://doi.org/10.1016/j.msea.2021.142101.

47 Carroll, B.E., Otis, R.A., Borgonia, J.P. et al. (2016). Functionally graded material of 304L stainless steel and inconel 625 fabricated by directed energy deposition: characterization and thermodynamic modeling. *Acta Materialia* 108: 46–54. https://doi.org/10.1016/j.actamat.2016.02.019.

48 Shah, K., ul Haq, I., Khan, A. et al. (2014). Parametric study of development of Inconel-steel functionally graded materials by laser direct metal deposition. *Materials and Design* 54: 531–538. https://doi.org/10.1016/j.matdes.2013.08.079.

49 Wu, D., Liang, X., Li, Q., and Jiang, L. (2010). Laser rapid manufacturing of stainless steel 316L/Inconel 718 functionally graded materials: microstructure evolution and mechanical properties. *International Journal of Optics* 2010: 802385. https://doi.org/10.1155/2010/802385.

50 Heer, B. and Bandyopadhyay, A. (2018). Compositionally graded magnetic-nonmagnetic bimetallic structure using laser engineered net shaping. *Materials Letters* 216: 16–19. https://doi.org/10.1016/j.matlet.2017.12.129.

51 Sahasrabudhe, H., Harrison, R., Carpenter, C., and Bandyopadhyay, A. (2015). Stainless steel to titanium bimetallic structure using LENS™. *Additive Manufacturing* 5: 1–8. https://doi.org/10.1016/j.addma.2014.10.002.

52 Reichardt, A., Dillon, R.P., Borgonia, J.P. et al. (2016). Development and characterization of Ti-6Al-4V to 304L stainless steel gradient components fabricated with laser deposition additive manufacturing. *Materials and Design* 104: 404–413. https://doi.org/10.1016/j.matdes.2016.05.016.

53 Bobbio, L.D., Bocklund, B., Otis, R. et al. (2018). Characterization of a functionally graded material of Ti-6Al-4V to 304L stainless steel with an intermediate V section. *Journal of Alloys and Compounds* 742: 1031–1036. https://doi.org/10.1016/j.jallcom.2018.01.156.

54 Li, W., Karnati, S., Kriewall, C. et al. (2017). Fabrication and characterization of a functionally graded material from Ti-6Al-4V to SS316 by laser metal deposition. *Additive Manufacturing* 14: 95–104. https://doi.org/10.1016/j.addma.2016.12.006.

55 Onuike, B. and Bandyopadhyay, A. (2020). Functional bimetallic joints of Ti6Al4V to SS410. *Additive Manufacturing* 31: 100931. https://doi.org/10.1016/j.addma.2019.100931.

56 Hofmann, D.C., Roberts, S., Otis, R. et al. (2014). Developing gradient metal alloys through radial deposition additive manufacturing. *Scientific Reports* 4: 5357. https://doi.org/10.1038/srep05357.

57 Durejko, T., Zietala, M., Polkowski, W., and Czujko, T. (2014). Thin wall tubes with Fe3Al/SS316L graded structure obtained by using laser engineered net shaping technology. *Materials and Design* 63: 766–774. https://doi.org/10.1016/j.matdes.2014.07.011.

58 Yan, L., Chen, X., Zhang, Y. et al. (2017). Fabrication of functionally graded Ti and γ-TiAl by laser metal deposition. *Journal of Metals* 69: 2756–2761. https://doi.org/10.1007/s11837-017-2582-5.

59 Schneider-Maunoury, C., Weiss, L., Acquier, P. et al. (2017). Functionally graded Ti6Al4V-Mo alloy manufactured with DED-CLAD® process. *Additive Manufacturing* 17: 55–66. https://doi.org/10.1016/j.addma.2017.07.008.

60 Gualtieri, T. and Bandyopadhyay, A. (2018). Additive manufacturing of compositionally gradient metal-ceramic structures: Stainless steel to vanadium carbide. *Materials and Design* 139: 419–428. https://doi.org/10.1016/j.matdes.2017.11.007.

61 Zhang, Y. and Bandyopadhyay, A. (2018). Direct fabrication of compositionally graded Ti-Al_2O_3 multi-material structures using laser engineered net shaping. *Additive Manufacturing* 21: 104–111. https://doi.org/10.1016/j.addma.2018.03.001.

62 Li, L., Wang, J., Lin, P., and Liu, H. (2017). Microstructure and mechanical properties of functionally graded TiCp/Ti_6Al_4V composite fabricated by laser melting deposition. *Ceramics International* 43: 16638–16651. https://doi.org/10.1016/j.ceramint.2017.09.054.

63 Liu, W. and DuPont, J.N. (2003). Fabrication of functionally graded TiC/Ti composites by laser engineered net shaping. *Scripta Materialia* 48: 1337–1342. https://doi.org/10.1016/S1359-6462(03)00020-4.

64 Gullino, A., Matteis, P., and Aiuto, F.D. (2019). Review of aluminum-to-steel welding technologies for car-body applications. *Metals* 10: 1–28. https://doi.org/10.3390/met9030315.

65 de Groh, H.C., Ellis, D.L., and Loewenthal, W.S. (2008). Comparison of GRCop-84 to other Cu alloys with high thermal conductivities. *Journal of Materials Engineering and Performance* 17: 594–606. https://doi.org/10.1007/s11665-007-9175-3.

66 Hinojos, A., Mireles, J., Reichardt, A. et al. (2016). Joining of Inconel 718 and 316 stainless steel using electron beam melting additive manufacturing technology. *Materials and Design* 94: 17–27. https://doi.org/10.1016/j.matdes.2016.01.041.

67 Yin, S., Yan, X., Chen, C. et al. (2018). Hybrid additive manufacturing of Al-Ti6Al4V functionally graded materials with selective laser melting and cold spraying. *Journal of Materials Processing Technology* 255: 650–655. https://doi.org/10.1016/j.jmatprotec.2018.01.015.

68 Wilde, G., Willnecker, R., Singh, R.N., and Sommer, F. (1997). The metastable miscibility gap in the system Fe–Cu. *International Journal of Materials Research* 88: 804–809. https://doi.org/10.3139/ijmr-1997-0151.

69 Gibson, I., Rosen, D.W., and Stucker, B. (2010). Powder bed fusion processes. In: *Additive Manufacturing Technologies Rapid Prototyping to Direct Digital Manufacturing* (ed. I. Gibson, D.W. Rosen and B. Stucker), 120–159. Boston, MA: Springer US https://doi.org/10.1007/978-1-4419-1120-9_5.

70 Kruth, J.P., Mercelis, P., Van Vaerenbergh, J. et al. (2005). Binding mechanisms in selective laser sintering and selective laser melting. *Rapid Prototyping Journal* 11: 26–36. https://doi.org/10.1108/13552540510573365.

71 Sun, S., Brandt, M., and Easton, M. (2017). 2 – Powder bed fusion processes: an overview. In: *Laser Additive Manufacturing Materials, Design, Technologies, and Applications* (ed. S. Sun, M. Brandt and M. Easton), 55–77. Woodhead Publishing https://doi.org/10.1016/B978-0-08-100433-3.00002-6.

72 King, W.E., Anderson, A.T., Ferencz, R.M. et al. (2015). Laser powder bed fusion additive manufacturing of metals; physics, computational, and materials challenges. *Applied Physics Reviews* 2: 041304. https://doi.org/10.1063/1.4937809.

73 Muller, P., Mognol, P., and Hascoet, J.-Y. (2012). Functionally graded material (FGM) parts: from design to the manufacturing simulation. In: *Proceedings: ASME 2012 11th Biennial Conference on Engineering Systems Design and Analysis*, Paper no. ESDA2012-82586.

74 Gibson, I., Rosen, D.W., and Stucker, B. (2010). *Additive Manufacturing Technologies* https://doi.org/10.1007/978-1-4419-1120-9.

75 Yan, L., Chen, Y., and Liou, F. (2020). Additive manufacturing of functionally graded metallic materials using laser metal deposition. *Additive Manufacturing* 31: https://doi.org/10.1016/j.addma.2019.100901.

76 Stichel, T., Laumer, T., Linnenweber, T. et al. (2016). Mass flow characterization of selective deposition of polymer powders with vibrating nozzles for laser beam melting of multi-material components. *Physics Procedia* 83: 947–953. https://doi.org/10.1016/j.phpro.2016.08.099.

77 Meng, W., Zhang, W., Zhang, W. et al. (2019). Additive fabrication of 316L/Inconel625/Ti6Al4V functionally graded materials by laser synchronous preheating. *International Journal of Advanced Manufacturing Technology* 104: 2525–2538. https://doi.org/10.1007/s00170-019-04061-x.

78 Pedersen, D.B. (2013). *Additive Manufacturing: Multi Material Processing and Part Quality Control*. The Technical University of Denmark.

79 Andriani, C. (2014). Review on melting of multiple metal materials in additive manufacturing. *Proceedings of the International Conference on Progress in Additive Manufacturing* 6: 139–144. https://doi.org/10.3850/978-981-09-0446-3.

80 Regenfuss, P., Streek, A., Hartwig, L. et al. (2007). Principles of laser micro sintering. *Rapid Prototyping Journal* 13: 204–212. https://doi.org/10.1108/13552540710776151.

81 Mahamood, R.M. and Akinlabi, E.T. (2015). Laser metal deposition of functionally graded Ti6Al4V/TiC. *Materials and Design* 84: 402–410. https://doi.org/10.1016/j.matdes.2015.06.135.

82 Zhang, B., Jaiswal, P., Rai, R., and Nelaturi, S. (2018). Additive manufacturing of functionally graded material objects: a review. *Journal of Computing and Information Science in Engineering* 18: 1215–1231. https://doi.org/10.1115/1.4039683.

83 Scaramuccia, M.G., Demir, A.G., Caprio, L. et al. (2020). Development of processing strategies for multigraded selective laser melting of Ti6Al4V and IN718. *Powder Technology* 367: 376–389. https://doi.org/10.1016/j.powtec.2020.04.010.

84 Demir, A.G. and Previtali, B. (2017). Multi-material selective laser melting of Fe/Al-12Si components. *Manufacturing Letters* 11: 8–11. https://doi.org/10.1016/j.mfglet.2017.01.002.

85 Guo, C., Ge, W., and Lin, F. (2015). 3D printing – Article dual-material electron beam selective melting: hardware development and validation studies. *Engineering* 1: 124–130. https://doi.org/10.15302/J-ENG-2015013.

86 Zhang, X., Chueh, Y., Wei, C. et al. (2020). Additive manufacturing of three-dimensional metal-glass functionally gradient material components by laser powder bed fusion with in situ powder mixing. *Additive Manufacturing* 33: 101113. https://doi.org/10.1016/j.addma.2020.101113.

87 Nadimpalli, V.K., Dahmen, T., Valente, E.H., Mohanty, S., and Pedersen, D.B. (2019). Multi-material additive manufacturing of steels using laser powder bed fusion. Proceedings: 19th International Conference & Exhibition at: Bilbao, ES.

88 Beal, V., Erasenthiran, P., Hopkinson, N., Dickens, P., and Ahrens, C. (2004). Fabrication of x-graded H13 and Cu powder mix using high power pulsed Nd:YAG laser. Conference Procedings: Solid Freeform Fabrication Symposium, Austin, TX, USA.

89 Lappo, K., Jackson, B., and Wood, D. (2003). Discrete multiple material selective laser sintering (M2SLS): experimental study of part processing. In: *Solid*

Freeform Fabrication Symposium, vol. 6, 109–119. Austin, TX: University of Texas http://edge.rit.edu/content/P10551/public/SFF/SFF 2003 Proceedings/2003 SFF Papers/10-Lappo.pdf.

90 Wu, W., Yongqiang, Y., Guishengi, M. et al. (2019). Free manufacturing of heterogeneous materials part by selective laser melting. *Optics and Precision Engineering* 27: 517–526.

91 Liu, X., Wang, X., Wang, E. et al. (2017). Study on ultrasonic response to mechanical structure of coal under loading and unloading condition. *Shock and Vibration* 2017: 7643451. https://doi.org/10.1155/2017/7643451.

92 Huang, H., Mills, J.K., Lu, C., and Sun, D. (2011). A universal piezo-driven ultrasonic cell microinjection system. *Biomedical Microdevices* 13: 743–752. https://doi.org/10.1007/s10544-011-9544-4.

93 Daenotes.com (2016). Ultrasonic waves, Daenotes.Com. http://www.daenotes.com/electronics/industrial-electronics/ultrasonic-waves.

94 Yang, S. and Evans, J.R.G. (2004). Acoustic control of powder dispensing in open tubes. *Powder Technology* 139: 55–60. https://doi.org/10.1016/j.powtec.2003.10.008.

95 Kumar, P., Santosa, J.K., Beck, E., and Das, S. (2004). Direct-write deposition of fine powders through miniature hopper-nozzles for multi-material solid freeform fabrication. *Rapid Prototyping Journal* 10: 14–23. https://doi.org/10.1108/13552540410512499.

96 Yang, Y. and Li, X. (2003). Experimental and analytical study of ultrasonic micro powder feeding. *Journal of Physics D: Applied Physics* 36: 1349–1354. https://doi.org/10.1088/0022-3727/36/11/316.

97 Chianrabutra, S., Mellor, B.G., and Yang, S. (n.d.). A dry powder material delivery device for multiple material additive manufacturing, pp. 36–48.

98 Chianrabutra, S. (2015). *A Dry Powder Micro Delivery Device for Multiple Material Additive Manufacturing*. University of Southampton.

99 Al-Jamal, O.M., Hinduja, S., and Li, L. (2008). Characteristics of the bond in Cu-H13 tool steel parts fabricated using SLM. *CIRP Annals - Manufacturing Technology* 57: 239–242. https://doi.org/10.1016/j.cirp.2008.03.010.

100 Wei, C., Zhang, Z., Cheng, D. et al. (2020). An overview of laser-based multiple metallic material additive manufacturing: from macro- to micro-scales. *International Journal of Extreme Manufacturing* 3 (1): 3. https://doi.org/10.1088/2631-7990/abce04.

101 Gradl, P.R., Protz, C.S., Ellis, D.L., and Greene, S.E. (2019). Progress in additively manufactured copper-alloy GRCOP-84, GRCOP-42, and bimetallic combustion chambers for liquid rocket engines. In: *Proceedings of the International Astronautical Congress, IAC 2019, October*, 21–25.

102 Schein, L.B. (1988). *Electrophotography and development physics*. Springer.

103 Kumar, A.V. and Zhang, H. (1999). Electrophotographic powder deposition for freeform fabrication. In: *10th Solid Freeform Fabrication Symposium*, 647–654.

104 Benning, M.J. and Dalgarno, K. (2018). Proof of concept of a novel combined consolidation and transfer mechanism for electrophotographic 3D printing.

Rapid Prototyping Journal 24: 1040–1048. https://doi.org/10.1108/RPJ-04-2018-0111.

105 Kumar, A.V., Anirban, D., and James, F. (2003). Solid freeform fabrcation by electrophotographic printing. In: *14th Solid Freeform Fabrication Symposium*, 39–49. http://utwired.engr.utexas.edu/lff/symposium/preceedingsArchive/Manuscripts/2003/2003-04-Kumar.pdf.

106 van der Eijk, C., Mugaas, T., and Karlsen, R. (2014). Metal printing process development of a new rapid manufacturing process for metal parts. In: *Proceedings: World PM2004 Conference 20*, vol. 20, 19860–19870. https://doi.org/10.1039/c8cp03103g.

107 Rafiee, M., Farahani, R.D., and Therriault, D. (2020). Multi-material 3D and 4D printing: a survey. *Advancement of Science* 1–26. https://doi.org/10.1002/advs.201902307.

108 Stichel, T., Geißler, B., Jander, J. et al. (2018). Electrophotographic multi-material powder deposition for additive manufacturing. *Journal of Laser Applications* 30: 032306. https://doi.org/10.2351/1.5040619.

109 Stichel, T., Brachmann, C., Raths, M. et al. (2020). Electrophotographic multilayer powder pattern deposition for additive manufacturing. *Journal of Metals* 72: 1366–1375. https://doi.org/10.1007/s11837-019-03965-z.

110 J. Santosa, D. Jing, S. Das, Experimental and numerical study on the flow of fine powders from small-scale hoppers applied to SLS multi-material deposition–part I, *Solid Freeform Fabrication Symposium* (2002) 620–627. https://doi.org/10.26153/tsw/5281, http://edge.rit.edu/edge/P10551/public/SFF/SFF 2002 Proceedings/2002 SFF Papers/70-Santos.pdf.

111 Lappo, K., Wood, K., Bourell, D., and Beaman, J. (2003). Discrete multiple material selective laser sintering (M2SLS): Nozzel design of powder design. pp. 93–108.

112 Pegna, J., Pattofatto, S., Berge, R. et al. (1999). E SAND-PAINTER: two-dimensional powder deposition. In: *Proceedings: Solid Freeform Fabrication Symposium*, 695–710.

113 Girnth, S., Koopmann, J., Klawitter, G. et al. (2019). 3D hybrid-material processing in selective laser melting: implementation of a selective coating system. *Progress in Additive Manufacturing* 4: 399–409. https://doi.org/10.1007/s40964-019-00082-w.

114 Wei, C., Sun, Z., Chen, Q. et al. (2019). Additive manufacturing of horizontal and 3D functionally graded 316L/Cu10Sn components via multiple material selective laser melting. *Journal of Manufacturing Science and Engineering* 141: 1–14. https://doi.org/10.1115/1.4043983.

115 Glasschroeder, J., Prager, E., and Zaeh, M.F. (2015). Powder-bed-based 3D-printing of function integrated parts. *Rapid Prototyping Journal* 21: 775–792.

116 Anstaett, C., Schafnitzel, M., Seidel, C., and Reinhart, G. (2017). Laser-based powder bed fusion of 3d-multi-material-parts of copper-chrome-zirconia and tool steel. Proceedings Euro PM 2017: International Powder Metallurgy Congress and Exhibition.

117 Wei, C., Gu, H., Zhang, X. et al. (2020). Hybrid ultrasonic and mini-motor vibration-induced irregularly shaped powder delivery for multiple materials additive manufacturing. *Additive Manufacturing* 33: 101138. https://doi.org/10.1016/j.addma.2020.101138.

118 Chueh, Y., Wei, C., Zhang, X., and Li, L. (2020). Integrated laser-based powder bed fusion and fused fi lament fabrication for three-dimensional printing of hybrid metal/polymer objects. *Additive Manufacturing* 31: 100928. https://doi.org/10.1016/j.addma.2019.100928.

119 Rittinghaus, S. and Wilms, M.B. (2020). *Laser Fusion of Powder and Foil (LFPF) – a multi material approach to AM*. https://doi.org/10.13140/RG.2.2.24648.96007.

120 Silva, M., Felismina, R., Mateus, A. et al. (2017). Application of a hybrid additive manufacturing methodology to produce a metal/polymer customized dental implant. In: *International Conference on Sustainable Intelligent Manufacturing*, 150–155. Leiria, Portugal: The Author(s) https://doi.org/10.1016/j.promfg.2017.08.019.

121 Nguyen, D.S., Park, H.S., and Lee, C.M. (2019). Applying selective laser melting to join Al and Fe: An investigation of dissimilar materials. *Applied Sciences* 9: 1. https://doi.org/10.3390/app9153031.

122 Tan, C., Zhou, K., and Kuang, T. (2019). Selective laser melting of tungsten-copper functionally graded material. *Materials Letters* 237: 328–331. https://doi.org/10.1016/j.matlet.2018.11.127.

123 Besnea, D., Dontu, O., Avram, M. et al. (2016). Study on laser welding of stainless steel/copper dissimilar materials. *IOP Conference Series: Materials Science and Engineering* 147: 1. https://doi.org/10.1088/1757-899X/147/1/012047.

124 Liu, Z.H., Zhang, D.Q., Sing, S.L. et al. (2014). Interfacial characterization of SLM parts in multi-material processing: Metallurgical diffusion between 316L stainless steel and C18400 copper alloy. *Materials Characterization* 94: 116–125.

125 Chen, J., Yang, Y., Song, C. et al. (2019). Interfacial microstructure and mechanical properties of 316L/CuSn10 multi-material bimetallic structure fabricated by selective laser melting. *Materials Science and Engineering A* 752: 75–85. https://doi.org/10.1016/j.msea.2019.02.097.

126 Chen, K., Wang, C., Hong, Q. et al. (2020). Selective laser melting 316L/CuSn10 multi-materials: Processing optimization, interfacial characterization and mechanical property. *Journal of Materials Processing Technology* 283: 116701. https://doi.org/10.1016/j.jmatprotec.2020.116701.

127 Zhang, M., Yang, Y., Wang, D. et al. (2019). Microstructure and mechanical properties of CuSn/18Ni300 bimetallic porous structures manufactured by selective laser melting. *Materials and Design* 165: 107583. https://doi.org/10.1016/j.matdes.2019.107583.

128 Sing, S.L., Lam, L.P., Zhang, D.Q. et al. (2015). Interfacial characterization of SLM parts in multi-material processing: Intermetallic phase formation between AlSi10Mg and C18400 copper alloy. *Materials Characterization* 107: 220–227.

129 Mhamdi, M.-B., Boujelbene, M., Bayraktar, E., and Zghal, A. (2012). Surface integrity of titanium alloy Ti-6Al-4V in ball end milling. *Physics Procedia* 25: 355–362. https://doi.org/10.1016/j.phpro.2012.03.096.

130 Wei, K., Zeng, X., Li, F. et al. (2020). Microstructure and mechanical property of Ti-5Al-2.5Sn/Ti-6Al-4V dissimilar titanium alloys integrally fabricated by selective laser melting. *Journal of Metals* 72: 1031–1038. https://doi.org/10.1007/s11837-019-03988-6.

131 Wang, P., Lao, C.S., Chen, Z.W. et al. (2020). Microstructure and mechanical properties of Al-12Si and Al-3.5Cu-1.5Mg-1Si bimetal fabricated by selective laser melting. *Journal of Materials Science and Technology* 36: 18–26. https://doi.org/10.1016/j.jmst.2019.03.047.

132 Hengsbach, F., Koppa, P., Joachim, M. et al. (2018). Inline additively manufactured functionally graded multi-materials: microstructural and mechanical characterization of 316L parts with H13 layers. *Progress in Additive Manufacturing* 3: 221–231. https://doi.org/10.1007/s40964-018-0044-4.

133 Bandyopadhyay, A. and Heer, B. (2018). Additive manufacturing of multi-material structures. *Materials Science & Engineering: R: Reports* 129: 1–16.

134 Tey, C.F., Tan, X., Sing, S.L., and Yeong, W.Y. (2020). Additive manufacturing of multiple materials by selective laser melting: Ti- alloy to stainless steel via a Cu-alloy interlayer. *Additive Manufacturing* 31: 100970. https://doi.org/10.1016/j.addma.2019.100970.

135 Beal, V.E., Erasenthiran, P., Hopkinson, N. et al. (2006). The effect of scanning strategy on laser fusion of functionally graded H13/Cu materials. *International Journal of Advanced Manufacturing Technology* 30: 844–852. https://doi.org/10.1007/s00170-005-0130-x.

136 Boley, C.D., Khairallah, S.A., and Rubenchik, A.M. (2014). Modeling of powder absorption in additive manufacturing. *Conference on Lasers and Electro-Optics/Europe – Technical Digest* 15–16. https://doi.org/10.1364/cleo_at.2014.am1l.5.

137 Wei, C., Gu, H., Li, Q. et al. (2020). Understanding of process and material behaviours in additive manufacturing of Invar36/Cu10Sn multiple material components via laser-based powder bed fusion. *Additive Manufacturing* 101683. https://doi.org/10.1016/j.addma.2020.101683.

138 Oliveira, J.P., Lalonde, A.D., and Ma, J. (2020). Processing parameters in laser powder bed fusion metal additive manufacturing. *Materials and Design* 193: 1–12. https://doi.org/10.1016/j.matdes.2020.108762.

139 Okamoto, H., Schlesinger, M.E., and Mueller, E.M. (ed.) (2016). *Binary Alloy Phase Diagrams*. ASM International https://doi.org/10.31399/asm.hb.v03.a0006247.

140 Shojaei Zoeram, A., Mousavi, A.A.A., and Mohsenifar, F. (2016). Effect of welding speed on mechanical and fracture behavior of nitinol laser-joints. *Welding in the World* 60: 11–19. https://doi.org/10.1007/s40194-015-0281-9.

141 Tomashchuk, I., Sallamand, P., Andrzejewski, H., and Grevey, D. (2011). Intermetallics The formation of intermetallics in dissimilar Ti6Al4V/copper/AISI 316

L electron beam and Nd: YAG laser joints. *Intermetallics* 19: 1466–1473. https://doi.org/10.1016/j.intermet.2011.05.016.

142 Mo, D., Song, T., Fang, Y. et al. (2018). A review on diffusion bonding between titanium alloys and stainless steels. *Advances in Materials Science and Engineering* 2018: 8701890. https://doi.org/10.1155/2018/8701890.

143 Ritchie, R.O. (1999). Mechanisms of fatigue-crack propagation in ductile and brittle solids. *International Journal of Fracture* 100: 55–83. https://doi.org/10.1023/A:1018655917051.

144 Hu, Y. and Cong, W. (2018). A review on laser deposition-additive manufacturing of ceramics and ceramic reinforced metal matrix composites. *Ceramics International* 44: 20599–20612. https://doi.org/10.1016/j.ceramint.2018.08.083.

145 Zhao, X., Wei, Q.S., Gao, N. et al. (2019). Rapid fabrication of TiN/AISI 420 stainless steel composite by selective laser melting additive manufacturing. *Journal of Materials Processing Technology* 270: 8–19. https://doi.org/10.1016/j.jmatprotec.2019.01.028.

146 Liu, Y., Tang, M., Hu, Q. et al. (2020). Densification behavior, microstructural evolution, and mechanical properties of TiC/AISI420 stainless steel composites fabricated by selective laser melting. *Materials and Design* 187: 1–13. https://doi.org/10.1016/j.matdes.2019.108381.

147 Trenke, D., Müller, N., and Rolshofen, W. (2006). Selective laser sintering of metal and ceramic compound structures. In: *Intelligent Production Machines and Systems 2nd I*PROMS Virtual International Conference 3–14 July 2006*, 198–203. https://doi.org/10.1016/B978-008045157-2/50039-0.

148 Koopmann, J., Voigt, J., and Niendorf, T. (2019). Additive manufacturing of a steel – ceramic multi-material by selective laser melting. *Metallurgical and Materials Transactions B: Process Metallurgy and Materials Processing Science* 50: 1042–1051. https://doi.org/10.1007/s11663-019-01523-1.

149 Ivers-Tiffée, E. and von Münch, W. (2007). Werkstoffe der Elektrotechnik. *Springer Nature* https://doi.org/10.1007/978-3-8351-9088-7.

150 Wang, R., Gu, D., Chen, C. et al. (2020). Formation mechanisms of TiB2 tracks on Ti6Al4V alloy during selective laser melting of ceramic-metal multi-material. *Powder Technology* 367: 597–607. https://doi.org/10.1016/j.powtec.2020.04.027.

151 Wang, R., Gu, D., Xi, L. et al. (2019). Selective laser melted TiB2/Ti6Al4V graded materials and first-principle calculations. *Materials Letters* 254: 33–36. https://doi.org/10.1016/j.matlet.2019.07.015.

152 Shishkovsky, I., Kakovkina, N., and Sherbakov, V. (2017). Graded layered titanium composite structures with TiB2 inclusions fabricated by selective laser melting. *Composite Structures* 169: 90–96. https://doi.org/10.1016/j.compstruct.2016.11.013.

153 Attar, H., Bönisch, M., Calin, M. et al. (2014). Selective laser melting of in situ titanium-titanium boride composites: processing, microstructure and mechanical properties. *Acta Materialia* 76: 13–22. https://doi.org/10.1016/j.actamat.2014.05.022.

154 Zhang, X., Wei, C., Chueh, Y.H., and Li, L. (2019). An integrated dual ultrasonic selective powder dispensing platform for three-dimensional printing of multiple material metal/glass objects in selective laser melting. *Journal of Manufacturing Science and Engineering, Transactions of the ASME* 141: 1–12. https://doi.org/10.1115/1.4041427.

155 Cheng, D., Wei, C., Huang, Y. et al. (2021). Additive manufacturing of lithium aluminosilicate glass-ceramic/metal 3D electronic components via multiple material laser powder bed fusion. *Additive Manufacturing* 102481. https://doi.org/10.1016/j.addma.2021.102481.

156 Wei, C. and Li, L. (2021). Recent progress and scientific challenges in multi-material additive manufacturing via laser-based powder bed fusion. *Virtual and Physical Prototyping* 16: 1–25. https://doi.org/10.1080/17452759.2021.1928520.

157 Chatham, C.A., Long, T.E., and Williams, C.B. (2019). A review of the process physics and material screening methods for polymer powder bed fusion additive manufacturing. *Progress in Polymer Science* 93: 68–95. https://doi.org/10.1016/j.progpolymsci.2019.03.003.

158 Verma, R. and Kaushal, G. (2019). State of the art of powder bed fusion additive manufacturing: a review BT – 3D printing and additive manufacturing technologies. In: (ed. L.J. Kumar, P.M. Pandey and D.I. Wimpenny), 269–279. Singapore: Springer Singapore https://doi.org/10.1007/978-981-13-0305-0_23.

159 Tang, S., Cheng, C., Yeh, R., and Hsu, R. (2018). Direct joining of 3D-printed thermoplastic parts to SLM-fabricated metal cellular structures by ultrasonic welding. *International Journal of Advanced Manufacturing Technology* 99: 729–736.

160 Hopmann, C., Kreimeier, S., Keseberg, J., and Wenzlau, C. (2016). Joining of metal-plastics-hybrid structures using laser radiation by considering the surface structure of the metal. *Journal of Polymers* 2016: 4734913. https://doi.org/10.1155/2016/4734913.

161 Chueh, Y., Zhang, X., Ke, J.C.-R. et al. (2020). Additive manufacturing of hybrid metal/polymer objects via multiple-material laser powder bed fusion. *Additive Manufacturing* 2.

162 Wang, X., Jiang, M., Zhou, Z. et al. (2017). 3D printing of polymer matrix composites: a review and prospective. *Composites. Part B, Engineering* 110: 442–458. https://doi.org/10.1016/j.compositesb.2016.11.034.

163 Parandoush, P. and Lin, D. (2017). A review on additive manufacturing of polymer-fiber composites. *Composite Structures* 182: 36–53. https://doi.org/10.1016/j.compstruct.2017.08.088.

164 Bekas, D.G., Hou, Y., Liu, Y., and Panesar, A. (2019). 3D printing to enable multifunctionality in polymer-based composites: a review. *Composites. Part B, Engineering* 179: 107540. https://doi.org/10.1016/j.compositesb.2019.107540.

165 Yuan, S., Shen, F., Chua, C.K., and Zhou, K. (2019). Polymeric composites for powder-based additive manufacturing: Materials and applications. *Progress in Polymer Science* 91: 141–168. https://doi.org/10.1016/j.progpolymsci.2018.11.001.

166 Shishkovsky, I.V., Scherbakov, V.I., Saraeva, I.N., and Ionin, A.A. (2017). Thermoelectric properties of gradient polymer composites with nano-inclusions fabricated by laser assisted sintering. *Laser Physics Letters* 14: 35601. https://doi.org/10.1088/1612-202x/aa58bf.

167 Shishkovsky, I. and Morozov, Y. (2013). Electrical and magnetic properties of multilayer polymer structures with nano inclusions as prepared by selective laser sintering. *Journal of Nanoscience and Nanotechnology* 13: 1440–1443. https://doi.org/10.1166/jnn.2013.6010.

168 Chung, H. and Das, S. (2006). Processing and properties of glass bead particulate-filled functionally graded Nylon-11 composites produced by selective laser sintering. *Materials Science and Engineering A* 437: 226–234. https://doi.org/10.1016/j.msea.2006.07.112.

169 Chung, H. and Das, S. (2008). Functionally graded Nylon-11/silica nanocomposites produced by selective laser sintering. *Materials Science and Engineering A* 487: 251–257. https://doi.org/10.1016/j.msea.2007.10.082.

170 Chueh, Y.-H., Zhang, X., Wei, C. et al. (2020). Additive manufacturing of polymer-metal/ceramic functionally graded composite components via multiple material laser powder bed fusion. *Journal of Manufacturing Science and Engineering* 142: 14. https://doi.org/10.1115/1.4046594.

171 Amoabeng, D. and Velankar, S.S. (2018). A review of conductive polymer composites filled with low melting point metal alloys. *Polymer Engineering and Science* 58: 1010–1019. https://doi.org/10.1002/pen.24774.

172 Kyogoku, H. and Ikeshoji, T.-T. (2020). A review of metal additive manufacturing technologies: Mechanism of defects formation and simulation of melting and solidification phenomena in laser powder bed fusion process. *Mechanical Engineering - Reviews* 7: 19–182. https://doi.org/10.1299/mer.19-00182.

173 Mohanty, S. and Hattel, J.H. (2017). Laser additive manufacturing of multimaterial tool inserts: a simulation-based optimization study. *Laser 3D Manufacturing IV* 10095: 100950G. https://doi.org/10.1117/12.2253600.

174 Tan, J.H.K., Sing, S.L., and Yeong, W.Y. (2020). Microstructure modelling for metallic additive manufacturing: a review. *Virtual and Physical Prototyping* https://doi.org/10.1080/17452759.2019.1677345.

175 Dai, D. and Gu, D. (2015). Tailoring surface quality through mass and momentum transfer modeling using a volume of fluid method in selective laser melting of TiC/AlSi10Mg powder. *International Journal of Machine Tools and Manufacture* 88: 95–107. https://doi.org/10.1016/j.ijmachtools.2014.09.010.

176 Dai, D. and Gu, D. (2016). Influence of thermodynamics within molten pool on migration and distribution state of reinforcement during selective laser melting of AlN/AlSi10Mg composites. *International Journal of Machine Tools and Manufacture* 100: 14–24. https://doi.org/10.1016/j.ijmachtools.2015.10.004.

177 Chen, C., Gu, D., Dai, D. et al. (2019). Laser additive manufacturing of layered TiB2/Ti6Al4V multi-material parts: understanding thermal behavior evolution. *Optics and Laser Technology* 119: 105666. https://doi.org/10.1016/j.optlastec.2019.105666.

178 Yuan, P. and Gu, D. (2015). Molten pool behaviour and its physical mechanism during selective laser melting of TiC/AlSi10Mg nanocomposites: simulation and experiments. *Journal of Physics D: Applied Physics* 48: 1. https://doi.org/10.1088/0022-3727/48/3/035303.

179 Sorkin, A., Tan, J.L., and Wong, C.H. (2017). Multi-material modelling for selective laser melting. *Procedia Engineering* 216: 51–57. https://doi.org/10.1016/j.proeng.2018.02.088.

180 Gu, H., Wei, C., Li, L. et al. (2020). Multi-physics modelling of molten pool development and track formation in multi-track, multi-layer and multi-material selective laser melting. *International Journal of Heat and Mass Transfer* 151: 119458. https://doi.org/10.1016/j.ijheatmasstransfer.2020.119458.

181 Sun, Z., Chueh, Y.-H., and Li, L. (2020). Multiphase mesoscopic simulation of multiple and functionally gradient materials laser powder bed fusion additive manufacturing processes. *Additive Manufacturing* 35: 101448. https://doi.org/10.1016/j.addma.2020.101448.

182 Gu, H., Wei, C., Li, L. et al. (n.d.). Numerical and experimental study of molten pool behaviour and defect formation in multi-material and functionally graded materials laser powder bed fusion. *Advanced Powder Technology*. ADVPT-D-21.

183 Dobosz, A. and Gancarz, T. (2018). Reference data for the density, viscosity, and surface tension of liquid Al–Zn, Ag–Sn, Bi–Sn, Cu–Sn, and Sn–Zn eutectic alloys. *Journal of Physical and Chemical Reference Data* 47: 13102. https://doi.org/10.1063/1.5010151.

184 Sato, Y., Sugisawa, K., Aoki, D., and Yamamura, T. (2005). Viscosities of Fe–Ni, Fe–Co and Ni–Co binary melts. *Measurement Science and Technology* 16: 363–371. https://doi.org/10.1088/0957-0233/16/2/006.

185 Brillo, J. and Egry, I. (2005). Surface tension of nickel, copper, iron and their binary alloys. *Journal of Materials Science* 40: 2213–2216. https://doi.org/10.1007/s10853-005-1935-6.

186 Boley, C.D., Khairallah, S.A., and Rubenchik, A.M. (2017). Calculation of laser absorption by metal powders in additive manufacturing. In: *Additive Manufacturing Handbook: Product Development for the Defense Industry*, vol. 54, 507–517. https://doi.org/10.1201/9781315119106.

187 Cunningham, C.R., Flynn, J.M., Shokrani, A. et al. (2018). Invited review article: strategies and processes for high quality wire arc additive manufacturing. *Additive Manufacturing* 22: 672–686. https://doi.org/10.1016/j.addma.2018.06.020.

188 McAndrew, A.R., Alvarez Rosales, M., Colegrove, P.A. et al. (2018). Interpass rolling of Ti-6Al-4V wire + arc additively manufactured features for microstructural refinement. *Additive Manufacturing* 21: 340–349: https://doi.org/10.1016/j.addma.2018.03.006.

189 Qi, Z., Cong, B., Qi, B. et al. (2018). Microstructure and mechanical properties of double-wire + arc additively manufactured Al-Cu-Mg alloys. *Journal of Materials Processing Technology* 255: 347–353. https://doi.org/10.1016/j.jmatprotec.2017.12.019.

190 Shen, Q., Kong, X., and Chen, X. (2021). Fabrication of bulk Al-Co-Cr-Fe-Ni high-entropy alloy using combined cable wire arc additive manufacturing

(CCW-AAM): microstructure and mechanical properties. *Journal of Materials Science and Technology* 74: 136–142. https://doi.org/10.1016/j.jmst.2020.10.037.

191 Buchanan, C. and Gardner, L. (2019). Metal 3D printing in construction: a review of methods, research, applications, opportunities and challenges. *Engineering Structures* 180: 332–348. https://doi.org/10.1016/j.engstruct.2018.11.045.

192 Chen, X., Su, C., Wang, Y. et al. (2018). Cold metal transfer (CMT) based wire and arc additive manufacture (WAAM) system. *Journal of Surface Investigation: X-ray, Synchrotron and Neutron Techniques* 12: 1278–1284.

193 Gordon, T.W., Bednarz, B., Neller, M.K., Fanning, M.P., Lathabai, S. (2008). *Consumable Electrode Arc Welding*

194 Shen, C., Hua, X., Reid, M. et al. (2020). Thermal induced phase evolution of Fe–Fe3Ni functionally graded material fabricated using the wire-arc additive manufacturing process: an in-situ neutron diffraction study. *Journal of Alloys and Compounds* 826: 154097.

195 Rodrigues, T.A., Duarte, V., Miranda, R.M. et al. (2019). Current status and perspectives on wire and arc additive manufacturing (WAAM). *Materials (Basel).* 12: https://doi.org/10.3390/ma12071121.

196 Wang, J., Pan, Z., Ma, Y. et al. (2018). Characterization of wire arc additively manufactured titanium aluminide functionally graded material: microstructure, mechanical properties and oxidation behaviour. *Materials Science and Engineering A* 734: 110–119.

197 Yang, Z., Liu, Q., Wang, Y. et al. (2020). Fabrication of multi-element alloys by twin wire arc additive manufacturing combined with in-situ alloying. *Materials Research Letters* 8: 477–482.

198 Tian, Y., Shen, J., Hu, S. et al. (2019). Wire and arc additive manufactured Ti–6Al–4V/Al–6.25 Cu dissimilar alloys by CMT-welding: effect of deposition order on reaction layer. *Science and Technology of Welding and Joining* .

199 Shen, C., Pan, Z., Cuiuri, D. et al. (2016). Fabrication of Fe-FeAl functionally graded material using the wire-arc additive manufacturing process. *Metallurgical and Materials Transactions B: Process Metallurgy and Materials Processing Science* 47: 763–772. https://doi.org/10.1007/s11663-015-0509-5.

200 Shen, C. (2016). *Application of Wire-Arc Additive Manufacturing (WAAM) Process in In-situ Fabrication of Iron Aluminide Structures*. University of Wollongong.

201 Dong, B., Pan, Z., Shen, C. et al. (2017). Fabrication of copper-rich Cu-Al alloy using the wire-arc additive manufacturing process. *Metallurgical and Materials Transactions B: Process Metallurgy and Materials Processing Science* 48: 3143–3151. https://doi.org/10.1007/s11663-017-1071-0.

202 Ahsan, M.R.U., Tanvir, A.N.M., Seo, G.J. et al. (2020). Heat-treatment effects on a bimetallic additively-manufactured structure (BAMS) of the low-carbon steel and austenitic-stainless steel. *Elsevier* https://doi.org/10.1016/j.addma.2020.101036.

203 Chandrasekaran, S., Hari, S., and Amirthalingam, M. (2020). Wire arc additive manufacturing of functionally graded material for marine risers. *Materials Science and Engineering A* 792: 139530.

204 Han, S., Zhang, Z., Liu, Z. et al. (2020). Investigation of the microstructure and mechanical performance of bimetal components fabricated using CMT-based wire arc additive manufacturing. *Materials Research Express* 7: 116525.

205 Liu, L., Zhuang, Z., Liu, F., and Zhu, M. (2013). Additive manufacturing of steel-bronze bimetal by shaped metal deposition: Interface characteristics and tensile properties. *International Journal of Advanced Manufacturing Technology* 69: 2131–2137. https://doi.org/10.1007/s00170-013-5191-7.

206 Kannan, A.R., Kumar, S.M., Kumar, N.P. et al. (2020). Process-microstructural features for tailoring fatigue strength of wire arc additive manufactured functionally graded material of SS904L and hastelloy C-276. *Materials Letters* 274: 127968.

207 Rajesh Kannan, A., Mohan Kumar, S., Pramod, R. et al. (2020). Microstructure and mechanical properties of wire arc additive manufactured bi-metallic structure. *Science and Technology of Welding and Joining* 26: 1–11.

208 Abe, T. and Sasahara, H. (2016). Dissimilar metal deposition with a stainless steel and nickel-based alloy using wire and arc-based additive manufacturing. *Precision Engineering* 45: 387–395.

209 Abe, T., Mori, D., Sonoya, K. et al. (2019). Control of the chemical composition distribution in deposited metal by wire and arc-based additive manufacturing. *Precision Engineering* 55: 231–239. https://doi.org/10.1016/j.precisioneng.2018.09.010.

210 Wu, B., Pan, Z., Qiu, Z. et al. (2020). Enhanced interface strength in steel-nickel bimetallic component fabricated using wire arc additive manufacturing with interweaving deposition strategy. *Journal of Materials Science and Technology* 52: 226–234. https://doi.org/10.1016/j.jmst.2020.04.019.

211 Shen, C., Liss, K.-D., Reid, M. et al. (2020). Fabrication of FeNi intermetallic using the wire-arc additive manufacturing process: A feasibility and neutron diffraction phase characterization study. *Journal of Manufacturing Processes* 57: 691–699.

212 Reisgen, U., Sharma, R., and Oster, L. (2019). Plasma Multiwire Technology with Alternating Wire Feed for Tailor-Made Material Properties in Wire and Arc Additive Manufacturing. *Metals (Basel)* 9: 14. https://doi.org/10.3390/met9070745.

213 Marinelli, G., Martina, F., Lewtas, H. et al. (2019). Functionally graded structures of refractory metals by wire arc additive manufacturing. *Science and Technology of Welding and Joining* 24: 495–503. https://doi.org/10.1080/13621718.2019.1586162.

214 Huckstepp, A. (2019). *Surface Roughness – A Guide to Metal Additive Manufacturing by Digital Alloys*. Manufactur3dmag. https://manufactur3dmag.com/surface-roughness-a-guide-to-metal-additive-manufacturing-by-digital-alloys/.

215 Zhang, Y., Wu, L., Guo, X. et al. (2018). Additive manufacturing of metallic materials: a review. *Journal of Materials Engineering and Performance* 27: 1–13. https://doi.org/10.1007/s11665-017-2747-y.

216 Luo, X., Chang, W., Zhong, W., and Wardle, F. (2018). Chapter 8 - Hybrid machine tool design. In: *Hybrid Mach* (ed. X. Luo and Y.B.T.-H.M. Qin), 197–214. Academic Press https://doi.org/10.1016/B978-0-12-813059-9.00008-7.

217 Wong, H., Dawson, K., Ravi, G.A. et al. (2019). Multi-laser powder bed fusion benchmarking—initial trials with Inconel 625. *International Journal of Advanced Manufacturing Technology* 105: 2891–2906. https://doi.org/10.1007/s00170-019-04417-3.

218 Khorasani, A.M., Gibson, I., Veetil, J.K., and Ghasemi, A.H. (2020). A review of technological improvements in laser-based powder bed fusion of metal printers. *International Journal of Advanced Manufacturing Technology* 108: 191–209. https://doi.org/10.1007/s00170-020-05361-3.

219 Silva, M., Mateus, A., Oliveira, D., and Malça, C. (2017). An alternative method to produce metal/plastic hybrid components for orthopedics applications. *Proceedings of the Institution of Mechanical Engineers, Part L: Journal of Materials: Design and Applications* 231: 179–186. https://doi.org/10.1177/1464420716664545.

220 Wei, C., Sun, Z., Huang, Y., and Li, L. (2018). Embedding anti-counterfeiting features in metallic components via multiple material additive manufacturing. *Additive Manufacturing* 24: 1–12. https://doi.org/10.1016/j.addma.2018.09.003.

221 Aerosint (n.d.). Aerosint, Aerosint.Com/. https://aerosint.com/.

222 Wei, C., Chueh, Y.-H., Zhang, X. et al. (2019). Easy-to-remove composite support material and procedure in additive manufacturing of metallic components using multiple material laser-based powder bed fusion. *Journal of Manufacturing Science and Engineering* 141: 1–18. https://doi.org/10.1115/1.4043536.

223 Hou, S., Qi, S., Hutt, D.A. et al. (2018). Three dimensional printed electronic devices realised by selective laser melting of copper/high-density-polyethylene powder mixtures. *Journal of Materials Processing Technology* 254: 310–324. https://doi.org/10.1016/j.jmatprotec.2017.11.020.

224 Löffelmann, F. and Šplíchal, J. (2018). Design study of the heat switch base plate with single and multi-material topology. *Research and Education in Aircraft Design* 1–12.

225 Suh, J., Shapiro, A.A., Liu, Z., and Borgonia, J. (2014). https://doi.org/10.1038/srep05357). Developing gradient metal alloys through radial deposition additive manufacturing. *Scientific Reports* 4: 5357.

226 Locci, I.E., Bowman, C.L., Gabbs, T.P. et al. (2009). Development of high temperature dissimilar joint technology for fission surface power systems. In: *4th International Brazing and Soldering Conference*, 165–175.

227 Akiba, M., Araki, M., and Yokoyama, K. (1992). Thermal cycling experiments of monoblock divertor modules for fusion experimental reactors AU. In: *Fusion Technology* (ed. S. Suzuki), 1858–1862. Taylor & Francis https://doi.org/10.13182/FST92-A29989.

228 Chopde, R.S. (2017). Study on laser beam welding of copper and aluminum joint. *Journal of Mechanical and Civil Engineering* 17: 65–74. https://doi.org/10.9790/1684-17010056574.

229 Terrazas, C.A., Gaytan, S.M., Rodriguez, E. et al. (2013). Multi-material metallic structure fabrication using electron beam melting. *International Journal of Advanced Manufacturing Technology* 71: 33–45. https://doi.org/10.1007/s00170-013-5449-0.

230 Bohidar, S.K., Sharma, R., and Mishra, P.R. (2014). Functionally graded materials: a critical review. *International Journal of Research* 1: 289–301.

231 Marcu, T., Todea, M., Maines, L. et al. (2012). Metallurgical and mechanical characterisation of titanium based materials for endosseous applications obtained by selective laser melting. *Powder Metallurgy* 55: 309–314. https://doi.org/10.1179/1743290112Y.0000000007.

232 Qian, X. and Dutta, D. (2003). Design of heterogeneous turbine blade. *CAD: Computed Aided Designing* 35: 319–329. https://doi.org/10.1016/S0010-4485(01)00219-6.

233 Lee, M. and Yoo, Y.H. (2001). Analysis of ceramic/metal armour systems. *International Journal of Impact Engineering* 25: 819–829. https://doi.org/10.1016/S0734-743X(01)00025-2.

234 Maciejewski, G. and Mróz, Z. (2013). Optimization of functionally gradient materials in valve design under cyclic thermal and mechanical loading. *Computer Assisted Methods in Engineering and Science* 20: 99–112.

235 Teo, E.Y., Ong, S.-Y., Chong, M.S.K. et al. (2011). Polycaprolactone-based fused deposition modeled mesh for delivery of antibacterial agents to infected wounds. *Biomaterials* 32: 279–287. https://doi.org/10.1016/j.biomaterials.2010.08.089.

236 Weisman, J.A., Nicholson, J.C., Tappa, K. et al. (2015). Antibiotic and chemotherapeutic enhanced three-dimensional printer filaments and constructs for biomedical applications. *International Journal of Nanomedicine* 10: 357–370. https://doi.org/10.2147/IJN.S74811.

237 David, C. and Weber, P. (2013). Ultrasonic bonding of discrete plastic parts to metal, US 8546702 B2.

238 Bai, Y., Zhang, J., Zhao, C. et al. (2020). Dual interfacial characterization and property in multi-material selective laser melting of 316L stainless steel and C52400 copper alloy. *Materials Characterization* 167: 110489. https://doi.org/10.1016/j.matchar.2020.110489.

239 Mohd Yusuf, S., Zhao, X., Yang, S., and Gao, N. (2021). Interfacial characterisation of multi-material 316L stainless steel/Inconel 718 fabricated by laser powder bed fusion. *Materials Letters* 284: 128928. https://doi.org/10.1016/j.matlet.2020.128928.

240 Shakerin, S., Hadadzadeh, A., Amirkhiz, B.S. et al. (2019). Additive manufacturing of maraging steel-H13 bimetals using laser powder bed fusion technique. *Additive Manufacturing* 29: 100797. https://doi.org/10.1016/j.addma.2019.100797.

241 Yuan-Hui, C., Xiaoji, Z., Chao, W. et al. (2020). Additive manufacturing of polymer-metal/ceramic functionally graded composite components via multiple material laser powder bed fusion. *Journal of Manufacturing Science and Engineering, Transactions of the ASME* DRAFT-MANU-19-1418-0.2019.

242 Grasso, M. and Colosimo, B.M. (2017). Process defects and in situ monitoring methods in metal powder bed fusion: a review. *Measurement Science and Technology* 28: 6–26. https://doi.org/10.1088/1361-6501/aa5c4f.

243 Bobbio, L.D., Otis, R.A., Paul, J. et al. (2017). Acta materialia additive manufacturing of a functionally graded material from Ti-6Al-4V to Invar: experimental characterization and thermodynamic calculations. *Acta Materialia* 127: 133–142. https://doi.org/10.1016/j.actamat.2016.12.070.

244 Bobbio, L.D., Bocklund, B., Otis, R. et al. (2018). Experimental analysis and thermodynamic calculations of an additively manufactured functionally graded material of v to Invar 36. *Journal of Materials Research* 33: 1642–1649. https://doi.org/10.1557/jmr.2018.92.

245 Atasoy, E. and Kahraman, N. (2008). Diffusion bonding of commercially pure titanium to low carbon steel using a silver interlayer. *Materials Characterization* 59: 1481–1490. https://doi.org/10.1016/j.matchar.2008.01.015.

246 Wang, T., Zhang, B., Chen, G. et al. (2010). Electron beam welding of Ti-15-3 titanium alloy to 304 stainless steel with copper interlayer sheet. *Transactions of the Nonferrous Metals Society of China* 20: 1829–1834. https://doi.org/10.1016/S1003-6326(09)60381-2.

247 Petrovskiy, P., Doubenskaia, M., Sova, A., and Travyanov, A. (2020). Analysis of copper-tungsten cold spray coating: kinetics of coating formation and its thermal properties. *Surface and Coatings Technology* 385: 125376. https://doi.org/10.1016/j.surfcoat.2020.125376.

248 Simonelli, M., Aboulkhair, N.T., Cohen, P. et al. (2018). A comparison of Ti-6Al-4V in-situ alloying in selective laser melting using simply-mixed and satellited powder blend feedstocks. *Materials Characterization* 143: 118–126. https://doi.org/10.1016/j.matchar.2018.05.039.

249 Vora, P., Mumtaz, K., Todd, I., and Hopkinson, N. (2015). AlSi12 in-situ alloy formation and residual stress reduction using anchorless selective laser melting. *Additive Manufacturing* 7: 12–19. https://doi.org/10.1016/j.addma.2015.06.003.

250 Yadroitsev, I., Krakhmalev, P., and Yadroitsava, I. (2017). Titanium alloys manufactured by in situ alloying during laser powder bed fusion. *Journal of Metals* 69: 2725–2730. https://doi.org/10.1007/s11837-017-2600-7.

251 Aboulkhair, N.T., Maskery, I., Tuck, C. et al. (2016). The microstructure and mechanical properties of selectively laser melted AlSi10Mg: The effect of a conventional T6-like heat treatment. *Materials Science and Engineering A* 667: 139–146. https://doi.org/10.1016/j.msea.2016.04.092.

252 C. Seidel, C. Anstaett, M. Horn, M. Binder, 2018 *Status-quo der metallischen Multimaterialverarbeitung mittels Laserstrahlschmelzen, in: DGM-Fachtagung Werkstoffe Und Addit*. Fert. - Conf.

253 Chivel, Y. (2016). New approach to multi-material processing in selective laser melting. *Physics Procedia* 83: 891–898. https://doi.org/10.1016/j.phpro.2016.08.093.

254 Woidasky, J. (2017). Recyclingtechnik – Fachbuch für Lehre und Praxis. Von H. Martens, D. Goldmann. *Chemie Ingenieur Technik* 89: 346. https://doi.org/10.1002/cite.201790016.

255 Ullrich, H. (1967). *Mechanische Verfahrenstechnik: Berechnung und Projektierung*. Heidelberg: https://books.google.co.uk/books?id=KTrrAAAAMAAJ.

256 Wu, B., Pan, Z., Ding, D. et al. (2018). A review of the wire arc additive manufacturing of metals: properties, defects and quality improvement. *Journal of Manufacturing Processes* 35: 127–139. https://doi.org/10.1016/j.jmapro.2018.08.001.

257 Wu, B., Ding, D., Pan, Z. et al. (2017). Effects of heat accumulation on the arc characteristics and metal transfer behavior in wire arc additive manufacturing of Ti6Al4V. *Journal of Materials Processing Technology* 250: 304–312. https://doi.org/10.1016/j.jmatprotec.2017.07.037.

5

Modified Inherent Strain Method for Predicting Residual Deformation and Stress in Metal Additive Manufacturing

Albert C. To, Xuan Liang, and Wen Dong

Department of Mechanical Engineering and Materials Science, University of Pittsburgh, 3700 O'Hara St, Pittsburgh PA, 15261, United States

5.1 Background

Part-scale thermomechanical simulations for the metal additive manufacturing (AM) process to compute residual stress and deformation can be very time-consuming, since it is a long-time scale problem involving transient heat transfer, nonlinear mechanical deformation, and addition of a large amount of material by melting and solidification [1–3]. In order to enable fast yet accurate process simulation, increasing attention has been focused on the inherent strain (IS) method. Several decades ago, the IS method was developed and established to allow for fast estimation of residual distortion/stress in the metal welding process [4–8]. Both simulation and experimental results validated the accuracy and efficiency of the IS method for predicting residual distortion/stress of simple metal welding. Another computationally efficient method inspired by the IS theory is the method of applied plastic strain [9, 10]. It utilized a two-dimensional (2D) elastoplastic simulation to calculate the plastic strain and then applied it as an equivalent mechanical load to the three-dimensional (3D) model to evaluate residual deformation/stress. Both the IS and applied plastic strain method are capable of significantly reducing the computational cost of thermomechanical simulation [10, 11].

However, the physical process of AM is much more complicated compared with the simple metal welding process. Attempts to directly apply the original IS method or the applied plastic strain method to the AM process have failed to obtain accurate residual distortion or stress of builds with multiple deposition layers. In recent years, some other approaches based on the IS theory have been developed, and the computed ISs are imposed onto a macroscale model to obtain mechanical deformation [12–14]. These methods are easy to implement and computationally efficient, but it is a challenge to achieve good accuracy in predicting residual distortion at part scale [15].

Therefore, the modified inherent strain (MIS) method is proposed to compute ISs accurately in a straightforward way from the detailed AM process simulation based

Additive Manufacturing Technology: Design, Optimization, and Modeling, First Edition. Edited by Kun Zhou.

on a moving point-wise heat source model and implement fast part-scale simulation for residual deformation and stress prediction. Accordingly, the entire workflow of the MIS method includes three steps. First, a small-scale detailed process simulation is performed to observe the typical stress and strain evolution history in the metal AM process. Second, ISs are extracted according to the MIS model proposed. Then finally, the extracted ISs are assigned to the part-scale model in the sequential layer-wise quasi-static analysis in order to obtain the residual stress and distortion in the final metal builds. There are two main differences between the original and modified IS method: (i) the original method extracts strains from the as-built part to get the IS, while the modified method calculates the IS by using strains not only from the as-built part but also from the part during deposition; (ii) the original method applied the IS to the numerical model with elastic material behavior, while the modified method adopts elastoplastic material behavior. Details regarding the modified method will be described later.

The goal of this book chapter is to review recent progress of the MIS method for residual deformation and stress modeling for the metal AM process using a heat source. The remainder of this chapter is organized as follows. In Section 5.2, the modified theory underlying the MIS method is introduced. Taking the laser powder bed fusion (L-PBF) process as an example, the approach in which the ISs are extracted from a small-scale detailed process simulation is presented in Section 5.3. The procedure for implementing the layer-wise MIS-based simulation of a part-scale model is given in Section 5.4. Experimental validations will be presented to illustrate the performance of the MIS method in Section 5.5. Some simulation-based design examples will be presented in Section 5.6. Finally, summary and outlook are given in Section 5.7.

5.2 Modified Inherent Strain (MIS) Method

5.2.1 Theory for Modification

In the original IS method proposed for fast simulation of the welding process, a key assumption is that the residual elastic strains are insignificant in large regions of the weld structures excluding the weld lines compared to the residual plastic strains in the weld-affected zones [6–8]. With respect to this assumption, all the residual elastic strains will vanish when the constraint like a four-corner-clamped boundary is removed for the weld structures. The residual plastic strains generated from the welding process will be kept in the final cooled welded parts. Since ISs are defined as the inelastic strains in the welded parts, the ISs become equal to the residual plastic strains in the original IS method [6–8]. For metal AM process simulation, however, the aforementioned assumption that the elastic strains in main regions are insignificant is invalid and inaccurate when modeling residual stress and deformation is considered for AM parts. This is because the AM process is much more complex than the welding process. On the theoretical side, the key assumption that the elastic strains in the main regions generated by the welding process vanish is generally

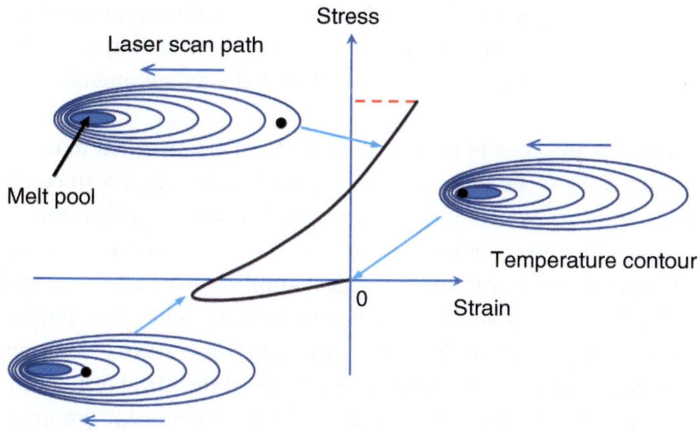

Figure 5.1 Mechanical strain induced at a concerned material point during a simple single-track welding process.

not valid in large AM parts. During the AM process, new depositions can become mechanically constrained by previous layers, and new mechanical boundaries continue to emerge with deposition of those new tracks and layers. After multiple layers of materials are printed, the mechanical constraints in the previously deposited layers can be considered as reaching a steady state. The residual elastic strains in the AM process will go into distortion of the remaining section of the AM build. Thus, to make the IS method valid for the AM process, the ISs should contain a term related to elastic strain evolution, in addition to the evolving plastic strains.

In order to overcome these issues, the MIS model is proposed to estimate the ISs based on detailed process simulation results [16]. Figure 5.1 schematically illustrates the stress–strain history induced at a concerned material point (see black dot), where the heat source passes through and creates a heat-affected zone (HAZ). The starting point indicates the moment when the material point of interest experiences the highest temperature due to the laser heating. The concerned material point is assumed to be both stress-free and strain-free in the beginning (right panel, Figure 5.1). A large temperature gradient will appear in front of the heat source center [17–19]. As the heat source moves away to the left side (lower panel, Figure 5.1), the material point of interest cools down and solidifies rapidly, experiencing a significant amount of shrinkage (compressive strain) but very small compressive stress because the yield strength is small at elevated temperatures.

As the heat source moves further away, the cooling process at the point of interest slows down, while the material ahead has just solidified and is undergoing large shrinkage, which as a result induces nonlinear tensile stress and strain. The tensile stress would reach a maximum point in the final state (upper panel, Figure 5.1). If this were a simple welding problem, the ISs can be obtained by relaxing the solid to a stress-free state and then the resulting compressive plastic strains are used as the ISs according to the original IS method [4, 6]. However, different from the simple welding problem, the elastic strains in each deposited layer in an AM process can be significantly affected by the evolving mechanical boundaries as additional materials are being deposited in a layer-wise manner. The effect due to the evolving mechanical boundaries on the ISs must be accounted for, leading to the development of the MIS method.

In the MIS method, for the purpose of computing the ISs at a concerned material point induced during the AM process, two distinct thermal states, an intermediate state and a steady state, are introduced and defined in the following. The intermediate state is defined as the state when the heat source just passes by. In this state, the local compressive mechanical strain reaches the largest amplitude (c.f., lower panel, Figure 5.1). Meanwhile, the steady state corresponds to the state when the temperature of the whole metal component cools to the ambient temperature (see upper panel of Figure 5.1), usually taken as the room temperature. In the MIS method, the ISs are defined as the difference between the total mechanical strains at the intermediate state and the residual elastic strains at the steady state. Therefore, the IS ε^* in the MIS method can be expressed mathematically by the following equation:

$$\varepsilon^* = \varepsilon_{t_I}^{\text{Mechanical}} - \varepsilon_{t_S}^{\text{Elastic}} \tag{5.1}$$

where t_I and t_S denote the intermediate and steady state, respectively. The two distinct states are the new features introduced for the proposed MIS method.

To facilitate understanding, another form of Eq. (5.1) can be written as:

$$\varepsilon^* = \varepsilon_{t_I}^{\text{Plastic}} + \varepsilon_{t_I}^{\text{Elastic}} - \varepsilon_{t_S}^{\text{Elastic}} \tag{5.2}$$

where $\varepsilon_{t_I}^{\text{Plastic}}$ and $\varepsilon_{t_I}^{\text{Elastic}}$ represent the plastic and elastic strains at the intermediate state, respectively, while $\varepsilon_{t_S}^{\text{Elastic}}$ represents the elastic strains at the steady state.

As the earlier-modified theory suggests, the residual stress and deformation are inherently related to heat input-induced strains and largely influenced by the interlayer effects from the evolving mechanical boundaries as new layers are deposited onto the already deposited layers. The ISs are employed to quantitatively capture the residual stress and deformation considering the thermal history and interlayer constraint effect in the AM process. Regarding numerical technique, the aforementioned MIS formulation will be employed to extract typical IS values from a small-scale detailed thermomechanical simulation for a specific AM process and material combination. And then the computed IS values are applied to the part-scale AM component to predict the residual stress/deformation profile. This method has been proven to be valid when the process has a laser scanning pattern that repeats itself periodically, as will be demonstrated for the L-PBF process next. Specifically, as a popular L-PBF process, the direct laser metal sintering (DMLS) technique is considered in this chapter for the model validation purpose.

5.2.2 Remarks on the IS Method

The original ISs are considered as the incompatible strains in a deformed structure, including plastic strains, creep strains, and strains caused by phase transformation [6]. These incompatible strains are the reason for the residual stress. In other words, from a mechanical viewpoint, if a metal part has no incompatible strains, there will be no residual stress inside. Consider a structure that undergoes multiple thermal cycles where only elastic and thermal strains are involved. Once the structure recovers its initial temperature, all the strains vanish, and no residual stress exists. However, the incompatible strains do not guarantee the formation of residual stress because it also depends on boundary conditions. For example, if a body somehow possesses creep strains or phase transformation-induced strains but can deform freely (no external and internal constraints), there should be no residual stress inside the body.

The basic idea of the original IS method is that by applying incompatible strains determined from a deformed structure to its undeformed counterpart with the same boundary conditions, the residual stress and deformation distribution caused by the welding process can be simulated through a static mechanical analysis. The successful application of the original IS method in the welding field is mainly based on two assumptions: (i) the weld is long enough (or the welding speed is sufficiently slow) so that the melt pool is stable when it moves, and the temperature distribution near the melt pool is uniform along the weld length; (ii) the cross section of the weld is way smaller than that of the base plate so that the melt pool dimension is relatively small compared to the base plate, and therefore the boundary conditions of the weld and surrounding HAZ can be considered unchanged along the weld length.

These two assumptions ensure that: (i) the ISs are uniform along the weld length; (ii) from a cross-sectional perspective, the ISs only exist in a small area around the weld; and (iii) the magnitude and distribution of the ISs are simple and mainly affected by the weld parameters, such as the heat source, the joint geometry, and material properties of the weld and base metal. Therefore, it is possible to mathematically express the distribution of the ISs for a welded structure based on the weld parameters. Further, these ISs are quite universal and applicable to predicting the residual stress distribution of welded structures involving similar weld parameters.

Compared with the success in the welding field, the original IS method exhibits poor performance in predicting residual stress and deformation in the AM field. Liang et al. [20] implemented the original IS method to a single-walled deposition with different layers manufactured by the laser engineered net shaping (LENS) process. The predicted maximum deformation of the as-built structure in the build direction is compared to the results obtained by the detailed process simulation, as listed in Table 5.1. Bugatti and Semeraro [21] employed the original IS method to predict the bending deformation of a double cantilever beam printed by the L-PBF process. The results are then compared with the experimental measurement in Table 5.2. There are six cases in total, involving three scanning strategies and two options of upper thickness. The large errors shown in both Tables 5.1 and 5.2 indicate that the original IS method cannot accurately compute the residual deformation of those AM metal parts.

The failure of using the original IS method to predict the residual deformation/stress distribution for AM metal parts is due to some significant differences between the welding and AM processes. For example, the fast, small, and intense heat source in the AM process leads to a relatively unstable melt pool which moves

Table 5.1 Maximum deformation (mm) in the build direction of the single-walled structure.

Number of deposition layers	Detailed process simulation	The original IS method	Error (%)
1	0.326	0.170	47.9
2	0.502	0.299	40.4
3	0.584	0.412	29.8
5	0.937	0.672	28.3
10	0.663	0.478	27.9

Source: Liang et al. [20]/with permission of Elsevier.

Table 5.2 Comparison of the bending deformation of the double cantilever beam [21].

Scanning strategy	Upper thickness (mm)	Bending measured in the experiment (μm)	Bending predicted by the original IS method (μm)	Error (%)
mx	1.1	−568.6	−576.9	1.5
mx	2.2	228.9	−453.8	298.2
my	1.1	−1294.3	−294.5	77.2
my	2.2	−315.4	−187.7	40.5
chess	1.1	−1235.7	−408.5	66.9
chess	2.2	−49.8	−307.4	517.3

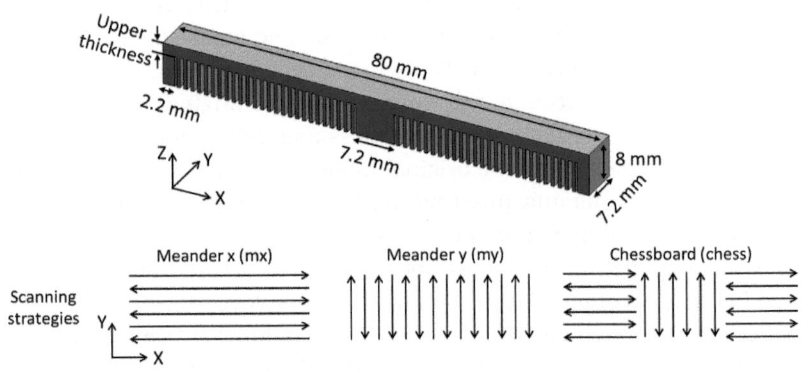

Source: Bugatti and Semeraro [21]/with permission of Elsevier.

over all material points in a large build. This suggests the entire build should be part of the HAZ which is much wider compared with the weld line in the welding process. During the layer-by-layer AM process, when a layer is being printed, all the previously deposited material, as well as the substrate, will work as thermal and mechanical boundaries. As the printing progresses, the height of the part increases, and the shape of each layer may also vary, resulting in the continuous change in boundary conditions during the printing. The different scanning strategies and remelting of the previously deposited material also increase the complexity of the residual deformation/stress distribution.

The aforementioned differences make the IS distribution in AM parts more complicated than that in welded structures. Therefore, in AM processes, it is difficult to use a uniform or simple expression to describe the IS distribution within the deposited material. Even if we successfully obtain the IS distribution in an AM part somehow, since it highly depends on the geometry of the part, using this distribution to predict the residual stress of other parts is impractical.

5.3 Extraction of ISs for L-PBF Process

The small-scale detailed process simulation of the L-PBF processing of Ti6Al4V (Ti64) is presented in the following section based on finite element analysis (FEA). Since the detailed process simulation for DMLS is adapted from that of the directed energy deposition (DED) process which was discussed in Ref. [22], only some pertinent details specific for the DMLS process are given in this section and only a brief description is provided hereby to present some necessary details. In the detailed process simulation, temperature-dependent material properties including the coefficient of thermal expansion (CTE), Young's modulus, and yield strength [22, 23] are involved. The L-PBF (DMLS) system can have different scanning modes involving different process parameters. The core-skin mode will be simulated as it is employed to print the core volume of any large metal component in the DMLS process. The major process parameters, such as laser power and scan velocity, of this laser scanning mode for Ti64 are listed in Table 5.3.

In the core-skin printing mode, a single layer is usually divided into multiple stripes, each having a constant width. The laser scanning path rotates by an angle of 67° when a new layer is deposited. This rotation angle is the default option for printing metal parts using Ti64 and Inconel 718 powders in the DMLS process. A reason for such an option is that this laser scanning strategy containing a rotation angle of 67° can provide the least directional stress difference inside the layers of the metal part [1]. Inside each stripe, the laser beam scans back and forth in an S-pattern, and those scan lines are perpendicular to the stripe edges. These detailed features are carefully considered in the small-scale simulation in order to better match the practical DMLS process. A small rectangular deposition is selected as the representative volume, as shown in Figure 5.2a, consisting of two layers with

Table 5.3 Default DMLS process parameters of the core-skin mode for Ti64.

Process parameters	Laser power (W)	Laser scan velocity ($m\,s^{-1}$)	Layer thickness (μm)	Laser beam diameter (μm)
	285	1.0	30	100

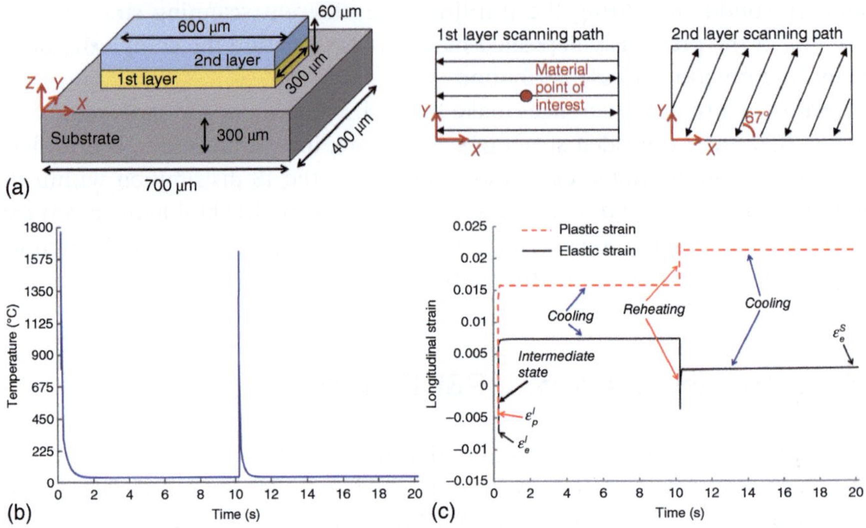

Figure 5.2 (a) Two-layer representative volume with the bottom-fixed substrate (the scanning paths only reflect the mode and do not indicate the actual number of tracks), (b) temperature history of the material point of interest, and (c) mechanical strain evolution history of the material point of interest.

a size of $0.6 \times 0.3 \times 0.06\,\text{mm}^3$ [24]. In the first layer, the laser beam moves along a straight line for three consecutive tracks. In the second layer, the laser path rotates by an angle of 67° and results in 11 line tracks. The two-layer model is printed onto a large substrate of the same metal material. The substrate bottom is fixed as the boundary condition. Choosing such a small-scale two-layer deposition to extract ISs is a trade-off between computational accuracy and efficiency. The two-layer model with different laser scanning directions is able to consider the influence of the scanning strategy and interlayer remelting. On the other hand, it is not necessary to have too many layers or a large computational domain because each line track in the L-PBF process only has a limited HAZ [24, 25].

The double-ellipsoidal heat source model [26] is adopted to simulate the laser beam in the small-scale process simulation. If needed, detailed mathematical equations of this heat source model can be found in Refs. [22, 26]. One major characteristic is that the respective values of the longitudinal, transverse, and through-the-depth semi-axes of the ellipsoidal heat source model can be different. Values of the semi-axes of the heat source model are determined based on the practical process parameters in Table 5.3. For Ti64 and Inconel 718, the values of longitudinal and transverse semi-axis are equal to the radius (50 μm) of the laser beam. The value of the depth is determined as 90 μm according to experimental measurement result.

The laser scanning strategy used in this chapter is a frequently used scanning strategy in the L-PBF process. In addition to the core-skin printing mode, there are some other laser scanning strategies such as the island scanning used by other AM systems. However, even though the scanning strategies are different for different AM systems, the proposed MIS method can still be applied. When a different scanning strategy is concerned, a small representative volume needs to be determined, which contains only a few layers for simulating the specific scanning paths for different layers. As long as the representative volume is established, the small-scale process simulation can be implemented. The MIS method can be implemented to extract the ISs and predict the residual stress and deformation. Nonetheless, the disadvantage is that the computational cost can be expensive because more layers have to be considered to capture the effects of different laser scanning paths for different layers.

The detailed process simulation includes a transient thermal analysis and a quasi-static mechanical analysis. The temperature data extracted from the thermal model is applied to the mechanical model as thermal loads to get the stress and strain distribution. Generally, the process parameters, such as laser power, absorptivity, and heat source geometry, are input variables in the thermal analysis and need calibration. Ref. [25] introduces a method for calibrating the process parameters using far-field thermocouple data and a line heat source model.

Based on the small-scale detailed process simulation of the two-layer model, one material point shown in Figure 5.2a is chosen to observe the temperature and strain histories. This point of interest is at the center of the first layer to avoid the boundary influence and consider the interlayer effect, that is, the remelting caused by the second-layer deposition. The temperature and strain histories at that point are shown in Figure 5.2b and c, respectively. Note that the strains are called longitudinal because they are in the direction of the laser scanning path. The sharp changes in temperature and strain histories are coincident. The first sharp change is due to the deposition of the material at that point of interest, involving heating, melting, and solidification. The temperature at that point surges when the heat source approaches and drops dramatically when the heat source moves away. The second sharp change is caused by the second-layer deposition above the point of interest, involving reheating, remelting, and resolidification. After the laser scanning of each layer, the temperature and strains are nearly constant for a while (about 10 s). It represents the dwell time between layers, accounting for the processes such as powder spreading and platform descending in reality. It has to be noted that the interlayer dwell time (about 10 s) is much longer than the laser scanning time (about 1 ms) due to the high scanning speed and the small computational domain. That is why, in Figure 5.2b,c, the sharp-change phases look like spikes compared with the nearly constant phases. The intermediate and steady states of a typical material point in the L-PBF process can be identified based on the strain evolution history as shown in Figure 5.2c. This figure displays the nodal elastic and plastic strain terms based on FEA to calculate the elemental ISs according to the MIS method.

For the lower layer in the two-layer deposition, it experiences complete melting, remelting, and cooling process. Thus, the lower layer can be considered as a representative layer to compute the ISs of a large build. If a material point in the lower layer is selected, the elastic and plastic strain evolution history can be observed directly based on the small-scale process simulation. In Figure 5.2c, the large compressive elastic and plastic strain is induced at the very beginning of the strain evolution history. This feature indicates the rapid melting and following solidification process, that is, the so-called intermediate state. Then the elastic and plastic strain will develop into tensile strain very rapidly and reach the steady state following the intermediate state. This feature also suggests that the rapid solidification finishes in a very short time. This feature is the most common characteristics of the mechanical strain evolution for different metal AM processes including the L-PBF process using a laser beam source.

As shown in Figure 5.2c, the compressive strains reach the maximum nearly at the same time as the material is deposited, which is defined as the intermediate state (annotated in Figure 5.2c). Then tensile strain starts to accumulate because of the rapid solidification. Regarding the first two terms in Eq. (5.2), the elastic and plastic strains at the intermediate state are first extracted from the small-scale simulation results. Then the elastic strain at the steady state is extracted as the third term in Eq. (5.2). By this way, the ISs of any material point in the first layer of the two-layer model can be obtained. For the convenience of numerical implementation, some simplifications have to be employed. Constitutively, the strain and stress terms are second-order tensors according to the continuum mechanics theory. For isotropic material, only six independent components in the strain–stress tensor are needed for three-dimensional problems. By rearranging the indices, the second-order tensors can be further reduced to lower-order forms such as vector and matrix. As for the IS vector, it consists of six components including three normal strain components (ε_{xx}, ε_{yy}, and ε_{zz}) and three shearing strain components (ε_{xy}, ε_{yz}, and ε_{xz}). The subscripts for each strain are consistent with the coordinates in Figure 5.2a. Based on the small-scale detailed process simulation, the shearing strains have a much smaller magnitude compared with those normal strain components [20]. A reasonable explanation is that the normal expansion and shrinkage in the in-plane dimension of a thin metal layer are the major cause of the residual deformation and stress instead of the shearing effect. Therefore, the three shearing strain components are neglected for simplicity, and the IS vector only contains three normal strain components in the MIS method.

For the L-PBF process, when the MIS method is applied, the obtained IS distribution curves have some similar characteristics as found in Ref. [20]. The IS components in three normal directions are continuous and nonuniform over the layer domain (Figure 5.3). The IS values are almost stable and flat in the middle region of the deposition layer. Whereas, the values show a drastic decrease to nearly zero when the curves get close to the two ends of the layer. This feature suggests the likely influence on the ISs by the thermal and mechanical boundaries in different locations of the deposition layer. The local materials at two ends of the layer are exposed to quite different thermal boundary conditions and also weaker mechanical constraint compared with those materials in the middle region of a large build.

In particular, the most important feature is that the computed IS vector consists of two compressive components in the in-plane direction and one tensile component in the build direction. The positive sign of the strain component in the build direction indicates the Poisson effect in the mechanical deformation. Given the small

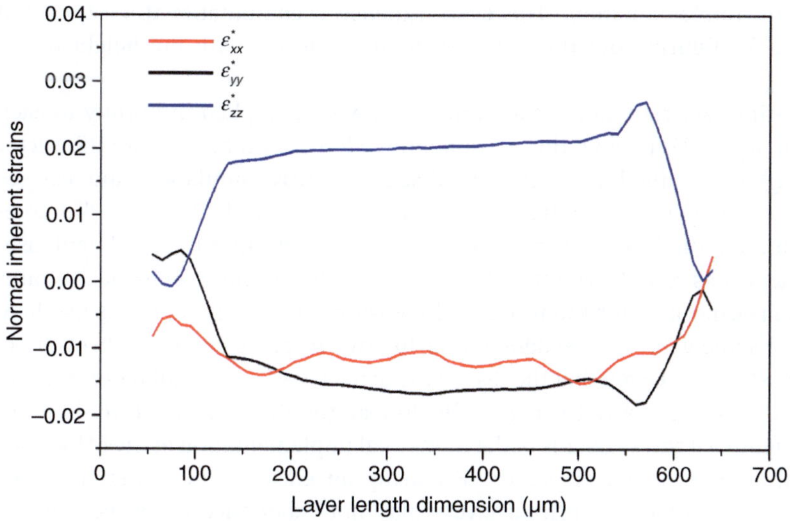

Figure 5.3 Normal IS component distribution for Inconel 718 builds by the L-PBF process. Source: Adapted from Chen et al. [25].

layer thickness in contrast to the other two dimensions of a single layer, temperature gradient in the in-plane directions is very different compared with that in the layer thickness dimension. As a result, the in-plane normal mechanical strains exert the dominant Poisson effect on the mechanical strain in the build direction. Moreover, the mechanical constraint for a layer in the build direction is relatively weaker than the in-plane directions. The thin layer is easier to expand and shrink normally in the build direction in the heating and cooling process. As a result of the coupled effect by the in-plane mechanical strains and the mechanical constraint, a positive normal strain in the build direction is obtained for the IS vector. The aforementioned discussion explains why in the simulation, it is found that the normal stress is quite small (not exactly zero) for the top layer compared with the lower layers, though the normal strain component in the IS vector is nonzero regarding the build direction.

5.4 Governing Equations for MIS-Based Sequential Analysis

In the L-PBF process, if the same laser scanning path is applied, the micro-welding phenomenon and mechanical strain history in each layer of the large deposition should be quite similar. As a result, the extracted ISs for each layer of a large metal build will be close. Thus, the same ISs can be used for each layer of the large build for simplification. Nonetheless, for a large and complex deposition containing thousands of thin layers, the thermal profile in the build direction can have some influence on the ISs of the deposition layers in different locations. It is reasonable that if varying IS values are applied to the layers in the build direction, accuracy of the MIS method can be improved compared to the current implementation. However, expensive computational cost cannot be avoided for figuring out the accurate trend of the ISs with the height of the build.

As a preliminary estimation, the average ISs will be applied uniformly to each deposition layer. Following the aforementioned discussion, a mean IS vector ($\varepsilon^*_{xx}, \varepsilon^*_{yy}, \varepsilon^*_{zz}$) is obtained from the small-scale process simulation and can be employed to predict the residual stress and deformation of a large L-PBF component. In addition, based on the simulation, it is found that the rotational laser scan strategy has a redistribution function and makes the mechanical strains homogenous especially for the in-plane dimensions. Therefore, it is reasonable to assume the same value can be adopted for the two in-plane components of the IS vector, i.e., $\varepsilon^*_{xx} = \varepsilon^*_{yy}$. Given that the ISs are extracted based on a small representative volume out of a large part, it is reasonable to load the IS vector in a layer-by-layer fashion. Figure 5.4 shows details of the numerical implementation of the MIS-based sequential analysis based on static FEA using an elastoplastic material model. Nonetheless, a couple of top/final layers may not experience the same thermal loading because they will not experience many remelting occurrences like the lower layers. Consequently, the ISs in the top/final layers can be a little varied from those in the lower deposition layers. This effect is ignored in this study, since the several top/final layers take up only a very small section of a large component. Regarding the several bottom layers, though they are exposed to strong constraint of a large build plate directly, it has been proved in the earlier work [20] that it is acceptable to assume the first few layers have the same ISs with the remaining layers in the bulk volume of a large build.

For the purpose of adapting to the layer-wise loading of the ISs, a large metal component needs to be meshed into thin slices to ensure the part-scale model shows a flat top surface at any height along the build direction, just like the scenario in a practical L-PBF process. The ISs extracted from the small-scale process simulation can be used in the MIS-based sequential analysis for any complex part if the identical process parameters are employed. In other words, we do not need to run the small-scale detailed thermomechanical simulation again. It should be

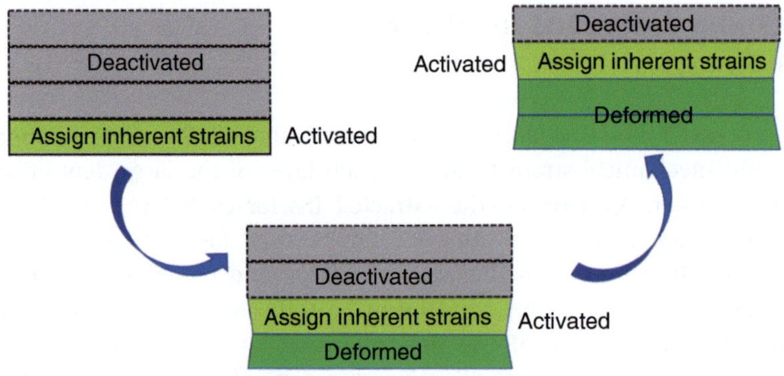

Figure 5.4 Applying the ISs to a part-scale model in a layer-by-layer fashion.

noted that the identical process parameters mainly suggest that the laser power, scanning velocity, and scanning strategy are identical. However, if different process parameters are considered, the small-scale thermomechanical simulation should be run again to extract accurate mechanical strain evolution history and compute the ISs accordingly.

Usually, a large component can contain as many as thousands of thin layers in the L-PBF process. Although the MIS method is efficient, it is not practical to fully simulate all the thin layers in the sequential analysis. Therefore, many thin layers are numerically merged into one equivalent layer (also called super layer [28]) in the layer-wise analysis. When the equivalent layer is adopted to represent many thin layers, the mean IS vector is applied as the orthotropic CTEs of the metal build. And a unit temperature change is loaded to the equivalent layers in a layer-wise manner to simulate the bottom-to-top feature of the AM process (see Figure 5.4). The body force is neglected in the MIS-based layer-wise simulation. For the large-scale L-PBF components, the governing equation for the sequential analysis can be formulated as:

$$\nabla \cdot \boldsymbol{\sigma}^i = 0 \tag{5.3}$$

where $\boldsymbol{\sigma}^i$ indicates the stress field in the part when the ith equivalent layer is added onto those already deposited layers. In other words, if each equivalent layer is simulated by a load step in the sequential FEA, the governing equation as shown in Eq. (5.3) indicates the analysis for the ith load step. In the numerical implementation, the element birth-and-death technique [22] is used to model the bottom-up material depositing process. All the equivalent layers will be killed/deactivated in the very beginning of the sequential analysis. Then each layer is activated and analyzed step by step in the build direction.

The elastic stress tensor involved in Eq. (5.3) is computed based on the elastic constitutive law as:

$$\boldsymbol{\sigma}^i = \boldsymbol{D} : \boldsymbol{\varepsilon}_e^i \tag{5.4}$$

where $\boldsymbol{\varepsilon}_e$ represents the elastic strain field and $\boldsymbol{D}$ is the constitutive elastic tensor. In contrast to $\boldsymbol{\sigma}^i$ in Eq. (5.3), $\boldsymbol{\varepsilon}_e^i$ represents the global elastic strain field of the large component when the ith equivalent layer is added onto those already activated layers.

Some details of how to load the ISs are provided here. Since the IS vector contains three normal strain components, the IS component values are defined as CTEs in the material property parameters. A unit temperature increase can be used as the thermal load to induce the thermal strain/stress related to the ISs. The total strain of the large-scale component can be computed as:

$$\boldsymbol{\varepsilon}_{tot}^i = \boldsymbol{\varepsilon}_e^i + \boldsymbol{\varepsilon}_p^i + \boldsymbol{\alpha}^* \Delta T^i \tag{5.5}$$

where $\boldsymbol{\varepsilon}$ and $\boldsymbol{\varepsilon}_p$ represent the total and plastic strain in the ith load step, respectively. The elastoplastic material model should be employed. The symbol $\boldsymbol{\alpha}^*$ denotes the equivalent CTEs in relation to three IS components, and ΔT denotes the unit temperature increase ($\Delta T = 1$). In addition to Eq. (5.5), the total strain ε can be written as:

$$\varepsilon^i = \frac{1}{2}[\nabla u^i + (\nabla u^i)^T] \tag{5.6}$$

where u^i represents the global displacement field in the ith load step. In particular, in the layer-wise simulation, the displacement field is updated in each load step. The displacement solution of the previous load step is naturally taken as the initial conditions of the following load step.

In addition to the governing equation, mechanical boundary conditions should be determined according to the practical manufacturing scenarios. For the L-PBF process, usually the metal depositions are printed on a fixed large and thick build plate. Therefore, in the MIS-based sequential simulation, the bottom surface of the build plate should be totally clamped as the boundary condition. However, it is obvious that the large and thick build plate can form a strong constraint to the bottom of the builds. Therefore, to reduce computational time, the build plate can be neglected. As the new boundary condition, the bottom surface of the large component is fixed accordingly. This boundary condition is adopted in this chapter.

The sequential solution process can be easily carried out in any commercial FEA software such as the ANSYS Mechanical Enterprise. As mentioned earlier, based on the element birth-and-death technique, the equivalent layers will be deactivated at first and then activated layer by layer with respect to the build direction (see Figure 5.4). In each load step, the contractive deformation of the activated region will be considered as the initial condition for the next layer waiting for activation. The experimental results have exhibited good accuracy of the aforementioned implementation procedure, which will be shown in the following section. In addition, the influence on simulation accuracy by the number of equivalent layers will also be studied in the following section. Since only static mechanical analysis is involved, the sequential layer-by-layer simulation can usually be finished in short time.

5.5 Experimental Validation: Double Cantilever Beam

In this section, a double cantilever beam is considered as the benchmark example to validate the MIS method. Profile of the double cantilever beam is illustrated in Figure 5.5. The teeth-like structures are employed to support the overhanging section so that it can be printed successfully by the L-PBF process. As for the size, the length is 76.2 mm, the width is 12.7 mm, and the height is 18.4 mm in total. After the deposition process is finished, the teeth-like structures are cut to allow the overhanging beams to deform due to stress release, as shown in Figure 5.5. The total deformation is measured experimentally, specifically on the two ends (Point #1 and #2) of the double cantilever beam.

The MIS-based sequential analysis for residual stress and deformation simulation is conducted in ANSYS v17.2. To set up the boundary condition, the bottom surface of the build plate is fixed in the mechanical analysis. For Ti64, the computed mean IS vector is (−0.012, −0.012, 0.01). The three normal IS components are, respectively,

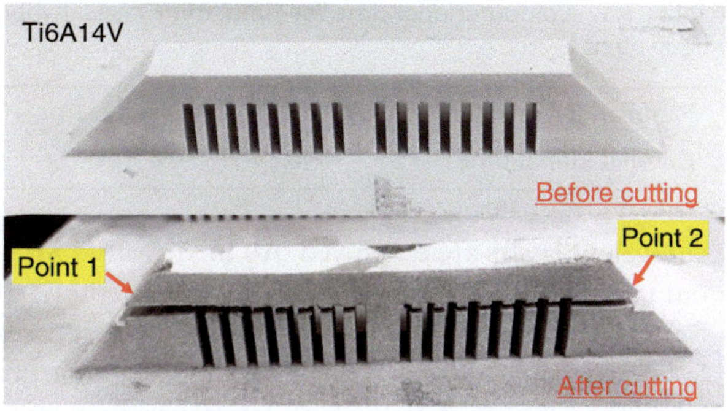

Figure 5.5 Double cantilever beam of Ti64 before and after cutting off the support structures.

parallel to the in-plane scan direction, transverse to the in-plane scan direction, and parallel to the vertical build direction. The IS values are assigned to the full-scale meshed double cantilever beam model in a layer-wise fashion. For the geometry as shown in Figure 5.5, 60 equivalent layers are contained in the entire deposition including support structures. The material properties of Ti64 employed in this validation example are Young's modulus of 105 GPa and yield strength of 768 MPa [24]. Note that in the MIS-based simulation, only material properties at ambient temperature are required.

Two end center points (#1 and #2, see Figure 5.5) of the double cantilever beam are specifically considered. The total deformation of each of these points, with vertical displacement as the dominant component, is measured and listed in Table 5.4. It is noted that the total deformations are a little different at the left and right end. A possible reason is provided for clarification. In the post-processing, those support structures in the right side were cut first. Consequently, the residual stress that remained in the cantilever beam deposition was largely relaxed. The relaxation also affects some residual stress in the left section of the deposition. When the remaining support structures were cut in the left side, the partially relaxed overall residual stress was released a second time. This time the stress relaxation effect resulted in a smaller vertical deformation at the left end of the beam deposition. According to Table 5.4, good accuracy (<10% error) can be observed. Therefore, the effectiveness of the MIS method is validated for predicting residual deformation of the L-PBF components.

In addition, the influence on the simulation accuracy based on the MIS method by the number of those equivalent layers is also studied. The 60-layer model is taken as the basic model. More simulations are implemented, where 30, 20, 15, 10, 6, and 3

Table 5.4 Total deformations at the two ends of the double cantilever beam by the MIS-based simulation and experiment.

Location	MIS-based simulation (mm)	Experiment (mm)	Error (%)
Point #1	0.87	0.94	7.4
Point #2	0.38	0.36	5.6

Table 5.5 Computational time for simulation cases involving different number of equivalent layers.

Number of equivalent layers	3	6	10	15	20	30	60
Computational time (h)	0.39	0.61	0.86	0.98	1.22	1.62	2.30

equivalent layers are employed. All the involved MIS-based simulations are carried out using a desktop computer with 8 Intel Xeon E5-1660 cores. The computational time is listed in Table 5.5 for each of the seven cases including the basic case (60 layers). It can be observed the computational time does not increase proportionally when the number of equivalent layers increases gradually. This is mainly related to the element activation process and the nonlinear characteristics of the calculation, e.g., the use of elastoplastic material models. As the number of equivalent layers increases, the number of elements for each equivalent layer decreases, and the total number of active elements for calculation in each step also changes. However, the computational time for each step may not change proportionally with the number of active elements due to material model nonlinearity. In addition, the element activation process itself takes time. In comparison with a detailed thermomechanical simulation where a point-wise heat source model is a must to simulate the complex laser scanning paths, the MIS-based sequential analysis can provide much better computational efficiency.

Additionally, the maximum total deformation values for Points #1 and #2 are shown in Figure 5.6 against the number of equivalent layers. It can be seen that the largest total deformation increases first and then decreases. It tends to be stable with the increasing number of equivalent layers. Some meaningful discussions for the important features of the plots (see Figure 5.6) are given as follows. In the extreme

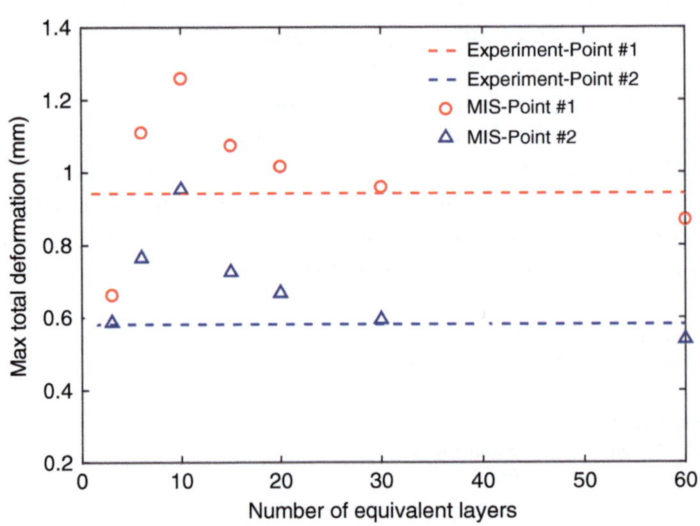

Figure 5.6 Total deformation of Points #1 and #2 (see Figure 5.5) obtained by experiment and MIS-based simulation using a different number of equivalent layers.

case, the simulation using only one equivalent layer gives the smallest total deformation in comparison with the remaining test cases. In that extreme case, the entire double cantilever beam is considered as only one equivalent layer. The unit temperature increase is applied to the entire model all at once, and accordingly, just one load step is needed to finish the mechanical analysis. The simulation result shows the residual stress is very insignificant in the top region of the deposition. Consequently, when the residual stress is released by removing the teeth-like support structures, the overhanging beam section does not show much flexural deformation. The conclusion is that the one-step modeling would result in underestimation of the residual stress and further underestimation of the residual deformation after stress release. Therefore, it is inaccurate to push the sequential MIS-based analysis to an extreme case where a large component is considered as only one equivalent layer.

5.6 Simulation-Driven Design for L-PBF Process

5.6.1 Support Structure Selection for Crack Prevention

A typical L-PBF build usually consists of substrates, support structures, and the desired components. Among them, the support structure plays a vital role in ensuring the continuity and stability of the printing process. First, the support structure makes overhangs printable. Second, it can fix and secure the component to the substrate. Third, the support structure allows for heat dissipation from the component to the substrate, reducing heat accumulation in the component. One of the most common support structures employed in the L-PBF process is the lattice structure comprised of thin walls with a thickness of around 100 μm [27]. In the design stage of the lattice support structure, one of the challenges is to avoid cracking at the solid/support interface within the printing process, which is mainly caused by the residual stress induced in the thermomechanical cycles [29–31]. Although residual stress occurs almost everywhere in the printed structure during processing, the solid/support interface is prone to cracking compared to other locations. This is because: (i) most of the residual stress is induced within the solid part; (ii) the geometric difference between the solid and lattice structure is significant, causing the residual stress to concentrate on the solid/support interface; and (iii) the large contrast between the effective stiffness of the solid and lattice structure makes the interface vulnerable [27]. To overcome this drawback, it is necessary to evaluate and reduce the possibility of cracking at the solid/support interface during the design stage. In this section, a novel approach for predicting cracks at the solid/lattice support interface will be introduced. It combines in situ experiments, global–local *J*-integral analysis, material homogenization, and MIS-based simulation.

5.6.1.1 Description of the Workflow

The possibility of cracking at the solid/support interface is evaluated based on a nonlinear elastic–plastic fracture parameter called the *J*-integral [27, 32–34].

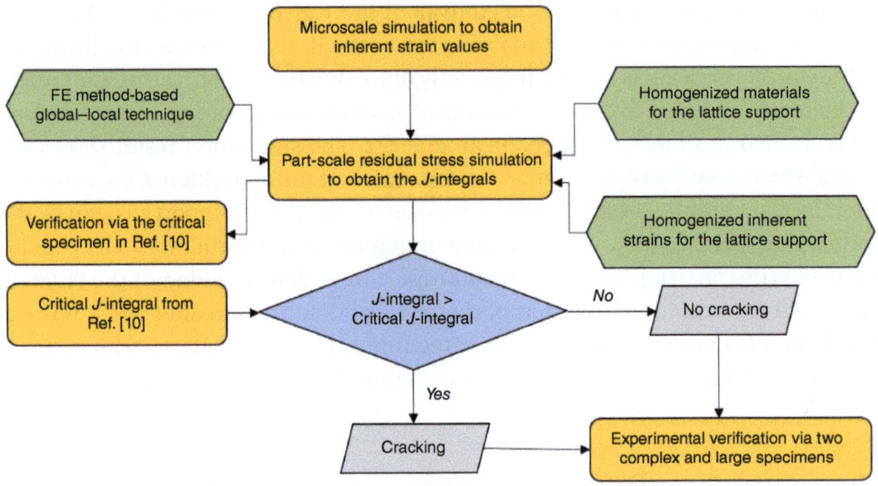

Figure 5.7 Workflow of the evaluation for cracking at solid/support interface [27].

As shown in Figure 5.7, the evaluation mainly has two steps. In the first step, the critical J-integral of the solid/support interface is determined based on a series of printing experiments and the MIS-based simulation. After that, in the second step, the J-integral of the solid/support interface at different locations of a real part is calculated according to the residual stress distribution obtained through the MIS-based simulation. If the calculated J-integral is smaller than the critical value, there will be no cracking at the corresponding location, and vice versa.

5.6.1.2 Determination of the Critical *J*-Integral for Solid/Support Interface

The J-integral is defined as a surface (or line) integral enclosing the crack front (or the crack tip) from one cracking surface to the other [35, 36]. It has the ability to characterize the local stress–strain field near the crack front (or the crack tip). The critical J-integral is the nonlinear fracture toughness or elastoplastic fracture toughness for a solid material or a bimaterial interface. The J-integral value of the material at the bound of cracking is considered as the critical one [32]. The interested reader is referred to Ref. [37] for a detailed description of the J-integral.

A common way to obtain the critical J-integral at the interface is to manufacture several compact tensile specimens and perform tests according to ASTM E1820-01 standard [32]. However, this method is impractical for the as-built AM solid/support interface because: (i) the production of the specimen that contains the solid/support interface may fail due to cracks occurring during the printing process; (ii) even if the specimen is successfully printed, the residual stress within the structure will affect the accuracy of the measured J-integral [32].

Therefore, instead of using the standard testing process, the MIS-based simulation combined with in situ experiments is adopted to determine the critical J-integral for the as-built AM solid/support interface [27, 32]. First, a series of fracture blocks (also called hybrid blocks), samples composed of half solid and half lattice support, are built as shown in Figure 5.8a. The height of the lattice support (or the thickness

Figure 5.8 (a) Geometry of the fracture block sample, and (b) samples w/ and w/o cracks. Source: Tran et al. [32]/with permission of Elsevier.

in the build direction) is constant for all the samples, while the height of the solid varies. As the solid portion of the block becomes higher, the residual stress concentrated at the interface becomes larger, thus increasing the possibility of cracking [32]. When the height of the solid is not large enough, the as-built samples will be intact, and no cracks will occur at the interface (Figure 5.8b, four samples to the right). For samples with relatively high solid, cracks at the interface can be observed after printing (Figure 5.8b, two samples to the left). Thus, there should exist a critical height, above which the block samples would crack at the interface and below which samples remain intact. In other words, the interfacial J-integral of the sample with the critical height is defined as the critical value when evaluating potential cracking at the solid/support interface. It may be necessary to print several sets of samples to accurately determine the critical height. For example, after printing a set of samples with a large height increment, say 0.5 mm, the first critical height is found. Then, another set of samples is printed with refinement, where the height of the second set of samples is around the first critical height and the height increment is set smaller, say 0.1 mm. After that, a new critical height with less uncertainty is found. This procedure can be repeated until the final critical height is sufficiently accurate. For Inconel 718 samples printed by the EOS M290 DMLS system with default process parameters, the critical height is found to be 8.2 mm [32].

After determining the critical height, the MIS-based simulation is employed to compute the residual stress of the as-built critical block sample. To ensure accuracy of the simulation, the lattice support structure is modeled explicitly, and the mesh around the crack front is refined to compute the J-integral accurately. Meanwhile, the solid part far from the interface is modeled by a coarser mesh so that the computational cost is affordable. On a workstation with 24 cores (Intel Xeon Gold 6136 CPU @3.00GHz), it takes about 98 hours to finish the simulation [32]. Based on the results of residual stress, the J-integral at the interface between the solid and support is evaluated. In ANSYS, the J-integral evaluation is achieved through the CINT command [38]. The critical J-integral at the solid/support interface is found to be $J_c = 7196\,\mathrm{J\,m^{-2}}$ for Inconel 718 processed by EOS DMLS system with default parameters [32].

5.6.1.3 Calculation of J-Integral at Solid/Support Interface for as-Built Part

Once the interfacial critical J-integral is determined, we can use it to predict cracking of a real part and improve the crack resistance design. First, the residual stress distribution of the as-built part is calculated by the MIS-based simulation. Afterward, the J-integrals of the solid/support interface are evaluated at multiple locations having especially high residual stress and compared to the critical J-integral. If the J-integral at a location is larger than the critical value, cracking is predicted to occur at this location during printing. In this case, the design of the part can be modified accordingly so that the J-integral will be lower than the critical value. It has to be noted that the critical J-integral at the solid/support interface depends on a lot of factors including the material, process parameters, and lattice structure geometries. To ensure that the critical J-integral determined by the hybrid block samples is applicable to the solid/support interface of the real part, the aforementioned factors should be the same for both the hybrid block samples and the real parts. This is true for all the samples mentioned in the rest of this section.

The procedure of implementing the MIS-based simulation for the real part is similar to that for the hybrid block samples. However, considering that the real part usually has a larger overall dimension and a more complex geometry than the hybrid block sample, explicitly creating the lattice support structure in the MIS-based simulation for the real part will drastically increase the computational cost and even make the simulation intractable. In order to reduce computational time and preserve the characteristics of the lattice structure as much as possible, the support of the real part is modeled by a block of solid elements with homogenized material properties in the MIS-based simulation. With the help of this material homogenization technique, the lattice structure consisting of many individual cells can be treated as a continuum solid with equivalent material properties (such as yield strength and Young's modulus) and can be meshed in the same way as the solid part above the support. This technique can reduce the modeling complexity and increase computational efficiency for full-scale and large finite element (FE) models. The homogenization of the material property is based on anisotropic elasticity and Hill's anisotropic plasticity model. More information regarding this material model can be found in Refs. [38–40]. In Ref. [41], this homogenization technique is validated by direct numerical simulation (i.e., modeling the lattice support explicitly) and experiments. A case study in Ref. [41] also shows that this technique can reduce the number of elements by 80% and achieve almost 10× in computational efficiency.

It has to be noted that each layer of the lattice structure printed by the EOS M290 DMLS system consists of multiple tracks with different (usually orthogonal) scanning directions. Since the ISs of the single-track exhibit anisotropy in the printing plane (i.e., the IS values along and perpendicular to the scanning direction are different), the ISs of the lattice structure should also be treated as material properties and need to be homogenized [41]. For example, for a thin-walled structure with a thickness of 0.1 mm, the in-plane IS along and perpendicular to the scanning direction is −0.01 and −0.003, respectively [41]. Considering that the ISs are implemented to the part-scale FE model as CTEs of the material, the homogenization of the ISs is mathematically equivalent to homogenization of CTEs. Interested readers are referred to Refs. [41–43] for more details about how to obtain the homogenized CTEs.

Although the material homogenization can improve efficiency of the MIS-based simulation, it may introduce some errors due to the large mesh size, especially when calculating the J-integral at the solid/support interface. Therefore, another technique called global–local analysis is employed to accurately evaluate the interfacial J-integral. With this technique, a real part is represented by two models: a global model including the entire system and a local model including only a small area cut from the system [38]. For static mechanical analysis, the global model is first used to obtain the strain and stress response of the entire structure. The local model is then employed to investigate the area that is mostly concerned.

Since the computational domain of the global model is large, in order to reduce the computational expense, the global model employs the homogenized properties for the lattice structure, ignores some geometrical details (e.g., replacing filets with sharp corners), and is meshed coarsely. The boundary conditions of the global model are either prescribed tractions or displacements, depending on the real build conditions. When the simulation based on the global model is complete, a local model representing the region of interest (ROI) is created. The boundary condition of the local model is the displacements on the cut boundary, which can be obtained from the nodal displacement results of the global model. The local model is developed based on the Saint-Venant's principle, which states that when replacing the force field with a static equivalent system, the change of the stress and strain field will be constrained near the load area [38]. In other words, as long as the cut boundary is far enough away from the location we are concerned with, the results obtained by the local model should be close to those from the global model. Of course, since a refined mesh is used, the local model can provide more accurate results within the ROI than the global model from the perspective of FEA. That is why this technique has been developed and widely used to reduce computing time and ensure the accuracy of results within the ROI.

After combining material homogenization and global–local analysis techniques, the evaluation of the J-integral at the solid/support interface for a real part includes the following steps. The first step is to perform the MIS-based simulation on the entire part using the global model where the support structure is homogenized. The second step is to create the local model on the ROI that includes the solid/support interface whose J-integral needs to be calculated. Note that the support structure can now be explicitly created for accuracy because the computational cost will not increase too much due to the limited size of the local model. The third step is to mesh the local model. The fourth step is to apply the boundary conditions to the local model, that is, extracting nodal displacements along the cut boundary from the global model and importing them to the local model. Interpolation algorithms may be involved during the importation of the nodal displacement because meshes along the cut boundary in these two models are usually incompatible. The fifth step is to solve the equilibrium problem to obtain the stress distribution near the solid/support interface. In the last step, the interfacial J-integral is calculated. The second step to the last step can be repeated to obtain the J-integral value at different locations of the solid/support interface.

Example 5.1 ***Fracture Sample with a Flat Interface***

In this example, the J-integrals at the solid/support interface are computed for two hybrid block designs (denoted by Sample A and Sample B) to evaluate the possibility of cracking. Note that the MIS-based simulation of these two samples does not consider the material homogenization and global–local techniques as all the lattice support is modeled explicitly. The configurations of the two samples are shown in Figure 5.9a,b. Qualitatively, for both samples, the tensile stress at the solid/support interface is accumulated as the solid deposition gets higher, resulting in the increase in both J-integral and cracking probability at that location. After obtaining the residual stress distribution of the as-built part (Figure 5.9c,d), the interfacial J-integrals at the end of the sample are calculated as $J_A = 7288\,\text{J m}^{-2}$ for Sample A and $J_B = 7167\,\text{J m}^{-2}$ for Sample B. The results are reasonable because the solid height in Sample A is 1 mm higher than that in Sample B. Compared with the critical J-integral $J_c = 7196\,\text{J m}^{-2}$, the J-integral of Sample A is larger, indicating that cracks will occur at the interface, while the smaller J-integral of Sample B implies there should be no crack at the interface. Both predictions are confirmed by printing experiments (Figure 5.9e,f).

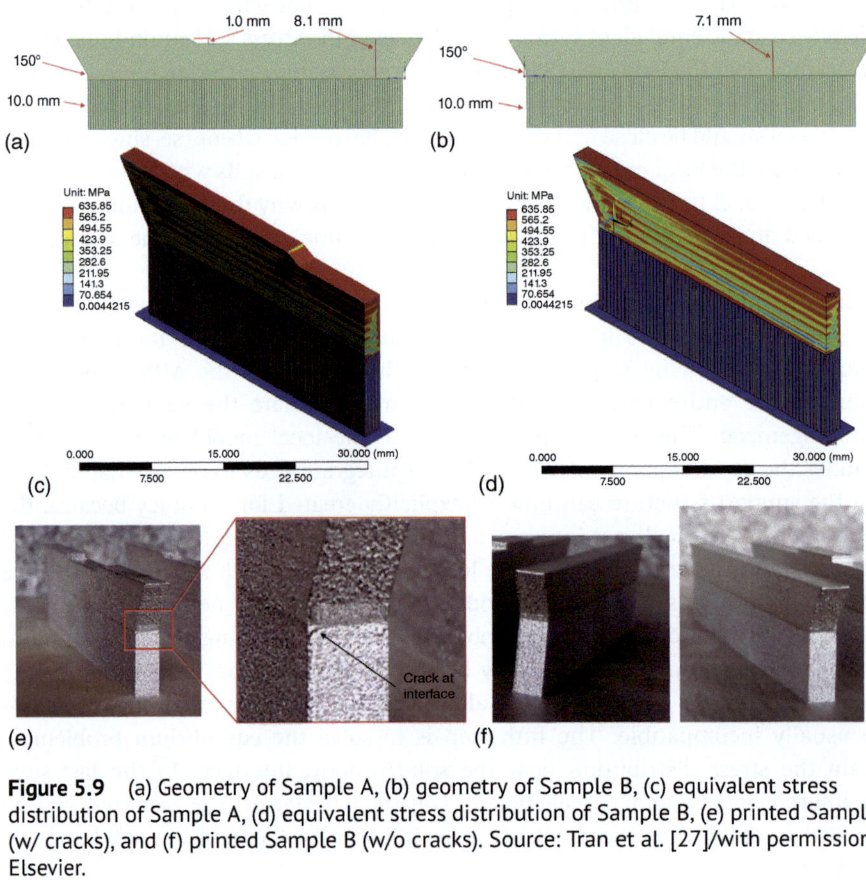

Figure 5.9 (a) Geometry of Sample A, (b) geometry of Sample B, (c) equivalent stress distribution of Sample A, (d) equivalent stress distribution of Sample B, (e) printed Sample A (w/ cracks), and (f) printed Sample B (w/o cracks). Source: Tran et al. [27]/with permission of Elsevier.

Example 5.2 ***Fracture Sample with a Curved Interface***

Given the geometry of a hybrid block sample ($63.6 \times 25.6 \times 26.2\,\text{mm}^3$) shown in Figure 5.10a, the J-integrals at the solid/support interface at the corners on the two ends (see Figure 5.10b) are calculated and compared to the critical value. Here, the material homogenization and global–local techniques are employed to save computational time. When evaluating the residual stress of the global model in MIS-based simulation, the lattice structure is modeled by the homogenized continuum material. When calculating the J-integral in the local model, the lattice structure is still created explicitly to obtain accuracy. Two local models with a dimension of $1 \times 2.1 \times 2\,\text{mm}^3$ are cut from the global model to compute the interfacial J-integral at the left and right ends of the sample. From the results shown in Figure 5.10c, the left end has a J-integral of $J_{\text{left}} = 8200\,\text{J m}^{-2}$, which is larger than the critical value of $J_c = 7196\,\text{J m}^{-2}$, while the right end has a J-integral of $J_{\text{right}} = 3500\,\text{J m}^{-2}$, which is approximately half of the critical value. These results suggest that the left interface of the sample will crack, and the right interface will be crack-free. The printed sample, as shown in Figure 5.10d, confirms the simulation predictions.

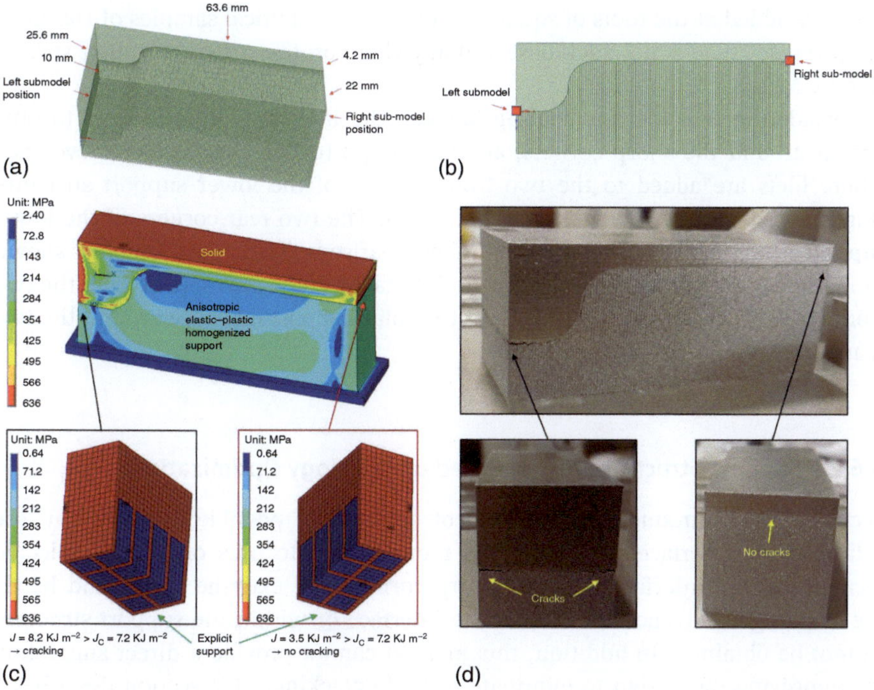

Figure 5.10 (a) Geometry of the fracture sample with a curved interface, (b) locations for calculating J-integrals, (c) evaluation of J-integrals, and (d) printed fracture sample with a curved interface (w/ cracks at the left interface and w/o cracks at the right interface). Source: Tran et al. [27]/with permission of Elsevier.

Example 5.3 ***Bearing Bracket***
In this example, a bearing bracket with a complex geometry (Figure 5.11a) is studied by using the MIS-based simulation coupled with material homogenization and global–local analysis techniques.

Two corners in the lower support of the bearing bracket, as shown in Figure 5.11b, are selected as the ROI for cutting the local model. The dimension of the local mode is $1\times2\times2\,\text{mm}^3$. The calculation results show that J-integrals at these two corner interfaces are 7600 and $9300\,\text{J m}^{-2}$, which are both higher than the critical value of $J_c = 7196\,\text{J m}^{-2}$. It implies that the solid/support interface at both corners will crack during the printing. Since cracks at both corners are observed in the printed sample, the prediction by the MIS-based simulation is confirmed. In addition, the significant difference in residual stress around the solid and support (Figure 5.11c) also explains why cracks are prone to occur at the solid/support interface.

Based on the experimental and numerical results of cracking evaluation, two new designs of the support structure are proposed to avoid cracking. As shown in Figure 5.12a, the lattice supports at four corners of the lower support structure are replaced by the solid square columns (called Design A). By doing so, the solid/support interfaces at these corners are eliminated, and the stress concentration there should also be reduced. In addition, to avoid possible delamination, filets are added at the roots of square columns. Two identical samples of Design A are printed. There is no crack observed anywhere on the structure's outer surface (Figure 5.12a).

Considering that for this bearing bracket, the interfacial crack seems to only appear around the sharp corners, another design (called Design B) is developed where filets are added to the two front corners of the lower support structure (Figure 5.12b) to see if cracks can be avoided. The two rear corners of the lower support structure are kept unchanged for comparison. The printed sample is shown in Figure 5.12b. As expected, there is no interfacial crack anymore around the two front corners, while cracks still exist at the solid/support interface around the two rear corners.

5.6.2 Support Structure Design Based on Topology Optimization

Section 5.6.1 introduces a method of evaluating possible cracking at the solid/support interface. However, this method only focuses on several selected locations at the interface (such as sharp corners and extreme ends), and hence the cracking occurrence over the entire interface or within the support structure cannot be obtained. In addition, this method cannot provide a direct suggestion for improving the design to eliminate possible cracking. This section describes a procedure for designing the graded lattice support structure for parts printed by L-PBF processes. The procedure combines topology optimization (TO), material homogenization, and MIS method. It is able to minimize the overall mass of the support structure while constraining the maximum residual stress over the entire deposition, including both the support structure and solid part.

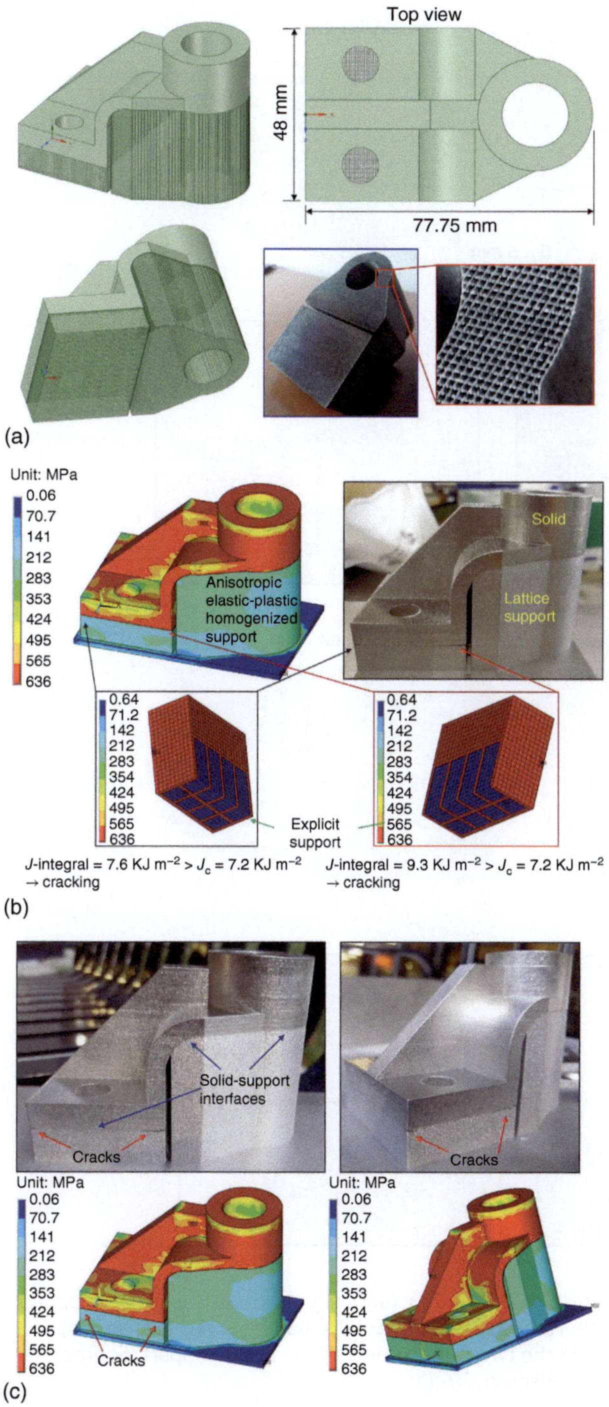

Figure 5.11 (a) Geometry of the bearing bracket, (b) evaluation of J-integrals, and (c) significant difference in residual stress around the solid and support interface. Source: Tran et al. [27]/with permission of Elsevier.

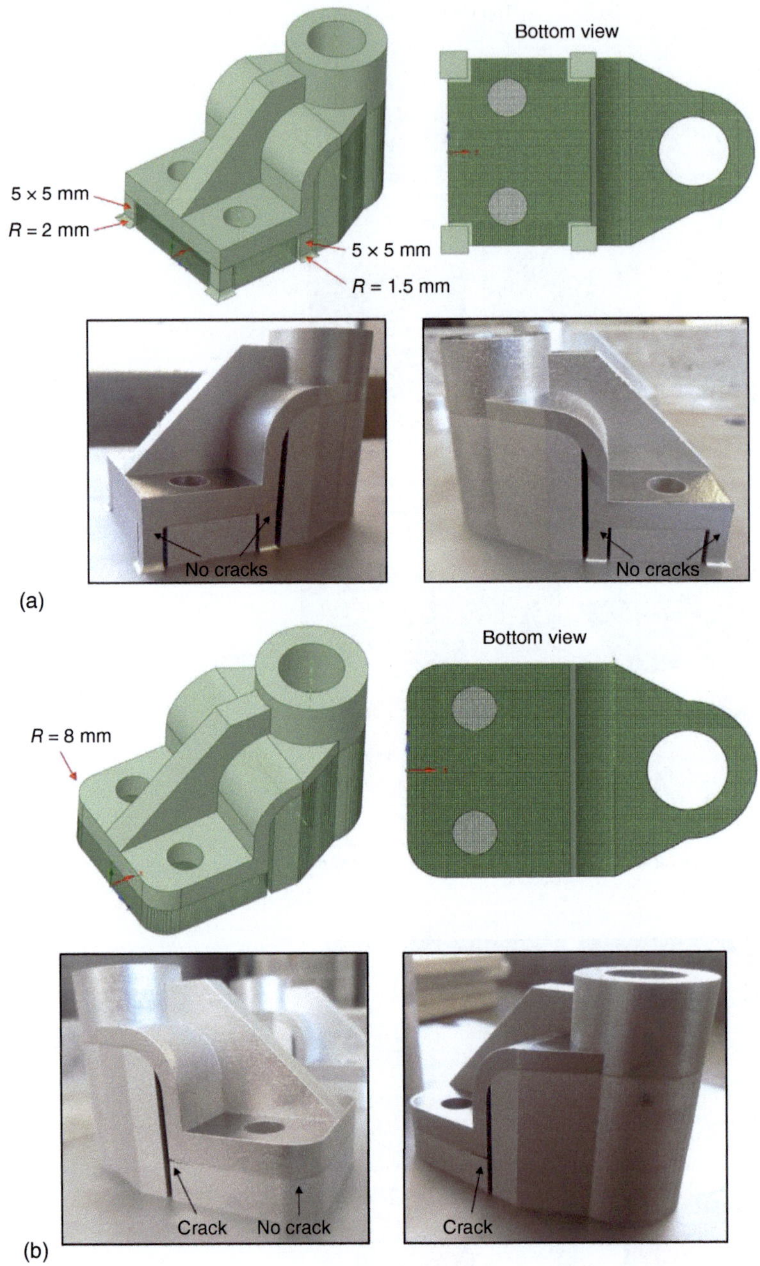

Figure 5.12 Two designs for cracking prevention: (a) Design A where the lattice supports at four corners of the lower support are replaced by solid square columns [27] and (b) Design B where filets are added to the two front corners of the lower support. Source: Tran et al. [27]/with permission of Elsevier.

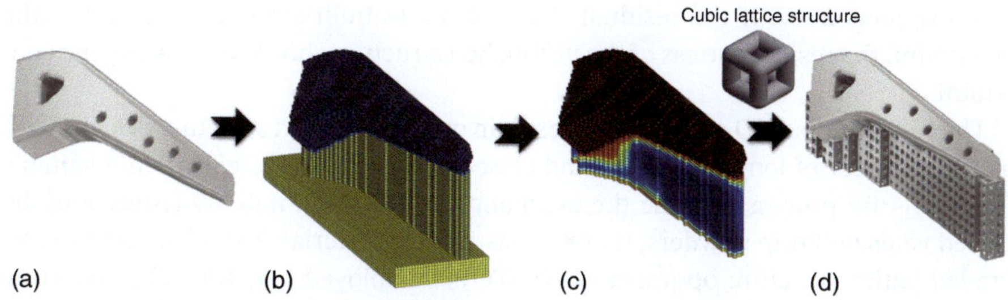

Figure 5.13 Workflow of the support structure design for a practical component: (a) geometry of the component, (b) initial structure meshed by the voxel elements, (c) optimal density profile of the support structure, and (d) CAD model of the optimal support structure. Source: Cheng et al. [46].

5.6.2.1 Description of the Workflow

The general process for support structure design of a component consisted of four steps is illustrated in Figure 5.13. First, for a given part shown in Figure 5.13a, the overhang is automatically detected. An initial support structure with uniform density is generated based on the detection results. After that, both component and support structure are meshed by voxel elements, as shown in Figure 5.13b. Then, TO is implemented and utilized to minimize the overall mass of the support structure under several constraints (such as the maximum residual stress and the density range). Figure 5.13c shows the optimal density profile of the support structure. Lastly, a CAD file is automatically generated, as shown in Figure 5.13d, where the continuous optimal density distribution is converted to the lattice support structure consisting of cubic cells with different relative densities. This CAD file can be further converted to the STL file and sent to the printer for manufacturing.

5.6.2.2 Topology Optimization of the Support Structure

Here, regarding the support structure design for L-PBF printed parts, the goal of the optimization is to reduce the overall residual stress within both the support structure and the desired component so that cracking can be prevented during the printing process. The computational domain of the optimization problem includes three sub-domains: the build substrate Ω_t, support structure region Ω_s, and the bulk component Ω_c. Here the support structure helps to anchor the component onto the substrate and make the overhang of the component printable.

The objective for support design optimization is to minimize the overall mass of the support structure, while the maximum residual stress in the support structure (domain Ω_s) and bulk component (domain Ω_c) is lower than the yield strength of the printing material. Note that after optimization, the mass of the support structure can only be minimized but should not vanish. That is because the support structure is essential for providing support to the overhang and fixing the component to the substrate. In order to reduce the complexity of the optimization problem, only the residual stress of the as-built support structure and component is considered. In other words, the residual stress variation during the printing process is ignored. This simplification is reasonable because the residual stress tends to increase as the printing progresses. If the residual stress of the as-built structure can satisfy the constraint, the residual stress of the unfinished structure should also satisfy the constraint.

The conventional TO method may result in optimal support structure designs with a large number of long overhangs and closed voids. These designs are not suitable for the L-PBF process because the overhang can cause printability issues, and the closed voids will trap powders, thereby wasting the material. To tackle these issues, graded lattice structure optimization (LSTO) is employed [39, 46]. After detecting the overhangs of the component and determining the support structure domain Ω_s, this domain will be infilled by the open-celled, self-supporting, and variable density lattice structures based on the LSTO instead of by the conventional block shell support. The optimization problem for support structure design can be expressed mathematically as [46]:

$$\text{Minimize } m(\rho) = \sum_{e=1}^{N} \rho_e v_e \tag{5.7}$$

such that

$$\begin{aligned} &\boldsymbol{KU} = \boldsymbol{F}^{\text{in}} \\ &\overline{C} = \overline{C}(\rho) \\ &\sigma_{\max}^{H}(\rho) \leq 1 \\ &\underline{\rho} \leq \rho_e \leq \overline{\rho} \end{aligned} \tag{5.8}$$

where $m(\rho)$ is the objective function, representing the total mass of the support structure. N denotes the total number of voxel elements in the support structure domain Ω_s. The symbols ρ_e and v_e denote the topological density and volume of the eth voxel element within Ω_s, respectively. The first constraint is the mechanical equilibrium equation for the entire computational domain, where $\boldsymbol{K}$ denotes the structural stiffness matrix, $\boldsymbol{U}$ denotes the displacement vector, and $\boldsymbol{F}^{\text{in}}$ denotes the external load. $\overline{C} = \overline{C}(\rho)$ is the elastic model of the support structure. Regarding the third constraint, $\sigma_{\max}^{H}$ denotes the maximum von Mises/equivalent stress of the computational domain based on the measure given by the modified Hill criterion [39]. The stress values are normalized by the yield strength of the metal material. In other words, when the normalized von Mises stress is smaller than 1, only elastic deformation occurs in the design domain. When the normalized von Mises stress is larger than 1, plastic deformation will also exist in addition to elastic deformation. By making the maximum stress smaller than 1, we ensure that only elastic deformation needs to be considered. No plasticity is considered in the optimization model.

As the design variable for the support structure optimization, ρ_e will be restricted to a range shown in the last constraint. The symbols $\underline{\rho}$ and $\overline{\rho}$, respectively, denote the lower and upper limit of the design variable, which corresponds to lattice structures with different volume densities. To solve the optimization in Eqs. (5.7) and (5.8), the p-norm stress is employed to approximate the maximum stress so that the sensitivity analysis can be performed analytically as maximum stress is not a differentiable function. Sensitivity analysis is performed with respect to the p-norm residual stress constraint and other constraints. The sensitivity information is employed in the optimization algorithm based on the method of moving asymptotes (MMA) [44]. By this iterative optimization process, the optimized layouts of the support structures can be found.

5.6.2.3 Residual Stress Estimation

For the optimization problem described in Eqs. (5.7) and (5.8), the optimal results may not be obtained until after hundreds of iterations. In each iteration, the mechanical equilibrium equation (the first constraint listed in Eq. (5.8)) of the as-built structure needs to be solved. Since the MIS method has been proven to be an efficient and accurate method for predicting the residual stress in the as-built metal AM parts, it is utilized to estimate the residual stress required in each iteration of the optimization process. The MIS method is usually implemented in the FEA. However, to further increase the computational efficiency and avoid the time-consuming and error-prone generation process of the FE mesh, the finite cell method (FCM) is employed here to solve the equilibrium equation. The main idea of the FCM is to embed the computational domain of interest $(\Omega_t + \Omega_c + \Omega_s)$ into an extended fictitious domain. The combined domain can then be easily meshed by structured finite cells, simplifying the mesh generation. When evaluating the residual stress, the material properties of the support structure are related to the relative density of the cubic lattice and homogenized as described in Ref. [45].

Example 5.4 ***Double Cantilever Beams***

To illustrate the aforementioned procedure for the support structure design, a double cantilever beam structure is taken as an example. The geometry of the double cantilever beam is shown in Figure 5.14a. Its overall size is $80 \times 20 \times 24\,\text{mm}^3$. The thickness of the build substrate is set to be 8 mm in the residual stress simulation. The support structure that needs to be designed is located immediately under the overhanging beams. The volume of the support structure domain on each side of the cantilever beams is $36 \times 20 \times 15\,\text{mm}^3$.

Both the beam and the lattice structure are printed with Ti64, with a Young's modulus of 110 GPa, a yield strength of 1140 MPa, and a Poisson's ratio of 0.3 [46]. It has to be noted that the material properties of Ti64 used here are different from those in

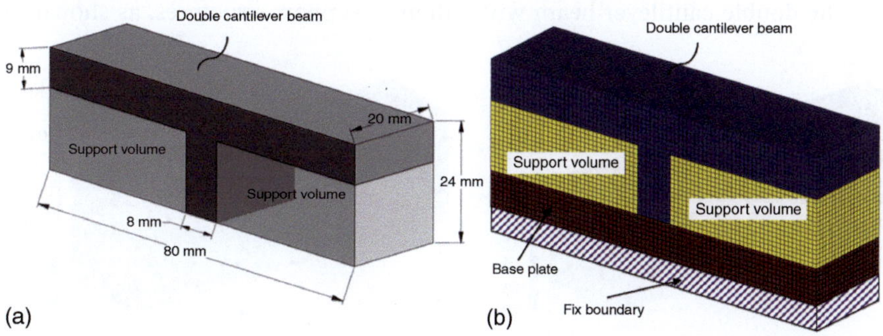

Figure 5.14 (a) Geometry of the double cantilever beam and (b) voxel-based mesh model. Source: Cheng et al. [46].

Section 5.5 (Ref. [24]). Commonly, material properties may vary due to machine variability, measurement devices, post-processing, etc. The IS vector for Ti64 is assumed to be (−0.012, −0.012, and 0.01). The entire computational domain is meshed with eight-node hexahedral elements, as shown in Figure 5.14b. The yellow voxels represent the support structure domain Ω_s, blue voxels are for the bulk component Ω_c, and red voxels denote the substrate domain Ω_t. When evaluating the residual stresses, the bottom of the substrate is fixed in all directions as boundary conditions, while ISs are assigned to both the component and support structure in the form of nodal forces.

The relative density of each element in the support structure domain Ω_s is initially set to 0.9 and constrained between 0.2 and 0.95 during the optimization. Meanwhile, the elements in the bulk component domain Ω_c and substrate domain Ω_t are kept to be solid ($\rho_e = 1$) during the optimization. Once the residual stress distribution in the computational domain is obtained, the optimization problem described in Eqs. (5.7) and (5.8) is solved iteratively to reduce the volume of the support structures until the p-norm residual stress constraint is satisfied. The optimization is terminated when the change of volume fraction of the support structure within three successive iterations is smaller than 0.001, while the global stress constraint is satisfied.

The optimization results are presented in Figure 5.15. As shown in Figure 5.15a, the lattice support structure has relatively high densities for the region close to the upper corner of the center pillar and low densities for the region far away from the center pillar. There exist several intermediate densities (i.e., the yellow voxels in Figure 5.15a) at the interface between the high-density and low-density regions. Figure 5.15b shows the normalized residual stress distribution of the optimal design. For most area of the support structure domain, the normalized residual stresses are smaller than 0.8 (i.e., the actual residual stress is 80% of the yield strength of Ti64). The large residual stress occurs at the root of the center pillar. In the iterative optimization process, after about 50 iterations, the overall density of the support structure decreases 54.2%, from the initial value of 0.9 to 0.412, and becomes stable. Meanwhile, the normalized maximum residual stress decreases from 1.1 to 0.99. Though the normalized maximum residual stress does not change a lot, the result of 0.99 (smaller than 1) indicates that all the material is in elasticity, which ensures manufacturability of the build in theory.

To validate the effectiveness of the optimized support structure design, four samples of the double cantilever beam with different support structures, as shown in

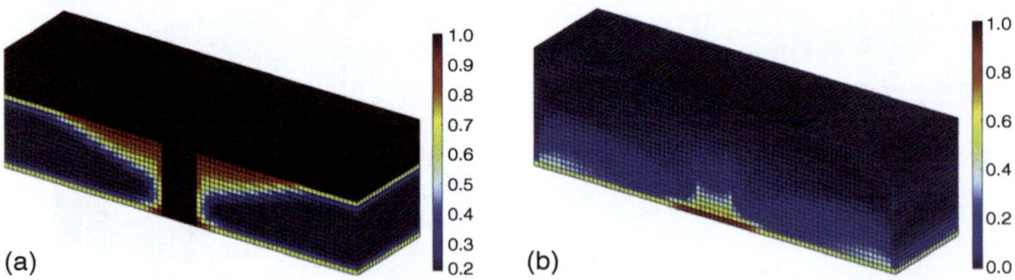

Figure 5.15 Optimization results of the double cantilever beam: (a) optimal density distribution and (b) normalized residual stress distribution. Source: Cheng et al. [46].

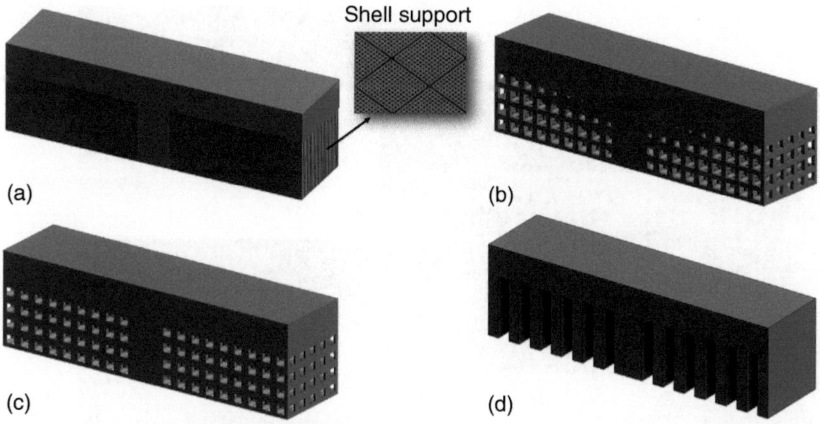

Figure 5.16 Four different support structure designs for double cantilever beam for the comparison purpose: (a) shell support from Magic software, (b) optimally variable density lattice structure support, (c) uniform lattice structure support, and (d) teeth-like support structure. Source: Cheng et al. [46].

Figure 5.16, are printed out by the EOS M290 DMLS system for residual deformation measurement. The support structure designs include (i) block shell structure generated by Materialize Magics software (widely used in L-PBF); (ii) optimized variable density lattice structure (overall relative density of 0.412); (iii) uniform density lattice structure (overall relative density of 0.45); and (iv) teeth-like structure (overall relative density of 0.5).

The original plan is to print these four samples simultaneously on a large substrate. However, the printing process is terminated abnormally due to cracking in the block shell structure sample. The cracking at the bottom of the support structure causes the top surface of the beam to bend upward, interfering with the powder recoater blade and resulting in a failed build. This example shows that the accumulation of residual stress in the sample may lead to severe problems. During the printing, one failed component can ruin the whole build and cause a loss of time and money. To avoid possible build failure, the block shell support design is discarded, and the remaining three cantilever beam samples are printed. As shown in Figure 5.17, although there are cracks observed in the teeth-like support sample, the build process for these three samples is successful. Compared to the block shell support and teeth-like support, the successfully manufactured (without observed cracks) uniform density and variable density support samples imply that the open-cell lattice structure can relieve residual stress effectively and can be used for the design of the support structure for components printed by the L-PBF process.

After printing, the support structure is cut from one side of the cantilever beam for all three samples using a wire-cut machine. When the connection between the support structure and the cantilever beam is cut off, the beam will deflect upward due to the residual stress relief. The bending distortion of the beam is measured by a 3D scanning device Faro Laser ScanArm V3 (FARO Technologies) with an accuracy of 0.065 mm, as shown in Figure 5.18. The maximum deflection occurs at the

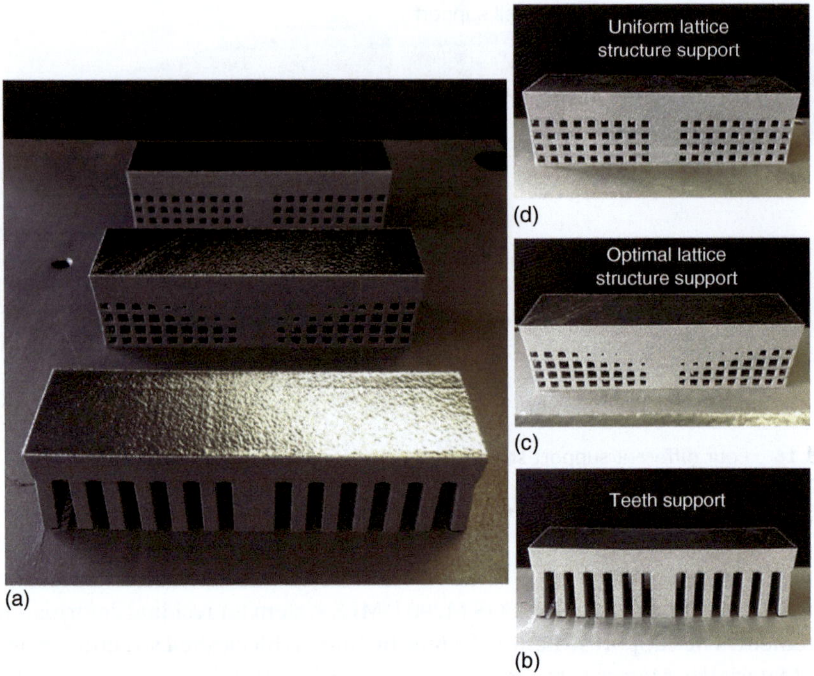

Figure 5.17 Build for samples of the double cantilever beam printed out by EOS M290 DMLS in Ti64: (a) the three printed out beams for validation, (b) the beam with teeth support, (c) the beam with optimal lattice structure support, and (d) the beam with uniform lattice structure support. Source: Cheng et al. [46].

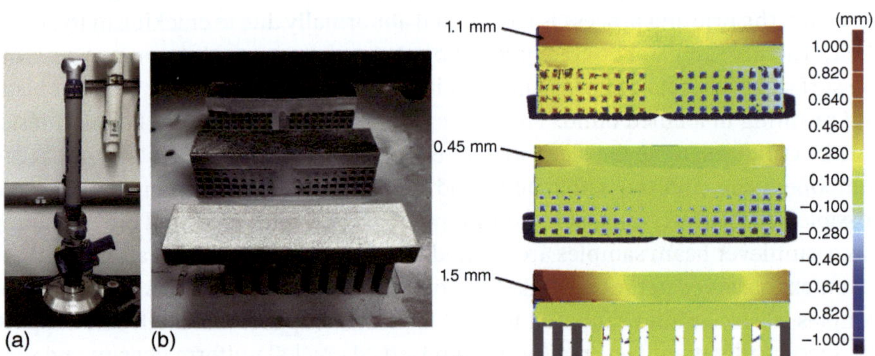

Figure 5.18 Experimental measurement for the distortion of the beams after cutting: (a) 3D laser scanning device, (b) three beams after cutting, and (c) residual deformation. Source: Cheng et al. [46].

end of the cantilever beam. It is 1.1 mm for the uniform density design, 0.45 mm for the optimized variable density design, and 1.5 mm for the teeth-like design. The results indicate that the proposed procedure for the support structure design can significantly reduce the residual stresses in metal parts manufactured by the L-PBF process. The optimized support design not only ensures the manufacturability of the

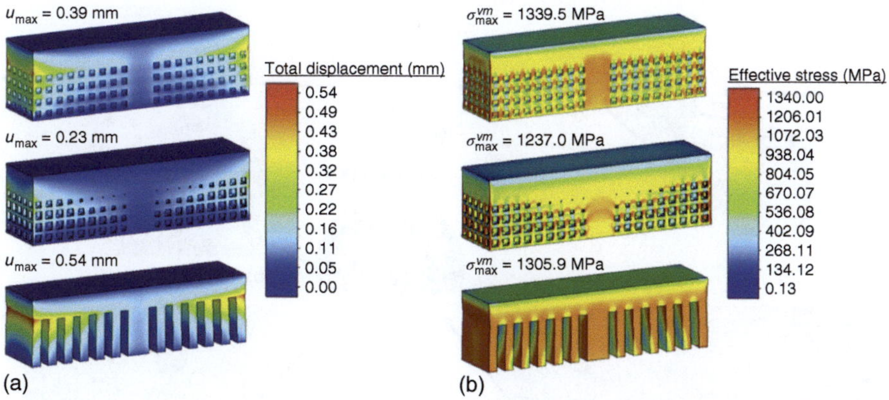

Figure 5.19 Full-scale simulation results of the beams before cutting: (a) displacement distribution of the beams and (b) residual stress distribution of the beams. Source: Cheng et al. [46].

component (i.e., avoid cracking during the printing process) but also considerably improves the quality of the bulk component (i.e., reduce residual distortion).

To further examine the mechanical performance of the optimized support structure design, the commercial software Simufact Additive v4.0 from MSC Inc. is used to implement the part-scale simulation for these three cantilever beams to evaluate the residual deformation and residual stress distribution before cutting. As shown in Figure 5.19a, the maximum distortions in all the samples before cutting are significantly smaller than those in the samples after cutting. In addition, the maximum distortion of the optimized variable density design is only 0.23 mm, which is smaller than that of the uniform density design (0.39 mm) and the teeth-like design (0.54 mm).

The residual stress distribution of samples obtained by Simufact is illustrated in Figure 5.19b. It can be observed that the residual stress is obviously reduced quite a lot in the case of having the optimized support structure when compared with those cases having uniform density and teeth-like support structures. Among the three support structure configurations, the optimized variable density lattice support structure shows the best overall control of residual stress distribution. This agrees well with the observation from the experiment and proves that the proposed methodology is efficient in constraining residual stresses in components manufactured by L-PBF.

Example 5.5 ***Hip Implants***

In this example, the proposed optimization procedure is implemented for the support structure design of a typical hip implant made of Ti64. The geometry of the implant is given in Figure 5.20a where the unit is mm. It has an overall dimension of $120 \times 35 \times 16\ \text{mm}^3$. Figures 5.20b–d show the cracks and delamination in a sample with conventional block shell support printed by the EOS M290 DMLS system. Although this sample is printed successfully, the cracking and delamination in the support structure indicate that the residual stress in the sample is relatively high,

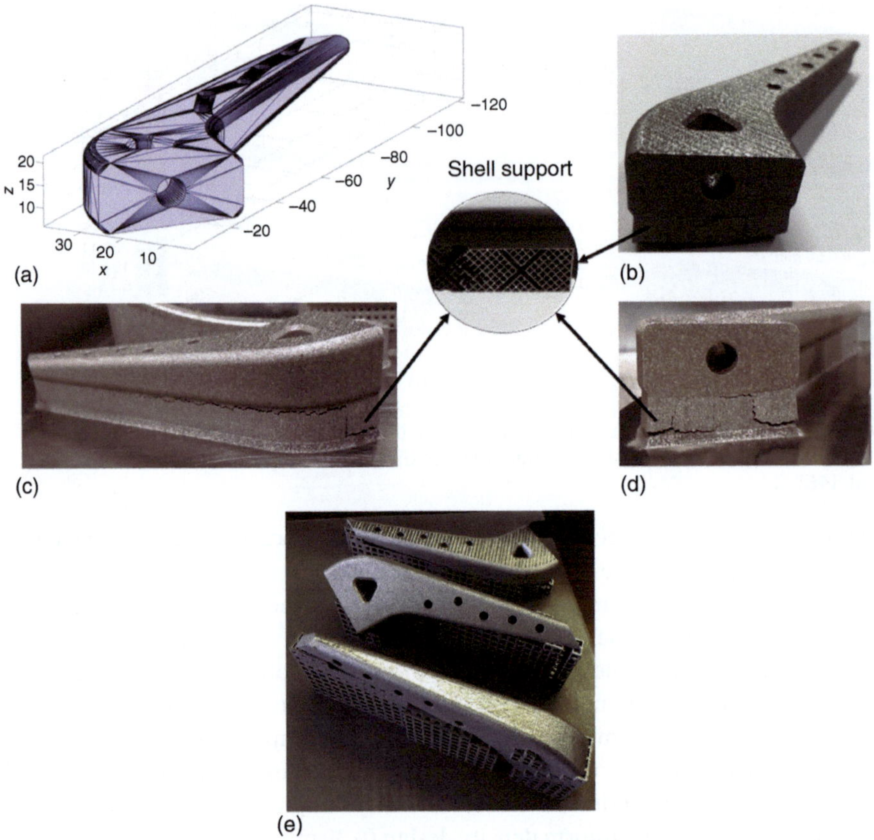

Figure 5.20 (a) Geometry of the hip implant, (b) printed sample after cutting off from the substrate, (c) left view of the implant before cutting, (d) front view of the implant before cutting, and (e) printed implants with optimized lattice support (in three different build directions). Source: Cheng et al. [46].

which increases the uncertainty of the printing process (e.g., the build may fail sometimes) and compromises the performance of the implant (e.g., low fatigue strength and poor accuracy). Further, using the block shell support structure leads to material waste, since a lot of powder is trapped in the closed voids of the support.

In contrast, the optimized open-celled variable density lattice support is able to effectively reduce the residual stress in the support and component, minimize the mass of the support, and also save the feedstock powder. Note that the optimization process depends on the build direction of the component because the volume and the outer surface area of the support structure domain may change with different build directions. For the hip implant shown in Figure 5.20a, three different build directions are selected. After the support structure optimization, the optimal designs obtained are printed out by the EOS M290 DMLS system in Ti64, as shown in Figure 5.20e. Compared with the sample containing the block shell support, all the optimal designs are successfully printed without obvious cracking, exhibiting the reliability of the optimization procedure for the support structure of the metal AM parts.

5.6.3 Laser Scanning Path Design

The residual stress and distortion that exist in the as-built metal AM parts are mainly caused by the large nonuniform thermal gradient induced during the printing by the heat source such as the L-PBF process. In addition to the support structure design, many heat-source-related factors can also affect the residual stress and distortion of the part. In this section, we will discuss scanning path design, which is able to influence the thermal gradient distribution by controlling the order and direction in which the heat source passes through the material points. A novel method [47] for reducing the residual stress and distortion will be presented here. It involves the level-set-based TO of the scanning path and the MIS method. Note that all the optimization calculations involved in this section are dimensionless. In other words, variables such as length and stress have units of 1.

5.6.3.1 Description of the Method

Given a part geometry, this method is able to optimize the scanning path of each layer within the part based on a level-set framework. The objective can be either compliance or residual stress minimization. Moreover, this method can be extended to achieve concurrent optimization by considering TO together, thus improving the performance of the method.

The residual stress/distortion of the as-built part needs to be obtained in each iteration of the optimization process. Considering that the part-scale thermomechanical simulation is computationally expensive, the MIS-based simulation is employed to ensure the optimization efficiency. One advantage of using the MIS method over other fast prediction methods is that the IS vector applied to the structure is related to the scanning direction. By assigning different IS vectors to the material points, the MIS method is able to capture the effect of the scanning direction of the heat source. However, since the IS vectors are applied to the entire layer simultaneously, the simulation cannot consider the order in which the heat source passes different material points.

The general steps of this method for scanning path optimization are described as follows:

(1) For a given part, generate the corresponding FE model and set initial scanning paths for each layer.
(2) Apply different IS vectors as nodal forces to each element based on the scanning direction.
(3) Set boundary conditions and solve the equilibrium equation for the FE model.
(4) Perform level-set-based optimization.
(5) Check if the convergence criterion is satisfied. If not, repeat steps (2)–(5).

Example 5.6 ***Compliance Minimization***
This example demonstrates the implementation of scanning path optimization for minimizing the compliance of a four-layer L-bracket. As shown in Figure 5.21a, the overall size of the L-bracket is 40 in length, 40 in width, and 4 in height (build direction). The structure is meshed by cubic elements (eight-node hexahedron) with an

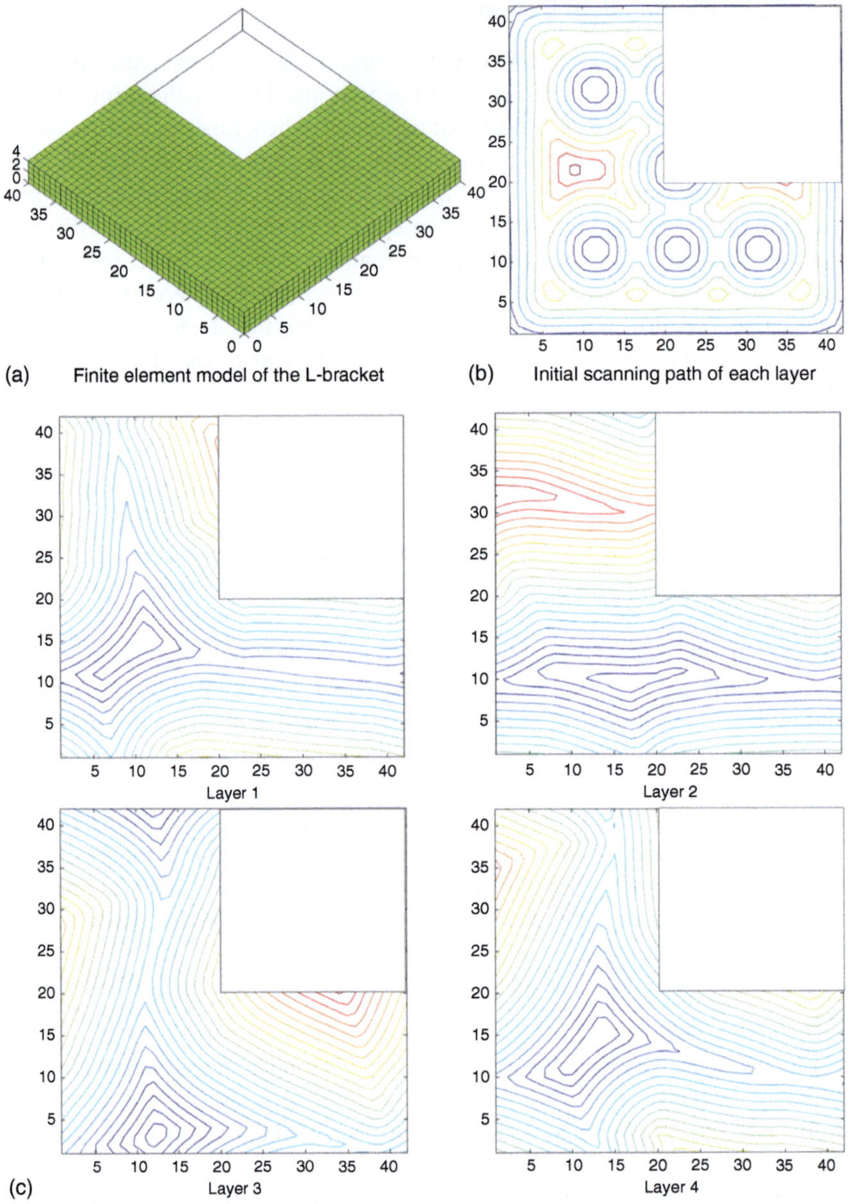

Figure 5.21 (a, b) Initial setup of the L-bracket optimization (Compliance 1147), and (c) layer-wise deposition optimization results. Source: Chen et al. [47].

edge length of one unit. For simplification, the substrate is not considered, and the bottom of the L-bracket is fixed in all directions as mechanical boundary conditions. The ISs along the scanning direction (in the printing plane), perpendicular to the scanning direction (in the printing plane), and along the build direction (out of the printing plane) are −0.0145, −0.0065, and 0.0120, respectively. The initial scanning

Table 5.6 Comparison of the compliance between the initial and optimized scanning paths of the L-bracket.

	Initial	Optimized	Reduction (%)
Layer 1	309.53	260.06	16.0
Layer 2	267.91	225.16	16.0
Layer 3	280.03	232.82	16.9
Layer 4	289.53	252.01	13.0
Total	1147	970	15.4

paths are the same for all the layers and are depicted in Figure 5.21b. For elements with different scanning directions, the coordinate transformation is employed to calculate their IS vectors with respect to the global coordinate system. Note that the IS vectors are layer-independent. In other words, elements in different layers will have the same IS vector if they have the same scanning direction.

Figure 5.21c presents the scanning paths for each layer after the optimization (Layer 1 is the bottom and Layer 4 is the top). As can be seen, although the structure has the same initial scanning paths and the layer-independent IS vectors, the optimized scanning paths in each layer are different due to the mechanical interaction between each layer. The compliance for the initial and optimized scanning paths is listed in Table 5.6. The results show that the optimization is effective, since the total compliance of the structure decreases by 15.4%, from 1147 to 970. Although the initial and optimized compliance values for each layer are not the same, the degrees of the compliance reduction for each layer are quite close (about 15%). It has to be noted that the scanning path optimization algorithm presented here is based on mathematics, and thus the resultant scanning paths may be irregular and counterintuitive.

In addition to the compliance comparison, the von Mises stress distribution in the L-bracket with the initial and optimized scanning paths is also compared (Figure 5.22). The maximum von Mises stress, which occurs at the corner of the bottom layer (Layer 1), is reduced by 7.8% (from 73.54 to 68.24) through compliance optimization. The maximum stress in each layer is reduced after the optimization, except for Layer 3. The results indicate that although the optimization aims to minimize structural compliance, it can also reduce the maximum stress in the structure.

Example 5.7 ***Stress Minimization***

This example performs the scanning path optimization to minimize the residual stress of the as-built part. The same L-bracket model and initial scanning paths shown in Figure 5.21 are adopted. The optimized scanning paths (Figure 5.23) for each layer are different, similar to the results of compliance minimization.

Figure 5.24 shows the von Mises stress distribution of each layer before and after the stress minimization. The stress results before the optimization are the same as the results shown in Figure 5.22 because of the same FE model and initial scanning paths used in these two examples. It can be seen that except for Layer 3,

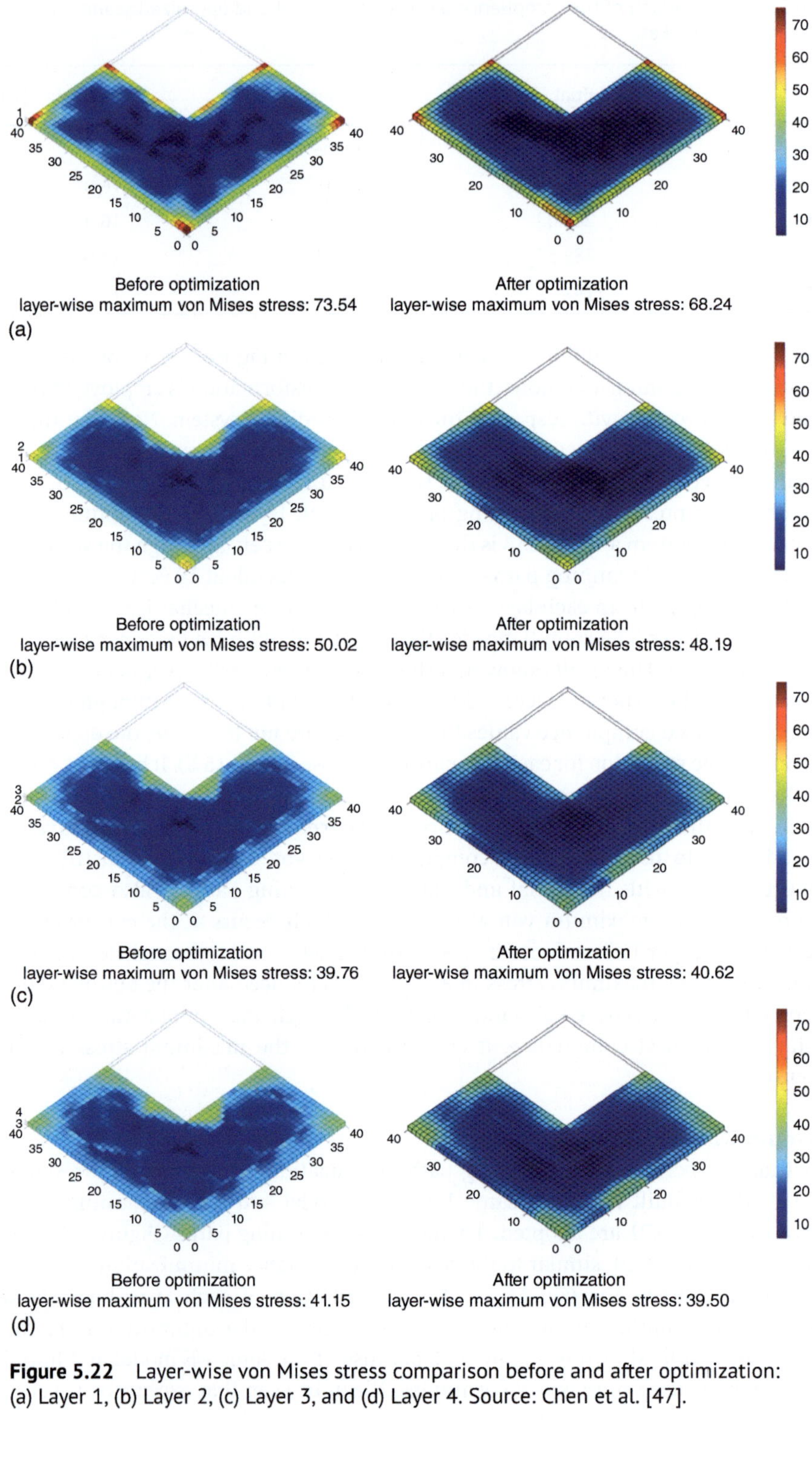

Figure 5.22 Layer-wise von Mises stress comparison before and after optimization: (a) Layer 1, (b) Layer 2, (c) Layer 3, and (d) Layer 4. Source: Chen et al. [47].

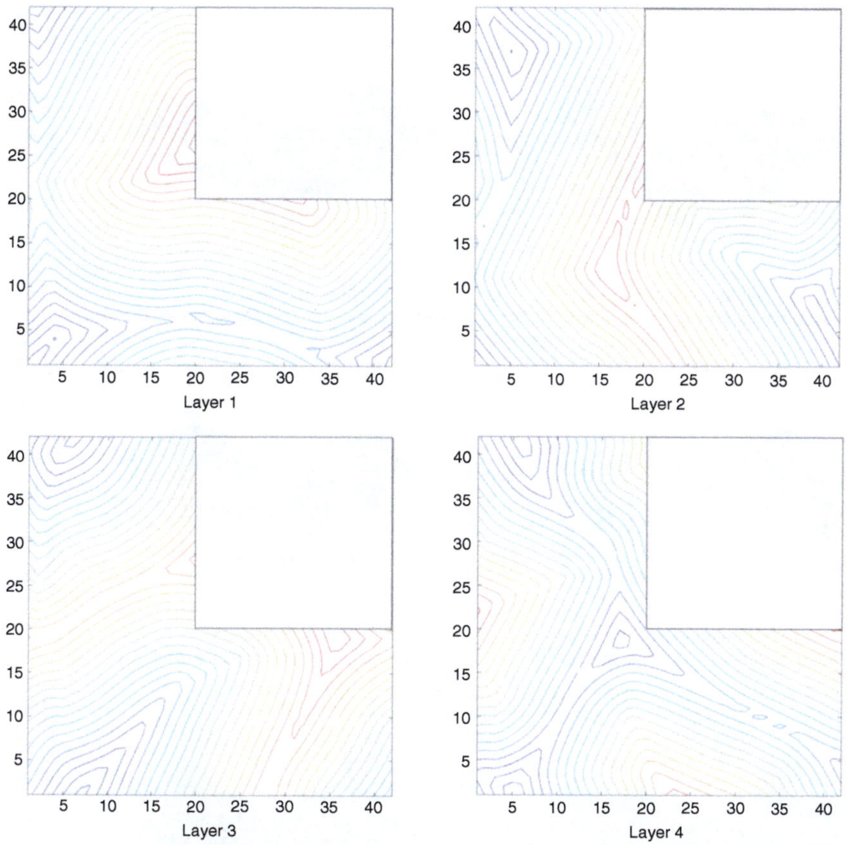

Figure 5.23 Layer-wise scanning path of the L-bracket for thermal stress minimization. Source: Chen et al. [47].

the maximum von Mises stress in each layer is reduced by optimization. Layer 1 has the greatest reduction from 73.54 to 65.24 (about 11.3%). The stress concentration at corners of Layer 1 is also significantly mitigated. The results prove that optimizing the scanning path can effectively reduce the structural residual stress and alleviate stress concentration, especially at the bottom corners, thereby reducing the likelihood of build failure such as cracking or delamination. By comparing the results of stress reduction between the compliance and stress minimization examples (Figures 5.22 and 5.24), it is found that the latter has a better performance in reducing the residual stress.

Example 5.8 ***Concurrent Optimization for Scanning Path and Structural Shape***

Considering that large residual stress usually occurs at the sharp corner of a structure due to the stress concentration, if the scanning path and structural shape optimizations are concurrently implemented, the residual stress level may be further reduced. In this example, the same four-layer L-bracket model and initial scanning paths (Figure 5.21) are employed to prove the effectiveness of the aforementioned

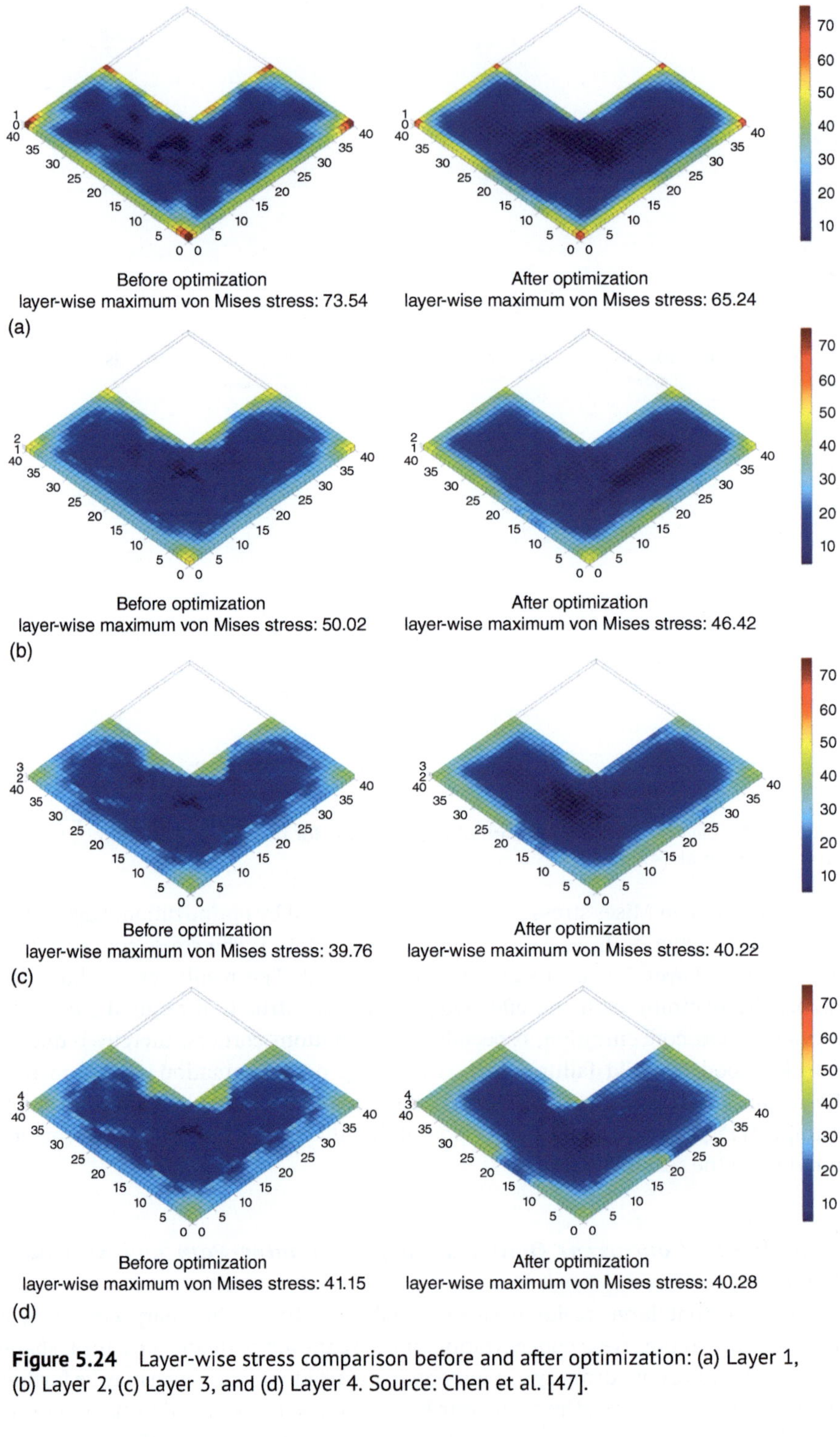

Figure 5.24 Layer-wise stress comparison before and after optimization: (a) Layer 1, (b) Layer 2, (c) Layer 3, and (d) Layer 4. Source: Chen et al. [47].

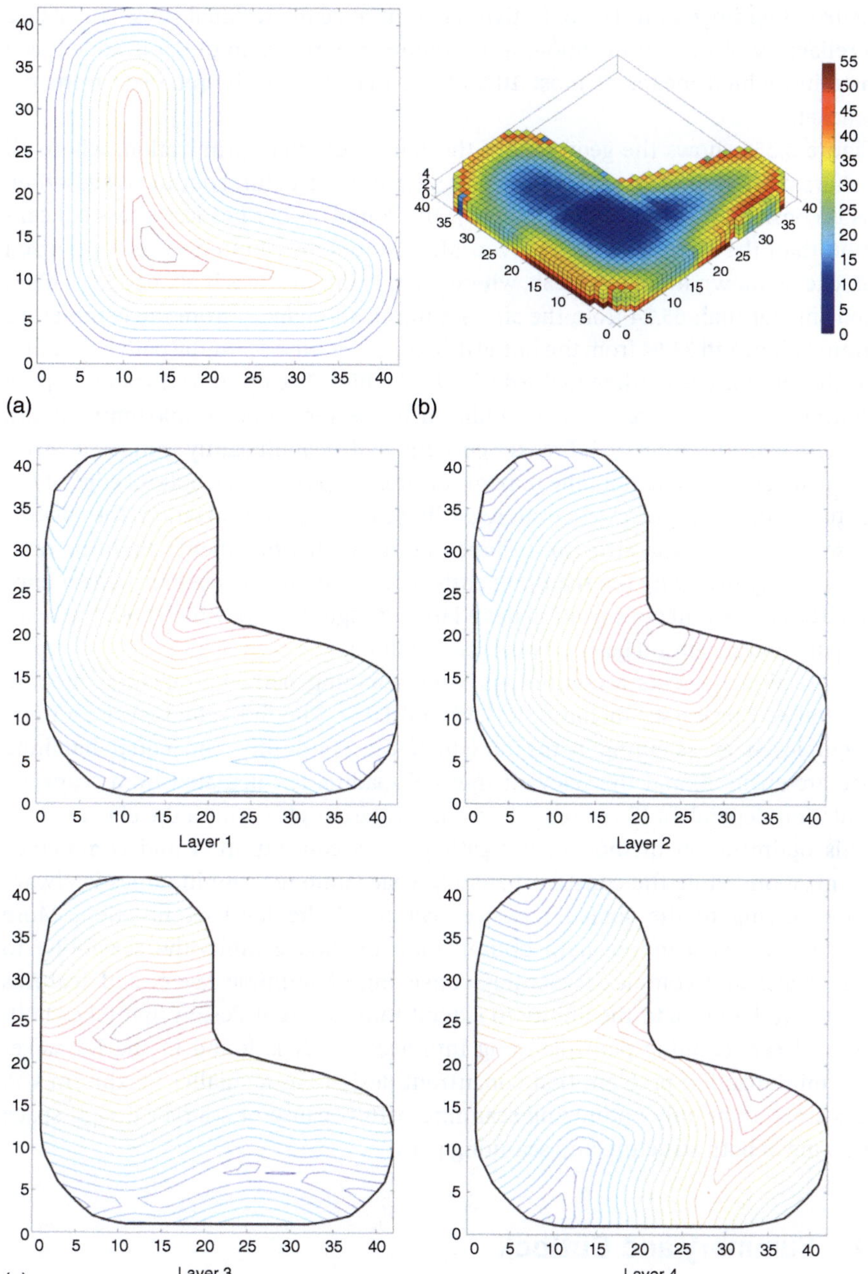

Figure 5.25 (a) Optimized structure, (b) bottom layer stress distribution, and (c) layer-wise scanning path after optimization for the concurrent optimization case. Source: Chen et al. [47].

concurrent optimization. The objective is to minimize the residual stress. To ensure the reliability of the optimization, a minimum material ratio of 0.9 is added as a constraint, which means at most 10% of the material can be removed from the L-bracket.

Figure 5.25a shows the geometry of the L-bracket after optimization, where all the sharp corners are replaced by the filets so that the structural outline is smoother. This is consistent with mechanics theory: the round corner has less stress concentration than the sharp corner. The von Mises stress distribution of the optimized L-bracket is shown in Figure 5.25b, where the maximum stress is about 55.5. This is much smaller than 65.24 from the stress minimization, 68.24 from the compliance minimization, and 73.54 from the initial design. Layer-wise scanning path is simultaneously optimized according to the derived velocities. The optimized scanning path is summarized in Figure 5.25c. In addition to the reduction in maximum stress, the stress concentration at L-bracket corners is also significantly mitigated. After the concurrent optimization, the maximum stress appears at the bottom edge of the structure. This example proves that, regarding the stress minimization, the concurrent scanning path and structural shape optimization works better than the optimization only involving the scanning path. The reason for this better performance is that the concurrent optimization has a larger design space in which both the scanning path and structural geometry can be modified.

This section presents a continuous laser scanning path optimization method for metal AM processes including L-PBF and DED. The individual level contours are employed as scanning paths of which the continuity and equal hatching space are ensured, and the relevant thermal loading is calculated from scanning orientation-dependent IS vectors. The numerical examples exhibit the effectiveness of this optimization method in mitigating stress concentration and compliance minimization, while the constant hatching space among scanning lines is always ensured owing to the signed distance feature of the level-set function. More importantly, this scanning path optimization method exhibits the scalability to much bigger and complex structures consisting of multiple layers and features such as overhangs, and the ability to extend to concurrent design that takes both TO and laser scanning path planning into account. It is found in the L-bracket stress minimization problem that concurrent design has a smaller maximum von Mises stress than only optimizing scanning path because of a larger design space consisting of both structure and scanning path.

5.7 Summary and Outlook

In this chapter, the MIS method is introduced which enables accurate and efficient simulation for residual stress and deformation prediction for a given metal AM process, especially for the L-PBF process. Then the way in which the IS values are extracted from detailed process simulation and applied to any part-scale model is discussed. The MIS method is fully validated by characterization experiments of as-built samples. Compared with the original IS method, the MIS method can provide better accuracy in residual stress and deformation simulation.

Based on the MIS method, some simulation-based design works are presented. First, the MIS-based simulation is used to perform support structure design to prevent cracking in the as-built metal components. The residual stress in the interface between the solid part and lattice support structures can be accurately simulated. The critical J-integral at the solid/support interface is finally computed. As a result, the cracking at the solid/support interface can be determined according to the critical J-integral. Moreover, in order to improve the efficiency, a global–local method is developed to simulate the residual stress at the solid/support interface. Only a very small volume needs to be considered in the global–local method. The accuracy of this global–local method has been validated by comparing to the full-scale method. As a benefit, a more efficient way is now accessible to perform the lattice support structure design for cracking prevention with respect to the layer-wise printing process.

Second, the MIS method is incorporated into the TO framework for support structure design in order to reduce the residual stress and deformation in the printed metal components. Variable density lattice support structures are used to fill in the design domain underneath the wanted bulk volume. The IS-induced force is considered as a thermal load to both the solid bulk volume and the support structures. By iterative TO, spatial density of the lattice support structures can be optimized, while the residual stress in the support structure and solid bulk volume is reduced largely. Experimental results reveal that the optimized support structure greatly decreases the residual stress in the as-built components by comparing with residual deformation in the deposition using non-optimized or teeth-like support structures. The experimental results fully demonstrate the necessity for support structure design based on TO.

Third, the MIS method is employed in the design optimization of scanning paths in the AM process. Directional IS values are dependent on the laser scan direction in the layer-wise printing process. Both compliance and thermal stress are taken as the objective function in the TO. According to the optimization result, the optimized scan paths can provide better mechanical performance and lighter residual stress for the as-built metal components.

With regard to the outlook for the MIS method, the accuracy and efficiency of this method will be further improved by studying how to further lump the layers reasonably in the layer-wise simulation. The potential influence on the MIS method by some critical factors such as laser scanning strategy will also be investigated. In addition, while past development of the MIS method has been on L-PBF and powder DED processes, future work will focus on the wire DED and binder jetting processes.

Acknowledgment

This work is supported by the Department of Energy's Nuclear Energy University Program (DE-NE0008813 and DE-NE0008994). Any opinions, findings, and conclusions or recommendations expressed in this material are those of the author(s) and do not necessarily reflect those of the Department of Energy.

References

1 Cheng, B., Shrestha, S., and Chou, K. (2016). Stress and deformation evaluations of scanning strategy effect in selective laser melting. *Additive Manufacturing* 12: 240–251.

2 Mukherjee, T., Zhang, W., and DebRoy, T. (2017). An improved prediction of residual stresses and distortion in additive manufacturing. *Computational Materials Science* 126: 360–372.

3 Vastola, G., Zhang, G., Pei, Q., and Zhang, Y.-W. (2016). Controlling of residual stress in additive manufacturing of Ti6Al4V by finite element modeling. *Additive Manufacturing* 12: 231–239.

4 Ueda, Y., Fukuda, K., and Tanigawa, M. (1979). New measuring method of three dimensional residual stresses based on theory of inherent strain (welding mechanics, strength & design). *Transactions of JWRI* 8 (2): 249–256.

5 Hill, M.R. and Nelson, D.V. (1995). The inherent strain method for residual stress determination and its application to a long welded joint. *ASME-Publications-PVP* 318: 343–352.

6 Yuan, M. and Ueda, Y. (1996). Prediction of residual stresses in welded T-and I-joints using inherent strains. *Journal of engineering materials and technology* 118 (2): 229–234.

7 Deng, D., Murakawa, H., and Liang, W. (2007). Numerical simulation of welding distortion in large structures. *Computer Methods in Applied Mechanics and Engineering* 196 (45-48): 4613–4627. https://doi.org/10.1016/j.cma.2007.05.023.

8 Murakawa, H., Deng, D., Ma, N., and Wang, J. (2012). Applications of inherent strain and interface element to simulation of welding deformation in thin plate structures. *Computational Materials Science* 51 (1): 43–52.

9 Zhang, L., Michaleris, P., and Marugabandhu, P. (2007). Evaluation of applied plastic strain methods for welding distortion prediction. *Journal of Manufacturing Science and Engineering* 129 (6): 1000–1010.

10 Michaleris, P., Zhang, L., Bhide, S., and Marugabandhu, P. (2006). Evaluation of 2D, 3D and applied plastic strain methods for predicting buckling welding distortion and residual stress. *Science and Technology of Welding & Joining* 11 (6): 707–716.

11 Wang, Y.-X., Zhang, P., Hou, Z.-G., and Li, C.-Z. (2008). Inherent strain method and thermal elastic-plastic analysis of welding deformation of a thin-wall beam. *Journal of Mechanics* 24 (04): 301–309.

12 Keller, N. and Ploshikhin, V. (2014). New method for fast predictions of residual stress and distortion of AM parts. In: *Proceedings of Solid Freeform Fabrication (SFF) Conference.*

13 Huang, G., Mehmert, P., Schafstall, H., and Buijk, A. (2016). *On The Creation of a New Finite Element Simulation Environment for Additive Manufacturing*, presented at the NAFEMS Americas Conference, Session Reviews, Session 7/1D.

14 Keller, N., Schlasche, J., Kober, C., Ploshikhin, V., Werner, C., and Domagala, T. (2015). *Distortion Compensation in Laser Beam Melting Processes*, presented at

the MSTAM 2014 - Proceedings of the 1st International Symposium Materials Science and Technology of Additive Manufacturing.

15 Afazov, S., Denmark, W.A., Toralles, B.L. et al. (2017). Distortion prediction and compensation in selective laser melting. *Additive Manufacturing* 17: 15–22.

16 Liang, X., Chen, Q., Cheng, L., and To, A. (2017) A Modified Inherent Strain Method for Fast Prediction of Residual Deformation in Additive Manufacturing of Metal Parts. In: *Proceedings of Solid Freeform Fabrication (SFF) Conference*, 2017.

17 Radaj, D. (2012). *Heat Effects of Welding: Temperature Field, Residual Stress, Distortion*. Springer Science & Business Media.

18 Frazier, W.E. (2014). Metal additive manufacturing: a review. *Journal of Materials Engineering and Performance* 23 (6): 1917–1928. https://doi.org/10.1007/s11665-014-0958-z.

19 Schoinochoritis, B., Chantzis, D., and Salonitis, K. (2017). Simulation of metallic powder bed additive manufacturing processes with the finite element method: a critical review. *Proceedings of the Institution of Mechanical Engineers, Part B: Journal of Engineering Manufacture* 231 (1): 96–117.

20 Liang, X., Cheng, L., Chen, Q. et al. (2018). A modified method for estimating inherent strains from detailed process simulation for fast residual distortion prediction of single-walled structures fabricated by directed energy deposition. *Additive Manufacturing* 23: 471–486. https://doi.org/10.1016/j.addma.2018.08.029.

21 Bugatti, M. and Semeraro, Q. (2018). Limitations of the inherent strain method in simulating powder bed fusion processes. *Additive Manufacturing* 23: 329–346. https://doi.org/10.1016/j.addma.2018.05.041.

22 Yang, Q., Zhang, P., Cheng, L. et al. (2016). Finite element modeling and validation of thermomechanical behavior of Ti-6Al-4V in directed energy deposition additive manufacturing. *Additive Manufacturing* 12: 169–177. https://doi.org/10.1016/j.addma.2016.06.012.

23 Heigel, J., Michaleris, P., and Reutzel, E. (2015). Thermo-mechanical model development and validation of directed energy deposition additive manufacturing of Ti–6Al–4V. *Additive manufacturing* 5: 9–19.

24 Liang, X., Chen, Q., Cheng, L. et al. (2019). Modified inherent strain method for efficient prediction of residual deformation in direct metal laser sintered components. *Computational Mechanics* 64 (6): 1719–1733. https://doi.org/10.1007/s00466-019-01748-6.

25 Chen, Q., Liang, X., Hayduke, D. et al. (2019). An inherent strain based multiscale modeling framework for simulating part-scale residual deformation for direct metal laser sintering. *Additive Manufacturing* 28: 406–418. https://doi.org/10.1016/j.addma.2019.05.021.

26 Goldak, J., Chakravarti, A., and Bibby, M. (1984). A new finite element model for welding heat sources. *Metallurgical transactions B* 15 (2): 299–305.

27 Tran, H.T., Liang, X., and To, A.C. (2020). Efficient prediction of cracking at solid-lattice support interface during laser powder bed fusion via global-local J-integral analysis based on modified inherent strain method and lattice support

homogenization. *Additive Manufacturing* 36: https://doi.org/10.1016/j.addma.2020.101590.

28 Peng, H. et al. (2018). Fast prediction of thermal distortion in metal powder bed fusion additive manufacturing: part 2, a quasi-static thermo-mechanical model. *Additive Manufacturing* 22: 869–882.

29 Mercelis, P. and Kruth, J.P. (2006). Residual stresses in selective laser sintering and selective laser melting. *Rapid prototyping journal.*

30 Colegrove, P.A., Donoghue, J., Martina, F. et al. (2017). Application of bulk deformation methods for microstructural and material property improvement and residual stress and distortion control in additively manufactured components. *Scripta Materialia* 135: 111–118.

31 DebRoy, T. et al. (2018). Additive manufacturing of metallic components–process, structure and properties. *Progress in Materials Science* 92: 112–224.

32 Tran, H.T., Chen, Q., Mohan, J., and To, A.C. (2020). A new method for predicting cracking at the interface between solid and lattice support during laser powder bed fusion additive manufacturing. *Additive Manufacturing* 32: https://doi.org/10.1016/j.addma.2020.101050.

33 Tran, H.T., Mraied, H., Iz0061di, S. et al. (2017). Optimizing ductility and fracture of amorphous metal thin films on polyimide using multilayers. *International Journal of Fracture* 204 (2): 129–142.

34 Tran, H.T., Shirangi, M.H., Pang, X., and Volinsky, A.A. (2014). Temperature, moisture and mode-mixity effects on copper leadframe/EMC interfacial fracture toughness. *International Journal of Fracture* 185 (1): 115–127.

35 Anderson, T.L. (2017). *Fracture Mechanics: Fundamentals and Applications.* CRC press.

36 Rice, J. and Rosengren, G.F. (1968). Plane strain deformation near a crack tip in a power-law hardening material. *Journal of the Mechanics and Physics of Solids* 16 (1): 1–12.

37 Short, J. and Hoeppner, D. (1989). A global/local theory of fatigue crack propagation. *Engineering fracture mechanics* 33 (2): 175–184.

38 Ansys (2019). *Ansys 2019 R1 User's Manual.* Canonsburg, Pennsylvania, USA: Ansys Inc.

39 Cheng, L., Bai, J., and To, A.C. (2019). Functionally graded lattice structure topology optimization for the design of additive manufactured components with stress constraints. *Computer Methods in Applied Mechanics and Engineering* 344: 334–359.

40 Hill, R. (1979). Theoretical plasticity of textured aggregates. In: *Mathematical Proceedings of the Cambridge Philosophical Society*, vol. 85, no. 1, 179–191. Cambridge University Press.

41 Liang, X. et al. (2020). Inherent strain homogenization for fast residual deformation simulation of thin-walled lattice support structures built by laser powder bed fusion additive manufacturing. *Additive Manufacturing* 32: 101091.

42 Zhang, Y., Shang, S., and Liu, S. (2017). A novel implementation algorithm of asymptotic homogenization for predicting the effective coefficient of thermal expansion of periodic composite materials. *Acta Mechanica Sinica* 33 (2): 368–381.

43 Cheng, G.-D., Cai, Y.-W., and Xu, L. (2013). Novel implementation of homogenization method to predict effective properties of periodic materials. *Acta Mechanica Sinica* 29 (4): 550–556.

44 Svanberg, K. (1987). The method of moving asymptotes—a new method for structural optimization. *International journal for numerical methods in engineering* 24 (2): 359–373.

45 Langelaar, M. (2018). Combined optimization of part topology, support structure layout and build orientation for additive manufacturing. *Structural and Multidisciplinary Optimization* 57 (5): 1985–2004.

46 Cheng, L., Liang, X., Bai, J. et al. (2019). On utilizing topology optimization to design support structure to prevent residual stress induced build failure in laser powder bed metal additive manufacturing. *Additive Manufacturing* 27: 290–304. https://doi.org/10.1016/j.addma.2019.03.001.

47 Chen, Q., Liu, J., Liang, X., and To, A.C. (2020). A level-set based continuous scanning path optimization method for reducing residual stress and deformation in metal additive manufacturing. *Computer Methods in Applied Mechanics and Engineering* 360: 112719.

6

High-Fidelity Modeling of Metal Additive Manufacturing

Hou Yi Chia and Wentao Yan

National University of Singapore, Department of Mechanical Engineering, 9 Engineering Drive 1, Singapore 117575, Singapore

6.1 Background

Metal additive manufacturing (AM) is a promising technology capable of producing complex-shaped parts without the need for special tooling [1, 2]. The unique manufacturing capability offered by metal AM has garnered interest from multiple industries such as the aerospace, biomedical, and automotive industries. However, as achieving high process reliability in metal AM remains a significant challenge despite more than 20 years of development since c. 2000 [3], AM as a manufacturing technology has seen limited adoption in these industries. In particular, defects that arise during the manufacturing process such as pores, insufficient bonding, delamination, and thermal cracking are detrimental to the part quality. As such, the comprehensive understanding of the mechanics and physics of the manufacturing process is crucial in guiding AM practitioners to reduce or eliminate defects and even aid in optimizing process parameter selection for production efficiency.

Depending on the metal AM method used, e.g. selective laser melting (SLM), electron beam melting (EBM), directed energy deposition (DED), and wire arc additive manufacturing (WAAM), different physical mechanisms will be encountered. In this chapter, the focus will be on the powder bed fusion (PBF) process, where many of the key physics in other metal AM processes can be illustrated due to their similarity. The aforementioned SLM and EBM techniques are PBF processes with different heat sources and will hereon be referred to as laser powder bed fusion (L-PBF) and electron-beam powder bed fusion (EB-PBF), respectively. In the PBF process, a thin layer of powder is mechanically spread onto a substrate, and a sliced region obtained from a three-dimensional computer-aided design (CAD) file is locally fused on the powder bed. Thereafter, the process repeats until the full part geometry is manufactured. Essentially, two main stages occur for each successive layer: powder spreading and powder melting.

Although the PBF process principle is seemingly simple, many complex physical mechanisms occur at each stage of the PBF process. Well-known characteristics of the PBF process include the dynamic powder behaviors, high heating and cooling

Additive Manufacturing Technology: Design, Optimization, and Modeling, First Edition. Edited by Kun Zhou.

rates, and high thermal gradients that have rarely been encountered in traditional metallurgical processes. Under these circumstances, the manufacturing defects are the result of the complex interplay between the hydrodynamic and phase-change effects. Therefore, it is imperative that we explore and understand the multi-physical mechanisms that govern the newly encountered physical phenomena in the PBF process. With a comprehensive understanding of the AM process, we can eliminate the sources of defect formation, construct reliable prediction capabilities, and eventually achieve optimal build.

The experimental challenges in understanding the AM process largely originate from the fact that the melting and solidification process occurs from mesoscopic to macroscopic spatiotemporal scales. Particularly, the short timescale of the process presents real physical challenges to experimental investigations of the AM process, despite the continued improvements in equipment and technique. Most literatures have reported the post-process characterization studies on AM materials. Several researchers have used high-speed cameras to observe the melt pool surface and powder dynamics [4–6]. Only recently, *in-situ* studies of the melt pool via ultrahigh-speed X-ray imaging have been reported [7–10].

Evidently, process characterization and understanding remain limited. While the *a posteriori* observation allows correlative relations to be drawn, it does not shed light on the actual mechanisms and melt pool dynamics that result in the observable melt pool dimensions and properties. Additionally, post-treatment of samples during experimental characterizations may alter the samples. Finally, *in-situ* observations and measurement of the melt pool dynamics are highly challenging and require specialized equipment which are usually expensive to acquire; even then, only 2D projections of the physical process can be obtained. Notably, experimentally tracing the thermal stress evolution during the melting process is inexpedient, if not impossible due to the high local temperature gradients and the fast heating and cooling rates. Therefore, the physical limitations of current experimental techniques require the aid of computational models for deeper understanding of the AM process.

Since the Digital Revolution, it has become feasible and even necessary to perform investigations with computational models. In the long history of metallurgy, trial-and-error experiment has been the only way to understand and optimize the manufacturing process. Now, researchers and engineers can use numerical modeling to mathematically describe the physical process and employ numerical techniques to solve for the evolution of the physical process. In particular, high-fidelity modeling is especially powerful for metal AM modeling, which allows us to capture unprecedented details at the micro- and mesoscale, a limitation of current experimental methodologies and investigations. From powder spreading, powder melting, and cooling of the melt pool, each process evolves from multi-physical interactions governed by their respective physical laws. It is therefore prudent to endeavor developing a description of the PBF process with well-known physics and possibly uncover new physical dynamics using high-fidelity modeling.

Furthermore, the validated models can be further used for optimization studies and guide the design and selection of process parameters for manufacturing. This approach saves time and material costs as it significantly reduces the number of repeated experiments to characterize the process with many manufacturing parameters. This will be especially useful and of interest for commercial companies, where minimum time and effort can be devoted to process characterization and optimization for different machines and powder sources.

In this chapter, we introduce a compilation of our works on high-fidelity modeling for PBF processes at various stages of processing. To strive for a comprehensive modeling approach, we attempt to incorporate almost every aspect and procedure of the PBF process using a laser or electron beam as the heat source (Figure 6.1). This framework includes powder spreading, powder preheating (for EB-PBF), powder melting, powder spattering and denudation (for L-PBF), and thermal-mechanical behaviors, which involves the physics of powder dynamics, molten pool dynamics, and thermal stress evolution. However, our works on the modeling of thermoelectric magnetohydrodynamics [17], keyhole pore formation [18], microstructure evolution [19–22], resultant mechanical properties [23, 24], and crystal plasticity for residual stresses in PBF [25, 26] will not be included in this chapter.

This chapter is segmented into four main sections: the powder spreading process is described in Section 6.2; models for powder melting are described in Section 6.3; a unique coupling method for evaluating thermal stress is proposed in Section 6.4; and other unique phenomena such as powder pre-sintering and powder spattering are elaborated in Section 6.5. In each section, the computational models used will be briefly introduced. The model is then verified with benchmarks (if applicable) and validated against experiments. The results are followed by the observations and insights attainable through the simulations at different processing conditions. Each section concludes with a summary and outlines future directions for AM modeling.

Overall, these simulations reveal multiple interesting and crucial multi-physics phenomena behind the AM process that influence the part quality. They also demonstrate the strength and value of modeling in aiding our understanding of the dynamic mechanisms behind the AM process, thereby providing insights for process improvements, parameter adjustments, and optimization.

6.2 Powder Spreading

Powder spreading is the process where a thin layer of powder is mechanically spread onto the substrate. The quality of the powder bed and the spreading efficiency are the main focuses in the powder spreading process. Specifically, a high packing density of the powder layer is favored for greater thermal conductivity and lower initial porosity. These conditions ensure sufficient bonding during the scanning process, thereby improving part quality. On the other hand, maximizing the spreading speed is also desired as it can significantly increase the production efficiency. Numerous factors influence the powder spreading mechanism, which can be revealed with high-fidelity simulations.

The powder spreading mechanism is investigated with the discrete element method (DEM) to include interactions at the microscopic particulate scale. A set of

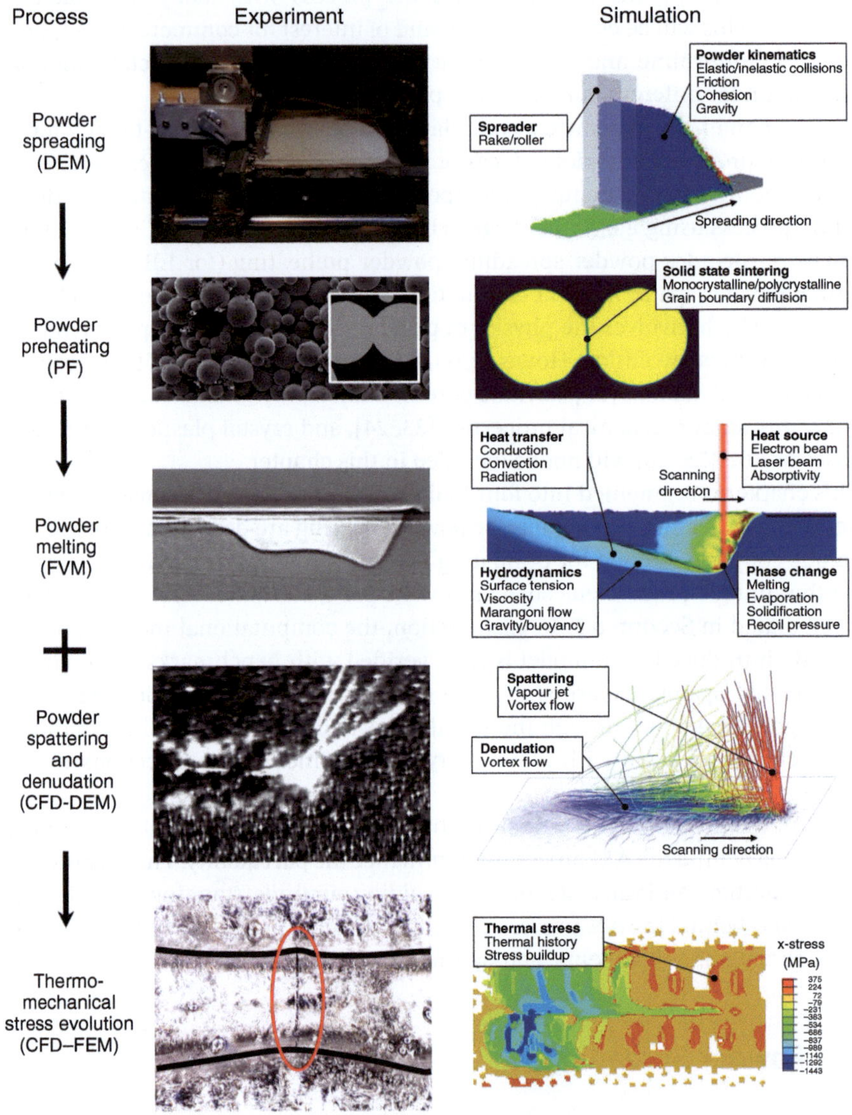

Figure 6.1 An overview of the overall framework for high-fidelity modeling of the PBF process. Below each process procedure, the computational method is stated in parenthesis (DEM: discrete element method, PF: phase-field, FVM: finite volume method, CFD: computational fluid dynamics, FEM: finite element method, CFD–DEM: coupling CFD and DEM for force interactions, and CFD–FEM: A novel one-way coupling of CFD and FEM for temperature input and morphology mapping). Source: Experiments (from top to bottom) – Yan et al. [11]/with permission of Elsevier; Yan et al. [12]/with permission of Elsevier; Wang et al. [13]; Bidare et al. [5]/with permission of Elsevier/CC BY 4.0; Chen and Yan [14]/Elsevier/CC BY 4.0. Simulations (from top to bottom) – Chen et al. [15], with permission from Elsevier; Yan et al. [12], with permission from Elsevier; Wang et al. [13]/ with permission of American Physical Society; Chen and Yan [16], with permission from Elsevier; Chen and Yan [14] under CC-BY 4.0 license.

powder particles with a preset size distribution is generated and randomly packed at the initial phase of the model. As the spreading tool moves and comes into contact with the powder particles, the kinetic behavior of each individual particle is calculated using classical equations of motions. At this fine level of resolution, cohesive van der Waals forces become significant, in addition to the tribological interactions described by contact mechanics. Two approaches for powder spreading commonly found in industrial equipment are the rake-type and roller-type spreading methods (Figure 6.2a). In this section, the model is firstly described. Then, the phenomena arising from these spreading methods as well as the influence of the spreading speed are presented, which are detailed in our previous studies [15, 27].

6.2.1 Governing Equations

In the DEM model, the powder particles are assumed to be spherical, and the mass, moment of inertia, radius, position vector, and the angular displacement of the particle i are denoted by m_i, I_i, $\mathbf{R}_i$, $\mathbf{r}_i$, and θ_i, respectively. The translational and rotational motions of each individual powder particle are calculated based on classical equations of motion as in Eq. (6.1):

$$\begin{cases} m_i \dfrac{\mathrm{d}^2 \mathbf{r}_i}{\mathrm{d}t^2} & = \mathbf{F}_\mathrm{g} + \sum \mathbf{F}_{\mathrm{ext},i} + \sum_j \left(\mathbf{F}_{\mathrm{n},ij} + \mathbf{F}_{\mathrm{t},ij} + \mathbf{F}_{\mathrm{c},ij} \right) \\ I_i \dfrac{\mathrm{d}^2 \theta_i}{\mathrm{d}t^2} & = \sum_j \mathbf{T}_{ij} = \sum_j \left(\mathbf{R}_i \times \mathbf{F}_{\mathrm{t},ij} \right) \end{cases} \tag{6.1}$$

where $\mathbf{F}_\mathrm{g}$ is the gravitational force, $\mathbf{F}_{\mathrm{ext},i}$ is any external force acting on particle i, and the normal contact force, tangential contact force, cohesion force, and moment acting on particle i due to particle j are denoted by $\mathbf{F}_{\mathrm{n},ij}$, $\mathbf{F}_{\mathrm{t},ij}$, $\mathbf{F}_{\mathrm{c},ij}$, and $\mathbf{T}_{ij}$, respectively. The moment is generated as a result of the forces applied.

Note that due to external influences, the gravitational force $\mathbf{F}_\mathrm{g}$ may be affected, and the external forces $\mathbf{F}_{\mathrm{ext},i}$ may arise (see Section 6.5.2.1). Nonetheless, under the usual circumstances with no external influences, the following conditions are applicable:

$$\begin{cases} \mathbf{F}_\mathrm{g} = m_i \mathbf{g} \\ \mathbf{F}_{\mathrm{ext},i} = 0 \end{cases} \tag{6.2}$$

where $\mathbf{g}$ is the gravitational acceleration.

The normal and tangential components of the contact force are calculated with the Hertz–Mindlin contact model (Figure 6.2b) [28, 29] with Eq. (6.3), which incorporates linear spring-damping contact forces of stiffness coefficient k and damping coefficient c:

$$\begin{cases} \mathbf{F}_{\mathrm{n},ij} & = k\delta_\mathrm{n} + c\dot{\delta}_\mathrm{n} \\ \mathbf{F}_{\mathrm{t},ij} & = \min\left(k\delta_\mathrm{t} + c\dot{\delta}_\mathrm{t}, \mu_s \left| \mathbf{F}_{\mathrm{n},ij} \right| \right) \\ \dot{\delta}_\mathrm{n} & = \left(\dot{\mathbf{r}}_i - \dot{\mathbf{r}}_j \right) \cdot \mathbf{n}_{ij} \\ \dot{\delta}_\mathrm{t} & = \left(\dot{\mathbf{r}}_i - \dot{\mathbf{r}}_j + \mathbf{R}_i \times \dot{\theta}_i + \mathbf{R}_j \times \dot{\theta}_j \right) \times \mathbf{n}_{ij} \end{cases} \tag{6.3}$$

where $\boldsymbol{\delta}_\mathrm{n}$ and $\boldsymbol{\delta}_\mathrm{t}$ are the normal and tangential displacements between the powder particles, $\mathbf{n}_{ij}$ is the normal unit vector between particles i and j; and μ_s is the sliding

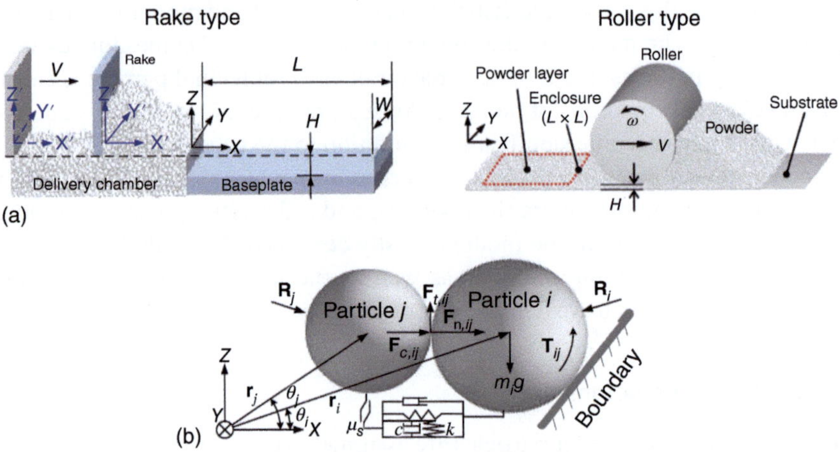

Figure 6.2 (a) Schematic of the two types of powder spreading mechanisms. The major physical factors affecting the powder spread includes powder size distribution, rake shape, spreading speed, and powder thickness. Source: Left image from Chen et al. [15], with permission from Elsevier. Right image Chen et al. [27], with permission from Elsevier. (b) Schematic of the Hertz–Mindlin model and the interaction forces between powder particles. Source: Chen et al. [15], with permission from Elsevier.

friction coefficient which can be adjusted to incorporate the influence of the shape irregularity of the powder particles. The values of k and c are calculated in Eq. (6.4):

$$\begin{cases} k &= \frac{16}{15}R^{*\frac{1}{2}}E^{*}\left(\frac{15m^{*}\eta_i^2}{16R^{*1/2}E^{*}}\right)^{\frac{1}{5}} \\ c &= \left(\frac{4m^{*}\,k\ln^2 e_{ij}}{\ln^2 e_{ij}+\pi^2}\right)^{\frac{1}{2}} \\ R^{*} &= \frac{R_iR_j}{R_i+R_j} \\ m^{*} &= \frac{m_im_j}{m_i+m_j} \\ E^{*} &= \frac{E_iE_j}{(1-\eta_j^2)E_i+(1-\eta_i^2)E_j} \end{cases} \tag{6.4}$$

where η_i and E_i are the Poisson's ratio and Young's modulus of the powder particle i, and e_{ij} is the coefficient of restitution.

Next, the van der Waals cohesive force $\mathbf{F}_{c,ij}$ is calculated with the Johnson-Kendall-Roberts (JKR) cohesion model [30, 31] described by the set of equations in Eq. (6.5):

$$\begin{cases} \mathbf{F}_{c,ij} = -4(\pi E^{*}\gamma)^{\frac{1}{2}}\xi^{\frac{3}{2}} + \frac{4E^{*}}{3R^{*}\xi^{3}} \\ \xi^4 - 2R^{*}\delta_n\xi^2 - \frac{4\pi\gamma R^{*2}}{E^{*}}\xi + R^{*2}\delta_n^2 = 0 \\ \delta_{cri} = \frac{\xi_{cri}^2}{R^{*}} - \sqrt{\frac{4\pi\gamma\xi_{cri}}{E^{*}}} \\ \xi_{cri} = \frac{9\left(3\sqrt{2}-4\right)\pi\gamma R^{*2}}{8\sqrt{2}E^{*}} \end{cases} \tag{6.5}$$

Table 6.1 Physical parameters of 316L stainless steel powder particle in the DEM model.

Material property	Value
Material density (ρ_O)	7.8×10^3 kg m^{-3}
Young's modulus (E)	2.2×10^3 MPa
Poisson's ratio (η)	0.3
Restitution coefficient (e)	0.9
Friction coefficient (μ_t)	0.62
Surface energy density (γ)	0.097 mJ m^{-2}

where γ is the surface energy density, ξ is an intermediate variable that defines the contact spot radius between particles, and δ_n is the distance between the particles. Note that it is possible for negative cohesion forces to exist. If the distance between any two powders exceed the critical value δ_{cri}, the cohesive force will be considered negligible.

Finally, with ρ_O being the material density, the stability of the simulation is ensured by using the Rayleigh critical time step criterion:

$$\Delta t < \min \left(\frac{\pi R}{0.163\eta + 0.877} \sqrt{\frac{\rho_O}{E}} \right) \tag{6.6}$$

6.2.2 Model Validation

The powder spreading simulations are compared with experimental results to validate the reliability of the DEM model. The material investigated is the 316L stainless steel powder, with the physical parameters of the DEM model given in Table 6.1. Note that the Young's modulus of stainless steel has been reduced by a factor of 100 to increase the time step as determined by Eq. (6.6), hence reducing simulation time. This adjustment is a common strategy in DEM to reduce the computational cost [32], provided that the numerical results are insensitive to the adjustment of the material properties, i.e. Young's modulus in this case [33].

For the rake-type spreading model, three groups of samples with different powder size distributions were spread at different layer thickness as shown in Figure 6.3a. In particular, the study includes a sample with a bimodal size distribution ($S_{c\text{-}2}$), as it is hypothesized that the finer powder particles can fill into the voids of its coarser counterparts, thereby improving the powder packing density. The other samples with size ratios of 1 : 5 and 1 : 3 were obtained from previous experimental studies [34, 35] and were used in the simulations shown here. For the roller-type spreading model, a common powder size distribution of 5-55 μm was used.

In both the rake-type and roller-type scenarios, the relative packing density of the DEM models are found to agree well with the experiments (Figure 6.3b–f). The rake-type spreader model is validated against the values of relative packing

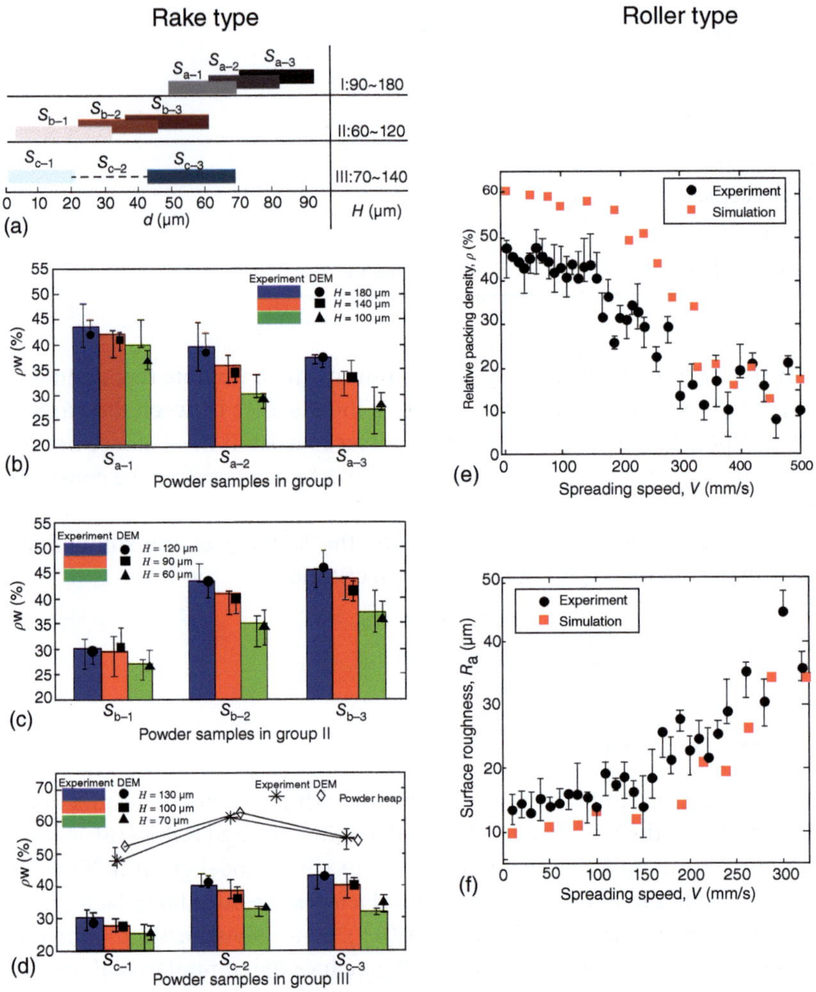

Figure 6.3 Experimental validation of the DEM model in the rake-type (a)–(d) and roller-type spreader (e) and (f). (a) Samples of different powder size distributions grouped in I–III depending on the spreading layer thickness; (b)–(d) comparison of the results of the relative packing density between the simulation and experiments; (e) comparison of the results of relative packing density between the experiment and simulation; and (f) comparison of the results of surface roughness between the experiment and simulation. The linear surface roughness was measured and calculated for comparison. In the experiments, laser profilometry was used to measure the linear profile of the powder layer at random zones after spreading. In the simulations, the height profile of the powder layer was obtained by sectioning along the deposited layer. Details of the experiments are elaborated in Refs. [15, 27]. Source: Figures (a)–(d) Chen et al. [15], with permission from Elsevier. Figures (e) and (f) Chen et al. [27], with permission from Elsevier.

density measured from experiments (Figure 6.3b–d). Furthermore, the dynamic response angle from the rake-type simulations agrees well with the experimental results conducted by Escano et al. [36], which measured the *in-situ* angles during the spreading process by employing a high-speed X-ray imaging technique. This agreement provides greater confidence to our rake-type powder spreading model. Similarly, results from the roller-type model are compared with the surface roughness and relative packing density obtained through experimental measurements (Figure 6.3e,f). The qualitative agreement of the results between the simulation and experiments shows that the simulation results are acceptable and reasonable. Overall, the validation from experiments conducted in the same study or independently by other researchers provides assurance to our powder spreading models.

Additionally, the influence of the particle size and layer thickness on the packing density can be observed from Figure 6.3. In the rake-type spreading scenario, there is a non-monotonic relationship between the particle size and the packing density: while packing density decreases as powder size increases in Group I, the opposite trend is observed in Group II. Next, the results show that the packing density increases with increasing layer thickness. Additionally, the results suggest that the use of bimodal-sized powders does not necessarily improve the packing density. Therefore, it is necessary to identify the optimal mixture ratio for the use of bimodal-sized powder distribution strategy to be feasible. The optimal size and mixture ratio are influenced by the material properties and powder particle characteristics, which could be obtained by performing a Design of Experiments study. On the other hand, the results from the study of the roller-type spreader show obvious trends: as the spreading speed increases, there is a deterioration in the packing quality and a greater rate of increase in the surface roughness.

6.2.3 Spreading and Deposition Mechanisms

By simulating the powder spread under different conditions using the DEM model described earlier, we have identified different mechanisms that affect the quality of the powder bed. The stress-dip regions and the powder flow characteristics can be observed, while the mechanisms such as the wall effect, cohesion effect, and percolation effect become apparent. These mechanisms are able to explain the results observed in Figure 6.3. Additionally, the influence of velocity on the powder bed quality is investigated in the counter-roller-type spreading.

6.2.3.1 Rake-Type Powder Spreading

The conventional definition of stress (in continuum mediums such as fluids and bulk solids), force per unit area, is not applicable for the discrete particles as the forces in the discrete medium are anisotropic and non-differentiable. Therefore, it will be prudent to define the stress σ as the total contact forces per unit volume within a cubic cell of unit volume l (Figure 6.4a):

$$\sigma = \frac{1}{l^3} \cdot \sum_{\text{cell}} \left| \mathrm{F}_{\mathrm{n},ij} + \mathrm{F}_{\mathrm{t},ij} \right| . \tag{6.7}$$

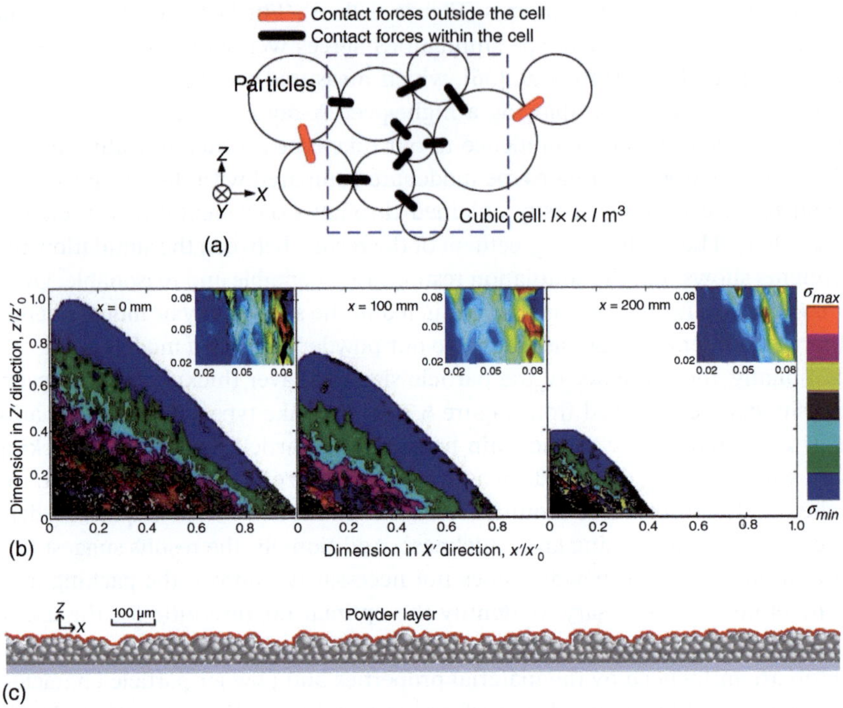

Figure 6.4 (a) The stress defined as the total contact forces per unit volume. (b) Stress contours of the powder pile, where the sudden dip (in blue) can be observed at the bottom region. (c) The height profile of the deposited powder layer. Source: Chen et al. [15], with permission from Elsevier.

The stress, defined as the total contact forces per unit volume for each particle in Eq. (6.7), dominates the dynamic behaviors and packing density. This is usually the case for most cases of particulate flow. However, if we examine the stress along the bottom of the pile, we find that the stress drops significantly under the apex of the pile. This "stress-dip" phenomenon exists at this region regardless of the variation of powder pile size (Figure 6.4b). While it is commonly known that the stress-dip phenomenon occurs in a static packed pile, this simulation shows that stress-dip also occurs in the dynamically moving powder pile during the powder spreading process, although this requires further experimental validations. In general, it can be observed that the stress along the bottom of the powder pile decreases with the decreasing powder pile size, excluding the stress-dip region. Additionally, due to the discrete nature of powder material and random collision between particles, local stress values fluctuate. Consequently, the stress fluctuation results in an unstable mass flow rate, which leads to noncontinuous or rough regions in the powder layer (Figure 6.4c).

Next, the dominant mechanisms are elaborated. Firstly, there are two types of "wall effects" that can be identified: the static and the dynamic wall effects. These are geometrical effects that induce voids and become increasingly apparent as the

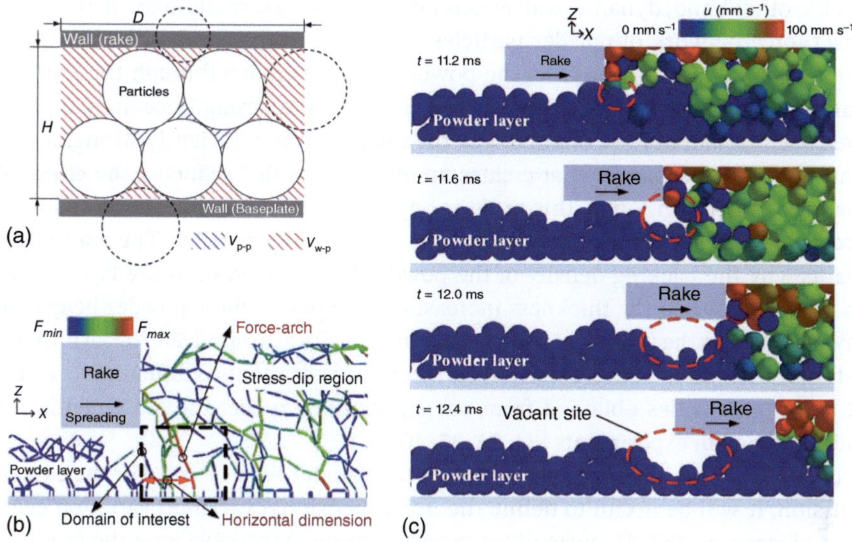

Figure 6.5 Static and dynamic wall effects reduce the packing layer density. (a) Schematic illustrating the static wall effect: greater packing cannot be achieved due to geometrical constraints; (b) a dynamic wall effect due to the transient force-arch that arise during powder spreading: this phenomenon creates an additional barrier that shortens the effective length in the static wall perspective; and (c) voids that are induced due to the dynamic wall effect. Source: Chen et al. [15], with permission from Elsevier.

layer thickness decreases. In the region between the rake and the base plate, the two barriers create a static wall effect. The static wall effect arises because of the geometrical constraints imposed on the packing of spherical powder particles between two flat plates. In essence, voids are induced because spaces that could have been filled by powder particles cannot exist due to the spatial constraints imposed by the walls (Figure 6.5a).

By considering the simple geometrical analysis of an orthorhombic packing of particles with a diameter d in a cylinder with a height H and diameter D, we can quantify the volume ratio between the wall-particle voids and the total void volume:

$$\frac{V_{\text{w-p}}}{V_{\text{total}}} \approx \frac{d\left(\frac{1}{H}+\frac{4}{D}\right)}{0.8+d\left(\frac{1}{H}+\frac{4}{D}\right)} \approx \begin{cases} 0, & \text{if } (D \gg d, H \gg d) \\ \frac{1}{0.8\cdot\left(\frac{H}{d}\right)+1}, & \text{if } (D \gg d) \end{cases} \tag{6.8}$$

The larger the value, the stronger the static wall effect is. If the dimensions of the container are several orders of magnitude larger than the particle diameter ($D \gg d$ *and* $H \gg d$), the static wall effect becomes negligible. In PBF AM, the horizontal dimension is much greater than the particle diameter ($D \gg d$). Effectively, it implies that the static wall effect in the thin powder layer becomes significant and increases if the particle size increases or if the layer thickness decreases. If $H = d$ (the critical height), approximately 56% of the total porosity is due to wall-particle voids, which demonstrates that the static wall effect will be a major reason for low packing density.

On the other hand, dynamic wall effects arise due to the formation of the transient contact force network of powder particles as illustrated in Figure 6.5b. This force network is broken by the rake as the powder particles move through the domain of interest and deposit onto the powder layer. Notably, a strong force-arch appears (in red), which can only be disrupted by the rake. As it is sufficiently strong, it will act as an additional barrier that creates an effective "wall," reducing the effective horizontal dimension. Thus, this transient dynamic effect will induce voids in the powder layer, further reducing the packing density (Figure 6.5c). The wall effect explains why the packing density of the powder layer increases as the layer thickness increases. When the thickness increases to the point where powder heaps are formed, it reaches the upper limit of the packing density at around 50–70%. Although the direct observation of these wall effects through experiments is difficult, the resultant packing densities obtained from our simulations reflect the aforementioned trends obtained in experiments [34, 35, 37–40].

Secondly, the cohesion effect arises due to the van der Waals force. For the ease of discussion, it will be useful to define the nondimensional variables variation coefficient of stress σ_V (VCS), normalized mass flow rate Q (NMFR), and the bonding number (K) with reference to the domain of interest in Figure 6.5b:

$$\begin{cases} \sigma_V &= \frac{\sigma_{c,st}}{\overline{\sigma}_c} \\ Q &= \frac{q_c}{q_{c,i}} = \frac{q_c}{\rho VHW} \\ K &= \frac{F_{c,max}}{m_i g} = \frac{9\gamma}{4\rho g} \cdot \frac{1}{d^2} \end{cases} \tag{6.9}$$

where $\sigma_{c,st}$ and $\overline{\sigma}_c$ are the standard deviation and the mean value of the stress σ_c, respectively; q_c and $q_{c,i}$ are the mass flow rate and ideal mass flow rate, respectively, through the domain of interest; and $F_{c,max}$ is the maximum van der Waals force. Physically, VCS measures the degree of fluctuation stress, NMFR represents the local relative packing density, and K is the ratio of van der Waals force to gravitational force which indicates degree of influence of the cohesive van der Waals force.

The graph of VCS and NMFR for the simulation of mono-sized powders with different particle diameters at a layer thickness of 150 μm is plotted (Figure 6.6). One can observe that a critical point exists. If the bonding number K is relatively small, it implies that the cohesion effect is insignificant and the wall effects will dominate. Thus, beyond the critical point, the decrease in powder particle size will therefore weaken the wall effects and reduce the fluctuation of stress and enhance mass flow, resulting in a denser powder layer with less voids. However, before the critical point, the cohesion effects dominate. As such, the decrease in the powder particle size will instead result in the decreased stability of powder flow, increasing the fluctuation of the stress and then reducing the mass flow rate, leading to a loose powder layer with powder agglomerations and voids. Note that these results pertain to the discussion of stainless steel powders, and the trends will necessarily change with the use of powders of different materials.

Thirdly, the percolation effect is the dynamic sieving effect when powder particles pass through the temporary void regions of moving powder bed while colliding with

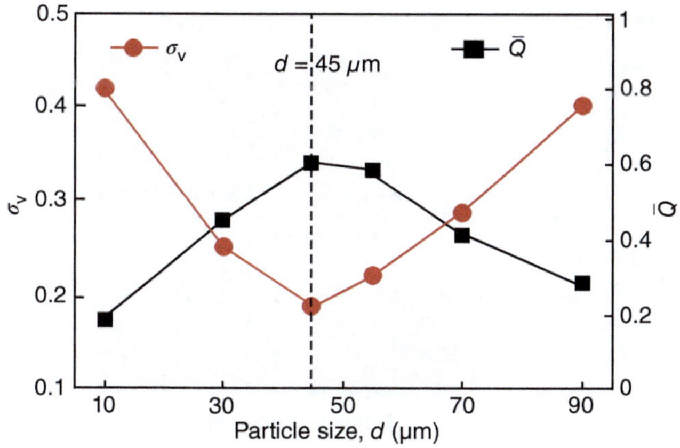

Figure 6.6 The plot of σ_V and $\overline{Q}$ against particle size, showing the optimal size for best powder bed quality. Source: Reproduced from Chen et al. [15], with permission from Elsevier.

each other randomly. This effect explains the reason for the apparent ineffectiveness of using a bimodal-sized powder distribution. Due to the percolation effect, powder size segregation occurs. As observed in the simulation of a powder spreading process of powder with bimodal-sized distribution $S_{c\text{-}2}$ (Figure 6.7), the segregation of the finer particles from the bulk resulted in more voids and reduced packing density. There is also no significant improvement over the powder flow rate or stress stability due to the percolation effect.

6.2.3.2 Roller-Type Powder Spreading

The roller in a roller-type spreader influences the spreading mechanisms slightly as compared to the rake-type spreader. In particular, the rolling motion coupled with the translational spreading motion, and the tendency for the particles to adhere to the roller affects the powder flow dynamics.

The powder particles close to the surface of the substrate have much lower speeds and are directly deposited onto the substrate. As such, the powder flow characteristics close to the substrate directly affects the packing quality. Therefore, the powder flow through the section between the roller and the powder bed (domain of interest and area of interest in Figure 6.8), and the velocities of the powder particles through the section are observed. On the other hand, the perspective can be shifted from the global reference to the frame attached to the bottom of the roller. By shifting the perspective, it can be observed that as the roller rotates and moves translationally, it imposes a moving boundary on the powder pile, leading to a vortex flow relative to the roller as observed in the powder movement.

Next, we demonstrate that the increase in the applied pressure on the powder bed by increasing the spreading speed is not necessarily favorable for the packing quality of the powder layer. By defining mass rate q and powder particle speed V in the area of interest, the ideal scenario would be that q is directly proportional to V. However, as the spreading speed increases, the mass flow rate increases at a decreasing

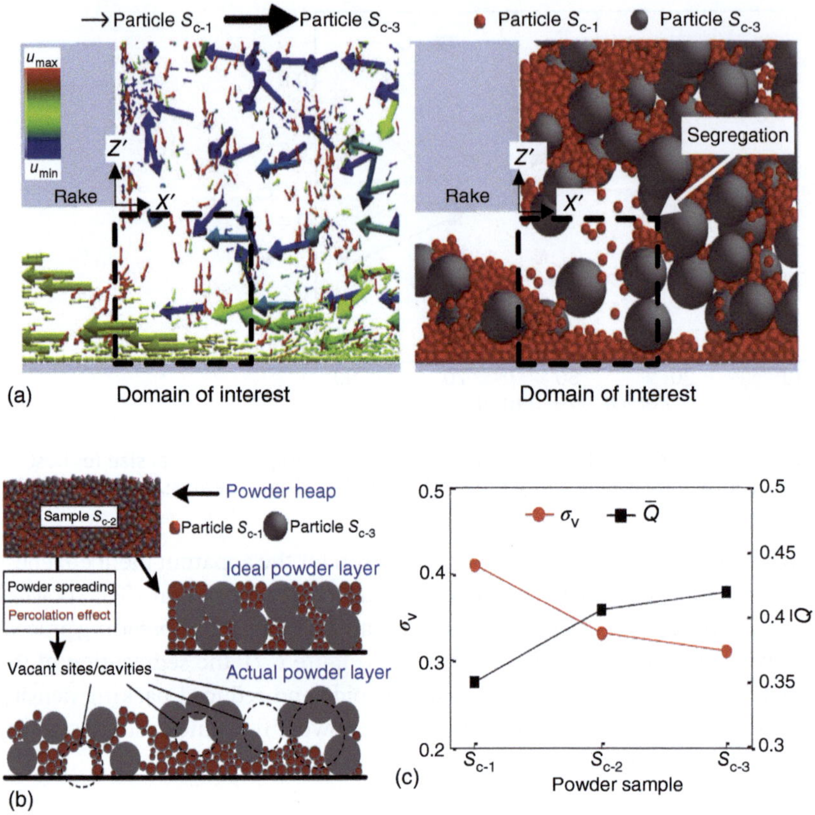

Figure 6.7 Percolation effect where $H = 140\,\mu\text{m}$. (a) Velocity field and particle morphology of the bimodal sample $S_{c\text{-}2}$, where the small and big arrows represent the velocity of particles $S_{c\text{-}1}$ and $S_{c\text{-}3}$, respectively, and their colors represent the magnitude of velocity; (b) difference between the ideal powder layer and the actual powder layer obtained from the spreading of bimodal sample $S_{c\text{-}2}$; and (c) values of σ_V and $\bar{Q}$ for samples in Group III. Source: Chen et al. [15], with permission from Elsevier.

rate (Figure 6.9). This implies that after a certain point, the mass flow rate suffers, resulting in the deterioration of the packing quality.

When the spreading speed is small ($50\,\text{mm}\,\text{s}^{-1}$), the frictional forces on the particles from the substrate are able to overcome the drag forces on the particles from the roller, allowing the particles to deposit properly onto the substrate. This is favorable for the formation of a dense powder layer. However, as the spreading speed increases, the drag force on the particles becomes stronger, eventually overwhelming the frictional forces on the particles from the substrate. Thus, the powder particles remain attached on the roller and are dragged along the surface of the roller rather than being deposited onto the substrate, which explains the reduced mass flow rate through the region. This effect results in gaps and voids on the powder bed (Figure 6.10a). The condition is further exacerbated by the fact that as the spreading speed increases, the particles behind the roller tend to undergo increasingly volatile

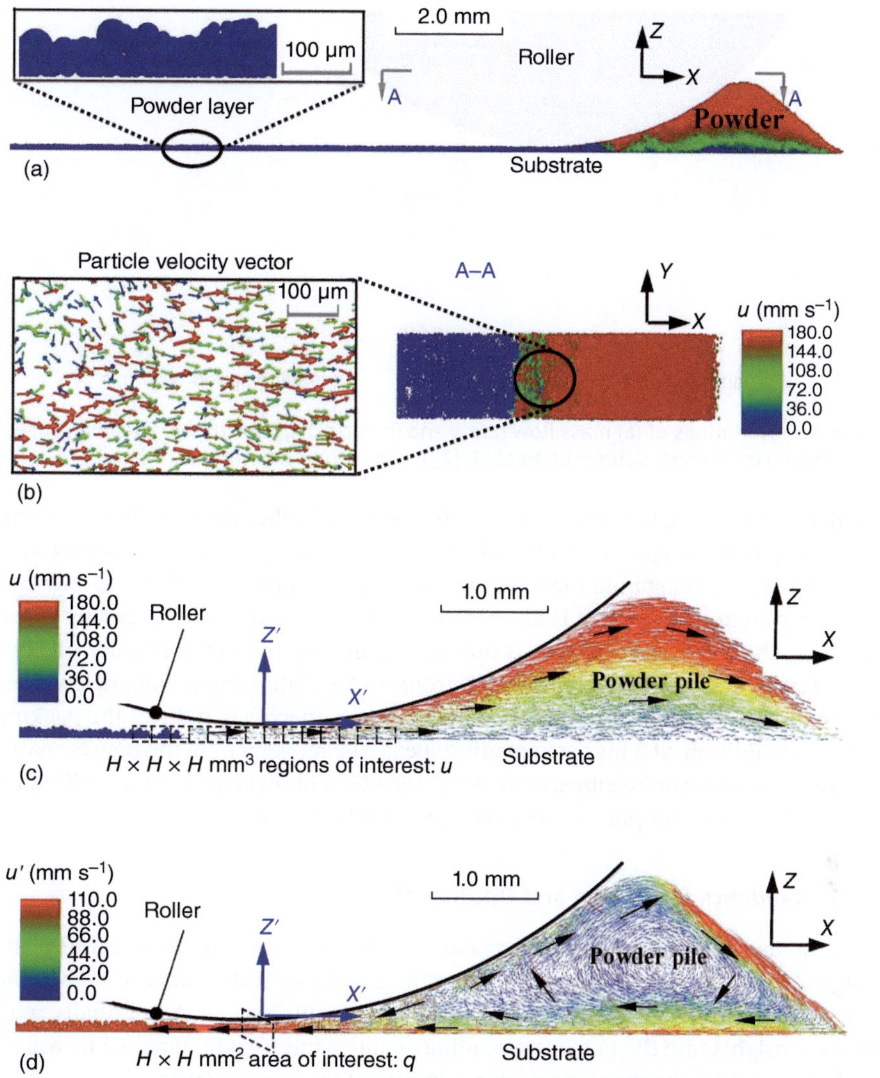

Figure 6.8 (a) Numerical simulation showing the powder dynamics during the spreading process at $V = 100\,\mathrm{mm\,s^{-1}}$. (a) Cross-sectional view (*XZ*). (b) Top view (*XY*) of the velocity profiles at section A–A of the powder particles. (c) The regions of interest and the powder flow velocity in the global *XYZ* frame of reference. (d) The area of interest and the powder flow velocity in the frame of reference $X'Y'Z'$ attached to the bottom of the roller. Source: Chen et al. [27], with permission from Elsevier.

motions before properly settling onto the substrate. This increases the susceptibility of vacancies on the substrate (Figure 6.10b).

Similar to the rake-type spreader, force-arches and transient contact force networks develop during the roller-type spreading. As the roller sweeps along the powder bed, the roller effectively compresses the incoming powder pile. The higher the

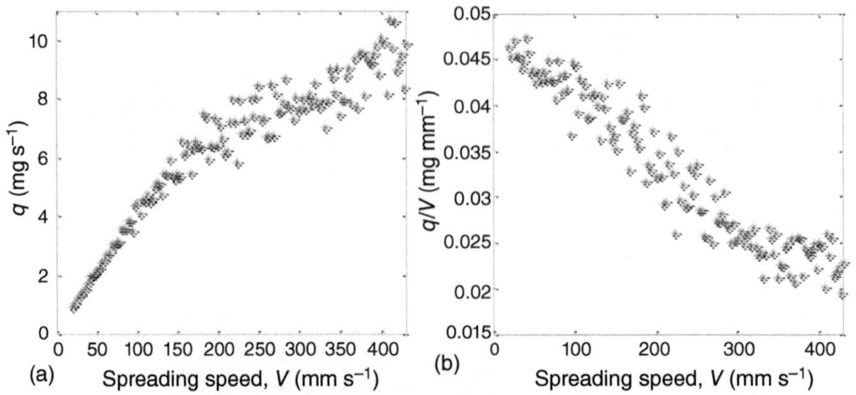

Figure 6.9 Variations of (a) mass flow rate q and (b) ratio of q/V with the spreading speed at the region of interest. Source: Chen et al. [27], with permission from Elsevier.

speed, the greater the strength of compaction and the higher the coordination number, leading to a stronger force network. Theoretically, the pressure imposed by such a compression effect should increase the coordination number and lead to a denser granular system. However, it is clear from Figure 6.11 that as the spreading speed increases, the increased pressure is not equally distributed to the powder pile but instead creates concentrated regions of "strong arches," allowing cavities of low pressure to exist between the arches. This phenomenon effectively reduces the packing density. In addition, at a higher spreading speed, the movement of the powder particles in the powder pile becomes increasingly erratic. This decreases the coordination number of the powder pile, thereby reducing the packing density.

6.2.4 Guidance for Design and Optimization

With the use of a validated DEM model, it is possible to guide and optimize the design of the key physical factors that affect the powder spreading quality, such as the rake shape, spreading velocity, layer thickness, and rolling speed. Additionally, with detailed insights into the powder spreading mechanisms, several targeted measures can be proposed to improve the packing density of the powder layer.

Firstly, for the rake-type spreader, the powder layer thickness should be increased to reduce the static and dynamic wall effects. However, a thicker layer would then require higher energy input to melt more material; the laser processing parameters need to be adjusted to account for the change. Secondly, the powder size distribution requires careful balancing to mitigate both the wall and cohesion effects. As powder size decreases, it will reduce the static and dynamic wall effects, which can increase powder packing density. However, it also leads to an increase of the cohesion effect which causes powder agglomeration. Hence, the powder size should be adjusted to the optimal point. In addition, one can still apply the bimodal-sized powder but modified so as to avoid the percolation effect, which requires adjustment of the size and density ratio simultaneously to improve the packing density of the powder layer. Finally, surface modification of the rake and powder particles can also be achieved

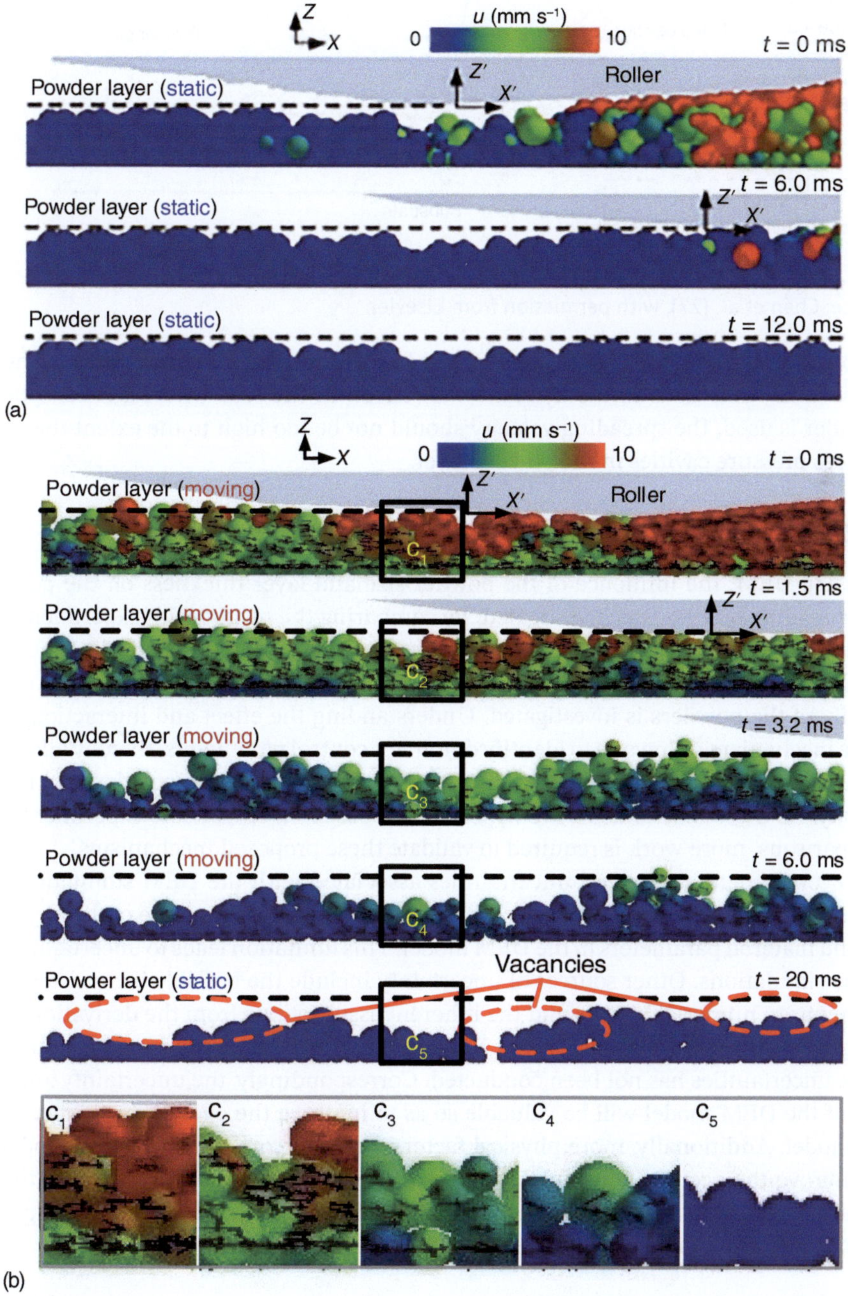

Figure 6.10 (a) The tendency for the particles to adhere to the roller leads to voids. (b) The volatile motions of the particles in the wake of the roller lead to voids. Source: Chen et al. [27], with permission from Elsevier.

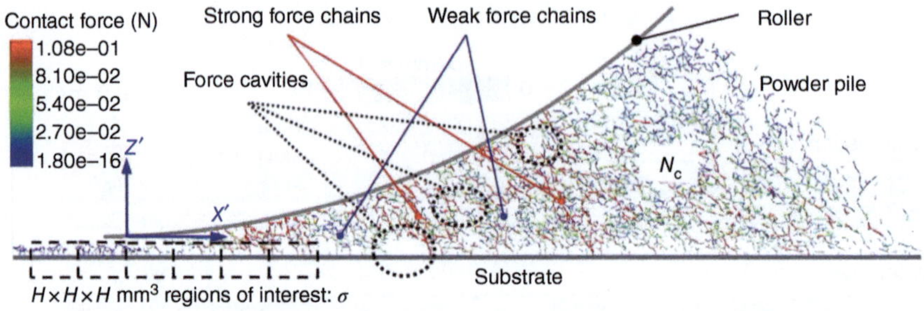

Figure 6.11 The contact force network of the powder pile at $V = 100$ mm s^{-1}. Source: Chen et al. [27], with permission from Elsevier.

by lowering the surface friction coefficient, so as to mitigate the formation of the "force-arch" that leads to the dynamic wall effect. In the case where a roller-type spreader is used, the spreading velocity should not be too high to the extent that it induces pressure cavities in the powder layer.

6.2.5 Summary and Outlook

By using DEM, the influence of the powder size and layer thickness on the powder spreading process was investigated. By comparing the results with experiments, three main deposition mechanisms were identified: the wall effect, cohesion effect, and percolation effect. Thereafter, the influence of spreading speed on the packing quality of the powders is investigated. Understanding the effect and interaction of these mechanisms allows us to identify dominant control parameters, thereby allowing us to further achieve a certain degree of optimization of the powder packing density. As these mechanisms are hypothesized based on reasonable experimental comparisons, more work is required to validate these proposed mechanisms.

However, there are several uncertainties associated with the DEM simulations. Notably, it is challenging, if not impossible to accurately measure some of the physical and material parameters in the DEM model. This limitation leads to uncertainties in the simulations. Other sources of uncertainty include the possible inconsistency arising from numerical errors and the inherent assumptions from the derivation of the governing equations used in the DEM model. The systematic investigation of these uncertainties has not been conducted. Correspondingly, the uncertainty analysis of the DEM model will be valuable so as to improve the overall confidence of the model. Additionally, more physical factors can be incorporated into the model to improve the accuracy and fidelity, such as the influence of the gas flow in the manufacturing chamber and the oxidation film on the powder surface which might be continuously changing with each powder reuse.

6.3 Powder Melting

Powder melting is the process where a heat source is applied to locally fuse material. In the available literature on numerical modeling of the powder melting process in AM, two main approaches prevail: the thermal model using the finite element method (FEM) approach and the thermal-fluid flow model using the computational fluid dynamics (CFD) approach. Notably, the thermal-fluid flow model captures the effects of molten pool flow dynamics and therefore possesses high fidelity, revealing intricate details arising from fluid phenomena that the FEM-based thermal model is unable to replicate (see Section 6.4.4.1). Indeed, this disparity between the models is clearly illustrated in the study by Yan et al. [41] which compared the accuracy of the CFD and FEM models against experimental results. The study showed that the FEM models are able to achieve sufficiently accurate results in thermal conduction mode by intentionally enhancing the effective thermal conductivity. However, the accuracy of the FEM models deteriorates in conditions where the keyhole mode is encountered (described in Section 6.3.3) due to the recoil pressure significantly influencing the molten pool behavior. In short, the CFD model will provide more accurate results as the molten pool dynamics are incorporated.

During the powder melting process, a heat source is applied to melt the deposited powders which subsequently form melt tracks. Here, a crucial information that can be obtained is the thermal history from the melt pool evolution as it will directly influence how the microstructure forms. Nonetheless, the influence of the process parameters on the geometry and dimensions of the melt pool are of primary interest. From a numerical perspective, the finite volume method (FVM) with appropriate surface tracking and reconstruction techniques is best suited to model the irregular and complex morphology and morphological evolution of the randomly packed powder particles and melt pool.

In this section, the general description of the thermal-fluid flow model is briefly presented. It is then further supplemented with the necessary high-fidelity heat source models and evaporation model for more accurate descriptions of the heat input and recoil pressures in the melt pool. Finally, the inclusion of all these subcomponents, along with the incorporation of the powder bed geometry obtained from the powder spreading model, elegantly completes a rudimentary high-fidelity model of the powder spreading and melting process.

6.3.1 Governing Equations

The thermal-fluid model is implemented in the CFD software FLOW-3D and has been experimentally validated to be accurate in a variety of AM cases [8, 11, 42, 43]. In the thermal-fluid simulation, the flow of the molten material is assumed to be incompressible and laminar. The governing equations are classical descriptions of mass continuity, momentum conservation, and energy conservation as given in Eq. (6.10) [42]:

$$\begin{cases} \nabla \cdot (\rho \mathbf{v}) &= 0 \\ \frac{\partial}{\partial t}(\rho \mathbf{v}) + \nabla \cdot (\rho \mathbf{v} \otimes \mathbf{v}) &= \nabla \cdot (\mu_{\text{vis}} \nabla \mathbf{v}) - \nabla p + \rho \mathbf{g} + \mathbf{f}_{\mathbf{B}} \\ \frac{\partial}{\partial t}(\rho h) + \nabla \cdot (\rho \mathbf{v} h) &= q_H + \nabla \cdot (\lambda \nabla T) \end{cases} \tag{6.10}$$

where **v** is the velocity vector, μ_{vis} is the viscosity, p is the pressure, λ is the thermal conductivity, and T is the temperature. The momentum equation incorporates gravity (**g**), buoyancy ($\mathbf{f_B}$, Boussinesq approximation), and viscosity (μ_{vis}). The boundary conditions for the momentum conservation equation incorporate surface tension, recoil pressure, and Marangoni forces, and the thermal boundary conditions incorporate the surface radiation and energy loss by evaporation. In the energy conservation equation, h represents the specific enthalpy calculated by $h = c_pT + (1 - f_s)\,L$, where c_p, f_s, and L denote the specific heat capacity, solid fraction, and latent heat of melting, respectively, i.e. the latent heat is linearly released with the temperature.

The volume of fluid (VOF) method [44] is applied to track the free surfaces at each time increment, given in Eq. (6.11):

$$\frac{\partial F}{\partial t} + \nabla \cdot (F\mathbf{v}) = 0 \tag{6.11}$$

where F denotes the fluid fraction. Further details about the model can be referred to [11, 42].

6.3.2 Heat Source Models

Heat source models are the basis for thermal simulations in AM, as they describe the energy input and interaction with the material. In PBF AM, two types of heat sources are often encountered: the electron beam and the laser beam. Since the pioneering works of Rosenthal, there have been different models proposed for the moving heat sources. The most common model is the Gaussian heat source model due to its relative accuracy and simplicity of the implementation. Yet, the direct implementation of these traditional heat source models fail to accurately predict the molten pool shape, temperature fields, and heat-affected zones (HAZs), as they describe the effective but not actual energy inputs. As such, more realistic and accurate heat source models are necessary for high-fidelity simulation in AM.

In this endeavor, the heating mechanisms due to material interaction with photons (laser beam source) and electrons (electron beam source) are considered at the particulate scale. The exact trajectories of the photons and electrons as they impinge on the material substrate (powder bed) are calculated, and the energy transfer due to the reflections of the photons or the collisions of the electrons are obtained. For the laser, photons are usually deflected away from the material due to relatively high reflectivity of metal surfaces. However, the multi-reflection of photons in the powder bed results in a higher than usual effective absorptivity. For the electron beam, electrons usually penetrate deep into the material with some backscattering effects.

6.3.2.1 Heat Source Model of Laser Beam

The heat source model of a laser beam is developed using a ray-tracing model. The laser beam source is initially modeled as a Gaussian surface heat source, where the cross section of the laser beam heat flux q_L in the x–y plane coordinates is given in Eq. (6.12):

$$q_L = \frac{Q}{2\pi R_b^2}\exp\left(-\frac{(x - x_o)^2 + (y - y_o)^2}{2R_b^2}\right) \tag{6.12}$$

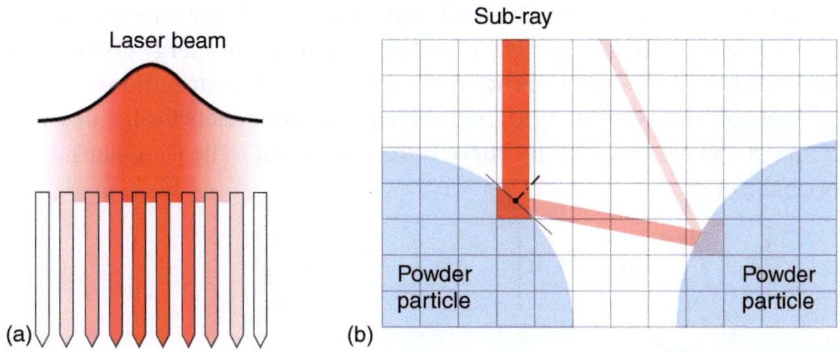

Figure 6.12 The schematic of the ray-tracing model. (a) The laser beam is divided into a finite number of sub-rays with constant energy levels, where the magnitude of the energy level corresponds to position of the sub-ray in the energy distribution (typically Gaussian). (b) A sub-ray loses energy at every instance of reflection as it encounters a surface. The energy is absorbed by the cell in contact.

where Q is the beam power, R_b is the beam radius, and (x_o, y_o) denotes the position of the laser beam. This heat source is then divided into a finite number of rays (Figure 6.12a). As the rays impinge onto the powder bed or melt pool surface, the multi-reflection routine and energy absorption of each ray are calculated (Figure 6.12b).

In the discretized cell mesh (Figure 6.12a), the laser sub-ray is allowed to move from one cell to another cell until the ray encounters a surface at an incidence point. At this point, the absorption and reflection are calculated based on the incidence angle on the surface normal plane using well-known Fresnel equations, Beer–Lambert's law, and Snell's law, where the details on the application of these equations can be found in [45]. The reflected ray continues its trajectory, and the process iterates until the ray travels out of the process domain. If the ray has dropped below a threshold energy after multiple reflections, the iteration stops as well (to save computational time).

6.3.2.2 Heat Source Model of Electron Beam

A more realistic heating mechanism of the material by the electron beam is considered here. In essence, the heating of the material arises due to the energy transfer from the electrons to the atoms in the material, which is the result of the elastic and inelastic collisions between the electrons and the atoms. This physical mechanism can be simulated by performing a Monte Carlo simulation of electron–material interaction on open-source software such as CASINO [46, 47]. The use of Monte Carlo simulation has been validated experimentally by the researchers in [48], and the results are shown to be in good agreement (Figure 6.13). The penetration trajectory of each individual electron is then calculated, and the spatial power density distribution can therefore be obtained. This yields the general description of the electron beam heat source $q(x, y, z)$ as presented in Eq. (6.13):

$$q(x, y, z) = \eta \cdot Q \cdot f(z) \cdot g(x, y) \tag{6.13}$$

where η is the energy absorptivity, Q is the beam power, $f(z)$ is the energy distribution in the transverse direction, and $g(x,y)$ is the radial in-plane energy distribution of the cross section of the beam. The energy absorptivity η and distributions $f(z)$ and $g(x,y)$ are obtained from Monte Carlo simulations. The parameters for the distributions $f(z)$ and $g(x,y)$ are obtained by curve fitting (assumed to be Gaussian), as in Eqs. (6.14) and (6.15):

$$f(z) = \left[\delta_\mathrm{d} \cdot \int_{\frac{z_0}{\delta_\mathrm{d}}}^{+\infty} \exp(-t^2)\,\mathrm{d}t\right]^{-1} \cdot \exp\left(-\frac{(z-z_0)^2}{{\delta_\mathrm{d}}^2}\right) \tag{6.14}$$

$$g(x,y) = \frac{N}{\pi R_\mathrm{b}^2} \exp\left(-N \cdot \frac{(x-x_s)^2 + (y-y_s)^2}{R_b^2}\right) \tag{6.15}$$

where in Eq. (6.14), δ_d is the characteristic depth, z_o is the location with the highest energy density, and t is the depth of penetration. The energy distribution in the penetration direction is determined by the initial kinetic energy of electrons and is therefore dependent only on the acceleration voltage. In Eq. (6.15), N is the concentration coefficient of the electron beam, R_b is the beam radius, and x_s and y_s are the coordinates of the center of the beam. For application of the model into PBF AM, the energy absorptivity takes into account the impinging angles at different locations within the powder.

6.3.3 Evaporation and Recoil Pressure

When a heat source is introduced onto the surface of a substrate, at sufficiently high thermal loads, rapid evaporation of material occurs, resulting in a strong recoil pressure that generates a deep cavity into the melt pool. In PBF AM, this effect leads to what is commonly known as the keyhole phenomenon and is speculated to be a source of porosities [9]. The objects of interest are the mass loss and recoil pressure that arise due to the evaporation of the material. In particular, the recoil pressure significantly affects the shape and stability of the keyhole structure, which influences the keyhole dynamics and pore formation dynamics. Hence, the accurate prediction of the recoil pressure is directly related to the need for an accurate evaporation model. Usually, the numerical simulations performed by various researchers employ Anisimov's model for the calculation of recoil pressures:

$$P_\mathrm{recoil} = \alpha P_\mathrm{b} \exp\left(\frac{L_\mathrm{v}}{R}\left(\frac{1}{T_\mathrm{b}} - \frac{1}{T}\right)\right) \tag{6.16}$$

where P_b is the saturated pressure at boiling temperature T_b, and $\alpha = 0.54$ based on Anisimov's assumption of bulk velocity conditions [49].

However, Anisimov's model does not consider the influences of the flow structure of the ambient gas and metal compositions, leading to inaccurate recoil pressure calculations. Owing to the insufficiencies of Anisimov's model, a new model is derived from thermodynamic principles of nonequilibrium evaporation. Only the key equations, concepts, and insights are elaborated in this section, while the rest of the derivations and discussions are detailed in the article [13].

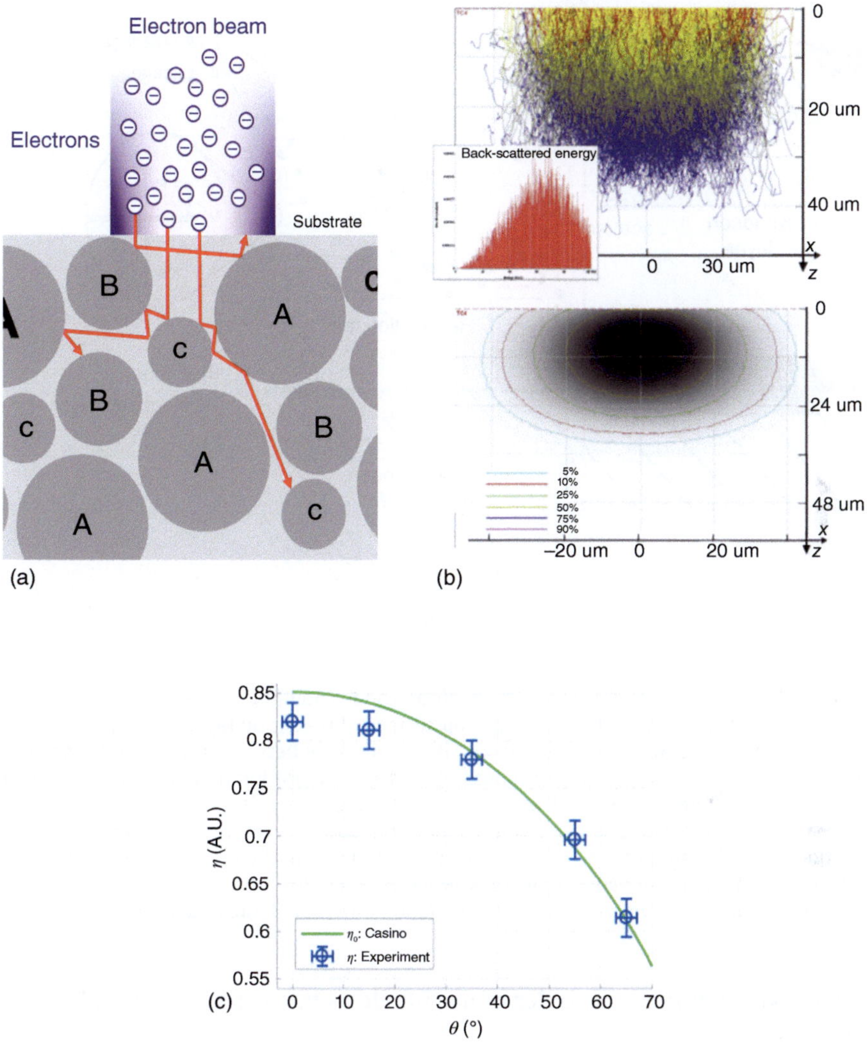

Figure 6.13 (a) Schematic of the electron–atom interaction model, with atoms of different sizes. (b) Result of the Monte Carlo simulation of the electron trajectories and the corresponding energy absorbed in the material substrate. Source: Yan et al. [47], with permission from Elsevier. (c) The comparison of the energy absorptivity at different incident angles between the CASINO model and experimental measurements performed by Carriere and Yue [48] shows good agreement. Source: Carriere and Yue [48], with permission from Elsevier.

6.3.3.1 Evaporation Model

During the evaporation of the material, the flow structure on the surface of evaporation consists of the transition layer, Knudsen layer, and vapor phase above the liquid surface (Figure 6.14). The effect of the transition layer is ignored and simply treated as a discontinuity. Here, we focus our attention to the Knudsen layer, which

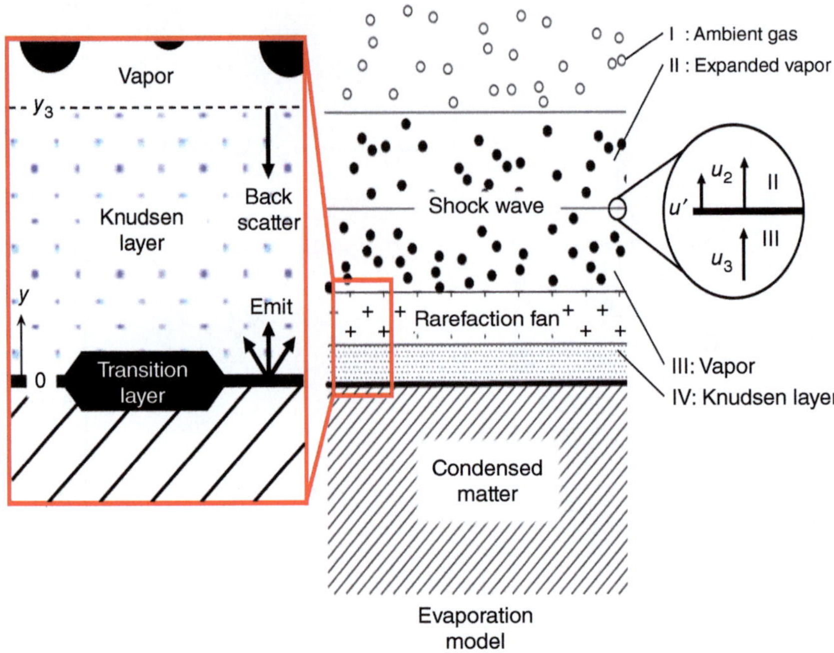

Figure 6.14 Schematic of the evaporation model, and schematic of the Knudsen layer shown in the red box. The origin is on the top surface of transition layer, and y-positive direction is the evaporating direction. $y = y_3$ is the interface between the vapor phase and Knudsen layer. Two flow structures are possible in this evaporation model: (i) flow structure in common atmosphere, in which the rarefaction fan does not exist, and (ii) flow structure in near-vacuum atmosphere which assumes the schematic as presented. The shock wave transition between region II and region III is illustrated within the circle on the right. Black solid circles represent the vapor molecules, while clear circles represent ambient gas molecules. Figure is drawn with reference to Wang et al. [13]. Source: Adapted from Wang et al. [13].

has a thickness in the same order of magnitude as the mean free path of the gas molecules.

At this scale, the vapor in this kinetic boundary layer exists in a nonequilibrium state, and the behavior of the molecules can no longer be described by continuum theory. Instead, the transport phenomenon in the Knudsen layer is best described by the Boltzmann equation. By considering different velocity distributions for the emitted molecules from the evaporating surface ($y = 0$) and backscattered molecules from the vapor phase layer ($y = y_3$), the mass, momentum, and energy conservation equations on both sides of the Knudsen layer can be derived as:

$$\begin{cases} \rho_e\sqrt{\frac{RT_e}{2\pi}} - \rho_3\sqrt{\frac{RT_3}{2\pi}}\beta F^- & = \rho_3 u_3 \\ \frac{1}{2}\rho_e RT_e + \frac{1}{2}\rho_3 RT_3\beta G^- & = \rho_3 u_3^2 + \rho_3 RT_3 \\ 2\rho_e RT_e\sqrt{\frac{RT_e}{2\pi}} - 2\rho_3 RT_3\sqrt{\frac{RT_3}{2\pi}}\beta H^- & = \frac{1}{2}\rho_3 u_3^2\left(u_3^2 + 5RT_3\right) \end{cases} \tag{6.17}$$

where ρ and T are the density and temperature of the vapor, respectively, $R = 8.314\text{J K}^{-1}\text{ mol}^{-1}$ is the gas constant, β is the condensation ratio, and the subscripts "e" and "3" denote the saturated and vapor phase layers, respectively. The variables F^-, G^- and H^- are defined with a dimensionless variable m:

$$\begin{cases} m &= \frac{u_3}{\sqrt{2RT_3}} = \sqrt{\frac{\gamma}{2}}\frac{u_3}{\sqrt{\gamma RT_3}} = \sqrt{\frac{\gamma}{2}}Ma \\ F^- &= -\sqrt{\pi}m\left[1-\text{erf}(m)\right] + \exp(-m^2) \\ G^- &= (2m^2+1)\left[1-\text{erf}(m)\right] - \frac{2}{\sqrt{\pi}}\exp(-m^2) \\ H^- &= -\frac{\pi}{2}m\left(m^2+\frac{5}{2}\right)\left[1-\text{erf}(m)\right] + \frac{1}{2}(m^2+2)\exp(-m^2) \end{cases} \tag{6.18}$$

where Ma is the Mach number at $y = y_3$ and $\text{erf}(m)$ is the Gaussian error function.

Essentially, the mass loss rate and recoil pressures are the mass flux and momentum flux on the liquid surface. Based on the conservation equations in Eq. (6.17), the mass loss m_{loss} and recoil pressure P_{recoil} can be obtained as:

$$\begin{cases} m_{\text{loss}} &= \rho_3 u_3 = m\sqrt{\frac{2}{R}} \cdot \frac{P_3}{P_e} \cdot \frac{T_e}{T_3} \cdot \frac{P_e}{T_e} \\ P_{\text{recoil}} &= \rho_3 u_3^2 + \rho_3 RT_3 = \frac{1}{2}\left[P_e \exp(m^2)\right](2m^2+1)\left(F^- + \sqrt{\frac{T_3}{T_e}}G^-\right) \end{cases} \tag{6.19}$$

where the saturated vapor pressure P_e can be obtained from the Clausius–Claperyon equation:

$$\ln\left(\frac{P_e}{P_{\text{ref}}}\right) = -\frac{L_V}{R}\left(\frac{1}{T_e} - \frac{1}{T_{\text{ref}}}\right) \tag{6.20}$$

As observed, the mass loss and recoil pressures can be fully expressed by the temperature and pressure ratios of the vapor regions. By substituting the equations from Eq. (6.18) into Eq. (6.17), one can obtain the temperature and pressure ratios for the mass loss and pressure recoil calculations in Eq. (6.19). More importantly, these ratios (as functions of m) can be fully expressed by a single independent variable: the Mach number.

6.3.3.2 Model of Flow in Common and Near-Vacuum Environments

From the model elaborated earlier, the properties of the mass loss and pressure recoil are entirely encapsulated by the Mach number. Hence, knowing the changes of the Mach number in its transition from the Knudsen layer to the ambient environment is crucial in describing the flow characteristics. Physically, two cases of the ambient environment can be considered: a common atmosphere and a near-vacuum atmosphere, as proposed by Knight (Figure 6.14) [50].

These flow structures are assumed to be continuous in regions II and III, implying that they are applicable until a certain threshold. By introducing the Knudsen number, which estimates the liquid rarefaction, the two cases of the ambient environment can be distinguished:

$$K_n = \frac{k_B T}{\sqrt{2}\pi d^2 PL}\begin{cases} < 0.1, & \text{common atmosphere} \\ \geq 0.1, & \text{near-vacuum atmosphere} \end{cases} \tag{6.21}$$

In the case where the common environment is considered, the pressure and temperature ratios can be obtained by assuming that the vapor phase in region II is fully expanded throughout the shock wave discontinuity. On the other hand, the case of near-vacuum environment presents more complications; usually, as the vapor flow transitions from the free molecular flow to gas dynamic flow regime, the Knudsen layer exists and $Ma = 1$. Yet, in a near-vacuum environment, the Knudsen layer barely exists as the vapor emitted from the liquid surface is quick to reach local equilibrium after several collisions. Therefore, an interpolation is performed in the region where $0.05 < Ma < 1$ to obtain the recoil pressure and mass loss values. The interpolation method is detailed in [13].

In addition, the effect of multi-element composition is accounted by considering that the overall saturated pressure is proportional to the linear combination of the saturation pressure of each component. For majority of the alloys in AM, this assumption is valid. It should be noted that the Knudsen number from Eq. (6.21) also sets the limiting criteria of the pressure-to-temperature ratio in each region:

$$\frac{P_i}{T_i} > \frac{k_B}{K_n\sqrt{2}\pi d^2 L} \tag{6.22}$$

Therefore, this model can be further extended to application in the AM of both pure metals and alloys, where d for alloys can be taken as the weighted average diameter of each vapor particle.

6.3.4 Model Verification and Validation

6.3.4.1 Realistic Heat Inputs

The mechanistically derived models in Section 6.3.2 are shown to be realistic descriptions of the heating physics, verified through both simulations and experiments. For the laser model, the ray-tracing simulation shows that ray tracking can be done accurately to reflect the correct calculation of energy absorptivity. At each incident spot, a fraction of the laser power is absorbed onto the plates, resulting in a decreasing laser power for each reflection point (Figure 6.15a).

For the electron beam model, a simulation of electron beam on a single spherical particle was conducted to verify the accuracy of the electron beam model (Figure 6.15b) [51]. The results show that the highest temperature occurs below the surface of the particle, correctly reflecting the effect of electron penetration and the accurate temperature distribution imposed by an electron beam. Moreover, the high-temperature region diminishes radially from the center of the particle, reflecting the influence of the incidence angle. Notably, this model shows that the location of the highest temperature exists below the surface of the substrate due to electron penetration. The seemingly trivial result not only implies that the melt pool formation initiates below the surface of the substrate but also potentially explains the physical phenomena of explosion and eruption observed in electron beam manufacturing.

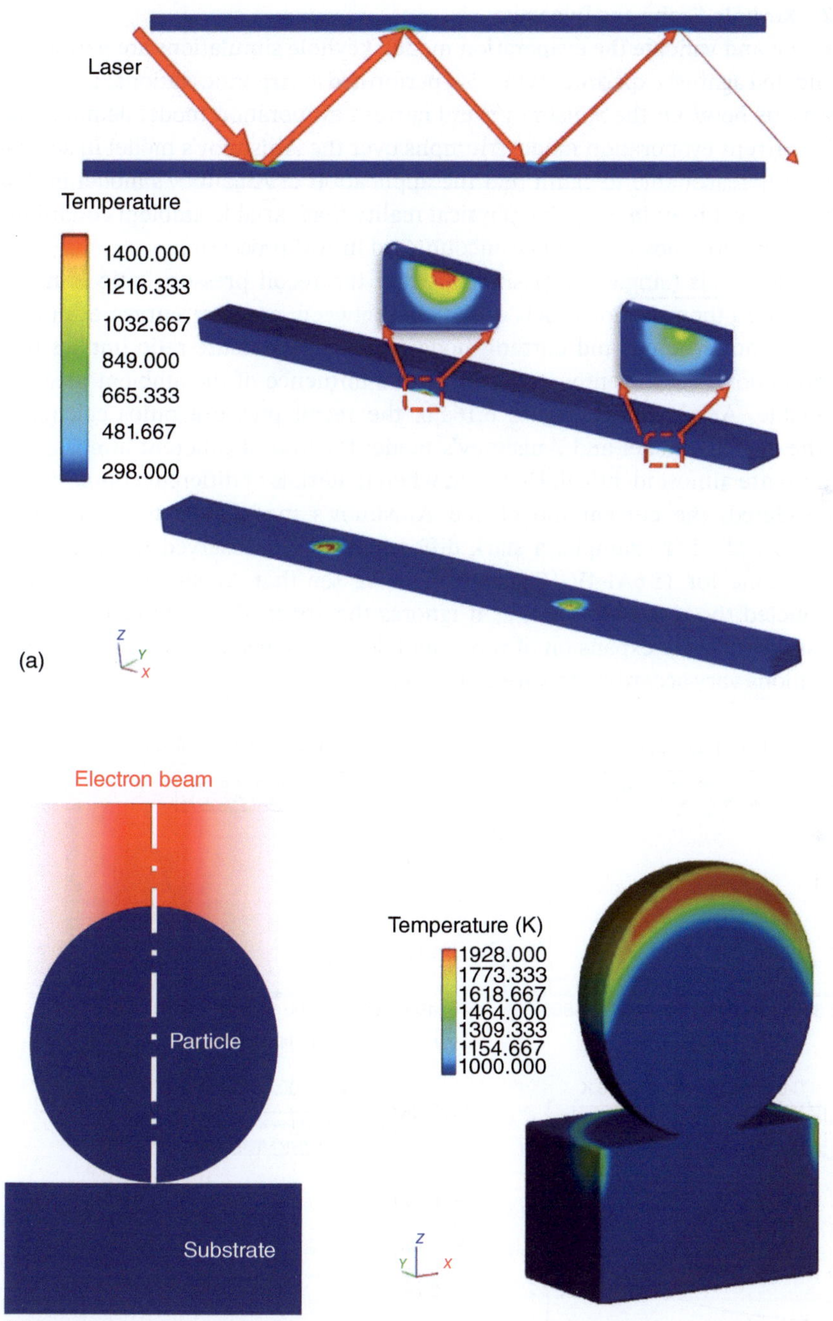

Figure 6.15 (a) The verification simulation for the ray-tracing model where the laser goes through four reflections between two flat plates spaced a distance apart. Source: Ge et al. [52], with permission from Elsevier. (b) Electron Beam Verification. Left: schematic of the simulation of an electron beam on a spherical powder particle; right: results of the electron beam simulation, showing clearly the heat distribution due to penetration of electrons that occurs below the surface of the particle. Source: Yan et al. [51], with permission from Elsevier.

6.3.4.2 Keyhole Shape and Dynamics

To evaluate and validate the evaporation model, keyhole simulations are performed and validated against experiments [53, 54] performed at Argonne National Labs. The comparisons between the Anisimov's and current evaporation model demonstrate that the current evaporation model triumphs over the Anisimov's model in several aspects. It is reasonable to claim that the application of Anisimov's model in AM is imprecise as it is an inescapable physical reality that variable ambient conditions and multi-material powders will be encountered in AM processing.

For pure metals (single composition), when the recoil pressure ratio is much larger than 1, there is not much difference between pressure curves evaluated using Anisimov's model and current model. The high-pressure ratio implies that the evaporation is sufficiently strong; thus, the influence of the ambient pressure is negligible. As shown in Figure 6.16a,b, the recoil pressure ratios calculated using the current model and Anisimov's model for iron at different atmospheric conditions are almost identical. However, when materials of different compositions are considered, the current model and Anisimov's model deviate significantly (Figure 6.16c,d). For example, a stark difference can be observed when calculations are done for Ti-6Al-4V (Figure 6.16c). We see that Anisimov's model has overpredicted the recoil pressure, as it ignores the effect of the ambient pressure which suppresses the expansion of vapor, and does not consider the fact that vapor compositions vary according to surface temperature.

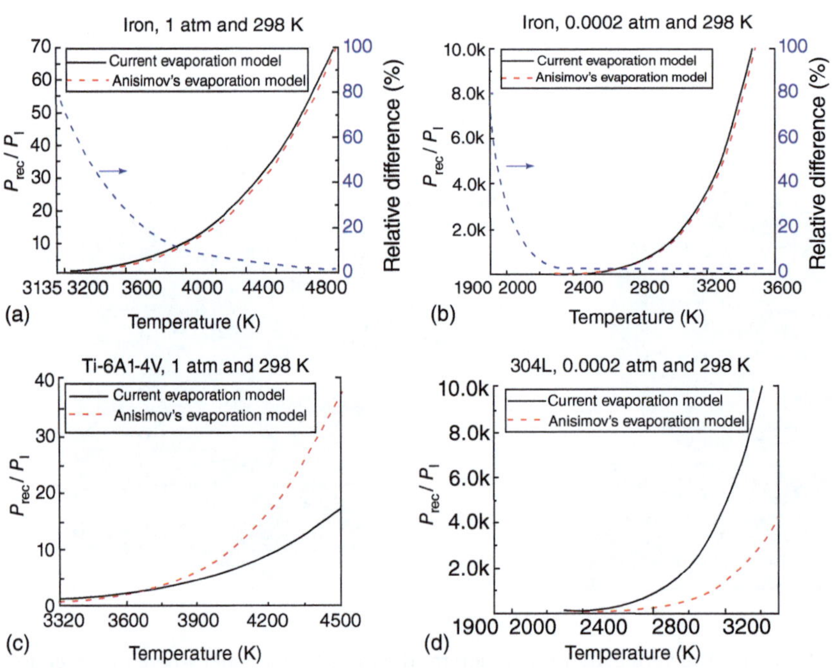

Figure 6.16 Comparison of the recoil pressure ratio at different molten pool surface temperatures between the current model and Anisimov's model. Source: Wang et al. [13], by the American Physical Society.

The overprediction of recoil pressure in Anisimov's model may explain why the current model matches experimental observations more closely. Firstly, the higher recoil pressure estimated by Anisimov's model may have suppressed fluctuations at the keyhole depth. Additionally, the rate of the depth penetrating process of Anisimov's model is far greater (Figure 6.17). Moreover, the increased recoil force results in larger aggregation of liquid metal overflowing from the keyhole in the common environment (Figure 6.18).

To further verify the model, a simulation of laser scanning in a common environment was performed. As shown clearly in Figure 6.19, the simulation results are consistent with the experimental results, even capturing the inclination of the melt pool front at the initial keyhole formation process.

6.3.4.3 Molten Track Profile

For an initial validation of the thermal-fluid flow model, it is judicious to first compare the simulation and experiment results of track melting on a bare plate to exclude the effects of the powder bed. Here, a good example is the comparison of the molten track profiles obtained from the simulations by Ge et al. [52] to the bare-plate studies conducted via the Additive Manufacturing Metrology Testbed (AMMT) at the National Institute of Standards and Technology (NIST).

Ge et al. [52] compared the simulation results using the ray-tracing model with the surface morphology of the experimental results from NIST. They found that the simulation results of the surface elevation and track morphology under different scan strategies agree well with the experimental results. The simulation results of different scan strategies (constant speed, exact stop, and continuous scan) show good agreement as observed in Figure 6.20, where a peak at the starting area and a valley at the end are faithfully reproduced in all cases, which have similar magnitudes to their experimental counterparts. Ge et al. [52] explained that these peaks and valleys form due to the Marangoni effect along the scan direction. The rear region of the metal fluid has lower temperature and correspondingly higher surface tension, which draws the fluid from the front region to the rear. After validation of the model with bare plate experiments, it provides greater confidence for incorporating a powder bed into this model for further simulations.

6.3.5 Coupling with Powder Spreading Model

A single melt track of good quality serves as the fundamental building unit of a high-quality AM product. Therefore, single-track experiments and simulations are often conducted to examine the range of processing parameters that yield no porosities. At this powder-scale (mesoscopic) simulation, the time step is typically in the order of nanoseconds to capture the relatively high-velocity movement of the fluid in the melt pool, governed by the hydrodynamic and thermodynamic effects. When discretizing the powder bed for thermal-fluid simulations, the mesh size is dependent on the size of the powders used in the simulation. In the subsequent thermal-fluid simulations, mesh convergence test has been conducted to exclude mesh sensitivity.

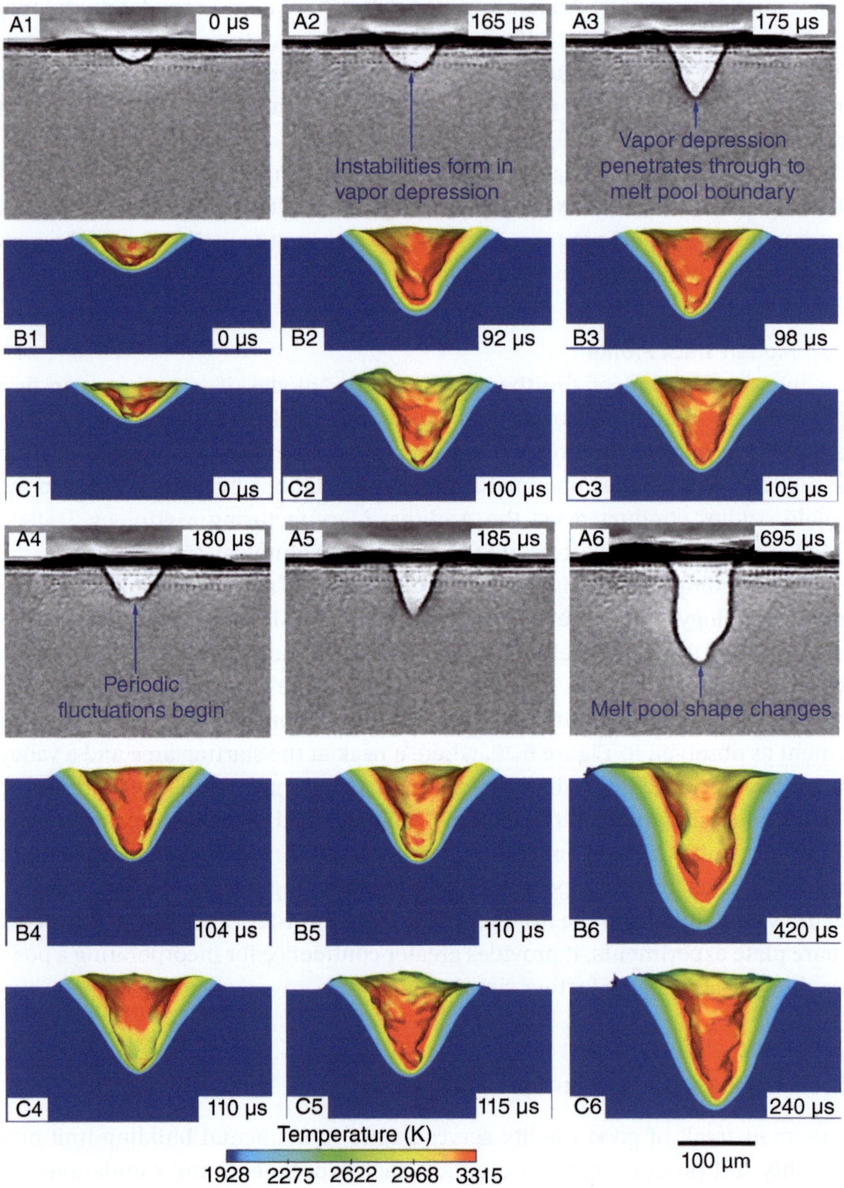

Figure 6.17 Comparison of molten pool and keyhole formation at different time frames in stationary melting at 1 atm and 298 K. (A1–A6) shows X-ray images of keyhole shapes; (B1–B6) shows the simulated keyhole shapes with the current evaporation model; (C1–C6) shows the simulated keyhole shapes with Anisimov's evaporation model. Source: Wang et al. [13]/with permission of American Physical Society.

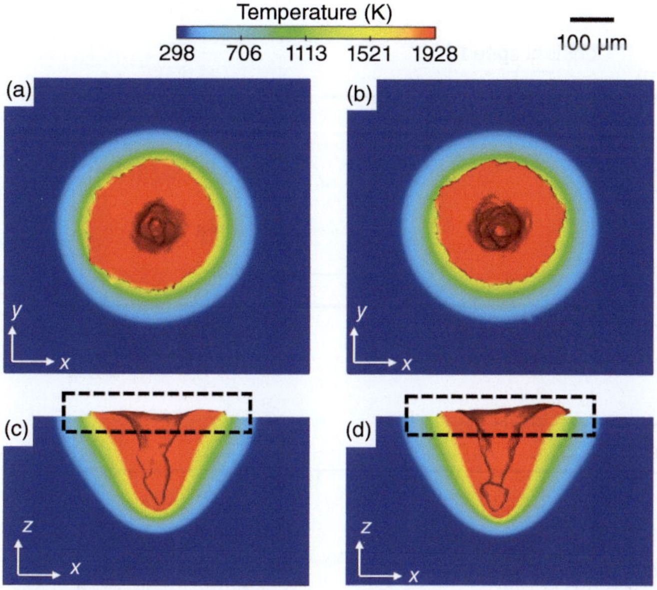

Figure 6.18 Shape of the molten pool at the same time frame (600 μs). (a) and (b) are the top views of the molten pool of the current and Anisimov's model; (c) and (d) are the cross section of the molten pool of the current and Anisimov's model. Source: Wang et al. [13], by the American Physical Society.

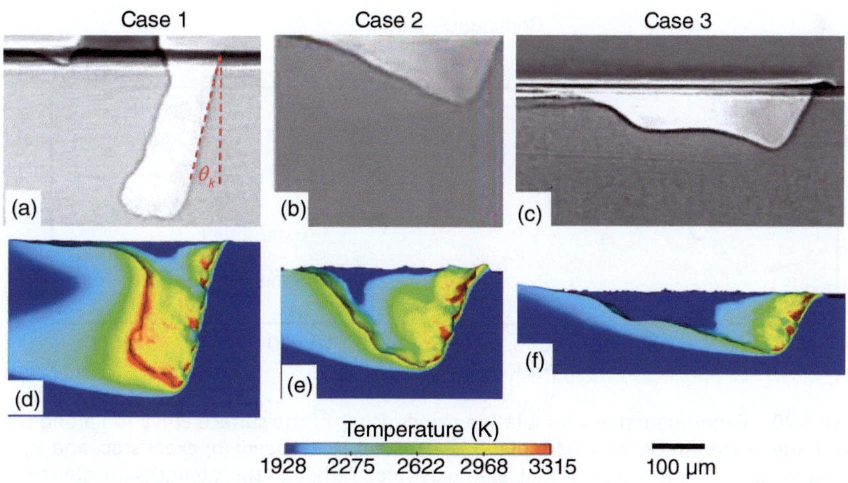

Figure 6.19 Comparison of the keyhole geometry feature of the current model with experimental results. (a)–(c) X-ray images of the keyhole shape; (d)–(f) simulated keyhole shapes and temperature distribution. Source: Wang et al. [13]/with permission of American Physical Society.

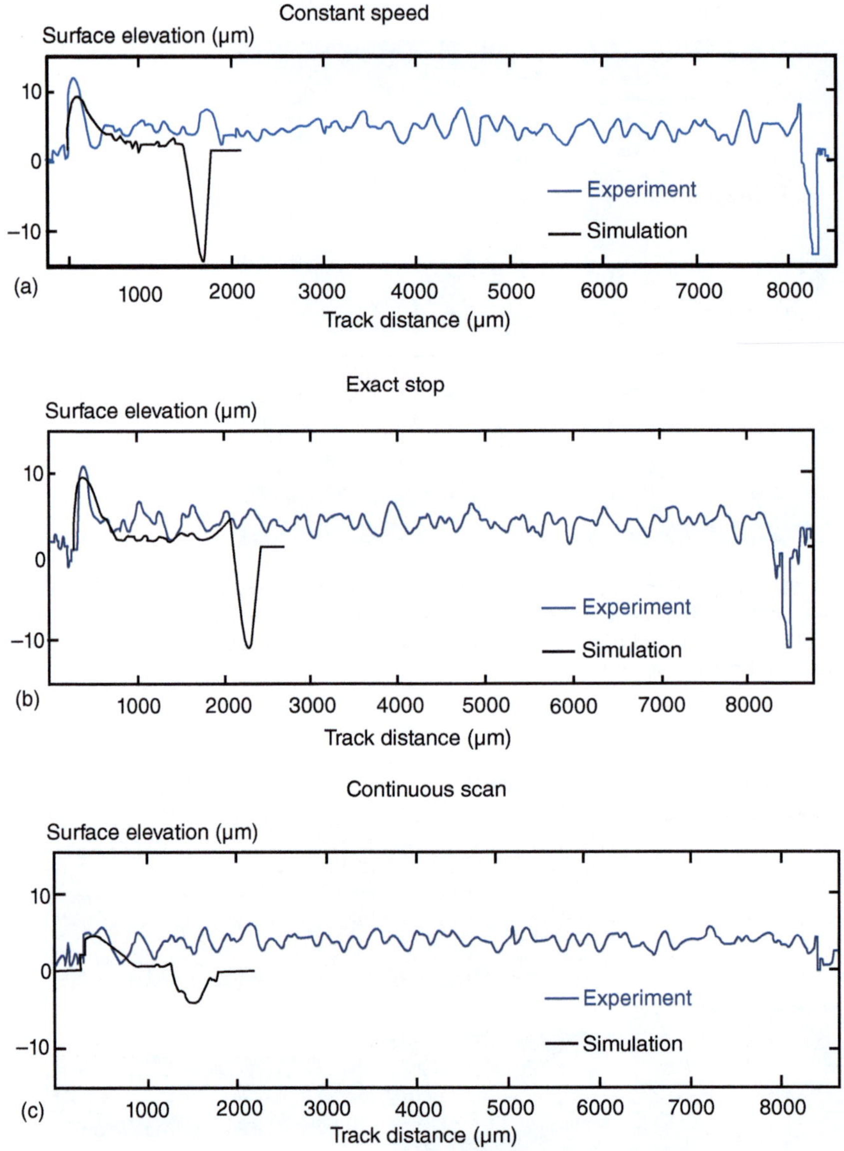

Figure 6.20 Experimental and simulation comparisons of the surface elevation along the central line of scan tracks on a bare plate: (a) constant scan speed; (b) exact stop, and (c) continuous scan. The material is 316L stainless steel. Although track lengths are shorter than those in the experiment to save computational time, the dip in surface elevation at the end of the track is faithfully reproduced. Source: Reproduced from Ge et al. [52], with permission from Elsevier.

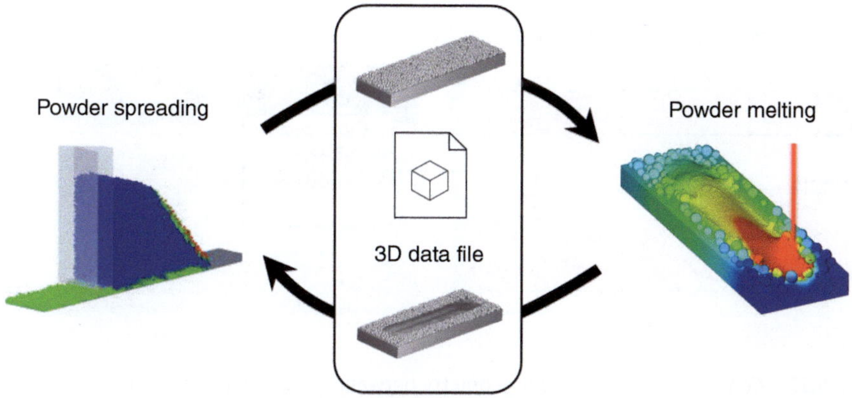

Figure 6.21 An integrated modeling framework to perform a simulation of a single-track or multilayered PBF process. The 3D data files (commonly STL) are exchanged between the DEM and CFD software for their respective simulations. Source: Figure of powder spreading adapted from Chen et al. [15], Yan et al. [11]/with permission from Elsevier. Figure of powder melting Yan et al. [42], with permission from Elsevier.

Typically, the powder particle with the smallest diameter should occupy at least three to five mesh cells in one dimension.

Before simulating the powder melting process, a high-resolution powder layer can be first generated with the DEM described in Section 6.2. The resultant powder layer model can be imported into the FVM model as a Standard Tessellation Language (STL) file to build the initial geometry during setup, for which the CFD simulation can be conducted with the application of the relevant heat source(s) (Figure 6.21).

To model a multilayered manufacturing process, after the simulation of the first layer is complete, the detailed solidified shape of the melt track is transferred into the DEM model as an STL file. Thereafter, the deposition of a new powder layer on the solidified tracks is simulated. The resultant combined geometry of both the second layer and the first layer is then transferred back into the CFD model as an STL file, of which a second CFD simulation of powder melting can be performed.

In this section, the validation of the thermal-fluid flow model coupled with a powder spreading model is first presented. Thereafter, the physical mechanisms of single-track, multi-track, and multilayer defects are clarified using high-fidelity powder-scale models, demonstrated in the works [11, 42] which simulate the EBM of titanium alloy Ti-6Al-4V in a vacuum environment. The material properties are given in Table 6.A.1.

6.3.5.1 Single-Track Cases

In Section 6.3.4.3, the thermal-fluid flow model has been validated with bare plate experiments. Hence, we can proceed to couple the model with a powder bed as described earlier and compare the simulation results with experiments. An example of such a comparison is illustrated in Figure 6.22. In simulations, when the historical peak temperature exceeds the liquidus temperature of the material, the cell will be considered within the interface of the fusion zone. On the other hand, the cross

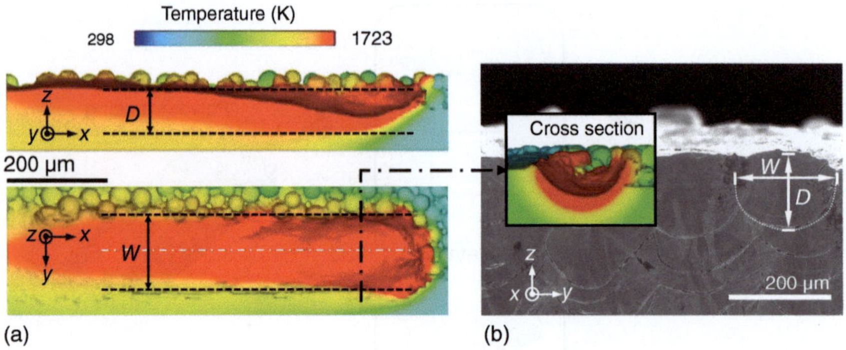

Figure 6.22 Comparison of melt pool geometry between (a) simulations and (b) experiments. The material is 316L stainless steel, with liquidus temperature set as 1723 K in this simulation. The white-dashed line indicates the slice position of the x–z plane for a side view of the melt pool. The cross section of the y–z plane is shown in the figure. The sample in the experiment was sectioned, polished, and etched before observing under SEM. Source: Yang et al. [21]/Springer Nature/CC BY 4.0.

Table 6.2 Results from Ge et al. [52], showing that the molten pool dimensions obtained from simulations (Sim.) and experimental (Exp.) measurements agree reasonably (material: 316L stainless steel, material properties listed in Table 6.A.1).

Power – Velocity		Width (μm)	Depth (μm)	Height (μm)
200 W – 0.8 m s^{-1}	Sim.	85	9	17
	Exp.	79	12	16
200 W – 1.0 m s^{-1}	Sim.	77	10	25
	Exp.	60	10	19
200 W – 1.2 m s^{-1}	Sim.	60	8	18
	Exp.	60	5	22

Source: Ge et al. [53]/with permission of Elsevier.

section of the melt pool in experiments is obtained after sectioning, polishing, and etching via conventional metallography techniques. Alternatively, the width of the melt pool from the top view can be directly compared with the melt tracks obtained from experiments.

Following the earlier validation of the base plate simulations, Ge et al. [52] subsequently compared simulation results of a powder bed with experiment results obtained an in-house SLM machine. The values of the melt pool width, depth, and height show a good agreement (Table 6.2). Similarly, Yan et al. [41] compared their simulation results with the single-track experiments with powder bed performed using the AMMT at NIST. The molten pool width at different laser powers and scanning velocities is found to agree well between simulation and experiments (Table 6.3).

Table 6.3 Results from Yan et al. [41], showing the molten pool width obtained from simulations and experimental measurements (material: IN625, material properties listed in Table 6.A.1).

Cases	Melt pool width	
Power – Velocity	**CFD (μm)**	**Experiment (μm)**
195 W – 0.8 m s^{-1}	117 ± 7	132.0 ± 14.1
195 W – 0.5 m s^{-1}	137 ± 8	149.7 ± 7.3
195 W – 0.2 m s^{-1}	216 ± 6	206.0 ± 20.9
122 W – 0.5 m s^{-1}	122 ± 6	126.9 ± 17.0
122 W – 0.2 m s^{-1}	170 ± 7	150.2 ± 8.2

Source: Adapted from Yan et al. [42].

6.3.5.2 Balling Phenomenon

It is observed in many experimental studies that a melt track is susceptible to breakup during its formation. Such discontinuities are a common effect observed in PBF AM, which can be primarily attributed to two "balling" regimes: low power balling and high power balling. Low power balling occurs when the input energy is insufficient, while high power balling occurs when the input energy is excessive. Undoubtedly, these balling effects arise due to the tendency of the melt track fluid to minimize its surface energy by forming surface curvatures. Yet, the exact nature of their causation is not necessarily clear, and conflicting mechanisms have been proposed for these balling effects. Therefore, simulations can be conducted to elucidate the mechanism for these track discontinuities.

Using the high-fidelity powder melting and spreading model described in Sections 6.2 and 6.3, a simulation study on EB-PBF conducted by Yan et al. [42] proposed a mechanism for low power balling. The investigation demonstrated that when insufficient heating occurs and the substrate is not melted, the molten metal fluid does not achieve sufficient wetting with the surface of the substrate. As such, the fluid tends to coalesce together into discontinuous groups to maximize the surface area of the fluid, instead of spreading on the solid substrate. When the coalescing effect is strong, balling occurs, where small clusters of molten fluid forms isolated balls. Subsequently, a study on L-PBF conducted by Ge at al. [52] further confirmed this mechanism with experimental validations. On the other hand, high power balling occurs when the heating effect is excessive, wherein the heat accumulates in the melt pool which requires more time to cool down. As such, the length of the melt pool tends to be relatively long. However, the long melt track is susceptible to breakup due to column-fluid instability (Plateau–Rayleigh instability), resulting in the track discontinuity.

In this section, balling effects observed in the L-PBF [52] and EB-PBF [42] simulations are presented. Regardless of the heating mechanism used (laser or electron beam), the balling effect that arises in either case is physically similar as balling is essentially a result of the thermo-capillary effects during melting. Therefore, the results from the L-PBF simulation will primarily be referred to for the discussions of the balling effect in both the L-PBF and EB-PBF simulations.

Firstly, by simulating conditions with constant laser power at increasing velocities, the surface morphology of the melt pool changes from a smooth track, discontinuous track, and finally a track with balling effects, similar to the experimental results obtained from scanning electron microscopy (SEM) in Figure 6.23. Comparison of the cross-sectional geometrical dimensions of the melt pool also show reasonable agreement (earlier presented in Table 6.2).

The E-PBF simulations demonstrated that it is crucial to achieve adequate wetting of the melt pool on the substrate, which can be ensured if the substrate under the particles is melted. This conclusion is corroborated in the L-PBF simulation. As observed in Figure 6.24, during the formation of the melt pool, the surface tension of the fluid in the melt pool causes the fluid from the fully or partially melted powder particles nearby to interact with the melt pool. The fluid of a fully melted powder particle will be entrained into the melt pool. Concurrently, the fluid of the partially melted powder particle with a relatively lower temperature will have a higher surface tension than the fluid in the melt pool. As such, the fluid from the molten pool

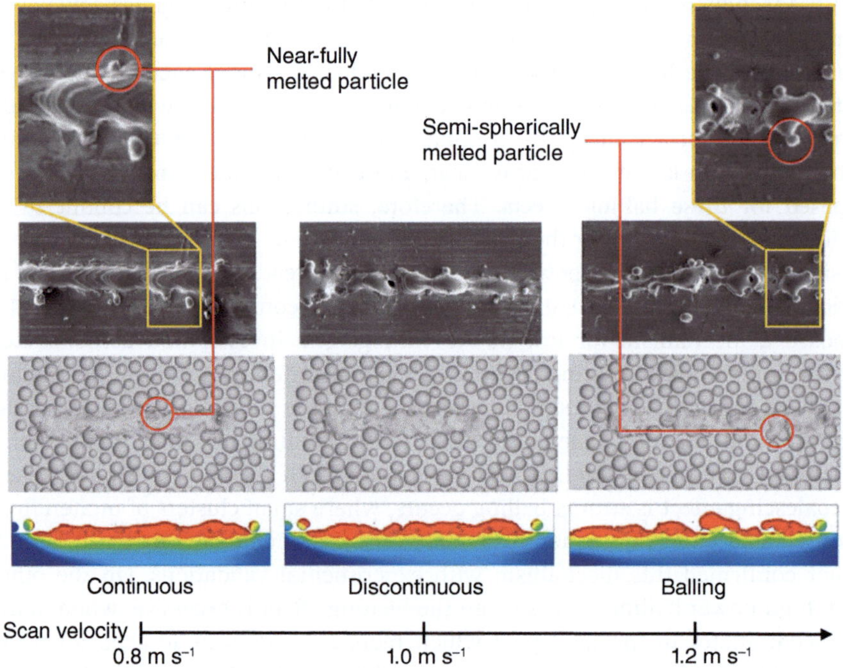

Figure 6.23 Transition from continuous melt track to balling effect. As scan speed increases, the melt pool becomes increasingly irregular, to the extent that it becomes discontinuous. Notably, balling effects can still occur even if smaller particles are used. From the SEM images, the morphology of the near-fully melted and semi-spherically melted particles can be observed near the melt pool, which are replicated in the simulations. Source: Ge et al. [52]/with permission of Elsevier.

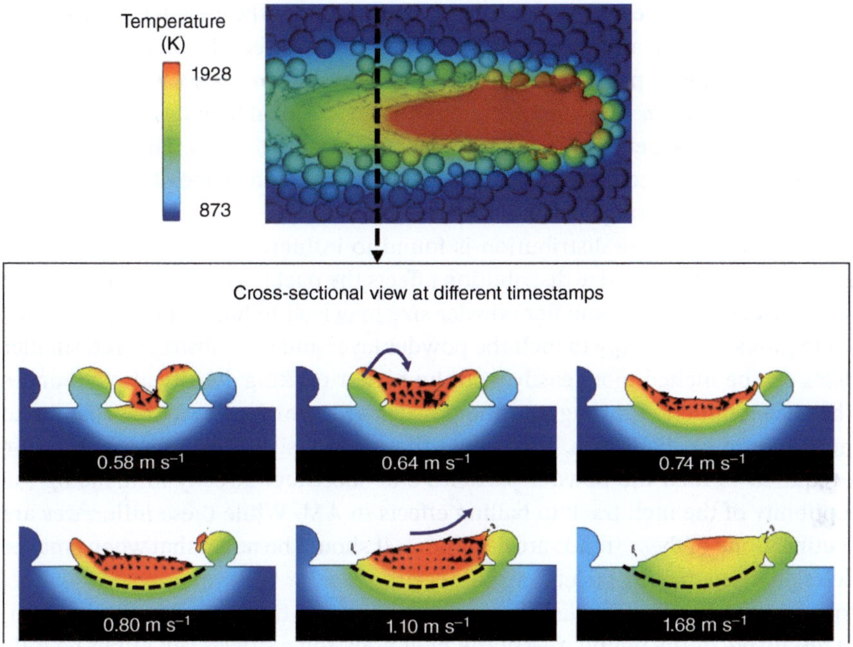

Figure 6.24 The time evolution of the flow pattern near a semi-spherically melted particle. The surface tension of the melt fluid causes two effects: the melted powder particle on the left is entrained into the melt pool, while the fluid adheres to the partially melted powder particle on the right. Source: Yan et al. [42], with permission from Elsevier.

tends to flow in the direction of the surface of the powder particle, adhering itself to the partially melted particle.

Therefore, as long as the substrate is melted, sufficient wetting and spreading can be achieved even if the initial region is bare (locations where powder particles are absent). Effectively, this observation demonstrates that insufficient energy input directly leads to the balling effect due to lack of melting of the substrate below the powder layer. Hence, the process parameters should be adjusted such that there is sufficient energy input to minimally melt the substrate beneath the powder layer.

Regardless, even if the balling effect is eliminated, the melt track will be continuous but irregular. As previously mentioned, powder particles near the melt pool will be dragged into the melt pool due to the effects of surface tension. The particles that are entrained into the melt pool are either almost fully melted or partially melted, as revealed in both the simulation and experiment (Figure 6.23). The particles that are almost fully melted will have relatively small effect on the molten pool flow and surface morphology. However, the partially melted particles will distort the molten pool morphology, decrease the size accuracy, and increase the surface roughness. This phenomenon may even result in a pore between the partially melted power particle and the melt pool, leading to a direct deterioration of the mechanical properties of the as-built metal parts.

Apart from the process parameters, the properties of the powder, namely the power size distribution and powder layer thickness also affect the melt pool quality. However, the random packing of powder particles does not have any significant influence. Indeed, from a macroscale perspective, the global behavior of the molten pool is expected to be generally similar with some local variations. Therefore, given a set of powder and process parameters, the global size and shape of the melt pool are expected to be generally consistent.

Firstly, the powder size distribution is found to influence the balling effect and track irregularity. Powder size distribution affects the packing density of the powder layer (see Section 6.2.3). A smaller powder size may lead to higher packing density, which requires more energy to melt the powder layer and the substrate. Yet, smaller particles can be melted more easily, and the higher packing density also enhances the thermal conduction throughout the powder bed, which allows the substrate to be melted more easily. Hence, changes to the powder size distribution affects the heat required to melt the powder particles and substrate, directly influencing the susceptibility of the melt track to balling effects in AM. While these influences are competing, none of these trends are dominant. It should be noted that when smaller particles are used, balling effects can still occur.

Moreover, the powder size distribution influences the track irregularity. As previously discussed, at the boundaries of the melt pool, the particles can either be fully or partially melted. If partial melting occurs, the melt pool adheres to the partially melted particles, giving a maximum possible width of w_{max}. On the other hand, if full melting occurs, the particles will be absorbed into the melt pool, giving a minimum possible width of w_{min}. Since the maximum width is a result of the partial melting of particles, the irregularity of the melt track and in terms of its amplitude increases with the powder size.

Finally, the powder layer thickness plays a significant role in the balling effect and single-track irregularity. Indeed, the thickness of the layer directly dictates the energy required to melt the substrate. In general, the thinner the layer, the less likely that balling effects will occur (given the same input power and scan speed). Similarly, the amplitude of irregularity increases with increasing layer thickness.

The simulation of single tracks reveals the different mechanisms that lead to discontinuous tracks, which can result in cracks, delamination, and even forced interruption of the entire manufacturing process (build failures). Based on these simulations, it can be observed that the balling effect can be simply prevented by ensuring a sufficient melting of the substrate regions below the melted particles, but adjusted such that it does not reach the high power balling regime. The most straightforward method to increase energy input is to reduce the powder layer thickness, but this approach will increase production time and reduce production efficiency. Hence, a careful selection of process parameters is required to optimize the multi-objective conflicts.

6.3.5.3 Multi-Track Cases

From the single-track investigations, we can proceed to the multi-track interactions. In the study of single-layer multi-track interactions, the primary factor is the distance

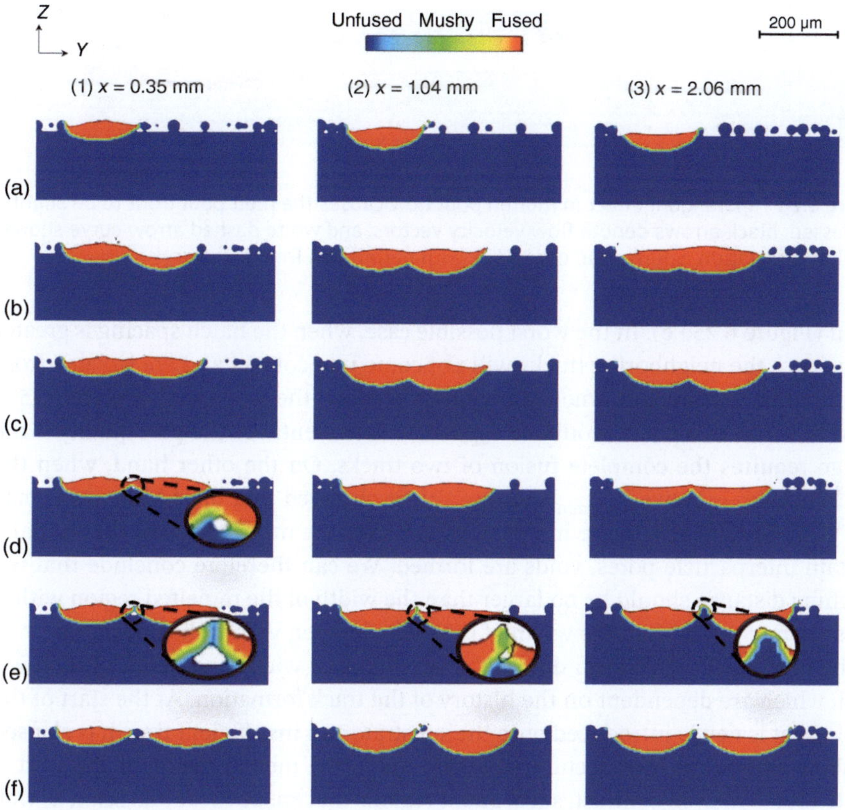

Figure 6.25 The fused zones with various scan paths of (a) single track, (b) S-shaped scan path with hatching distance of 0.2 mm, and (c)–(f) Z-shaped scan path with hatching distance of 0.20, 0.24, 0.26, and 0.32 mm, respectively, at various cross-sectional views at a given *x*-position, denoted by (1) 0.35 mm, (2) 1.04 mm, and (3) 2.06 mm. Source: Yan et al. [42]/with permission from Elsevier.

between the tracks, known as the hatching distance. Another factor is the scan path, which directly influences the energy input contours on the substrate and hence affects the heat flow characteristics.

Firstly, single layer S-shaped (zigzag, anti-parallel, and bidirectional) and Z-shaped (zigzig, parallel, and unidirectional) scan paths are investigated (Figure 6.25a). When simulated under the same hatching distances and processing parameters, a noticeable height difference can be observed in Figure 6.25b,c. The major difference between the two cases are caused by a height difference between the start and end regions for each individual track. This difference is attributed to the Marangoni effect in the molten pool flow as shown in Figure 6.26. The front part of the molten pool has a lower surface tension due to the higher temperatures; therefore, the material flows from the front to the rear at the surface.

Next, the influence of the hatch spacing is investigated. Understandably, the hatch spacing should be less than the minimum width of the melt pool ($w_{\min}$) to ensure sufficient overlap. However, when that is not the case, pores and voids are likely to

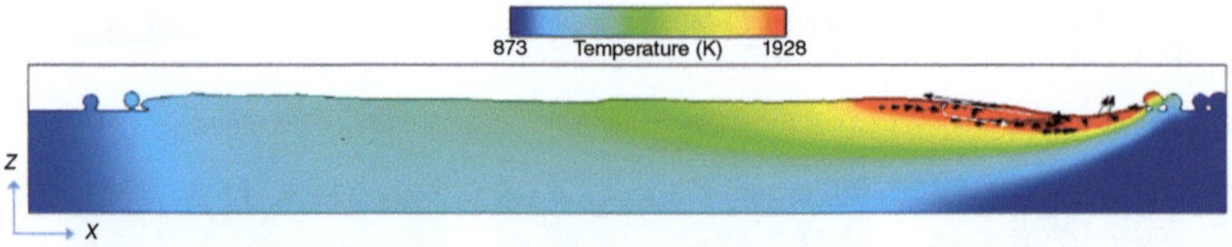

Figure 6.26 Marangoni effect in molten pool flow causes the melt pool front to be slightly depressed; black arrows denote flow velocity vectors, and white dashed arrow curve shows the flow circulation. Source: Yan et al. [42], with permission from Elsevier.

occur (Figure 6.25d,e). In the worst possible case, when the hatch spacing is greater than w_{max}, the neighboring tracks will not come into contact and are isolated from one another, leaving an unmelted region in between the two tracks (Figure 6.25f). In most cases, this result would be especially detrimental to the part quality if the design requires the complete fusion of two tracks. On the other hand, when the hatch spacing is between w_{min} and w_{max}, it is observed that inter-track voids may still form. Specifically, in the inter-track regions of the unmelted zone that initially contain interparticle pores, voids are formed. We can therefore conclude that the hatching distance should be no larger than the width of the remelted region within the substrate rather than the width of the melted region within the powder layer.

The inter-track voids form due to the varying melt width and depth of the melt pool, which are dependent on the history of the track formation. At the start of the track, heat is newly introduced onto the substrate and insufficient time has elapsed to allow significant heat accumulation. As such, the melted region at the start is relatively smaller. In contrast, subsequent regions that ensue the initial scan, as well as the tracks adjacent to newly laid tracks have larger melted regions. This result is observed because the subsequent regions to be scanned have been "preheated" by the heat conduction from the previously formed tracks, hence resulting in larger melt width and depth. Depending on the current position of the melt pool formation relative to the previously formed track, the lack of a "preheating" effect will lead to smaller cross-sectional regions and hence result in incomplete overlap.

Although it seems difficult to predict and control the melt region between tracks, it is almost certain that the melted regions at the start of the tracks will always be smaller. Hence, it can be hypothesized that edge-scanning strategies (such as "contouring") can be used to ensure the starting regions are melted and consolidated before scanning the inner regions, so as to mitigate inter-track voids.

6.3.5.4 Multilayer Cases

Next, by following the modeling process described earlier in Figure 6.21, multilayer simulations can be conducted for multilayer investigations. To demonstrate this simulation approach, a 2-layer-2-track simulation of the powder melting process of a Z-shaped track is conducted. It is assumed that a 2-layer-2-track study would be sufficient due to the repeated pattern that will arise between the interaction of the first and second track. Two scanning strategies are investigated: the layer-wise rotating strategy and the layer-wise interlacing strategy (Figure 6.27). In summary, the

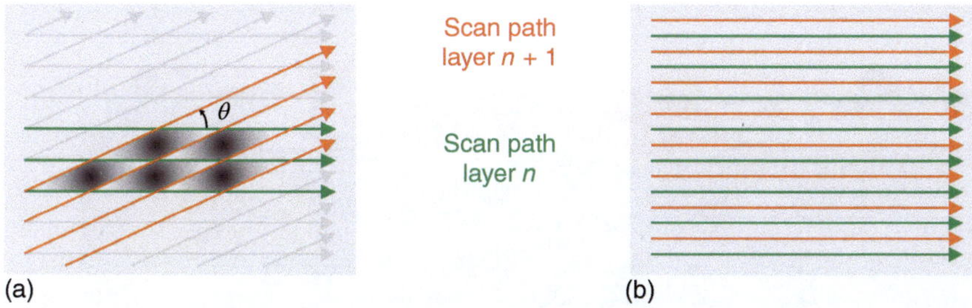

Figure 6.27 Schematic of the layer-wise scan strategy of (a) rotated by θ, and (b) parallel and interlaced scans. The arrows indicate the centrelines of the scan paths at different layers. The shaded regions in (a) highlight the positions where voids are likely to form.

results show that the interlaced scanning strategy is more beneficial in reducing the porosity.

Under the layer-wise rotating scanning strategy, the current layer is rotated by an angle θ ($0° \leq \theta \leq 180°$) relative to the preceding layer. Four different cases were considered in the investigation [11]; (i) $\theta = 0°$, unidirectional; (ii) $\theta = 180°$, counter-directional; (iii) $\theta = 45°$; and (iv) $\theta = 90°$. In all of these cases, it was evident that pores are present between the melt tracks.

These voids were observed to occur between the centerlines of adjacent scan paths in both layers as a result of the variation of the melt pool cross section as it forms along the powder bed. An example of the voids observed between the centerlines of the melt tracks is shown in the unidirectional case (Figure 6.28a). The cumulative effect is that the pores will most likely form in the region contained by the centerlines of the melt track from each layer (Figure 6.27). As these voids arise due to lack of fusion, the size of such voids are proportional to the powder size distribution and the difference between the hatch distance and the melting width. Certain randomness of the pore location and size will be observed due to the variability of the powder packing arrangement.

In addition, as the interparticle voids consolidate during melting, the fused surface of the first powder layer is lower than the height of the original powder layer. As such, when the second layer of powder is deposited, the effective thickness of the new layer varies from the original powder layer thickness to twice the powder layer thickness.

Moreover, in the unidirectional case ($\theta = 0°$), a void can be observed which is likely due to the weaker energy intensity and thicker powder layer, as it is in the middle region between two fused tracks. In the counter-directional case ($\theta = 180°$), a cluster of voids are observed near the start of the track of the second layer. This effect arises because after the first layer is formed, a dip exists at the end of the melt track due to the Marangoni effect, as elaborated previously (Figure 6.26). Subsequently during the powder spreading process, as the powder spreads onto this location, the resultant powder layer is effectively thicker. Therefore, more energy is required to melt an effectively thicker powder layer. Hence, by comparison of the contrasting observations, the voids near the edge are more pronounced in counter-directional cases than in unidirectional cases.

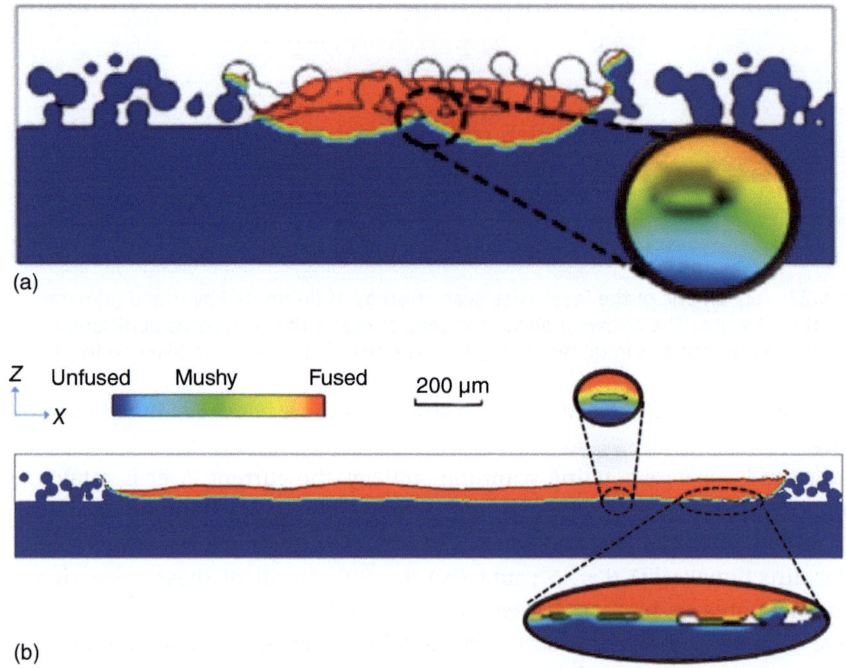

Figure 6.28 (a) Unidirectional case: A pore is observed between the centerlines of the melt track in the second layer. (b) Counter-directional case: A cluster of pores are observed at the end of the melt track in the second layer of melting. Source: Yan et al. [42], with permission from Elsevier.

Finally, the interlacing strategy was observed to contain no pores. Therefore, the layer-wise interlaced scan strategy was observed to be more beneficial for reducing the porosity in as-built products than the layer-wise rotated scan strategy. However, other voids may arise due to trapped gas into the melt pool but are currently not included in this model, as the incorporation of gas transport into the model remains challenging. This inadequacy may also complicate the comparison of pores between experiments and simulations, as the nature of pore formation in either case must be properly distinguished for a fair and accurate comparison.

6.3.6 Porosity Reduction and Optimization

The formation and interaction of the melt pools in different cases provide a slight insight into the possible directions and methods to reduce the porosity within and between the melt pools, and therefore, the porosity of the as-built part.

From the single-layer single-track studies, the clear strategy is to ensure that the process parameters are adjusted to provide sufficient melting of the substrate regions below the melted particles to prevent balling effects and formation of discontinuous droplets. However, there is no one-size-fits-all solution, and optimization studies are needed. In general, track uniformity can be improved by having smaller powder thickness, but the better quality print is obtained at the expense of build time. From the single-layer multi-track studies, the interlayer voids can be reduced by adopting contouring scanning strategies. Moreover, the hatching distance should be no larger than the width of the melted region within the substrate.

From the multilayer multi-track studies, the optimal fabrication parameters should ensure the sufficient fusion of a powder layer equivalent to more than twice the designated powder thickness. Additionally, the layer-wise interlace scan strategy is better than the layer-wise rotated strategies as the voids will likely form in the middle of adjacent tracks where the powder layer is relatively thicker at these regions.

6.3.7 Summary and Outlook

In this section, the heat source models, the evaporation model, and the coupling of FVM with DEM model have been presented along with the experimental comparisons. These examples illustrate the need for more accurate models in the context of AM. The heat source models described in this section show that the traditional description of a Gaussian heat distribution is insufficient. One needs to consider the ray-tracing in the powder bed for an accurate energy absorptivity and electron penetration mechanism for an accurate temperature distribution in the depth. On the other hand, the evaporation models show that Anisimov's evaporation model is only applicable for pure metal compositions, but in the case of AM most metal powders have alloy compositions.

While much effort has been devoted to derive this high-fidelity model, there is still much room for progress. Similar to any numerical model, it is susceptible to inaccuracies that may arise from reliable material data. In this regard, the high-temperature physics of liquid and vapor metal and its associated material properties will become increasingly valuable as modeling becomes increasingly relevant and feasible.

Moreover, other crucial physical phenomena have yet to be incorporated. As previously mentioned, the incorporation of gas models into the melting model with accurate physics is challenging. Furthermore, the multiphase modeling for vaporization and melt pool interaction prevalent in laser based melting systems remains a significant challenge. Very recently, several research groups [55, 56] developed preliminary multiphase flow models incorporating both molten pool flow and the gas flow interacting with powder particle motion. Additionally, there is currently no model on the effect of oxidation and oxides on PBF processes; only experimental investigations are available. Finally, burgeoning experimental investigations, such as the introduction of nanoparticles into the powder bed, introduce additional challenges for computational modeling. Regardless, the powder spreading and melting models established earlier serve as the groundwork for further modeling of more intricate physical phenomena.

6.4 Thermal Stress

Due to the high cooling rates in AM, it is highly desirable to predict the thermal stress and distortion during AM. Yet, existing literature employing thermomechanical models such as the FEM approach are oversimplified, which simply model the melt tracks as smooth and continuous regions. However, such a simplification ignores the effects of thermal-fluid flow and the actual morphology of the melt pool. In reality, as a new melt pool track forms, the solidified regions do not influence the stress states of subsequent regions. In addition, the remelting of previously deposited regions releases the residual stress in previous tracks. Finally, the tracks formed are highly irregular, have high surface roughness, and may contain porosities, in which such defects can result in local stress concentrations. The aforementioned effects are not captured with conventional FEM models, and these subtle intricacies ignored by conventional FEM approach can lead to profoundly different results.

In striving for a high-fidelity model for thermal stress simulation, we leverage on the precise temperature profiles and detailed track geometry obtained from CFD, and application of the quiet element model (QEM) for element suppression in FEM to address the shortcomings of the common FEM approach. By a one-way coupling of the thermal-fluid (CFD) and thermomechanical (FEM) models via interpolation between CFD cells and FEM nodes, accurate evolution of the thermal stress in the AM process in single- and multi-track cases can be obtained. For a detailed investigation into this methodology, the reader is referred to the reading [14].

6.4.1 Model Construction

The modeling approach presented here is applicable to a wide variety of fusion-based manufacturing processes. This section elaborates three essential parts of the modeling in the following order: (i) governing equations; (ii) interpolation equations for coupling; and (iii) state determination for QEM.

Firstly, the temperature history obtained from the CFD simulation (using the methods presented in Section 6.3) is directly imposed onto the FEM model via a one-way coupling. Therefore, the FEM simulation is a pure mechanical analysis with no transient heat transfer. Due to the lack of reliable material parameters for the thermal-elastic-plastic constitutive model, the material is currently assumed to be linear elastic. The governing equations applied to correlate the thermal strain, local temperature, and thermal stress are shown in Eq. (6.23) [57, 58].

$$\begin{cases} \varepsilon^{\text{th}} = \alpha\left(T,f_\beta\right)\left(T - T^0\right) - \alpha\left(T^{\text{I}},f_\beta^{\text{I}}\right)\left(T^{\text{I}} - T^0\right) \\ \boldsymbol{\sigma} = \mathbf{D}\boldsymbol{\varepsilon} \end{cases} \tag{6.23}$$

where $\alpha\left(T,f_\beta\right)$ represents the thermal expansion coefficient, T is the current temperature, T^{I} is the initial temperature, f_β represents the current material state, f_β^{I} represents the initial material state, and T^0 represents the reference temperature for the thermal expansion coefficient. Additionally, $\boldsymbol{\sigma}$ denotes the stress matrix, $\boldsymbol{D}$ denotes the elasticity tensor, and $\boldsymbol{\varepsilon}$ is the elastic strain matrix.

Next, a 3D interpolation is implemented to map the temperature stored in the center of Eulerian control volume (CV) of the CFD model onto the node of the Lagrangian element of the FEM model (Figure 6.29). Given the Eulerian structural mesh in the CFD model, we utilize the shape function of the brick element in the

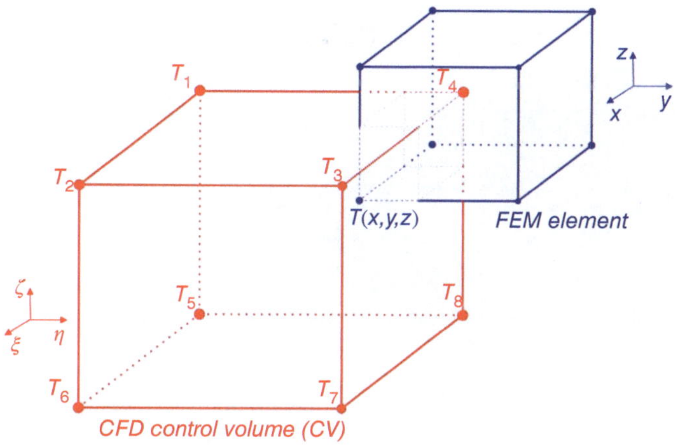

Figure 6.29 Schematic of the temperature interpolation between the CFD control volume centers (red) and FEM nodes (blue). Source: Chen and Yan [14]/Elsevier/CC-BY 4.0.

FEM as the interpolation function, given in Eq. (6.24) [59],

$$\begin{cases} T = \sum_{i=1}^{8} N_i(\xi, \eta, \zeta) T_i \\ N_i(\xi, \eta, \zeta) = \frac{1}{8} \left(1 - \xi\xi_i\right) \left(1 - \eta\eta_i\right) \left(1 - \zeta\zeta_i\right), \ i = 1,2, \dots, 8 \end{cases} \tag{6.24}$$

where ξ, η, and ζ are the local coordinates in the CFD coordinate system, ξ_i, η_i, and ζ_i are the natural coordinates of the CV centers surrounding the FEM node, and $T_i(\xi, \eta, \zeta)$ denotes the temperature values of eight CV centers surrounding the element node. Moreover, T is the interpolated temperature value for the FEM node inside the CV, and $N_i(\xi, \eta, \zeta)$ is the corresponding interpolation function.

This mapping scheme and the interpolation function is generally applicable to any 3D element shapes in FEM. After obtaining the temperature $T\left(t_1\right)$ and $T\left(t_2\right)$ by Eq. (6.24) on two adjacent instants t_1 and t_2 specified in CFD simulation, the temperature value $T\left(t'\right)$ between $\left(t_1 \leq t' \leq t_2\right)$ can be approximated by the 1D interpolation given in Eq. (6.25):

$$T(t') = T\left(t_1\right) + \frac{T\left(t_2\right) - T\left(t_1\right)}{t_2 - t_1} \left(t' - t_1\right) \tag{6.25}$$

The time step size in the FEM model, $\Delta t'$, is different from the data output time interval of the CFD model, Δt. To fully exploit the output of the CFD temperature profiles, $\Delta t'$ is smaller than Δt to allow a temperature interpolation between any two time-adjacent CFD temperature profiles.

Finally, the QEM is implemented by assigning null material properties to non-deposited elements, as their presence do not affect the deposited elements (Figure 6.30). The material properties in each element node change according to its own corresponding "state," dependent on the evolution of the morphology and temperature of the melt pool. The application of QEM gives rise to three possible states: gas, liquid, and solid (Figure 6.30b).

Therefore, the material state of each element depends not only its current temperature but also on its temperature history. Within a time increment Δt, the interpolated

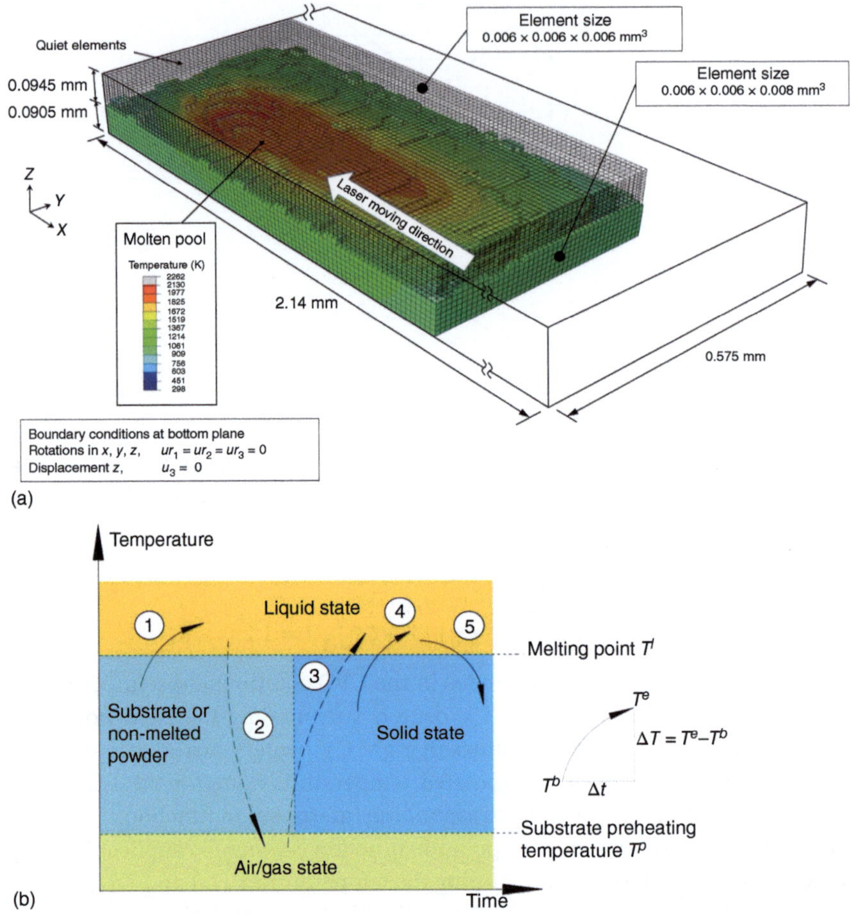

Figure 6.30 (a) Schematic of the quiet element model, within the defined domain, with mesh sizes and boundary conditions described in Section 6.4.2. (b) The material state changes during the melting and solidification process, which includes ① the substrate and powders are melted into liquid metal, ② the liquid material flows out of the cell and the state changes to air, ③ the state changes to liquid when liquid flows in, ④ the solidified material may be remelted into liquid, and ⑤ the liquid metal solidifies. It is noted that (i) there is no state transfer between solid and gas; (ii) since the effect of the unmelted powders on the stress distribution can be ignored, the unmelted powders are treated to be in the gas state. Source: Chen and Yan [14]/Elsevier/CC-BY 4.0.

temperature value for a node is T^b at the beginning and becomes T^e after Δt, so the temperature change is $\Delta T = T^e - T^b$. The melting point (the liquidus temperature for an alloy) and initial temperature (i.e. the preheating temperature) are constant and denoted by T^l and T^p respectively. Hence, the temperature criteria for the material state transformation can be defined as follows:

- **Melting.** if $\Delta T > 0$, $T^b < T^l$ and $T^e > T^l$, the node undergoes melting process and the material state for the node changes from solid to liquid;

Table 6.4 Material parameters for different material states in the FEM model.

Material states	Young's modulus (Pa)	Poisson's ratio	Thermal expansion coefficient (1/K)
Solid	1.32×10^{11}	0.310	9.2×10^{-6}
Liquid	1×10^{4}	0.001	9.2×10^{-6}
Air	1×10^{4}	0.001	0

- **Solidification**. if $\Delta T < 0$, $T^b > T^l$ and $T^e < T^l$, the node is solidifying and the material state changes from liquid to solid;
- if $T^e < T^p$, it is the environmental temperature and the material state is set as air.

6.4.2 Simulation Case

A simulation of two-layer, two-track EBM process is conducted. The electron beam diameter, scanning speed, power, and layer thickness are 0.4 mm, 0.5 m s^{-1}, 60 W, and 0.05 mm, respectively. The preheating temperature is 873 K. The thermal-fluid modeling and material properties (Ti-6Al-4V) are similar to that as listed in Section 6.3.5.

In addition to the material parameters, the material states in the FEM model, i.e. solid, liquid, and air, change depending on the temperature profile as shown in Figure 6.30b. The material properties for different states are listed in Table 6.4, in which the solidus elements are assigned with the material properties of Ti-6Al-4V. Note that the stiffness of the quiet elements (air and liquid elements) are assigned a nonzero value. Theoretically, the stiffness values should be set to zero, but assigning a zero Young's modulus to the air and liquid metal will result in numerical divergence and serious element deformation near the interface with the solid phases. Therefore, arbitrarily small values are assigned to the zero-valued factors, which are chosen such that it is sufficiently large to avoid numerical divergence but adequately small to be "quiet" for the stress analysis. Additional simulations have also been conducted to show that the results of the stress distribution are not affected by the arbitrarily assigned values if the values are sufficiently small.

As a result of the CFD-FEM coupling, the geometrical features (e.g. rough surfaces and voids) as well as the remelting of the previously deposited track and layer are successfully incorporated into the FEM model, as elaborated in Section 6.2 and 6.3. Naturally, the computational requirement for this approach is of interest and is briefly elaborated here.

The temperature profiles from the thermal-fluid model are output in an interval of 0.2 ms, while the time step size of the FEM model is set to be 0.1 ~ 0.2 ms. The mesh size in the thermal-fluid model is 0.005 mm. Figure 6.30 shows that in the FEM model, there are 922 752 elements and 969 612 nodes; the element dimension is $0.006 \times 0.006 \times 0.008$ mm^3 in the substrate and $0.006 \times 0.006 \times 0.006$ mm^3 in the region above the substrate. A mesh convergence test has been conducted to exclude mesh sensitivity.

The resultant computation time for the setup is around 700 hours for the thermal-fluid simulation on an Intel Core i7-2600 CPU (4 cores, 8 threads, and 3.40 GHz for the Processor Base Frequency) and 330 hours for the FEM simulation on an Intel X5650 CPU (6 cores, 12 threads, and 2.66 GHz for the Processor Base Frequency). The computational cost for the FEM model can be greatly reduced by optimizing the temperature loading scheme.

6.4.3 Stress Concentrations

The sectional views of the mechanical and thermal-fluid simulation results in Figure 6.31a reveal partially the voids, rough surface, and molten pool morphology. Aside from the terraced effect caused by the voxel mesh, the geometrical features and the temperature distribution between the FEM and CFD models show a good agreement. More importantly, the voids from the CFD model have also been mapped, which induces the stress concentration in the z direction (Figure 6.31b). The fidelity of the FEM temperature distribution can be further improved by refining the mesh and reducing the time step size of the FEM simulation.

The temperature gradient along the x–y plane in the cross section of the first layer is plotted at the end of the fabrication of the first layer ($t = 0.008$ seconds) as shown in Figure 6.31c. The temperature gradient at the start of the melt track after heating is larger than that of the middle region (circled in red), leading to a higher level of compression at this region.

6.4.4 Model Comparison and Application

6.4.4.1 Thermomechanical Model for Cross Comparison

The CFD-FEM modeling approach leverages on the high-fidelity thermal-fluid simulation to provide physically informed temperature profiles and accurate geometrical features for the mechanical simulation. This approach compensates for the oversimplification of conventional FEM models. To illustrate the superiority of the CFD-FEM model, a thermomechanical simulation is conducted for cross comparison with the CFD-FEM simulation.

To provide a fair comparison, a curved track shape resembling that from the thermal-fluid simulation was constructed (Figure 6.32a). However, the fine geometrical features of the rough surface are not modeled into FEM as it will lead to problems in mesh generation and heat source input. Hence, the surface needs to be modified to be smooth (Figure 6.32b). The scanning parameters and the material properties are identical to those listed in Section 6.4.2. Then, the input energy is modeled by Eq. (6.13) (elaborated in Section 6.3.2.2) and is effectively two-dimensional as no electron penetration is considered, since the element size is larger than the maximum penetration depth of the electron beam. The eight-node thermally coupled brick element (C3D8T) is used, and the mesh contains 209 304

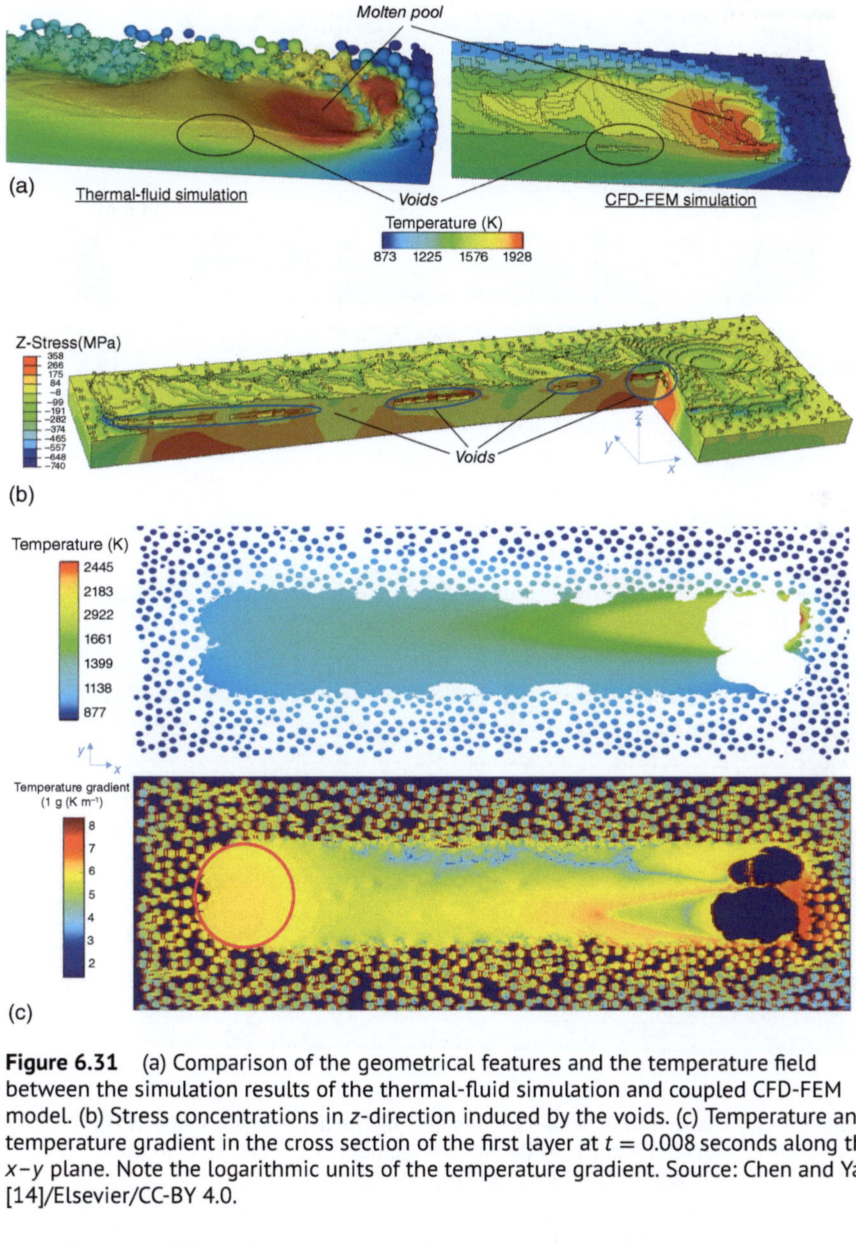

Figure 6.31 (a) Comparison of the geometrical features and the temperature field between the simulation results of the thermal-fluid simulation and coupled CFD-FEM model. (b) Stress concentrations in z-direction induced by the voids. (c) Temperature and temperature gradient in the cross section of the first layer at $t = 0.008$ seconds along the x–y plane. Note the logarithmic units of the temperature gradient. Source: Chen and Yan [14]/Elsevier/CC-BY 4.0.

hexahedral elements and 234 866 nodes. The computational cost for this FEM simulation is about 200 hours on an Intel X5650 CPU.

The comparison of the results for the thermal stress and temperature distribution between the thermomechanical and coupled CFD-FEM simulation are shown in Figure 6.33, where only the first layer is included. Several physical effects that occur in the thermal-fluid (FVM) model are not captured in the thermomechanical (FEM) model. For example in Figure 6.33a, energy loss due to evaporation is not

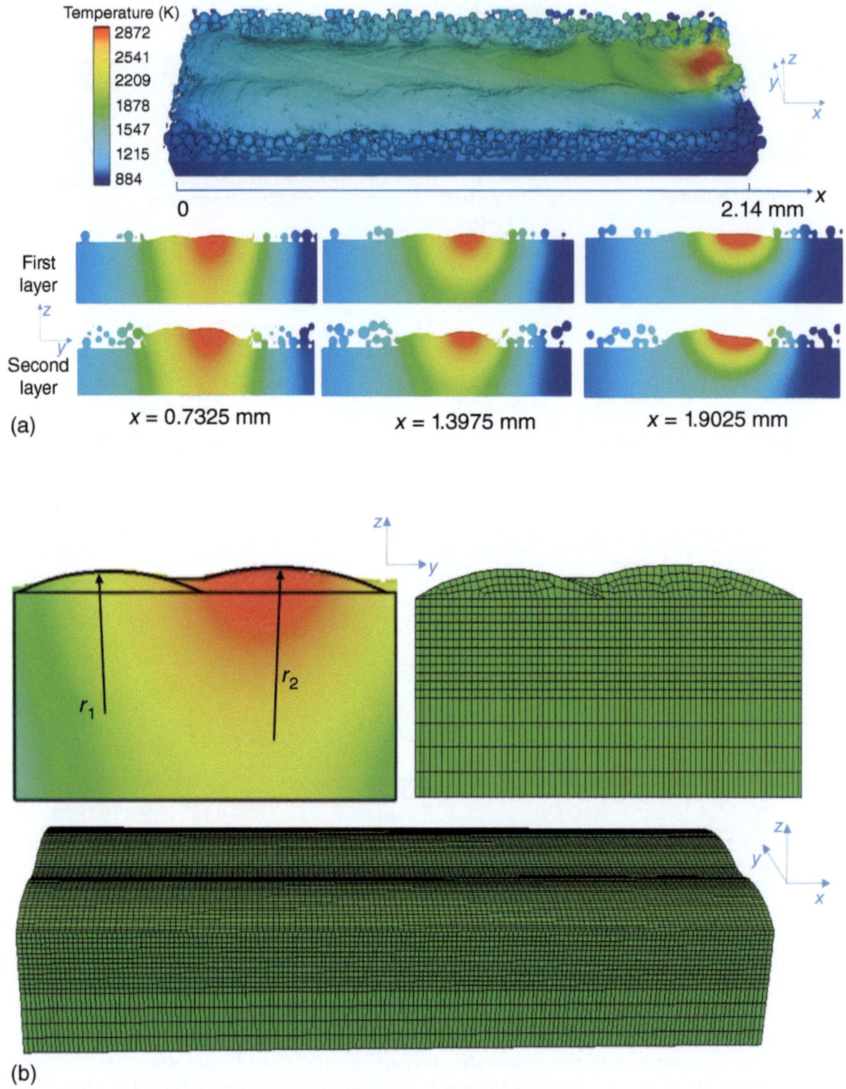

Figure 6.32 (a) Thermal-fluid simulation results. (b) The geometrical shape of the thermomechanical model mimicking the middle segment of the first layer in the thermal-fluid model. Source: Chen and Yan [14]/Elsevier/CC-BY 4.0.

accounted for in the FEM model, leading to a higher peak temperature (4228 K) than that obtained by the CFD-FEM model (2382 K). Additionally, the molten pool size in the FEM simulation is smaller than that of the CFD simulation. This discrepancy is observed because the hydrodynamic effects in the molten pool are not captured in the FEM model. For example, the Marangoni effect enlarges the molten pool dimension by convection due to thermal gradients. Finally, the CFD-FEM simulation is able to capture induced stress concentration regions, which are observed on the processed track of the CFD-FEM model. However, the stress

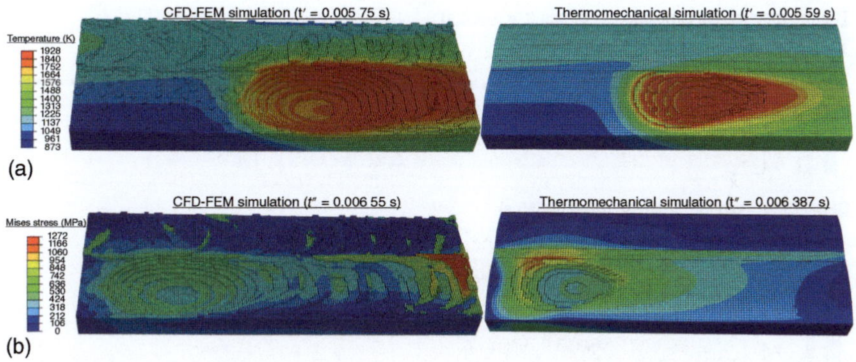

Figure 6.33 The comparison of (a) the temperature and (b) stress distribution between the CFD-FEM and thermo-mechanical simulation. Source: Chen and Yan [14]/Elsevier/CC-BY 4.0.

in the thermomechanical model is mostly released after the heating as the track surface is smooth (Figure 6.33b).

6.4.4.2 Thermal Cracking

The CFD-FEM model is able to offer a potential explanation for the mechanism of crack generation and other defects due to thermal effects [60,61], a key aspect where the thermomechanical model fails. The results of the CFD-FEM model are compared with three studies [14,62,63]. Normally, the deposited elements bear large compressive thermal stress in the heated regions. As the heat source moves away, the elements start to shrink. Thus, we focus on the cooling region where the residual stress concentration is most likely to occur.

As shown in Figure 6.34, the red regions of all the stress distribution plots highlight the tensile stress concentration regions where the cracks are more likely to occur. The single-track experiments by the SLM of the Fe-based metallic glass is conducted on Aconity Mini (Aconity GmbH, Germany) with the heat power of 100 W, scanning speed of 200 mm s^{-1}, and the layer thickness of 100 μm as shown in Figure 6.34a.

Note that the surface crack observed in the experiment propagates perpendicularly to the scanning path. Usually, such cracks are caused by the large stress components along the scan path. In the CFD-FEM simulation, the X-component stress concentration regions are found primarily along the edge of the surface wrinkles. The following mechanism is possible: these regions of relatively small stress concentrations are crack initiation sites, which then propagate and amalgamate into the long cracks vertical to the scanning path. However, in the thermomechanical simulation, the X-stress is distributed quite smoothly on the track surface, and no regions of tensile stress concentration can be observed.

Moreover, the stress component distributions in y and z directions are shown in Figure 6.34b and 6.34c, respectively. The long stress (Y-component) concentration region along the inter-track gap may result in cracks parallel to the scanning path

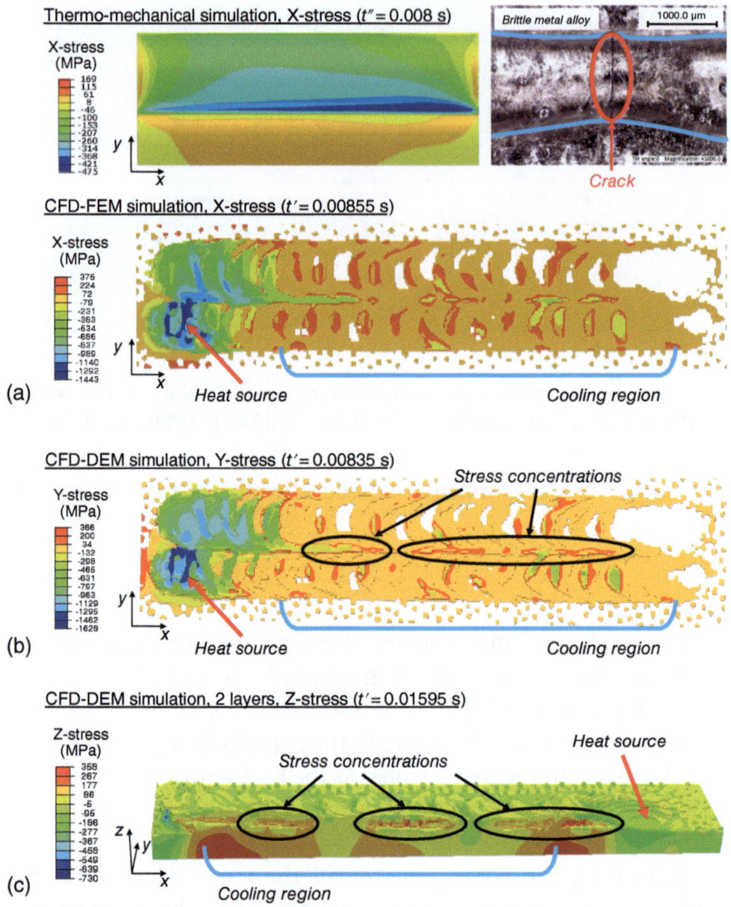

Figure 6.34 Comparison with crack observed in the SLM of Fe-based metallic glass by Aconity Mini (Aconity GmbH, Germany). (a) The crack occurring on the single-track sample; the X-stress distribution in the thermo-mechanical and CFD-FEM simulation at $t_1'' = 0.008$ seconds, $t_1' = 0.00855$ seconds, respectively; (b) the Y-stress distribution at $t_1' = 0.00835$ seconds; (c) the Z-stress distribution at $t_1' = 0.01595$ seconds for two layers. Source: Chen and Yan [14]/Elsevier/CC-BY 4.0.

direction. On the other hand, the stress (Z-component) concentration occurs mainly at the interface between different layers, especially in the regions around the voids, which may lead to delamination. To show the defects between two layers, we specifically observed the Z-stress distribution at the instant when two layers are fully deposited. On the other hand, as the X and Y stress components mainly concentrate on the layer surface, both of them are output at the instant when the first layer has been fully formed and cooled down.

In addition, another comparison between the single-track CFD-FEM simulation results with the experimental results by Vrancken et al. [63] is performed. As shown in Figure 6.35, the experimental cases conducted with different sets of laser power and scan speed are shown at different instants, i.e. t_1', t_2', t_3', and t_4'. Cracks can be

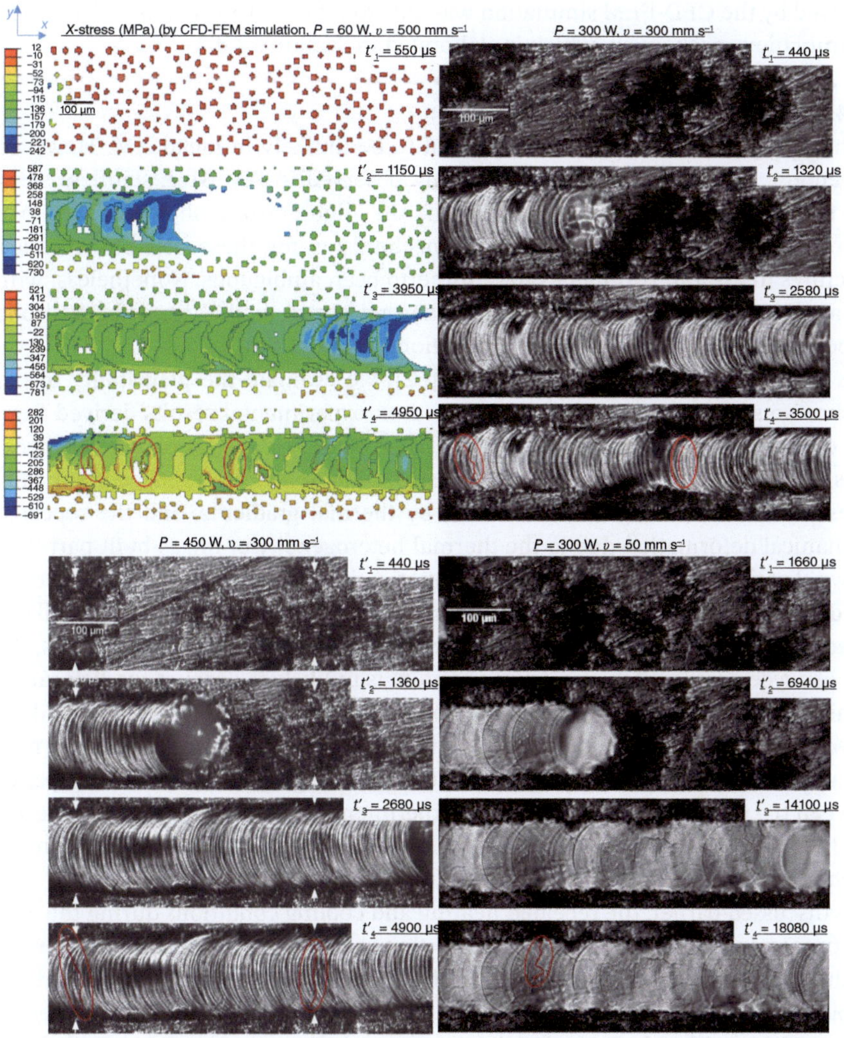

Figure 6.35 Comparison with microcracks in AM tungsten by Vrancken et al. [63]. The evolution of the X-stress by CFD-FEM simulation compared with the observed cracking in the single-track experiments. Source: Chen and Yan [14]/Elsevier/CC BY 4.0. Vrancken et al. [63]/Elsevier/CC BY 4.0.

observed at t_4' where the laser has traversed far from its previously scanned region. The evolution of the X-stress from the CFD-FEM simulation is reasoned here: the track segment is under compression at t_3', so no cracks can be found in the experimental results. Yet, when the laser passes far away at t_4', the track segment cools down further and shrinks. The compressive X-stress becomes tensile under this shrinkage, causing stress concentration regions perpendicular to the track. This mechanism explains the cracks generation in the single-track experiments.

Finally, this approach has also been applied in the investigation of high-density dislocations in additively manufactured copper [62]. The residual stress distribution achieved by the CFD-FEM simulation was able to rationally explain the dislocation accumulation mechanism, which is elaborated in Section 6.4.4.3.

6.4.4.3 Thermal Stress-Induced Dislocation

High-density dislocations were considered to be unlikely in metal AM. However, low-energy dislocation cells are commonly observed in AM 316L stainless steel, and sophisticated dislocation activities in the high-entropy alloy CoCrFeNiMn are observed. The cause and physical mechanisms for these dense dislocation structures in as-built AM metals have hitherto been ambiguous. Nonetheless, with the application of the aforementioned thermal stress model, we offer a strong proposition for the origin of dense dislocations in metal AM.

As demonstrated in our investigation [62], we evaluate the plausibility of the mechanisms of dislocation formation based on experimental results derived from PBF of AM copper. Among the many speculated causes for the formation of dislocations, we consider three possible mechanisms: cellular solidification, geometrically necessary dislocations (GNDs) rendered by thermal gradients, and the repeated mechanical deformation due to the thermal heterogeneity of the as-built part. It is shown that the repeated mechanical deformation is the most probable cause. Firstly, the dislocation formation due to cellular solidification of impurities is unlikely. Based on the EDS maps and elemental composition analysis of the Cu samples, the lack of oxygen segregation and oxygen content is unlikely to account for dislocation pinning and accumulation. Secondly, dislocation formation due to GNDs is unlikely as well. The orientation mapping with nanoscale resolution reveals no grain misorientations across the dislocation cell boundaries (Figure 6.36a). The discrepancy between the proposed mechanisms and experimental results leds us to hypothesize that the repeated mechanical deformation due to thermal heterogeneity is the main mechanism for dislocation formation.

As discussed earlier, the repeated heating and cooling conditions during the PBF AM process induces expansion and contraction in the melt pool, generating thermal residual stresses as the melt pool solidifies (Figure 6.34). Consequently, microscopic compression–tension cycles arise from the local thermal heterogeneities (Figure 6.36c). The dynamic loading and the deformation cycles are likely to affect the dislocation nucleation and accumulation. In this study, the simulation conducted for L-PBF Cu material revealed peak residual stresses in the order of tens of MPa (Figure 6.36b,c). It is reasonable to suggest that these high residual stresses activate plastic deformation in the as-fabricated regions that subsequently experience reheating. The dislocation patterns are self-organized into low-energy configurations via dislocation cross-slip and climb to minimize local strain energies, resulting in marginal or no local misorientations as observed in Figure 6.36a. These heterogeneous dislocation assemblies facilitate the development of high long-range internal stresses (Type-III residual stress), which concurs with widely observed microscale residual stresses in AM metals. Unless other deformation mechanisms are preferentially activated to release the residual stress, the dislocation structures are expected to be a characteristic feature of PBF metals.

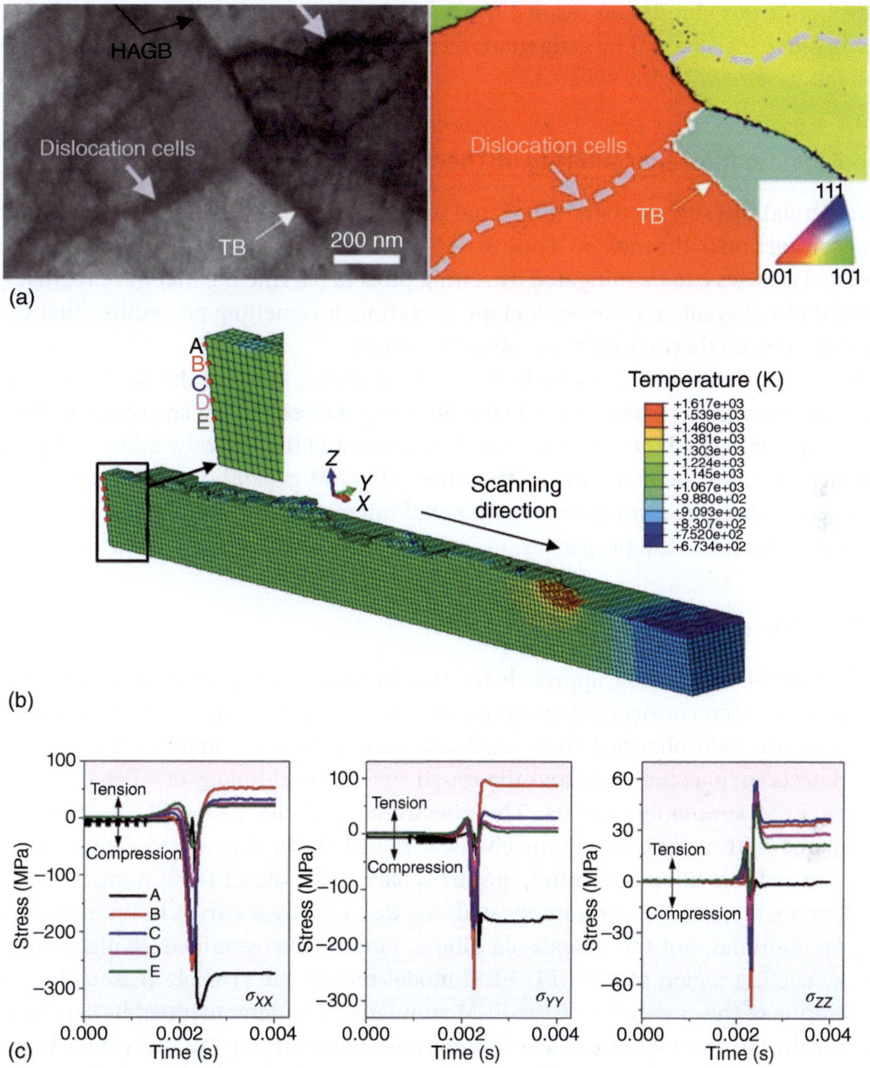

Figure 6.36 (a) The Bright-Field Scanning Transmission Electron Microscopy image (Left) and Inverse Pole Figure map (Right). The dislocation cell boundaries and twin boundary (TB) are indicated in the figures. (b) Thermal stress simulation of Cu with the CFD-FEM model described earlier. Five points with an average spacing of 10 μm in the as-built part were deliberately selected for local stress analysis. (c) Stress evolution over time of the respective components (σ_{XX}, σ_{YY}, and σ_{ZZ}) at the selected points shows clearly the presence of compression–tension cycles. Source: Wang et al. [62]/Informa UK Limited/CC BY 4.0.

Hence, the experimental evidence and comprehensive analysis of possible mechanisms strongly support the theory that the deformation cycles due to residual thermal stresses are the origin of the formation of dislocations in as-built AM metal. Unfortunately, it is experimentally challenging to capture this phenomenon to corroborate our simulation results. Until novel imaging techniques are developed, the *in-situ* experimental investigations to probe for effects of thermal stress and its validation remain highly coveted.

6.4.5 Mitigation and Tailoring of Thermal Stress

The simulations suggest that the residual stresses are induced by the rapid expansion and contraction of the melt pool due to high thermal gradients. Where necessary, the thermal stresses can be mitigated by setting process parameters that have relatively tame thermal gradients or even include reheating or remelting procedures that can release induced thermal stresses that are too high.

Moreover, there is the possibility of using this model to aid in the design and tailoring of the dislocations by varying the process parameters. The layer-wise melting procedure with local thermomechanical control could theoretically allow AM practitioners to tailor the metal microstructures. This AM capability offers the exciting prospect of direct control over the AM metal microstructure, and hence, the direct tailoring of mechanical properties.

6.4.6 Summary and Outlook

A high-fidelity modeling approach for the thermal stress prediction of the PBF process has been constructed and presented in this section. Based on the accurate temperature field obtained from CFD simulations and the application of QEM, the defects such as the voids and the rough surface morphology can been mapped to the FEM simulation results. The effectiveness of the CFD-FEM approach is compared with a pure thermomechanical model. With the consideration of the melting and remelting procedure, the stress pattern for the CFD-FEM model in the molten pool region is more complicated: the trend of stress curves in the unmelted region is similar, but the magnitude differs. Finally, the overall stress distribution in the cooling region of the CFD-FEM model reveals the possible reasons for the generation of the cracks. The CFD-FEM simulation has demonstrated its capability to explain the thermal stress distribution and evolution on the individual tracks with accurate temperature fields and few geometrical simplifications. The qualitative results and its relative magnitudes remain relevant insofar as to explain the stress concentrations induced by residual heat. However, without data of accurate temperature-dependent thermal expansion coefficients and thermo-elastic-plastic constitutive relations, the current model is not able to provide accurate quantitative prediction of thermal stress.

Moreover, the quantitative validation through experimental measurement of thermal stress on individual tracks is highly difficult. Nonetheless, high-quality experimental measurements of this nature are still possible, although they require the highly sophisticated facilities at US national labs using X-ray [64] or neutron diffraction [65].

Currently, due to the large computational cost, the CFD-FEM model faces difficulty to simulate the fabrication of a whole part. However, once the thermo-elastic-plastic constitutive law and the temperature-dependent material properties are accurately measured in experiments and incorporated to the model, coupled with proper multiscale modeling approaches, e.g. macroscale modeling of the sequential accumulation of melt tracks with the accurate thermal strains from this mesoscale CFD-FEM model, a whole part can be simulated in an efficient and accurate way, and the experimental validation at the part scale can be further conducted.

6.5 Modeling of Other Unique Phenomena

In the modeling of AM processes, most efforts have been devoted to the modeling of powder spreading and powder melting mechanisms. In this section, the modeling of two other phenomena are presented. The first is the simulation of sintering mechanisms in EBM processes using FVM and phase-field models [12]. The second is the coupling of DEM and FVM (a multiphase model) to simulate the powder spatter and denudation due to a vapor jet during AM [16].

6.5.1 Powder Sintering in EB-PBF

It is common to incorporate a preheating procedure in EB-PBF processes. During this procedure, the electron beam slightly sinters the powder bed prior to processing. Here, the sintering of Ti-6Al-4V powders in EB-PBF is presented. There are two possible sintering mechanisms: liquid-state and solid-state sintering. The liquid-state sintering is simulated using the same thermal-fluid model described in Section 6.3. On the other hand, a phase-field model is used to simulate the solid-state sintering.

6.5.1.1 Liquid-State Sintering

The simulation of powder particles with different diameters (20 μm, 28 μm, 40 μm) is conducted, and the maximum temperature of each particle is acquired as shown in Figure 6.37. Notice that the temperature profile at the highest point barely reaches the solidus temperature. The partial melting of a powder particle will occur, but the particle maintains its spherical shape due to surface tension. Moreover, the Marangoni effects are negligible due to minor temperature gradients. The results indicate that liquid-state sintering should be minimal or nonexistent in the absence of very fine particles. This indication is pertinently the case when the powder gets increasingly reused, resulting in fewer fine particles in the mix over time [66].

6.5.1.2 Phase-Field Model

Due to constraints of computational cost, 2D simulations are performed. Two kinds of fields are used, and the evolution of solid-state sintering process is driven by minimizing the total free energy F:

$$F = \int_v \left[f(c;\eta_\alpha) + \frac{12}{\kappa_c}|\nabla c|^2 + \sum_\alpha \frac{12}{\kappa_\eta}|\nabla \alpha|^2 \right] \, dv \tag{6.26}$$

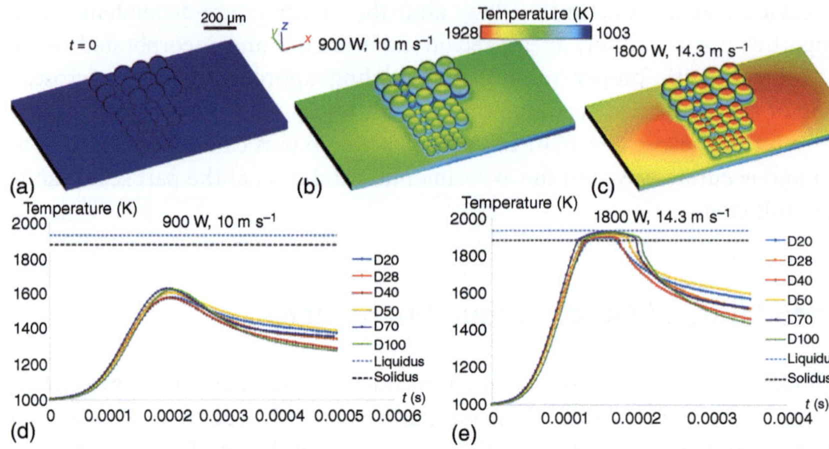

Figure 6.37 Thermal-fluid flow simulation of liquid-state sintering. (a) initial condition; (b) snapshot of temperature field under the power of 900 W and scan speed of 10 m s^{-1}; (c) snapshot of temperature field under 1800 W and scan speed of 14.3 m s^{-1}; (d) temperature profiles of powder particles at the top points for 900 W and scan speed of 10 m s^{-1}; (e) temperature profiles of the top points for 1800 W and scan speed of 14.3 m s^{-1}. Source: Yan et al. [12], with permission from Elsevier.

where c is the volume of solid, η_α is the order parameter field, f is the bulk free energy, and κ_c and κ_η are the gradient energy parameters. Furthermore, the Cahn–Hilliard equation is used for the conserved field c according to mass conservation principle, and the Allen–Cahn equation is used to solve for the nonconservative order parameter field η_α:

$$\left\{\begin{aligned} \frac{\partial c}{\partial t} &= \nabla \cdot \left(\mathbf{M}\nabla \frac{\delta F}{\delta c}\right) = \nabla \cdot \left[\mathbf{M}\nabla \left(\frac{\partial f}{\partial c} - \kappa_c \nabla^2 c\right)\right] \\ \frac{\partial \eta_\alpha}{\partial t} &= -L\frac{\delta F}{\delta \eta_\alpha} = -L\left(\frac{\partial f}{\partial \eta_\alpha} - \kappa_\eta \nabla^2 \eta_\alpha\right) \end{aligned}\right. \tag{6.27}$$

where $\mathbf{M}$ is the concentration mobility tensor dependent on the diffusion coefficients, and L is the order parameter scalar mobility dependent on the grain boundary mobility and energies. For a detailed elaboration of these equations, the reader is referred to the reading [12].

6.5.1.3 Solid-State Sintering

Three cases of simulation studies are performed: the necking of equal-sized monocrystalline powder particles, unequal-sized monocrystalline powder particles, and equal-sized polycrystalline powder particles are simulated (Figures 6.38 and 6.39). Generally, the neck grows rapidly in the first five seconds and slowly approaches a stable value. It should be noted that while the preheating procedure lasts only a few seconds, the particles remain at the preheating temperature during the whole manufacturing process. As such, the solid-state sintering mechanism would theoretically occur over the entire manufacturing duration until the temperature reaches a stable state. However, due to limitations of computational cost, it is impractical to explicitly simulate the process over the entire time span.

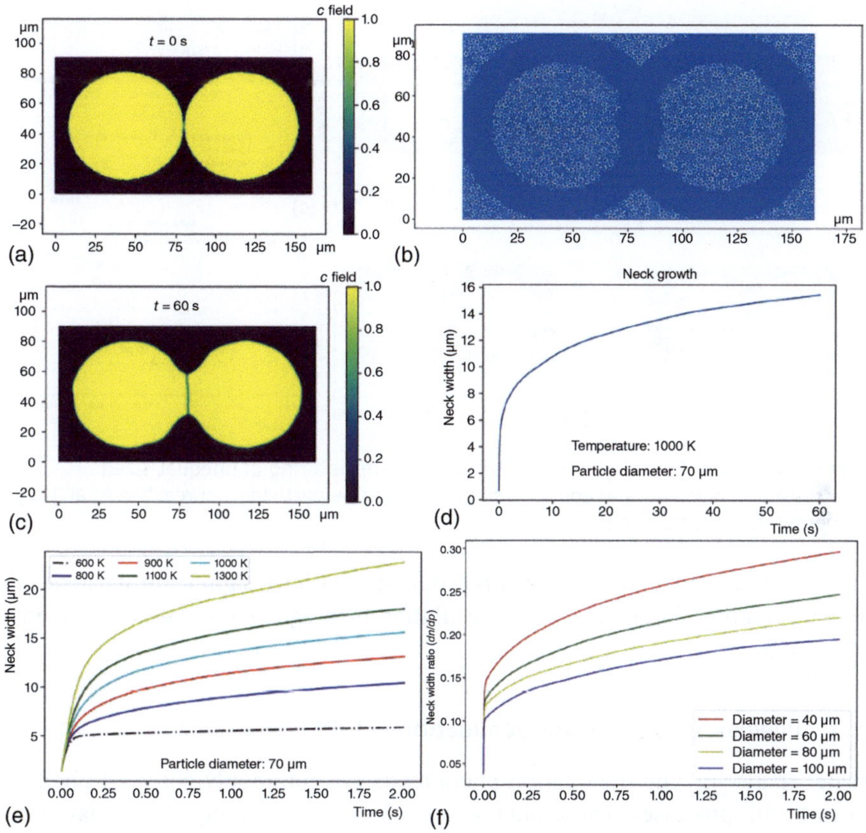

Figure 6.38 Phase-field simulation of the solid-state sintering of equal-sized monocrystalline particles. Source: Yan et al. [12], with permission from Elsevier.

Fortunately, the result of the simulation can still be compared to an independent study conducted by Leung et al. [67], where it is demonstrated that the neck diameter usually reaches about 30% of its diameter. This comparison shows that the simulation results agree well with the experimental results.

In addition, the preheating temperature has remarkable influence on neck growth. In general, the neck grows faster at higher temperatures, and when the temperature is 600 K or lower, no solid-state sintering occurs, which is consistent with experimental results. In addition, smaller particle diameters lead to faster neck growth rates. Unfortunately, the critical temperature for solid-state sintering was not identified due to lack of reliable material parameters. In the case of unequally sized monocrystalline powder particles, the results also agree qualitatively with the experimental observations.

In the polycrystalline powder particles, a simpler four-grain particle is considered. The results in Figure 6.39 indicate that the neck growth is similar to that of monocrystalline particles, although the grain boundary between the two particles becomes twisted due to grain boundary diffusion. Therefore, it is speculated that the

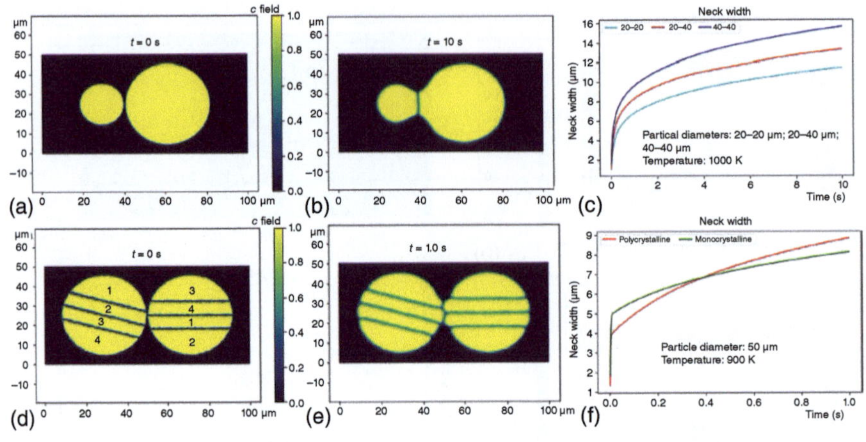

Figure 6.39 Phase-field simulation of the solid-state sintering of unequal-sized monocrystalline particles and equal-sized polycrystalline particles. Source: Yan et al. [12], with permission from Elsevier.

initial grain structures of the virgin powders do not have significant influence on the powder sintering process. This speculation is in agreement with experimental observations.

6.5.2 Powder Spattering and Denudation in L-PBF

Powder spattering and denudation phenomena are commonly observed phenomena in L-PBF processes. These phenomena disrupt the existing powder layer and introduce potential sources of defects. Powder spattering occurs when powder particles near a scan track are blown up, while powder denudation occurs when powder particles immediately surrounding the scan track are apparently depleted during powder melting. It has been hypothesized that these phenomena are caused by the metal vapor jet generated due to the vaporization of material, which induces strong fluid dynamic effects on the powder bed. However, most numerical models currently ignore these effects by assuming that the powder bed is static. Hence, this section presents a first attempt of a model that couples the dynamic behaviors of the powder bed and metal vapor jet, which corroborates well with the findings of the experimental measurements of powder spatter with state-of-the-art equipment.

6.5.2.1 Multiphase Flow Model

The multiphase flow model employs a bidirectional coupling between the DEM and FVM, using the same governing equations listed in Sections 6.2 and 6.3, respectively, where in this case, the FVM model is for the fluid simulation of the vapor jet. Due to the existence of this vapor jet, the effects of the drag force and energy exchange due to heat flux need to be included into the respective governing equations. The additional equations added to the governing equations of DEM and FVM, along with the coupling equations, are listed here. For the full list of equations and treatment of the metal vapor and ambient gas in the fluid phase, the reader is referred to the article [16].

Essentially, a two-way coupling for the momentum and energy equations in both FVM and DEM is implemented. For direct exchange of data, the FVM and DEM modules share the same Eulerian mesh. First, the FVM module iterates to convergence for a single time step, and the following values are exported from fluid mesh cells to the DEM mesh:

$$\begin{cases} \mathbf{F}_{\mathrm{g}} &= m_P \mathbf{g}\left(1 - \frac{\rho_{\mathrm{F}}}{\rho_{\mathrm{P}}}\right) \\ \mathbf{F}_{\mathrm{P-F}} &= 6\pi R\left(\mathbf{u}_{\mathrm{F}} - \mathbf{u}_{\mathrm{P}}\right)\mu_{\mathrm{F}} \\ Q_{\mathrm{P-F}} &= h_{\mathrm{P-F}} A_{\mathrm{P}}\left(T_{\mathrm{F}} - T_{\mathrm{P}}\right) \end{cases} \tag{6.28}$$

where ρ_{F} and ρ_{P} are the density of the fluid and powder phase, $\mathbf{F}_{\mathrm{P-F}}$ is the drag force from the gas phase to the powder, $\mathbf{u}_{\mathrm{P}}$ is the velocity of the powder particle, $\mathbf{u}_{\mathrm{F}}$ is the fluid velocity of the CV (where the particle center stays), and $Q_{\mathrm{P-F}}$ is the heat flux from the gas phase to the powder. In the DEM module, the values in Eq. (6.28) are input into the equations listed in Eq. (6.29):

$$\begin{cases} \mathbf{F}_{\mathrm{ext},i} &= \mathbf{F}_{\mathrm{P-F}} \\ m_{\mathrm{P}} c_p \frac{\mathrm{d}T_P}{\mathrm{d}t} &= Q_{\mathrm{P-F}} + \sum Q_{\mathrm{P-P}} \\ Q_{\mathrm{P-P}} &= 2\lambda_P\left[\frac{3F_n(1-\eta^2)a}{2E}\right]^{\frac{1}{3}} \end{cases} \tag{6.29}$$

where the external force $F_{\mathrm{ext},i}$ is equal to $F_{\mathrm{P-F}}$. The effect of heat transfer is included, where T_{P} is the temperature of the powders and $Q_{\mathrm{P-P}}$ is the heat flux between two contacting powder particles. The heat flux $Q_{\mathrm{P-P}}$ is calculated with the thermal conductivity λ_{P}, radius of the contact area a, Young's modulus E, and Poisson's ratio η.

After the DEM module runs for a time step, the information of the particles are input to the FVM module to calculate the forces on the gas phase by introducing momentum sinks $\boldsymbol{f}_{\mathrm{P-F}}$ and heat losses $q_{\mathrm{P-F}}$:

$$\begin{cases} \boldsymbol{f}_{\mathrm{P-F}} &= \frac{-\sum_j \boldsymbol{F}_{\mathrm{P-F}}}{V_{\mathrm{mesh}}} \\ q_{\mathrm{P-F}} &= \frac{-\sum_j Q_{\mathrm{P-F}}}{V_{\mathrm{mesh}}} \end{cases} \tag{6.30}$$

where V_{mesh} is the volume of the local mesh cell in the simulation domain and j is the total number of powder particles in the local mesh cell. The calculation process repeats again until the end of simulation time.

6.5.2.2 Multiphase Flow Behaviors

With the flow model established earlier, the simulation of an iron (Fe) vapor jet on a stainless steel powder bed was conducted. The dynamics of the multiphase flow have been observed to produce the powder denudation and powder spattering phenomena observed in experiments. In addition, the temperature profiles surrounding the vapor jet provide additional insights and reveal the dominant mechanism for the observed spatter in experiments.

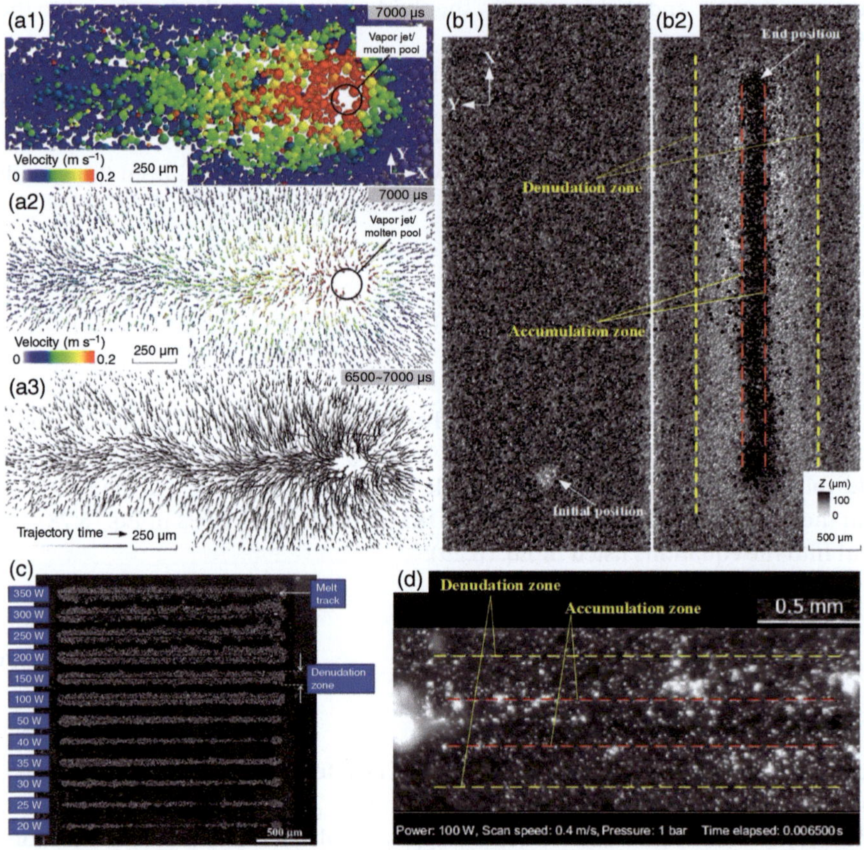

Figure 6.40 Denudation zones in simulations and experiments. Simulation: (a1) particle motions, (a2) particle velocity vectors, and (a3) trajectories of particles within the powder bed ($Z < 90\,\mu m$). Height maps of the powder bed are shown in (b1) before and (b2) after the scanning. The optical observations in experiments are shown in (c) by Mathews et al. [4] and (d) by Bidare et al. [5]. Source: Chen and Yan [16]/with permission of Elsevier.

As the vapor jet moves along the powder bed, particles in the powder layer are entrained toward the metal vapor due to vortex flows near the jet, resulting in a denudation zone near the scanning track (Figure 6.40). The denudation zone is relatively wide (400 μm) compared to the hatch spacing (∼100 μm) used in conventional AM. Such a denudation effect results in either thinner powder layers or exposed areas with no powders, which will deteriorate the quality of the subsequent melt tracks. Additionally, as the powders accumulate, they may gather onto the track and be partially melted or get attached, worsening the surface roughness (as discussed in Section 6.3.5).

Furthermore, the particles entrained into the vapor jet lead to powder spattering. The extent of the spatter depends on the momentum of the powder particles. Depending on the trajectories of the powder particles, these spattered particles may deteriorate the layer quality (Figure 6.41). Most of the powder particles are ejected far away from the powder layer at high velocities (exceeding the simulation domain)

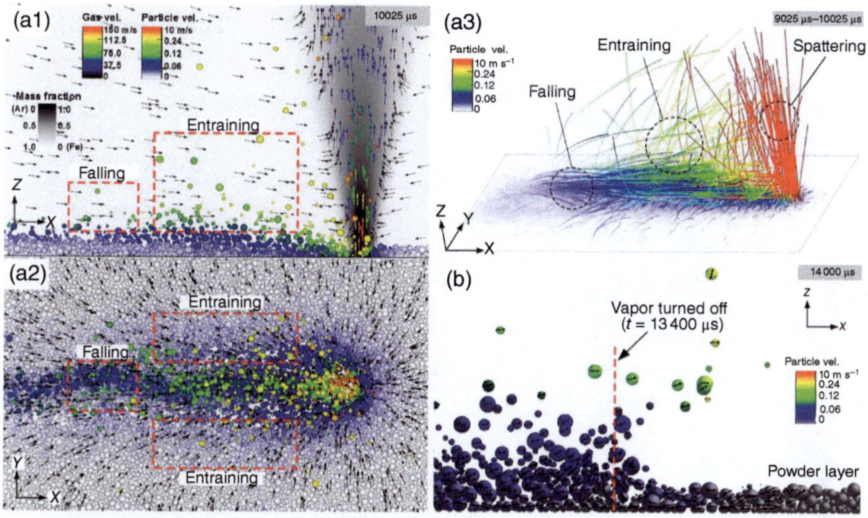

Figure 6.41 Spattering phenomenon: (a1) side view and (a2) top view of the velocity field; (a3) particle trajectories; (b) particle motions after the metal vapor is off, and the powder particles settle onto the powder bed.

and are likely to settle at some distance away from the vapor jet source, disrupting the existing powder bed. Due to the radial expansion of the vapor jet, only a few powder particles settle into the denudation zones. However, it is worth mentioning that in practice, the powders may get entrained or denuded into the melt pool, which is not incorporated here.

While the metal vapor has high temperatures of up to 3080 K, the quick heat dissipation due to rapid expansion and convective heat transfer under turbulent conditions result in extreme temperature drop from the vapor jet source to the ambient environment (2000 K over 1.6 mm) as seen in Figure 6.42. In addition, the inward

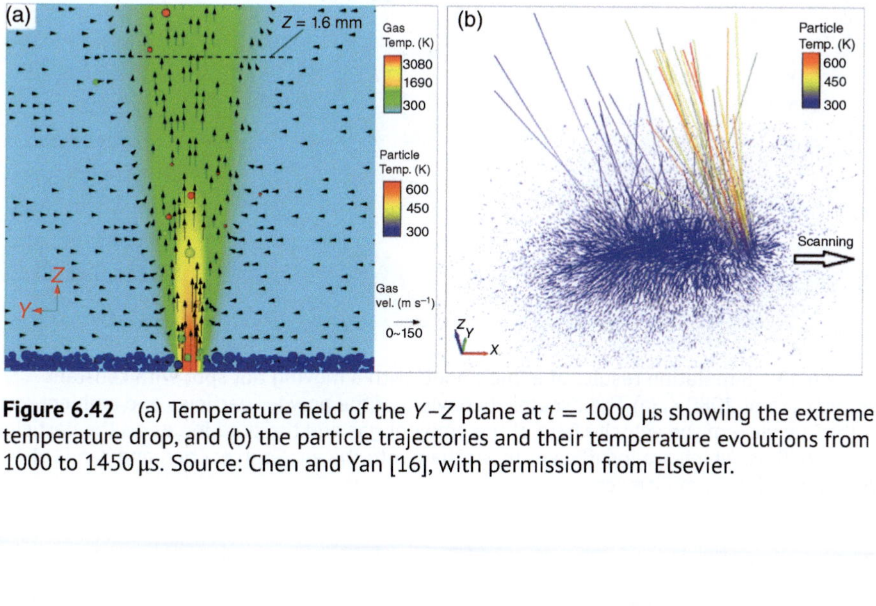

Figure 6.42 (a) Temperature field of the Y–Z plane at $t = 1000$ μs showing the extreme temperature drop, and (b) the particle trajectories and their temperature evolutions from 1000 to 1450 μs. Source: Chen and Yan [16], with permission from Elsevier.

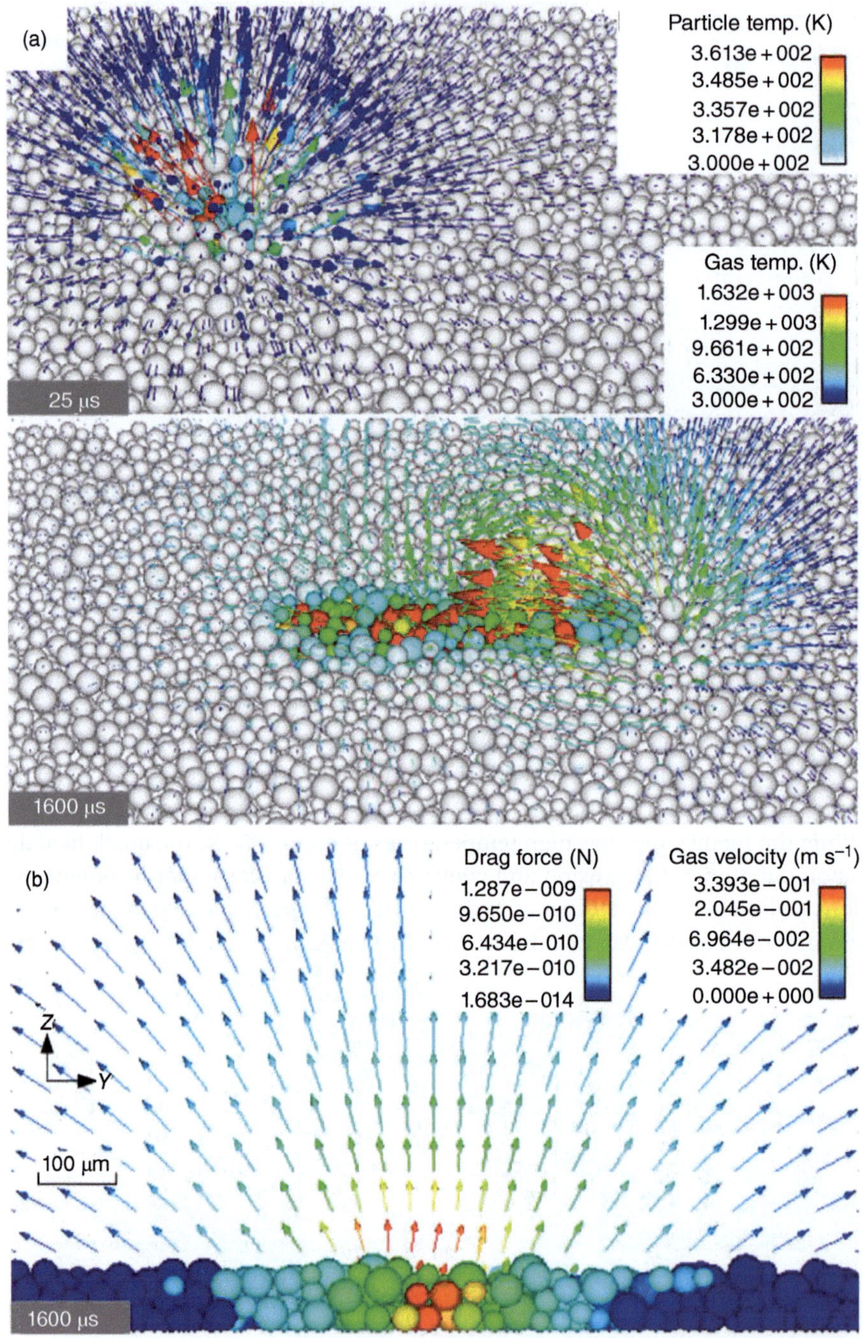

Figure 6.43 Simulation results of a "melt pool" with a moving hot spot with constant temperature of 2080 K. (a) The dynamic behaviors of the powder particles and ambient gas Ar. (b) Side view of the velocity field of the ambient gas and the drag forces on the particles at $t = 1600\ \mu s$, which are insufficient to lift them into the air. Source: Chen and Yan [16], with permission from Elsevier.

vortex circulation results in constant supply of cold ambient gas which ensures only moderate heating of the surrounding gas occurs near the vapor jet. Hence, the rapid heating of the entrained particles do not result in melting. We can therefore eliminate the possibility that the entrained particles are a source of liquid spattering droplets observed in experiments, and that the observed liquid spatter most likely originate from the molten pool.

Finally, a simulation case was conducted to mimic a melt pool under thermal-conduction mode of PBF process, where there is almost no metal vapor. Nonetheless, a thermal buoyancy effect is observed: the molten pool heats up its surrounding ambient gas, resulting in a convection effects similar to the metal vapor. However, the effects are relatively tame (Figure 6.43) due to small magnitudes of air velocity. No vortex flow is observed, and consequently, drag forces are insufficient to overcome gravity and friction to result in spattering or denudation phenomena. Hence, we conclude that the vapor jetting is the dominant mechanism for powder spattering and denudation phenomena, and the thermal buoyancy effect on powder spatter and denudation is negligible.

6.5.2.3 Influence of Jetting Angle

The jetting angle is determined by the laser power and scanning speed. In brief, there are no significant differences when the jetting angle is in the following range $30° < \theta_{jet} < 150°$. However, when the jetting angle is more than 150°, there will be no vortex flow downstream of the vapor jet. Due to the small angle between the jetting vapor and powder layer, the radial expansion of the vapor acts on the powder layer and blows away most of the particles within the trajectory. As such, there is no accumulation zone, resulting in a completely exposed denudation zone (Figure 6.44). Similarly, one can conclude that a jetting angle of less than 30° can lead to exposed denudation zones in front of the laser spot, resulting in too few particles to be melted to form a complete melt track. However, it should be noted that the range of angles are not a universal value, as the jetting speed and geometrical expansion are determined by the combined effects of laser power, scan speed, and powder layer properties.

6.5.3 Summary and Outlook

Firstly, based on the high-fidelity model of the preheating procedure, the sintering process is determined to be dominated by solid-state sintering mechanism, which is determined by the material properties, designated preheating temperature, and particle size. However, the model can be further improved by additional considerations such as movement of particles, and more importantly impurities and oxidation film on the surfaces. This improvement will be especially crucial since powders will be constantly reused, wherein such a practice may influence the powder quality and sintering behavior.

Next, the multiphase model demonstrates the first simulation, verified with experimental studies to replicate the denudation and spattering phenomena of powder particles. However, there are several factors that were not included in the

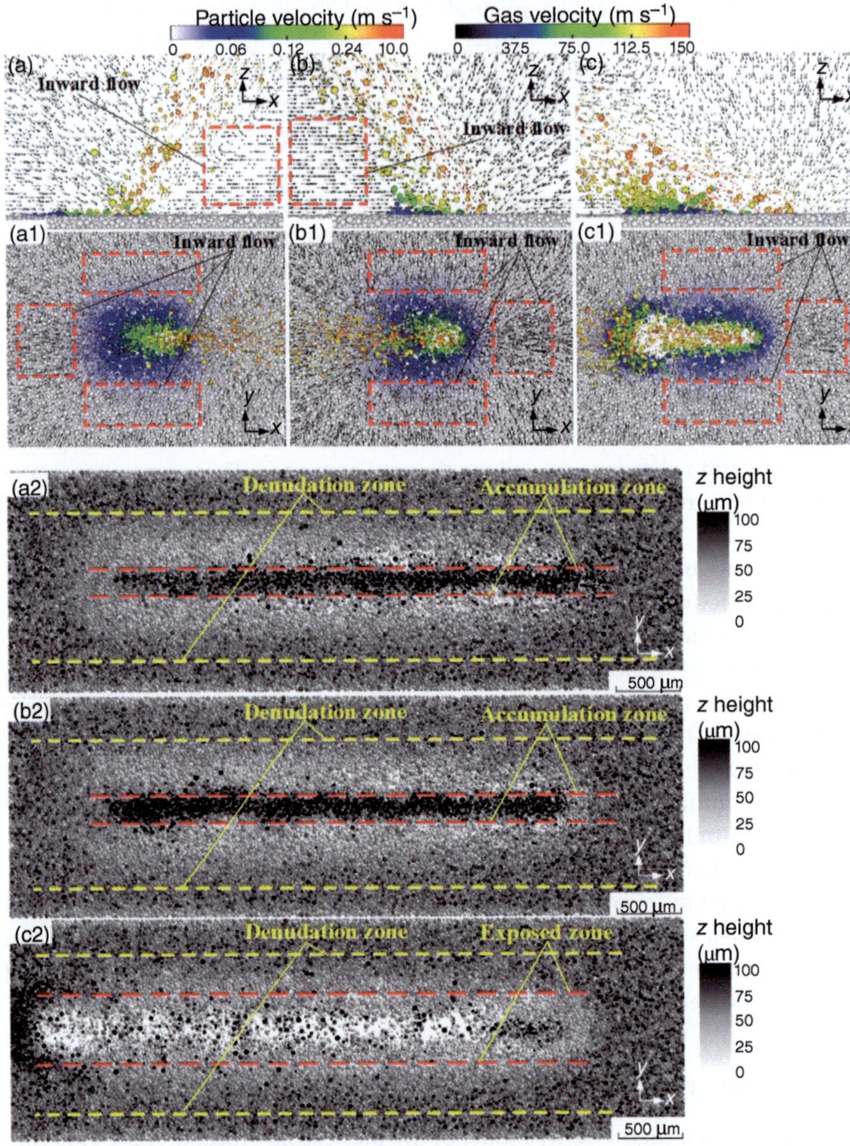

Figure 6.44 Simulation results of the spattering and denudation phenomena at different jetting angles: (a) 60°, (b) 120°, and (c) 150°. (1) shows the simulated velocity fields of the gas phase and particle, while (2) shows the resultant powder layer. Source: Chen and Yan [16]/with permission of Elsevier.

model, such as the influence of the laser beam, molten pool, and laser-induced plasma. In particular, the vapor jet is likely to be dynamic as the molten pool is constantly oscillating. The dynamic nature of the vapor jet results in a more complex powder dynamic behavior. Moreover, a flat plate is used in the simulation but in reality, the previously melted layer is likely to be irregular and rough.

Therefore, in the powder denudation phenomenon, adjustment of laser parameters and ambient pressure conditions can be further investigated. In the powder spattering phenomenon, systematic studies on the complex spattering behavior with incorporation of the molten pool and shielding gas flow can be conducted. Moreover, additional influences due to laser irradiation and laser-induced plasma may result in more complex temperature profiles. Looking further, one can consider the effective implementation of a vapor jet on the molten pool and successive powder layers.

6.6 Conclusions

In this chapter, we have presented our works on high-fidelity modeling encompassing a wide range of the AM process. These results become especially valuable when detailed experimental observations are difficult or unfeasible. When employed properly, simulations are powerful tools that can aid researchers and engineers to reliably predict and therefore design optimal processes. Nonetheless, simulations still require experimental validation to be accurate. The research in AM and its manufacturing dynamics require the combined effort of high-fidelity modeling and high-quality experiments.

Much effort hitherto has been directed at uncovering the underlying physical mechanisms during powder spreading and powder melting. However, many other physical processes such as multiphase fluid interactions, oxidation, and effect of element compositions have yet to be incorporated, which present further opportunities for modeling. The modeling of these physics are challenging, but it only highlights the interdisciplinary nature of research in AM and the exciting areas that can be explored in the future of computational modeling for AM.

The investigations of these physical mechanisms are more inclined to academic pursuits and scientific inquiry of physics. From the industrial perspective, the greatest value these computational models can bring is the efficient predictive capability that it can offer. Once these models are validated with a high degree of confidence, it will allow us to have full prediction capability with data-driven techniques further supplemented with uncertainty quantification methodologies. The final goal would be to achieve a thorough optimization of the AM manufacturing chain: from part modeling, building, and post-processing.

As the requirement for product quality in AM becomes increasingly demanding, computational modeling will play an increasingly powerful role in both AM prediction and design. In the future, with established and verified models, a large portion of trial-and-error experiments can be entirely replaced by virtual experiments conducted through simulations to obtain the optimal process parameters. Furthermore, the validated models can even aid design at the microlevel, such as tailored microstructures. This exciting vision can only be realized with a strong foundation of AM modeling that will allow further extension into process optimization.

6.A Material Properties for Thermal-Fluid Simulation

Table 6.A.1 Material parameters of 316L stainless steel (SS316L) [52], Ti-6Al-4V (Ti64) [13], and Inconel 625 (IN625) [41] used in the CFD simulation studies.

Material property	SS316L	Ti64	IN625	Units
Specific heat, c_p	730	872	TD[a)]	$J\ K^{-1}\ kg^{-1}$
Latent heat of melting, L_m	3.30×10^5	2.86×10^5	2.72×10^5	$J\ kg^{-1}$
Latent heat of evaporation, L_v	6.34×10^6	9.70×10^6	9.70×10^6	$J\ kg^{-1}$
Density, ρ_l	6900	4000	8440	$kg\ m^{-3}$
Surface radiation coefficient, α_b	0.4	0.4	0.4	—
Surface tension coefficient, σ	1.943	1.680	1.680	$N\ m^{-1}$
Viscosity, μ	0.005	0.005	0.005	Pa s
Solidus thermal conductivity, k	32.4	16.0	TD[a)]	$W\ m^{-1}\ K^{-1}$
Liquidus thermal conductivity, k	36.6	32.0	TD[a)]	$W\ m^{-1}\ K^{-1}$
Temperature sensitivity, σ_s^T	4.3×10^{-4}	2.6×10^{-4}	2.6×10^{-4}	$N\ m^{-1}\ K^{-1}$
Solidus temperature, T^s	1673	1878	1563	K
Liquidus temperature, T^l	1732	1928	1623	K

a) TD denotes that temperature dependent properties were used.

References

1 Smith, J., Xiong, W., Yan, W. et al. (2016). Linking process, structure, property, and performance for metal-based additive manufacturing: computational approaches with experimental support. *Computational Mechanics* 57 (4): 583–610.

2 Li, C., Liu, Z., Fang, X., and Guo, Y. (2018). Residual stress in metal additive manufacturing. *Procedia CIRP* 71: 348–353.

3 Körner, C. (2016). Additive manufacturing of metallic components by selective electron beam melting-a review. *International Materials Reviews* 61 (5): 361–377.

4 Matthews, M.J., Guss, G., Khairallah, S.A. et al. (2016). Denudation of metal powder layers in laser powder bed fusion processes. *Acta Materialia* 114: 33–42.

5 Bidare, P., Bitharas, I., Ward, R. et al. (2018). Fluid and particle dynamics in laser powder bed fusion. *Acta Materialia* 142: 107–120.

6 Heigel, J.C. and Lane, B.M. (2018). Measurement of the melt pool length during single scan tracks in a commercial laser powder bed fusion process. *Journal of Manufacturing Science and Engineering* 140 (5). 051012.

7 Zhao, C., Fezzaa, K., Cunningham, R.W. et al. (2017). Real-time monitoring of laser powder bed fusion process using high-speed X-ray imaging and diffraction. *Scientific reports* 7 (1): 1–11.

8 Hojjatzadeh, S.M.H., Parab, N.D., Yan, W. et al. (2019). Pore elimination mechanisms during 3D printing of metals. *Nature Communications* 10 (1): 1–8.

9 Zhao, C., Parab, N.D., Li, X. et al. (2020). Critical instability at moving keyhole tip generates porosity in laser melting. *Science* 370 (6520): 1080–1086.

10 Leung, C.L.A., Marussi, S., Towrie, M. et al. (2019). The effect of powder oxidation on defect formation in laser additive manufacturing. *Acta Materialia* 166: 294–305.

11 Yan, W., Qian, Y., Ge, W. et al. (2018). MESO-scale modeling of multiple-layer fabrication process in selective electron beam melting: inter-layer/track voids formation. *Materials and Design* 141: 210–219.

12 Yan, W., Ma, W., and Shen, Y. (2020). Powder sintering mechanisms during the pre-heating procedure of electron beam additive manufacturing. *Materials Today Communications* 25: 101579.

13 Wang, L., Zhang, Y., and Yan, W. (2020). Evaporation model for keyhole dynamics during additive manufacturing of metal. *Physical Review Applied* 14: 064039.

14 Chen, F. and Yan, W. (2020). High-fidelity modeling of thermal stress for additive manufacturing by linking thermal-fluid and mechanical models. *Materials and Design* 196: 109185.

15 Chen, H., Wei, Q., Zhang, Y. et al. (2019). Powder-spreading mechanisms in powder-bed-based additive manufacturing: experiments and computational modelling. *Acta Materialia* 179: 158–171.

16 Chen, H. and Yan, W. (2020). Spattering and denudation in laser powder bed fusion process: multiphase flow modelling. *Acta Materialia* 196: 154–167.

17 Wang, L. and Yan, W. (2021). Thermoelectric magnetohydrodynamic model for laser-based metal additive manufacturing. *Physical Review Applied* 15 (6): 064051.

18 Wang, L., Zhang, Y., Chia, H.Y., and Yan, W. (2022). Mechanism of keyhole pore formation in metal additive manufacturing. *npj Computational Materials* 8 (1): 1–11.

19 Yu, Y., Kenevisi, M.S., Yan, W., and Lin, F. (2020). Modeling precipitation process of Al-Cu alloy in electron beam selective melting with a 3D cellular automaton model. *Additive Manufacturing* 36: 101423.

20 Liu, D.R., Wang, S., and Yan, W. (2020). Grain structure evolution in transition-mode melting in direct energy deposition. *Materials and Design* 194: 108919.

21 Yang, M., Wang, L., and Yan, W. (2021). Phase-field modeling of grain evolutions in additive manufacturing from nucleation, growth, to coarsening. *npj Computational Materials* 7 (1): 1–12.

22 Yang, M., Wang, L., and Yan, W. (2021). Phase-field modeling of grain evolution in additive manufacturing with addition of reinforcing particles. *Additive Manufacturing* 47: 102286.

23 Yan, W., Lian, Y., Yu, C. et al. (2018). An integrated process–structure–property modeling framework for additive manufacturing. *Computer Methods in Applied Mechanics and Engineering* 339: 184–204.

24 Tang, H., Huang, H., Liu, C. et al. (2021). Multi-scale modelling of structure-property relationship in additively manufactured metallic materials. *International Journal of Mechanical Sciences* 194: 106185.

25 Grilli, N., Hu, D., Yushu, D. et al. (2021). Crystal plasticity model of residual stress in additive manufacturing using the element elimination and reactivation method. *Computational Mechanics* 69: 825–845.

26 Hu, D., Grilli, N., Wang, L. et al. (2022). Microscale residual stresses in additively manufactured stainless steel: computational simulation. *Journal of the Mechanics and Physics of Solids* 161: 104822.

27 Chen, H., Chen, Y., Liu, Y. et al. (2020). Packing quality of powder layer during counter-rolling type powder spreading process in additive manufacturing. *International Journal of Machine Tools and Manufacture* 153: 103553.

28 Di Renzo, A. and Di Maio, F.P. (2004). Comparison of contact-force models for the simulation of collisions in DEM-based granular flow codes. *Chemical Engineering Science* 59: 525–541.

29 Thornton, C., Cummins, S.J., and Cleary, P.W. (2013). An investigation of the comparative behaviour of alternative contact force models during inelastic collisions. *Powder Technology* 233: 30–46.

30 Barthel, E. (2008). Adhesive elastic contacts: JKR and more. *Journal of Physics D: Applied Physics* 41 (16): 163001.

31 Johnson, K. and Greenwood, J. (1997). An adhesion map for the contact of elastic spheres. *Journal of Colloid and Interface Science* 192 (2): 326–333.

32 J. Hanley, K. and O'Sullivan, C. (2017). Analytical study of the accuracy of discrete element simulations. *International Journal for Numerical Methods in Engineering* 109 (1): 29–51.

33 Alizadeh, E., Bertrand, F., and Chaouki, J. (2014). Comparison of DEM results and Lagrangian experimental data for the flow and mixing of granules in a rotating drum. *AIChE Journal* 60 (1): 60–75.

34 Karapatis, N., Egger, G., Gygax, P., and Glardon, R. (1999). Optimization of powder layer density in selective laser sintering. *1999 International Solid Freeform Fabrication Symposium.*

35 Bai, Y., Wagner, G., and Williams, C.B. (2017). Effect of particle size distribution on powder packing and sintering in binder jetting additive manufacturing of metals. *Journal of Manufacturing Science and Engineering* 139 (8). 081019.

36 Escano, L.I., Parab, N.D., Xiong, L. et al. (2018). Revealing particle-scale powder spreading dynamics in powder-bed-based additive manufacturing process by high-speed X-ray imaging. *Scientific Reports* 8 (1): 1–11.

37 Elliott, A.M., Nandwana, P., Siddel, D.H., and Compton B. (2016). A Method for Measuring Powder Bed Density in Binder Jet Additive Manufacturing Process and the Powder Feedstock Characteristics Influencing the Powder Bed Density. *Tech. Rep.* Oak Ridge, TN (United States): Oak Ridge National Lab.(ORNL). High …; 2016.

38 Wouterse, A., Williams, S.R., and Philipse, A.P. (2007). Effect of particle shape on the density and microstructure of random packings. *Journal of Physics: Condensed Matter* 19 (40): 406215.

39 Kiss, N., Brenn, G., Pucher, H. et al. (2011). Formation of O/W emulsions by static mixers for pharmaceutical applications. *Chemical Engineering Science* 66 (21): 5084–5094.

40 Xiang, Z., Yin, M., Deng, Z. et al. (2016). Simulation of forming process of powder bed for additive manufacturing. *Journal of Manufacturing Science and Engineering* 138 (8). 081002.

41 Yan, W., Lu, Y., Jones, K. et al. (2020). Data-driven characterization of thermal models for powder-bed-fusion additive manufacturing. *Additive Manufacturing* 36: 101503.

42 Yan, W., Ge, W., Qian, Y. et al. (2017). Multi-physics modeling of single/multiple-track defect mechanisms in electron beam selective melting. *Acta Materialia* 134: 324–333.

43 Yan, W., Lin, S., Kafka, O.L. et al. (2018). Data-driven multi-scale multi-physics models to derive process–structure–property relationships for additive manufacturing. *Computational Mechanics* 61 (5): 521–541.

44 Hirt, C.W. and Nichols, B.D. (1981). Volume of fluid (VOF) method for the dynamics of free boundaries. *Journal of Computational Physics* 39 (1): 201–225.

45 Han, S.W., Ahn, J., and Na, S.J. (2016). A study on ray tracing method for cfd simulations of laser keyhole welding: a progressive search method. *Welding in the World* 60: 247–258.

46 Yan, W., Smith, J., Ge, W. et al. (2015). Multiscale modeling of electron beam and substrate interaction: a new heat source model. *Computational Mechanics* 56: 265–276.

47 Yan, W., Ge, W., Smith, J. et al. (2016). Multi-scale modeling of electron beam melting of functionally graded materials. *Acta Materialia* 115: 403–412.

48 Carriere, P. and Yue, S. (2017). Energy absorption during pulsed electron beam spot melting of 304 stainless steel: Monte–Carlo simulations and in-situ temperature measurements. *Vacuum* 142: 114–122.

49 Khalatnikov, I.M. and Mineev, V.P. (1996). *30 Years Of The Landau Institute-Selected Papers*, vol. 11. World Scientific.

50 Knight, C.J. (1979). Theoretical modeling of rapid surface vaporization with back pressure. *AIAA journal* 17 (5): 519–523.

51 Yan, W., Qian, Y., Ma, W. et al. (2017). Modeling and experimental validation of the electron beam selective melting process. *Engineering* 3 (5): 701–707.

52 Ge, W., Han, S., Na, S.J., and Fuh, J.Y.H. (2019). Numerical modelling of surface morphology in selective laser melting. *Computational Materials Science* 186: 110062.

53 Cunningham, R., Zhao, C., Parab, N. et al. (2019). Keyhole threshold and morphology in laser melting revealed by ultrahigh-speed X-ray imaging. *Science* 363 (6429): 849–852.

54 Guo, Q., Zhao, C., Escano, L.I. et al. (2018). Transient dynamics of powder spattering in laser powder bed fusion additive manufacturing process revealed by in-situ high-speed high-energy X-ray imaging. *Acta Materialia* 151: 169–180.

55 Li, X., Zhao, C., Sun, T., and Tan, W. (2020). Revealing transient powder-gas interaction in laser powder bed fusion process through multi-physics modeling and high-speed synchrotron X-ray imaging. *Additive Manufacturing* 35: 101362.

56 Yu, t. and Zhao, J. (2021). Semi-coupled resolved CFD-DEM simulation of powder-based selective laser melting for additive manufacturing. *Computer Methods in Applied Mechanics and Engineering*. 377: 113707.

57 Cheng, B., Shrestha, S., and Chou, K. (2016). Stress and deformation evaluations of scanning strategy effect in selective laser melting. *Additive Manufacturing* 12: 240–251.

58 Dassault Systèmes D (2016). Abaqus Analysis User's Guide. *Tech. Rep.* Technical Report Abaqus 6.14 Documentation. Simulia Corp.

59 Reddy, J.N. (1993). *An Introduction to the Finite Element Method*, vol. 27. New York. McGraw-Hill Education.

60 Yadroitsev, I. and Yadroitsava, I. (2015). Evaluation of residual stress in stainless steel 316L and Ti6Al4V samples produced by selective laser melting. *Virtual and Physical Prototyping* 10 (2): 67–76.

61 Vrancken, B., Wauthlé, R., Kruth, J.P., and Van Humbeeck, J. (2013). Study of the influence of material properties on residual stress in selective laser melting. *Proceedings of the Solid Freeform Fabrication Symposium*, pp. 393–407.

62 Wang, G., Ouyang, H., Fan, C. et al. (2020). The origin of high-density dislocations in additively manufactured metals. *Materials Research Letters* 8 (8): 283–290.

63 Vrancken, B., Ganeriwala, R.K., and Matthews, M.J. (2020). Analysis of laser-induced microcracking in tungsten under additive manufacturing conditions: experiment and simulation. *Acta Materialia*. 194: 464–472.

64 Ganeriwala, R., Strantza, M., King, W. et al. (2019). Evaluation of a thermomechanical model for prediction of residual stress during laser powder bed fusion of Ti-6Al-4V. *Additive Manufacturing* 27: 489–502.

65 An, K., Yuan, L., Dial, L. et al. (2017). Neutron residual stress measurement and numerical modeling in a curved thin-walled structure by laser powder bed fusion additive manufacturing. *Materials and Design* 135: 122–132.

66 Tang, H., Qian, M., Liu, N. et al. (2015). Effect of powder reuse times on additive manufacturing of Ti-6Al-4V by selective electron beam melting. *JOM* 67 (3): 555–563.

67 Leung, C.L.A., Tosi, R., Muzangaza, E. et al. (2019). Effect of preheating on the thermal, microstructural and mechanical properties of selective electron beam melted Ti-6Al-4V components. *Materials and Design* 174: 107792.

7

Modeling of Polymer Powder-Based Additive Manufacturing

Fei Shen[1], *Pengfei Tan*[2], *and Kun Zhou*[2]

[1] *Tianjin University, School of Mechanical Engineering, Department of Mechanics, No.135 Yaguan Road, Haihe Education Park, Tianjin 300350, China*
[2] *Nanyang Technological University, School of Mechanical and Aerospace Engineering, 50 Nanyang Avenue, Singapore 639798, Singapore*

7.1 Background

Polymers and their composites for additive manufacturing (AM) have attracted great attention due to their advantages of the high strength-to-weight ratio and corrosion resistance [1, 2]. Polymer powder-based AM technologies, including binder jetting, selective laser sintering (SLS), and multi-jet fusion (MJF), have been widely used for fabricating polymeric materials. The AM of the polymeric powder consists of three processes: powder recoating, energy input (lasers for SLS and infrared lamps for MJF) or binder deposition, and cooling and coalescence or bonding [3]. Although great advances in AM of polymeric materials have been made recently, there exist several challenges such as pores, debonding and delamination, thermal distortion, and poor surface finish, which are closely associated with the mechanical properties of printed components. To circumvent these challenges and achieve a satisfactory performance of the printed products, numerical modeling of the AM process provides a feasible and efficient way to study the underlying mechanisms of defect formation and optimize the printing process parameters.

Powder recoating is a process to spread powder particles to form a powder bed using a blade or a roller. A counterrotating roller-type recoating system can produce a denser and smoother powder bed than a blade-type system [4]. The recoating characteristics of the powder bed are in high correlation with the mechanical properties and dimensional accuracy of printed components. A powder bed with low packing density and large surface roughness may lead to the emergence of more pores and delamination in the printed components. Ziegelmeier et al. [5] reported that the increasing packing density of the powder bed could lead to the enhancement in the ultimate tensile strength, the decrease in the surface roughness, and the increase in the density of SLS-printed parts. The recoating quality of the powder bed is dependent on numerous factors such as the morphological properties of powder particles (geometrical shape, size, and size distribution), material properties

Additive Manufacturing Technology: Design, Optimization, and Modeling, First Edition. Edited by Kun Zhou.

(Young's modulus, Poisson's ratio, density, and surface energy), recoating parameters (powder layer thickness, recoating velocity, and geometry of the spreader), and composition of composite powder particles [3]. These factors mutually determine the powder flow dynamic behavior in the recoating process, thus affecting the recoating density and surface morphology of the powder bed. Particularly, for fiber-reinforced polymer composite powder particles, properly chosen processing parameters can help to improve fiber homogeneity and control fiber orientation for obtaining anisotropic mechanical properties in the printed components.

Experiments have been conducted to investigate the powder flow behavior and the recoating characteristics of the powder bed. Escano et al. [6] used *in situ* high-speed high-energy X-ray imaging to study the powder recoating behavior in powder bed–based AM. They found that powder particles with an average diameter of 67 μm exhibited a larger average dynamic angle of repose (AOR) and a higher average slope surface flow speed than the powder particles with an average diameter of 23 μm. Ruggi et al. [7] investigated the flowability of polymer powder particles for SLS at different temperatures. Chen et al. [8] used the packing density and surface roughness measured from a metal powder bed to validate their numerical model. Recently, the discrete element method (DEM) that is proposed to describe the movement and mechanical behavior of granular materials and geomaterials has been applied to simulate the powder recoating process in AM [9–11]. This method can help to analyze the powder flow dynamics and explore the underlying mechanisms of defect formation in the powder recoating process [12–16].

Powder particles are fused by heat energy from the laser beam for SLS or the infrared lamps for MJF. The molten polymer then cools down to form desired parts. The solid-state polymer undergoes nonuniform thermal contraction during the temperature variation. Moreover, some molecular chains fold together in an orderly arrangement to form a crystalline phase for semicrystalline polymers. This recrystallization process causes a significant shrinkage in the polymer volume. Extensive efforts are devoted to determining the AM processability of different polymers [17–23]. The optimization of processing parameters is an important topic to acquire the high thermal and mechanical performances of printed parts [24–26]. However, tuning the parameters with the trial and error method in the experiment is time-consuming and costly. Numerical approaches offer a great alternative to model the sintering and cooling processes for various polymers and process parameters. According to their simulation scales, the process modeling methods can be classified into two types: mesoscopic and macroscopic scales. The mesoscopic method involves the melting of individual particles and the formation of a melt pool with high fidelity. Körner et al. [27, 28] used a two-dimensional (2D) lattice Boltzmann model for the process modeling. Three-dimensional (3D) mesoscopic models were also developed to simulate the melting and solidification of powder [29–31]. As a macroscopic method that treats the powder bed as a continuum medium, the finite element method is widely used to simulate the temperature, stress, and strain distribution in AM processes [20, 32–35]. The finite element method is suitable for large-scale process modeling.

This chapter focuses on the modeling of powder recoating and sintering processes for polymer powder-based AM. In Section 7.2, a DEM model is presented for simulating the recoating process of polymer powder and its composite powder, and the effects of recoating parameters on the recoating characteristics of the powder bed are investigated. In Section 7.3, a finite element model is proposed for simulating the sintering process of polymer powder and its composite powder, which is used to study the thermal and mechanical behavior of printed polymeric materials. Finally, Section 7.4 summarizes the main contents of this chapter and gives an outlook on future works.

7.2 Discrete Element Modeling of the Powder Recoating Process

A DEM model is presented for simulating the powder recoating process of polymer powder-based AM. The Hertz–Mindlin theory that describes the spring–dashpot interaction between particles is used, and the van der Waals force is also taken into account. To precisely simulate the recoating process, the geometrical features of powder particles such as the shape and size distribution that have an important effect on the powder flowability are properly described. Based on the DEM model, the powder dynamic behavior in the recoating process is investigated. The effects of recoating parameters (e.g. powder layer thickness and recoating velocity) and powder properties (e.g. geometrical shape and composition of composite powder particles) on the recoating characteristics of the powder bed are studied and used to provide guidance for the parameter optimization.

7.2.1 Discrete Element Model

The DEM is first proposed by Cundall and Strack for analyzing rock mechanics [36, 37]. In DEM, the translational motion and rotational motion of the particle i are derived from Newton's second law [36]:

$$\begin{cases} m_i \dfrac{\mathrm{d}\mathbf{v}_i}{\mathrm{d}t} = \boldsymbol{F}_\mathrm{c} + m_i\mathbf{g} + \boldsymbol{F}_\mathrm{v} \\ I_i \dfrac{\mathrm{d}\boldsymbol{\omega}_i}{\mathrm{d}t} = \boldsymbol{M} \end{cases} \tag{7.1}$$

where m_i is the mass of the particle, $\boldsymbol{v}_i$ is the velocity, $\boldsymbol{F}_\mathrm{c}$ is the contact force, $\mathbf{g}$ is the gravitational acceleration, $\boldsymbol{F}_\mathrm{v}$ is the van der Waals force, I_i is the moment of inertia, $\boldsymbol{\omega}_i$ is the angular velocity, and $\boldsymbol{M}$ is the total moment.

Considering two contact particles i and j, as shown in Figure 7.1, the relative velocity $\boldsymbol{v}_{ij}$ of the two particles at the contact point is calculated as follows [40]:

$$\boldsymbol{v}_{ij} = \boldsymbol{v}_i - \boldsymbol{v}_j + (R_i\boldsymbol{\omega}_i + R_j\boldsymbol{\omega}_j) \times \boldsymbol{n} \tag{7.2}$$

where the subscripts i and j denote the two interacting particles, $\boldsymbol{v}$ and $\boldsymbol{\omega}$ are the translational velocity and angular velocity of the particle, respectively, R is the

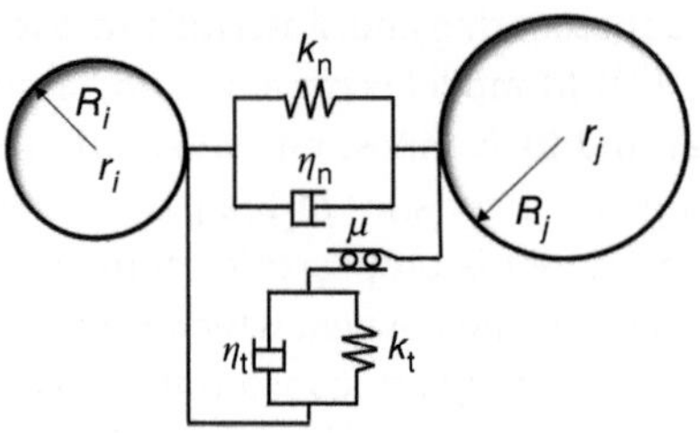

Figure 7.1 Contact force model of two particles *i* and *j* [38, 39].

radius of the particle, and $\boldsymbol{n}$ is the normal unit vector of the contact surface. The relative normal velocity $\boldsymbol{v}_\text{n}$ and the relative tangential velocity $\boldsymbol{v}_\text{t}$ are, respectively, expressed as:

$$\begin{cases} \boldsymbol{v}_\text{n} = (\boldsymbol{v}_{ij} \cdot \boldsymbol{n})\boldsymbol{n} \\ \boldsymbol{v}_\text{t} = \boldsymbol{v}_{ij} - \boldsymbol{v}_\text{n} \end{cases} \tag{7.3}$$

The displacement in the normal direction u_n is calculated as:

$$u_\text{n} = R_i + R_j - (\boldsymbol{r}_j - \boldsymbol{r}_i) \cdot \boldsymbol{n} \tag{7.4}$$

where $\boldsymbol{r}_i$ and $\boldsymbol{r}_j$ are the position vectors of the two particles, respectively.

The contact force $\boldsymbol{F}_\text{c}$ consists of two components: the normal force $\boldsymbol{F}_\text{n}$ and the tangential force $\boldsymbol{F}_\text{t}$ which are expressed as [39, 41]:

$$\begin{cases} \boldsymbol{F}_\text{n} = k_\text{n} u_\text{n} \boldsymbol{n} + \eta_\text{n} \boldsymbol{v}_\text{n} \\ \boldsymbol{F}_\text{t} = \min(k_\text{t} u_\text{t} \boldsymbol{n}_\text{t} + \eta_\text{t} \boldsymbol{v}_\text{t}, \mu |\boldsymbol{F}_\text{n}| \boldsymbol{n}_\text{t}) \end{cases} \tag{7.5}$$

where k_n and k_t are the normal and tangential stiffness, respectively; η_n and η_t are the normal and tangential damping coefficients, respectively; μ is the coefficient of friction; $\boldsymbol{n}_\text{t}$ is the tangential unit vector. In Eq. (7.5), the contact energy dissipation is considered by adding viscous damping forces to the normal and tangential components of the contact force. The maximum tangential force is determined according to the Mohr–Coulomb failure criterion.

The commonly used method to describe the contact behavior is the nonlinear Hertz–Mindlin theory [39], and the normal stiffness and shear stiffness are expressed as [39, 42]:

$$\begin{cases} k_\text{n} = 2E^* \sqrt{R^* u_\text{n}} \\ k_\text{t} = 8G^* \sqrt{R^* u_\text{n}} \\ E^* = \left(\dfrac{1 - \upsilon_i^2}{E_i} + \dfrac{1 - \upsilon_j^2}{E_j} \right)^{-1} \\ R^* = \left(\dfrac{1}{R_i} + \dfrac{1}{R_j} \right)^{-1} \\ G^* = \left(\dfrac{2 - \upsilon_i}{G_i} + \dfrac{2 - \upsilon_j}{G_j} \right)^{-1} \end{cases} \tag{7.6}$$

where the notation * represents the equivalent properties, E is the Young's modulus, and the shear modulus $G = \frac{E}{2(1+\upsilon)}$ with υ being the Poisson's ratio.

The damping coefficients in the normal and tangential directions are calculated as:

$$\begin{cases} \eta_n = 2\beta\sqrt{m^* k_n} \\ \eta_t = 2\beta\sqrt{m^* k_t} \\ \beta = \dfrac{|\ln \lambda|}{\sqrt{\ln^2\lambda + \pi^2}} \\ m^* = \dfrac{m_i m_j}{m_i + m_j} \end{cases} \tag{7.7}$$

where λ is the coefficient of restitution.

The van der Waals force is usually considered using the Johnson–Kendall–Roberts (JKR) [43] model, which is expressed as:

$$\begin{cases} \boldsymbol{F}_v = -4\sqrt{\pi a^3 \gamma E^*}\boldsymbol{n} \\ a^4 - 2R^* u_n a^2 - \frac{4\pi\gamma R^{*2}}{E^*} a + R^{*2} u_n^2 = 0 \end{cases} \tag{7.8}$$

where a is the contact radius and γ is the surface energy density. Furthermore, the Derjaguin–Muller–Toporov (DMT) model [38, 44] is an alternative used to calculate the van der Waals force:

$$\boldsymbol{F}_v = -4\pi R^* \gamma \boldsymbol{n} \tag{7.9}$$

7.2.2 Polymer and Composite Powder Particles

The commonly used and commercially available polymers for powder-based AM are categorized into semicrystalline thermoplastics (e.g. polyamide 11 [PA11] and polyamide 12 [PA12]), amorphous thermoplastics (e.g. polystyrene and polycarbonate), and thermoplastic elastomers (e.g. thermoplastic polyurethane) [1, 45]. Reinforcement materials in a polymeric matrix include metal, ceramic/glass, carbon, and organics [1]. Among these reinforcements, glass fibers and carbon fibers have been widely used for developing composite powder due to their advantages of high mechanical strength and light weight.

7.2.2.1 Polymer Powder Particles

The geometrical shapes of polymer and composite powder particles play a crucial role in the powder flow during the recoating process. Many researchers have focused on investigating the effects of powder shapes on the recoating characteristics of the powder bed and the quality of the printed parts. Shi et al. [46] studied the effects of material properties and powder sizes on the quality of SLS-printed parts and found that the particle size not only influences the dimensional accuracy but also the density. Berretta et al. [47] studied morphology, flowability, and interparticle interactions between two kinds of polymer powder, that is, polyether ether ketone (PEEK) and PA12, and revealed that nanosized fillers can significantly improve the flowability of the materials.

Figure 7.2 shows the morphology of different kinds of polymer powder particles. Polymer powder particles usually have complex nonspherical shapes.

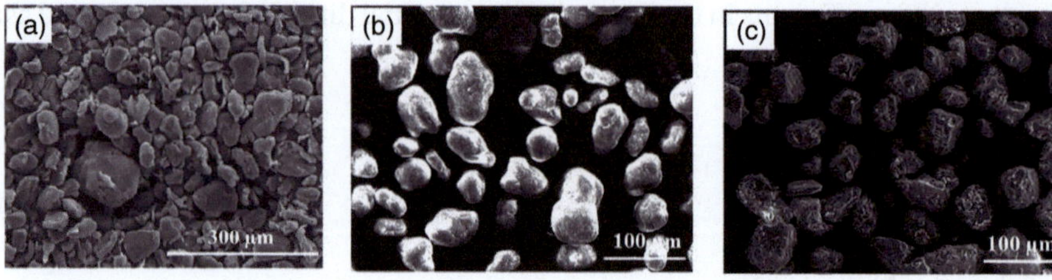

Figure 7.2 Morphology of polymer powder particles: (a) polyetherketone. Source: Berretta et al. [47]/with permission of Elsevier, (b) PA12. Source: Schmid et al. [48]/with permission of Elsevier, (c) thermoplastic polyurethane. Source: Ziegelmeier et al. [5]/with permission of Elsevier.

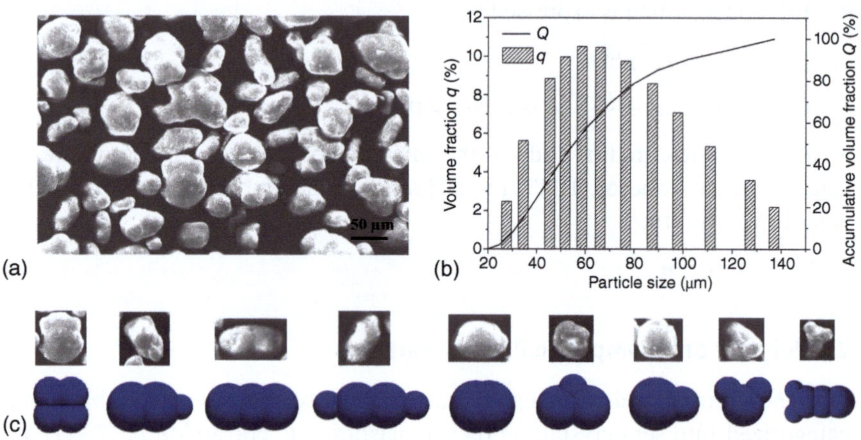

Figure 7.3 (a) Microscopic image of PA12 powder, (b) the particle size distribution of PA12 powder, and (c) several representative multi-sphere particles. Source: Tan et al. [50]/with permission of Informa UK Limited.

The multi-sphere method has been a common and versatile method to represent complex shapes by clumping a collection of spheres [12, 13, 49]. For example, the scanning electron microscope (SEM) image of commercially available PA12 powder particles from HP Inc. is shown in Figure 7.3a. The particle size distribution for the PA12 powder particles [51] is shown in Figure 7.3b. Several representative particles are chosen and approximated using the multi-sphere method, as shown in Figure 7.3c. The particle size is assumed to be the minimum diameter of a sphere that can enclose the multi-sphere particle. Parteli et al. [12] also used the multi-sphere method to represent the PA12 powder particles using the 2D shapes of powder particles that are extracted from light microscopic images.

Furthermore, new multi-sphere approximation approaches have been proposed. Haeri et al. [4, 13] developed a DEM model for the recoating process of PEEK powder particles that are approximated as roll-shaped particles. They also developed a new algorithm to generate multi-sphere approximations from 3D standard triangle language (STL) models of the real powder particles based on a greedy heuristic method, which is suitable for large-scale DEM simulations, as shown in Figure 7.4. Nasato and Pöschel [52] proposed a novel multi-sphere representation method for

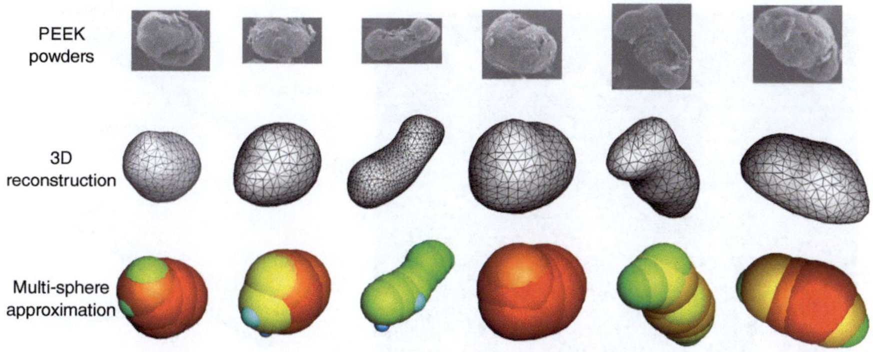

Figure 7.4 Different types of PEEK powder particles approximated from STL models using the multi-sphere approximation strategy. Source: Haeri [4]/Elsevier/CC BY 4.0.

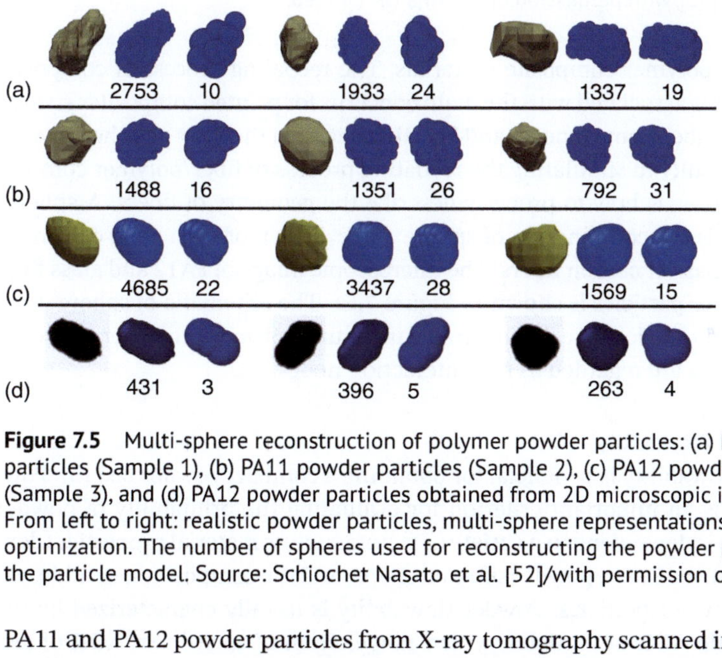

Figure 7.5 Multi-sphere reconstruction of polymer powder particles: (a) PA11 powder particles (Sample 1), (b) PA11 powder particles (Sample 2), (c) PA12 powder particles (Sample 3), and (d) PA12 powder particles obtained from 2D microscopic images (Sample 4). From left to right: realistic powder particles, multi-sphere representations before and after optimization. The number of spheres used for reconstructing the powder particles is below the particle model. Source: Schiochet Nasato et al. [52]/with permission of Elsevier.

PA11 and PA12 powder particles from X-ray tomography scanned images, as shown in Figure 7.5. The approximated particles are able to retain the geometrical features such as the aspect ratio, flatness ratio, and elongation ratio, which are closely associated with the powder flowability. They found that elongated particles with low aspect ratios are in favor of a dense powder bed at low recoating velocities, but spherical particles are beneficial to achieving a dense powder bed at high recoating velocities.

7.2.2.2 Composite Powder Particles

As one of the emerging materials for powder-based AM, fiber-reinforced polymer composites have attracted extensive focus recently [1, 2, 53]. However, the study on the recoating process of composite powder particles is rarely reported. The mechanical mixing of polymer powder particles and fibers is a common way to develop

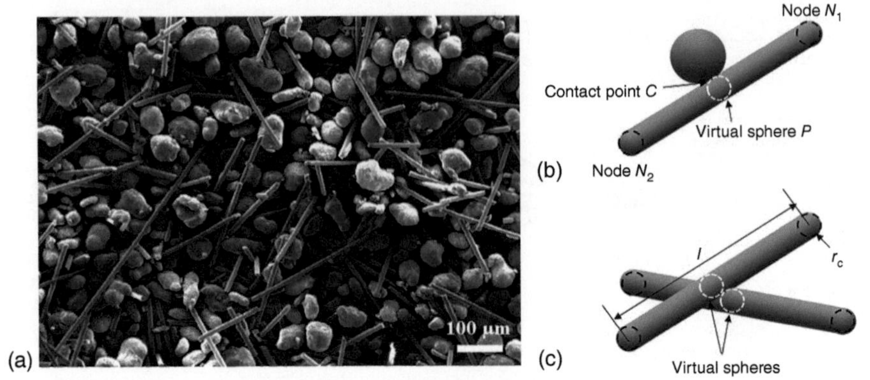

Figure 7.6 (a) SEM image of PA12 and glass fiber-reinforced PA12 composite powder particles, and the schematic of (b) sphere–fiber and (c) fiber–fiber interactions. Source: Tan et al. [50]/with permission of Informa UK Limited.

fiber-reinforced polymer composite materials. The recoating process of composite powder particles is associated with the main concerns for printed composites such as fiber alignment, fiber homogeneity, and bonding between the fiber and the polymer matrix. The difficulty in simulating the recoating process of fiber/polymer composite powder particles is how to properly describe the geometry of fibers. A spherocylinder, which is the combination of sphere and cylinder, offers a good option for approximating glass or carbon fibers. The microscopic image of PA12 and glass fiber composite powder particles is shown in Figure 7.6a. The schematic of sphere–fiber and fiber–fiber interactions are demonstrated in Figure 7.6b,c. For the fiber, the semispherical two ends are regarded as two interaction nodes.

7.2.2.3 Powder Flowability

Good powder flowability is beneficial for obtaining a compact powder bed [3]. Powder flowability is an important criterion for evaluating the printability of powder particles, which is dependent on a variety of factors such as material properties (density and friction), moisture, powder size and shape, and the composition or additives of composite powder particles. Powder flowability is usually characterized by the static or dynamic AOR of particles, which ranges from 0° to 90°. An AOR of less than 30° suggests good flowability, while an AOR above 55° indicates very limited flowability [54, 55]. The definitions of the AOR are dependent on specific applications. Generally, the static AOR is defined as the angle between the steepest slope of piled granular material without sliding and the horizontal plane [56]. The common methods to measure the AOR include the tilting box method [57], fixed funnel method [54, 58], revolving drum method [59, 60], hollow cylinder method [61–63], and tilting cylinder method [56].

The AOR of powder particles can be employed to calibrate the material properties used in the DEM model. Nasato and Pöschel [52] found that the increase in the coefficient of friction led to the increase in the AOR of virgin PA12 powder particles. As the coefficient of friction equaled to 0.5, the AOR of PA12 powder (32.66° ± 0.94°)

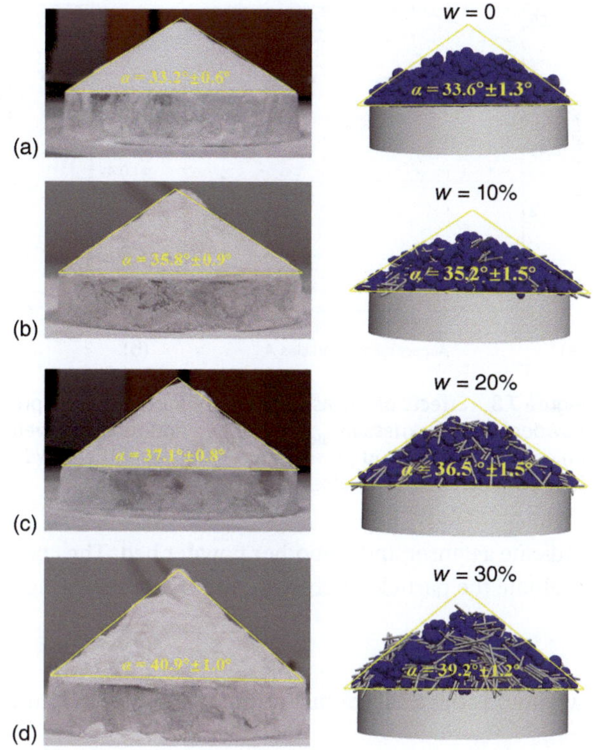

Figure 7.7 AOR of the powder particles with different fiber loading w: (a) 0%, (b) 10%, (c) 20%, and (d) 30%. Source: Tan et al. [50]/with permission of Informa UK Limited.

from the simulation matched well with the experimental value ($33.1° \pm 1.5°$) [47]. Tan et al. [50] investigated the AOR of PA12 and its glass fiber-reinforced composite by experiment and simulation, as shown in Figure 7.7. They found that the AOR increased with the increase in fiber loading in the composite powder. The increase in the AOR indicated a decrease in the flowability of composite powder particles.

7.2.3 Recoating Quality of the Powder Bed

In this section, the effects of particle properties and recoating parameters on the recoating quality of the powder bed are studied. The recoating quality is commonly assessed by the packing density φ and surface roughness ε:

$$\begin{cases} \varphi = \dfrac{V_{\mathrm{p}} + V_{\mathrm{f}}}{LWH} \\ \varepsilon = \sqrt{\dfrac{\sum_{i=1}^{N} (h_i - h_{\mathrm{a}})^2}{N-1}} \end{cases} \tag{7.10}$$

where V_{p} and V_{f} are the volumes of fibers and polymer powder, respectively; L, W, and H are the length, width, and thickness of the powder bed, respectively; h_i is the height of the powder bed at the point i; N is the number of points; h_{a} is the average height of the powder bed. Larger packing density and smaller surface roughness

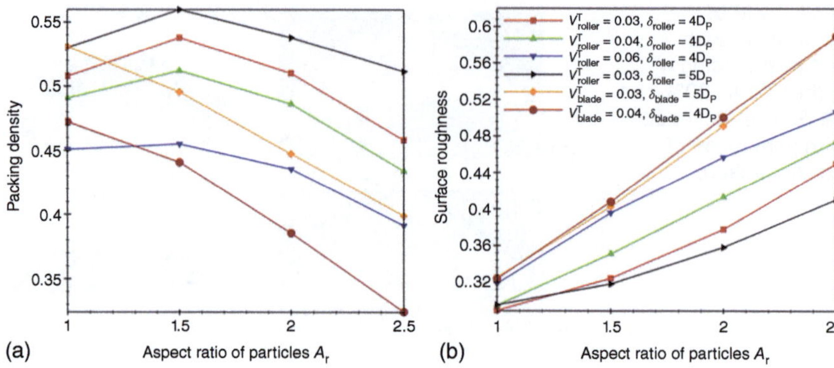

Figure 7.8 Effects of the aspect ratio of particles (A_r), spreader type (roller and blade), powder layer thickness (δ_{roller} and δ_{blade}), and recoating velocity (V^T_{roller} and V^T_{blade}) on the (a) packing density and (b) surface roughness of the powder bed. Source: Haeri et al. [13]/With permission of Elsevier.

indicate a denser and smoother powder bed. The coordination number N_c is used to evaluate the particle microstructure in the recoating process:

$$N_c = \frac{2C}{N} \tag{7.11}$$

where C is the total contact number and N is the number of particles.

7.2.3.1 Layer Thickness Effect

The powder layer thickness H is an important parameter to influence the powder flow behavior in the recoating process. Haeri et al. [13] studied the effects of the powder shape, spreader type, powder layer thickness, and recoating velocity on the packing density and surface roughness of the powder bed, as given in Figure 7.8. The PEEK polymer powder particles used in the simulation were approximated as rod-shaped particles. For different particle aspect ratios, the packing density of the powder bed at a higher powder layer thickness is larger than that at a lower powder layer thickness. For elongated particles (aspect ratio $A_r > 1$), the increase in the powder layer thickness reduces the surface roughness of the powder bed. Moreover, the powder bed spread by a roller can obtain better quality than that by a blade because the roller can offer more contact areas for particle rearrangement.

The top view and x–z cross section of the powder beds for glass fiber/PA12 composite powder particles at H = 100, 120, 150, and 170 μm are demonstrated in Figure 7.9. When H = 100 or 120 μm, large voids are observed in the powder bed, as shown in Figure 7.9a,b. At H = 100 μm, a thin and sparse powder bed is formed. In Figure 7.10, the effects of the powder layer thickness on the powder bed are demonstrated. As shown in Figure 7.10a, the increase in the powder layer thickness increases the packing density and reduces the surface roughness of the powder bed. Figure 7.10b shows that the height of the powder bed generally increases with the increase in the layer thickness. The height fluctuation becomes smaller at larger powder layer thicknesses.

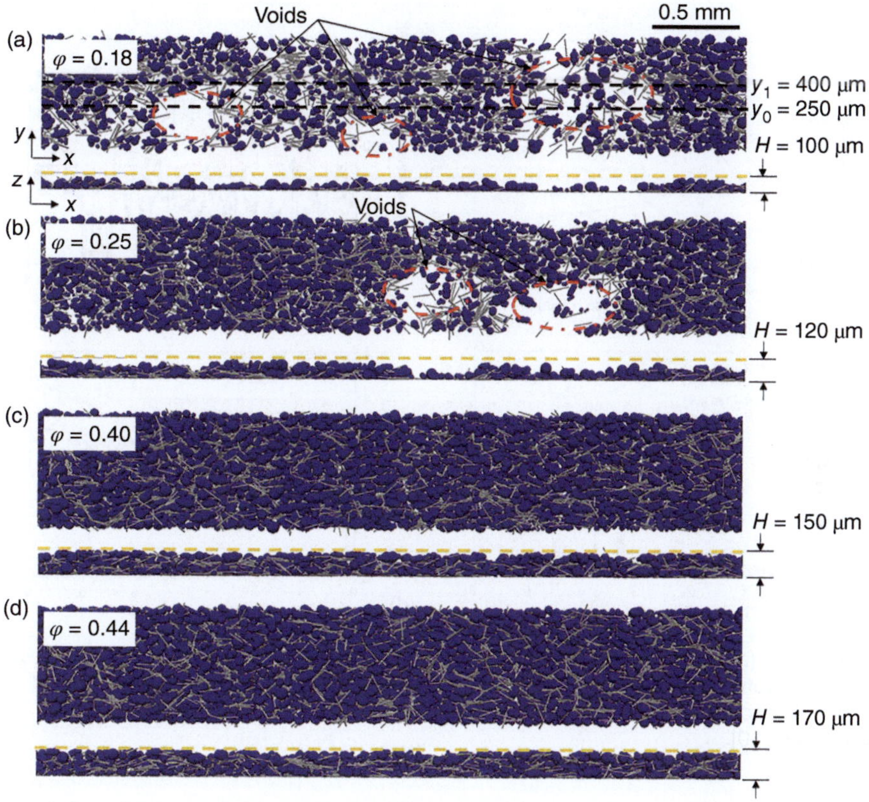

Figure 7.9 Powder beds of different powder layer thicknesses H: (a) 100 μm, (b) 120 μm, (c) 150 μm, and (d) 170 μm. Source: Tan et al. [50]/with permission of Informa UK Limited.

The jamming phenomenon in the powder recoating process is one of the important reasons that give rise to the formation of voids in the powder bed [49]. At a narrower gap between the roller and the substrate, jamming may occur more frequently, particularly for the particles with complex shapes and large sizes. The contact forces of the roller at different powder layer thicknesses are illustrated in Figure 7.10c. Strong contact interactions between the roller and the powder particles are observed at lower powder layer thicknesses $H = 100$ and $120\,\mu m$. At the peak points of the roller contact force, the powder particles may be dragged forward along the spreading direction due to their strong interaction, which results in void formation.

7.2.3.2 Recoating Velocity Effect

Parteli and Pöschel [12] found that the increase in the recoating velocity may lead to a decrease in the particle distribution homogeneity. In Figure 7.11a, the side and top views of the powder beds on a built part at the recoating velocities $v = 20$ and $180\,mm\,s^{-1}$ are demonstrated. The powder bed at $v = 20\,mm\,s^{-1}$ has a smoother top surface and is more compact than that at $v = 180\,mm\,s^{-1}$. As shown in Figure 7.11b, the surface roughness of the powder bed increases as the recoating velocity increases.

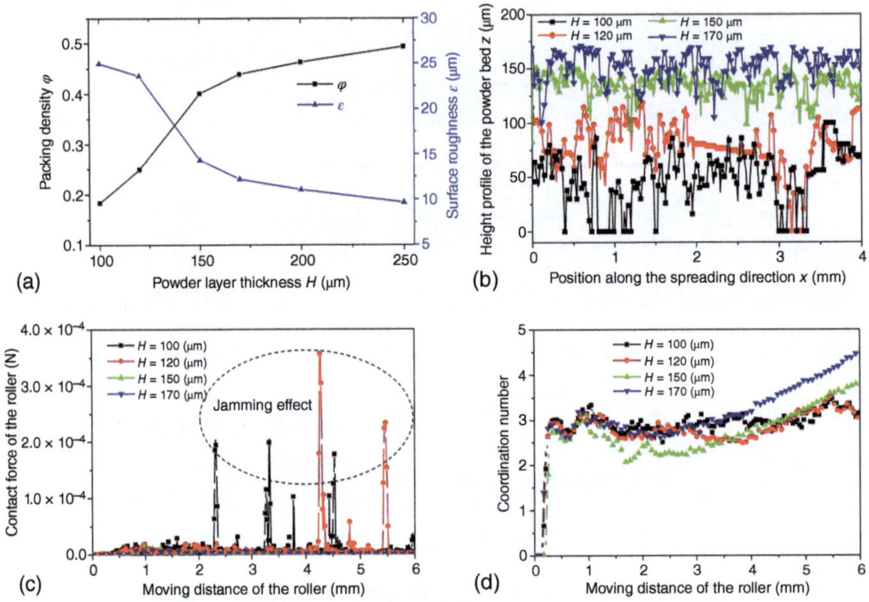

Figure 7.10 (a) Packing density and surface roughness versus the powder layer thickness. (b) Height profiles of the powder bed at different powder layer thicknesses. Evolution of (c) the roller contact force and (d) the coordination number at different powder layer thicknesses [50].

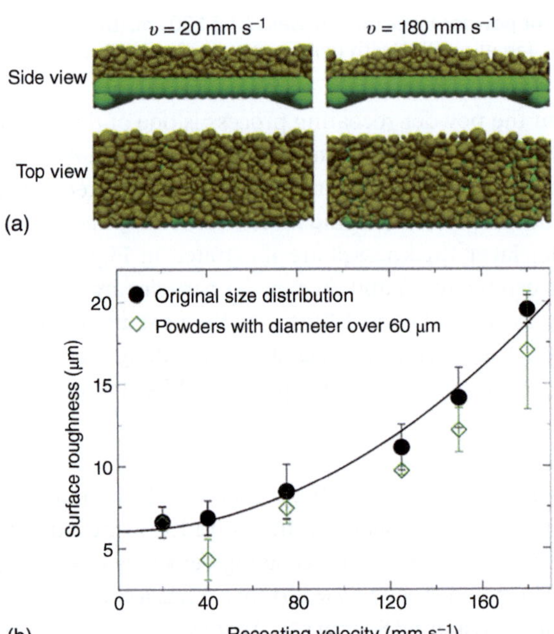

Figure 7.11 (a) Side and top views of the powder bed at different recoating velocities and (b) the surface roughness of the powder bed versus the recoating velocity. Source: Parteli and Pöschel [12]/With permission of Elsevier.

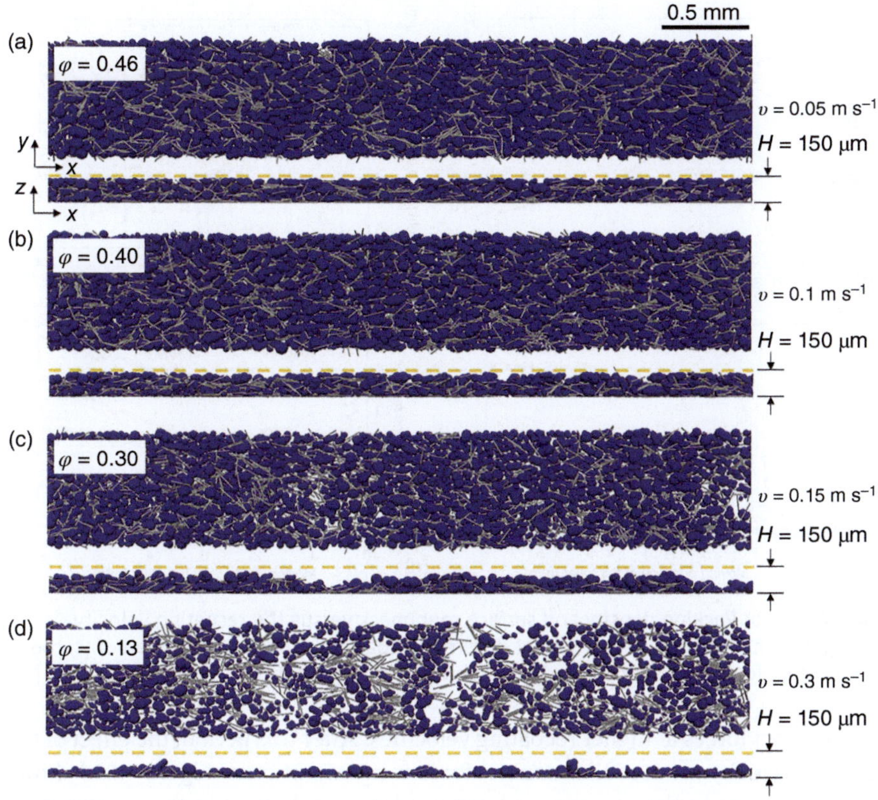

Figure 7.12 Powder beds with different recoating velocities v: (a) 0.05 m s^{-1}, (b) 0.1 m s^{-1}, (c) 0.15 m s^{-1}, and (d) 0.3 m s^{-1}. Source: Tan et al. [50]/with permission of Informa UK Limited.

After the powder particles with a diameter below 60 μm are removed, the powder particles have a new size distribution and the powder bed has smaller surface roughness. This is because the removed small particles tend to develop large agglomerates in the recoating process due to the increasing effect of the van der Waals force. Haeri et al. [13] also studied the effect of the recoating velocity on the packing quality of the powder bed for spherical and elongated particles. They concluded that the increase in the recoating velocity would lead to a decrease in the packing density and an increase in the surface roughness, as shown in Figure 7.8.

For the recoating process of composite powder particles, the increase in the recoating velocity also has an adverse effect on the recoating quality of the powder bed. The powder beds at the recoating velocities v = 0.05, 0.1, 0.15, and 0.3 m s^{-1} are demonstrated in Figure 7.12. With the increase in the recoating velocity, the powder bed becomes sparser and thinner. From Figure 7.13a, the increase in the recoating velocity leads to the decrease in the packing density and the increase in the surface roughness. As shown in Figure 7.13b, the average height of the powder bed increases with the reduction of the recoating velocity. In Figure 7.13, at the beginning of the period (moving distance between 0.5 and 2 mm), the roller experiences

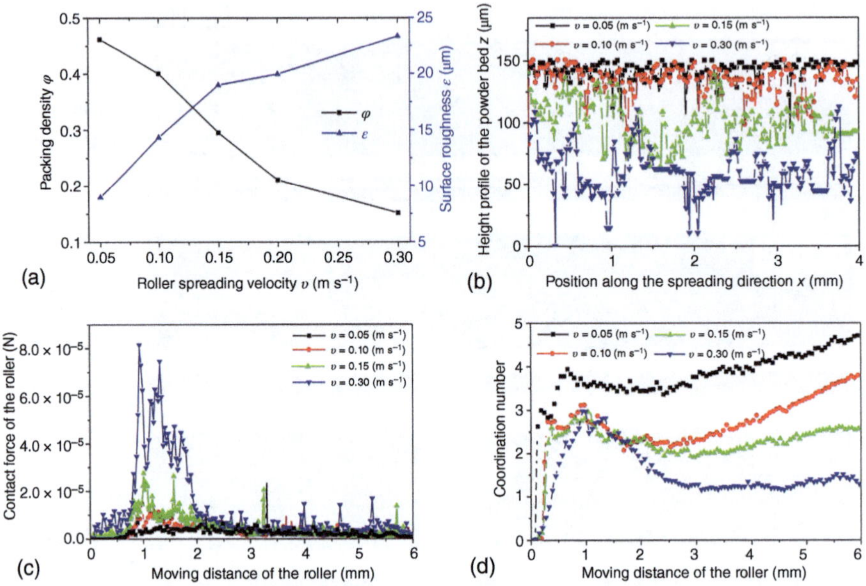

Figure 7.13 (a) Packing density and surface roughness versus the recoating velocity. (b) Height profiles of the powder bed at different recoating velocities. Evolution of (c) the roller contact force and (d) the coordination number at different recoating velocities [50].

larger contact forces at higher recoating velocities. After this period, the effect of the recoating velocity on the roller contact force is not significant. Figure 7.13d shows that the increase in the recoating velocity leads to a reduction in the coordination number during the recoating process, which means insufficient contact between the particles.

7.2.3.3 Fiber Loading Effect

The effect of fiber loading in composite powder on the packing quality is studied. The recoated powder beds at $w = 0$, 10%, 20%, and 40% are demonstrated in Figure 7.14. As depicted in Figure 7.15a, mixing a small number of fibers ($w = 5\%$ and 10%) with PA12 powder results in a slight increase in the packing density and a decrease in the surface roughness. As the fiber loading is more than 10%, the increase in the fiber loading leads to a decrease in the packing density. This can be ascribed to two reasons: the reduction of the particle flowability in the composite powder and the agglomeration of fibers. As shown in Figure 7.7, the composite powder particles with higher fiber loading have a larger AOR, which indicates poorer particle flowability. As shown in Figure 7.15d, fiber agglomerates may be developed as more fibers are mixed with PA12 powder particles due to the van der Waals force.

7.2.3.4 Particle Shape Effect

The powder shape is strongly associated with the powder flowability in the recoating process. Figure 7.8 shows that the particles with larger aspect ratios have higher surface roughness. For the roller-type recoating process, the maximum packing density

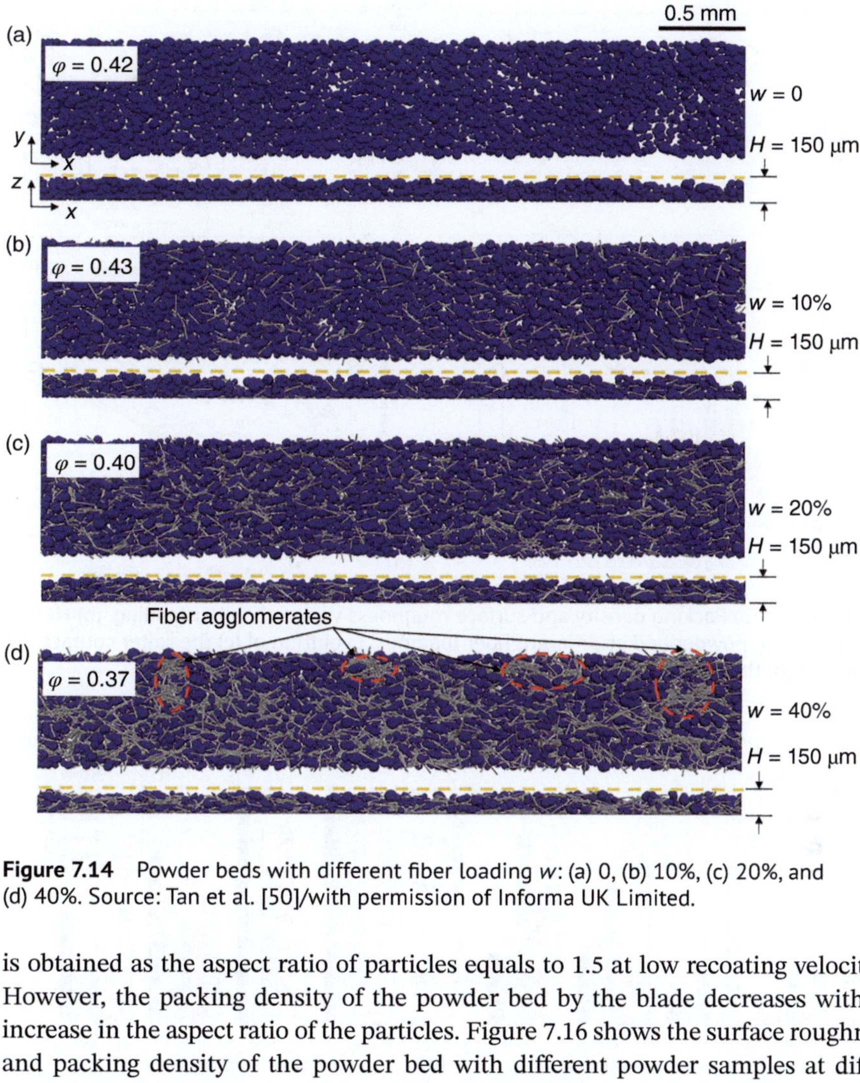

Figure 7.14 Powder beds with different fiber loading w: (a) 0, (b) 10%, (c) 20%, and (d) 40%. Source: Tan et al. [50]/with permission of Informa UK Limited.

is obtained as the aspect ratio of particles equals to 1.5 at low recoating velocities. However, the packing density of the powder bed by the blade decreases with an increase in the aspect ratio of the particles. Figure 7.16 shows the surface roughness and packing density of the powder bed with different powder samples at different recoating velocities. Samples 1–4 represent the different reconstruction methods of PA11 and PA12 powder particles (see Figure 7.5). As shown in Figure 7.16, the particles with low aspect ratios are beneficial to the formation of a dense powder bed at low recoating velocities. For higher recoating velocities, spherical particles can result in a denser powder bed as compared with elongated particles because spherical particles possess better flowability.

7.3 Finite Element Modeling of the SLS Process

The general steps for the finite element modeling of the SLS process include preprocessing, thermal analysis, and mechanical analysis [64]. As shown in Figure 7.17,

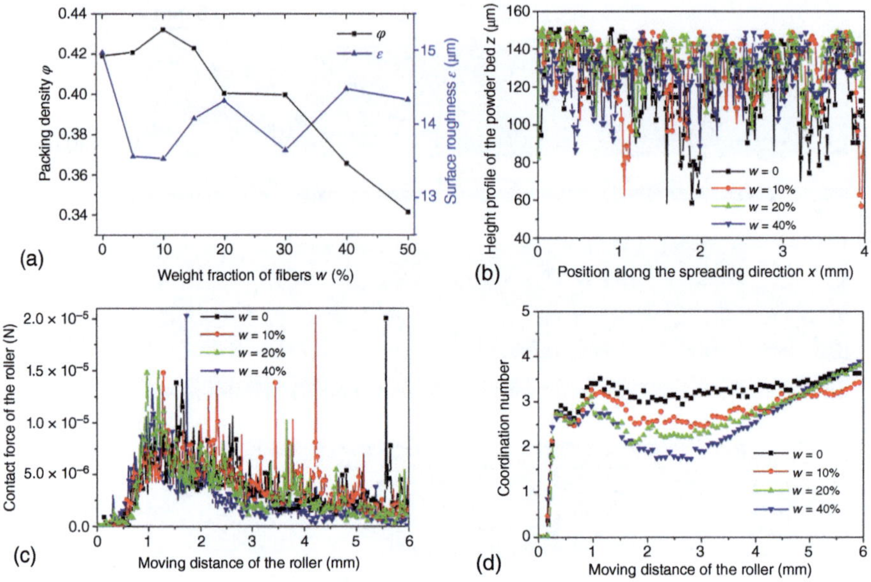

Figure 7.15 (a) Packing density and surface roughness versus the fiber loading. (b) Height profiles of the powder bed at different fiber loadings. Evolution of (c) the roller contact force and (d) the coordination number at different fiber loading [50].

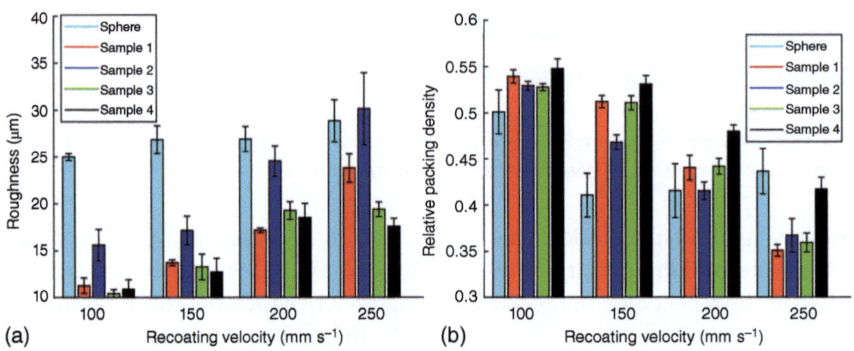

Figure 7.16 (a) Surface roughness and (b) packing density of the powder beds with different powder samples at different recoating velocities. Source: Schiochet Nasato et al. [52]/With permission of Elsevier.

the main purpose of the preprocessing is to build a finite element model for the SLS process. The finite element model defines the computational domain of the thermal and mechanical analyses. The domain usually includes the substrate and desired printing part. Because of the high computational cost of the 3D thermal and mechanical analyses, it is necessary to consider the dimension of the substrate and part carefully. Furthermore, the temperature-dependent physical, thermal, and mechanical properties of the material should be adopted, which is critical to the accurate simulation of the SLS process. These material properties include

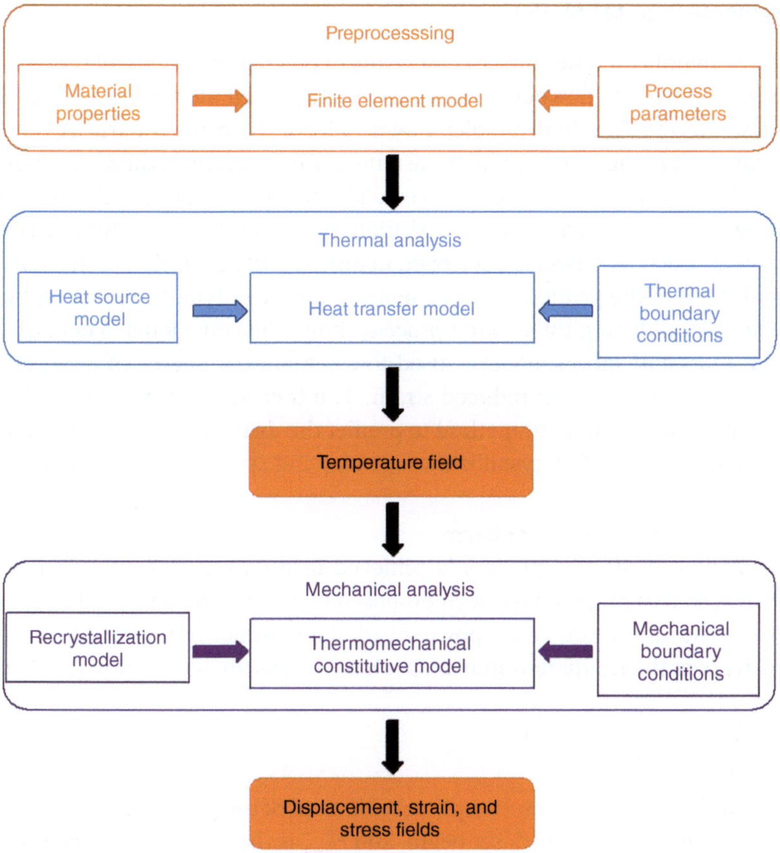

Figure 7.17 Flowchart of the thermo-mechanical finite element modeling for the SLS process.

the density, specific heat, thermal conductivity, coefficient of thermal expansion (CTE), elastic modulus, yield stress, etc. Several approaches have been developed to describe these temperature-dependent properties [32, 33, 35, 65–68]. The processing parameters in SLS include the laser power, laser radius, scanning speed, scanning pattern, hatching space, and layer thickness.

The thermal analysis is to predict the temperature field during the SLS process. The governing equation is based on the transient heat conduction, in which the heat energy is provided by the laser beam. A reasonable heat source model is necessary to describe the distribution of the laser energy.

The mechanical analysis to predict the displacement, strain, and stress of the printed part is based on the temperature field. To describe the mechanical behavior of the polymer material during the SLS process, a thermomechanical constitutive model is needed. For semicrystalline polymers, the recrystallization phenomenon should be considered in the modeling. Therefore, a recrystallization model is required to calculate the degree of recrystallization and the recrystallization-induced deformation.

7.3.1 Thermomechanical Model

The SLS process mainly includes the laser sintering of powder particles and the cooling of the printed part. The powder particles are rapidly fused after the scanning of the laser beam and then gradually cooled down to form the printed parts. A thermomechanical model is needed to predict the temperature, thermal strain, mechanical strain, and stress fields in the SLS process. The transient heat transfer model with a laser heat source model is used to calculate the temperature evolution. The thermo-elasto-viscoplastic model is developed to describe the evolution of mechanical strain and stress of the printed part. For semicrystalline polymers, the recrystallization phenomenon during the cooling process should be considered. Therefore, a nonuniform recrystallization model is used to calculate the degree of recrystallization and the recrystallization-induced strain. The thermomechanical model is implemented in the finite element method to predict the distribution and evolution of the temperature, degree of recrystallization, strain, and stress in the SLS process.

7.3.1.1 Heat Source Model for Laser Beam

The polymer fusion in the SLS process is achieved using a CO_2 laser as the heat source. Two heat source models have been proposed to describe the energy distribution provided by the laser beam on the powder bed. It is commonly assumed that the energy intensity over the irradiated circular spot has a Gaussian distribution [69–72]:

$$Q_0 = \frac{2AP}{\pi r_{\text{laser}}^2} \exp\left(-2\frac{r^2}{r_{\text{laser}}^2}\right) \tag{7.12}$$

where P is the laser power, A is the absorption coefficient of the powder material, r is the distance from the center of the laser beam, and r_{laser} is the laser radius. The laser energy is treated as a surface energy input over the irradiated spot on the powder bed. This heat source model has been used in the reported works on SLS modeling [32, 33].

The model ignores the interaction between the laser beam and the powder particles. The laser beam may experience multiple reflections and absorption when transmitting in the porous powder bed. The laser energy can penetrate a certain depth beneath the top surface of the powder bed. Laumer et al. [18] conducted an optical analysis of polymer powder materials for SLS. The reflectance, absorptance, and transmittance of polymer powder under different layer thicknesses at a laser wavelength of 10.6 μm were measured. Peyre et al. [20] further measured the laser beam transmission at various powder layer thicknesses. Results indicated a fast in-depth attenuation of the transmitted laser energy, which follows the Beer–Lambert's equation:

$$Q(z) = Q_0 \eta \exp(-\eta z) \tag{7.13}$$

where z is the depth from the top surface of the powder bed and η is the extinction coefficient. For PA12 and polyaryletherketone (PAEK) powder, η equals to 9000 and 7500 m^{-1}, respectively [20]. The term $\exp(-\eta z)$ represents the exponential decay of the laser power as a function of the depth from the top surface [73]. Therefore, the laser beam is treated as a volumetric heat source [35, 74]:

$$Q = \frac{2AP}{\pi r_{\text{laser}}^2} \exp\left(-2\frac{r^2}{r_{\text{laser}}^2}\right) \eta \exp(-\eta z) \tag{7.14}$$

This heat source model provides a more realistic description of the laser beam and its interaction with the powder bed.

7.3.1.2 Transient Heat Transfer Model

Polymer powder particles are preheated to the desired temperature (also called the bed temperature T_b) before the SLS process. The temperature rapidly increases up to the melting temperature after the laser irradiation. Because the powder bed is porous, the homogenization method is needed to model the transient heat transfer in the SLS process. The powder layer is treated as the continuum medium with homogenized thermal and mechanical properties. The powder particles and their interaction are not considered in this macroscopic modeling. The heat transfer model with the laser heat source is written as [75]:

$$\rho c \frac{\partial T}{\partial t} = \nabla \cdot (k \nabla T) + Q \tag{7.15}$$

where ρ, c, and k are the density, heat capacity, and heat conductivity, respectively. These three material properties are temperature-dependent. Moreover, their changes due to the transitions between powder, liquid, and solid phases are significant and should be considered in the process modeling of SLS. The initial temperature of the powder layer is set as:

$$T(t = 0) = T_b \tag{7.16}$$

Because the melting temperature of polymer powder is much lower than that of the metal powder, the thermal irradiation boundary condition is not considered in the modeling. The heat convection boundary condition is given as:

$$\begin{cases} -k(\boldsymbol{n} \cdot \nabla T) = h_h(T_b - T) \\ -k(\boldsymbol{n} \cdot \nabla T) = h_c(T_{rt} - T) \end{cases} \tag{7.17}$$

where $\boldsymbol{n}$ is the surface normal unit vector, T_{rt} represents the room temperature, and h_h and h_c are the convection coefficients during the heating and cooling processes, respectively.

The differential scanning calorimetry (DSC) analysis is conducted to investigate the melting and solidification behavior of the polymer powder [22, 76]. The start and end temperatures of melting and solidification (or recrystallization) are characterized through the DSC diagram. The heat capacity and heat conductivity at the powder and liquid phases are also obtained by DSC. The porosity of the powder bed is used to determine the homogenized density in the powder phase. The powder density is expressed as:

$$\rho_p = \rho_s(1 - \phi) \tag{7.18}$$

where ρ_p and ρ_s are the density at the powder state and the solid state, respectively, and ϕ is the powder porosity. Therefore, a significant increase in the density occurs at the transition from the powder to the liquid phase.

7.3.1.3 Thermo-Elasto-Viscoplastic Constitutive Model

Polymer materials experience thermal expansion and contraction in the heating and cooling processes, respectively. The thermal expansion and contraction induce the mechanical deformation of materials, i.e. the elastic (or viscoelastic) deformation and the viscoplastic deformation, resulting in the occurrence of internal stresses in the SLS-printed parts. For semicrystalline polymers, all molecular chains are amorphous after powder melting. However, parts of chains begin to recrystallize and form the crystalline phase as the temperature reaches the recrystallization point during the cooling process [77]. The recrystallization process induces a significant reduction in the polymer volume [17, 78]. The deformation can be classified into thermal, mechanical, and recrystallization-induced components. The thermal expansion and contraction and the recrystallization-induced volume shrinkage are accommodated by the mechanical deformation of polymer materials.

A thermo-elastic-viscoplastic constitutive model is developed to describe the relationship of the stress, mechanical strain, and thermal strain in the heating and cooling processes. Because the material deformation in the SLS process is small, the constitutive model is proposed in the small deformation framework. The total strain $\boldsymbol{\varepsilon}$ is written as:

$$\boldsymbol{\varepsilon} = \boldsymbol{\varepsilon}_{\mathrm{e}} + \boldsymbol{\varepsilon}_{\mathrm{vp}} + \boldsymbol{\varepsilon}_{\mathrm{th}} + \boldsymbol{\varepsilon}_{\mathrm{re}} \tag{7.19}$$

where $\boldsymbol{\varepsilon}_{\mathrm{e}}$, $\boldsymbol{\varepsilon}_{\mathrm{vp}}$, $\boldsymbol{\varepsilon}_{\mathrm{th}}$, and $\boldsymbol{\varepsilon}_{\mathrm{re}}$ are the elastic strain, viscoplastic strain, thermal strain, and recrystallization-induced strain, respectively. The Cauchy stress $\boldsymbol{\sigma}$ is related to the elastic strain $\boldsymbol{\varepsilon}_{\mathrm{e}}$:

$$\boldsymbol{\sigma} = \boldsymbol{C}(T) : \boldsymbol{\varepsilon}_{\mathrm{e}} \tag{7.20}$$

where $\boldsymbol{C}$ is the temperature-dependent fourth-order elastic tensor.

The viscoplastic strain rate is usually determined by the plastic dissipation potential. If the yield function is taken as the plastic dissipation potential, the viscoplastic strain rate is a tensor normal to the yield surface in the space of stresses [79]. If the plastic dissipation potential is a function of the equivalent stress and the hydrostatic stress, the viscoplastic strain rate is no longer normal to the yield surface [80–82]. The obtained viscoplastic model considers the contribution of the hydrostatic stress on the viscoplastic flow. However, this contribution is not significant in the SLS process because the stress and strain are relatively small compared to the case of the tension loading on polymer materials. Therefore, the von Mises yield function is used in the plastic dissipation potential [83]:

$$\phi = \sqrt{3\boldsymbol{s} : \boldsymbol{s}/2} - \sigma_{\mathrm{y}}(T) \tag{7.21}$$

where $\boldsymbol{s}$ is the deviatoric component of $\boldsymbol{\sigma}$ and σ_{y} is the temperature-dependent yield stress. The viscoplastic strain rate $\dot{\boldsymbol{\varepsilon}}_{\mathrm{vp}}$ is derived as:

$$\dot{\boldsymbol{\varepsilon}}_{\mathrm{vp}} = \dot{\lambda}_{\mathrm{vp}} \frac{\partial \phi}{\partial \boldsymbol{\sigma}} = \dot{\lambda}_{\mathrm{vp}} \frac{3\boldsymbol{s}}{2\sqrt{3\boldsymbol{s} : \boldsymbol{s}/2}} \tag{7.22}$$

where $\dot{\lambda}_{\mathrm{vp}}$ is the viscoplastic multiplier and $\partial\phi/\partial\boldsymbol{\sigma}$ is the direction of the viscoplastic flow. The viscoplastic multiplier is not obtained according to the consistency condition under the plastic yielding. It is commonly defined as a function of the von Mises stress, yield stress, and equivalent viscoplastic strain [80]:

$$\dot{\lambda}_{\text{vp}} = \dot{\lambda}_0 \left(\frac{\sqrt{3\boldsymbol{s} : \boldsymbol{s}/2}}{\sigma_y(T)(1 + \delta - \delta \exp(-p/p_y))} \right)^m \tag{7.23}$$

where $\dot{\lambda}_0$, δ, p_y. and m are material parameters and p is the equivalent viscoplastic strain expressed as $p = \sqrt{2\boldsymbol{\varepsilon}_{\text{vp}} : \boldsymbol{\varepsilon}_{\text{vp}}/3}$. The term $1 + \delta - \delta \exp(-p/p_y)$ is used to describe the smooth increase of the stress–strain curve at material yielding, and its value ranges from 1 to $1 + \delta$ as the equivalent viscoplastic strain increases.

The thermal strain $\Delta\boldsymbol{\varepsilon}_{\text{th}}$ is calculated according to the temperature increment and the CTE:

$$\Delta\boldsymbol{\varepsilon}_{\text{th}} = \alpha(T)\Delta T \cdot \boldsymbol{I} \tag{7.24}$$

where α is the temperature-dependent CTE. The unit second-order tensor **I** indicates that the isotropic thermal expansion is used in the modeling.

7.3.1.4 Recrystallization Model

For semicrystalline polymers, the recrystallization phenomenon occurs in the cooling process as the polymer temperature is lower than the onset point of recrystallization. The recrystallization process ends when the temperature reaches the end point. Therefore, the length of the time span of the recrystallization determines the degree of crystallization in polymer materials. If the time span is long enough for the recrystallization process to be completed, the degree of crystallization can reach 1. Otherwise, the degree of crystallization is less than 1, which is determined by the cooling rate of the SLS process. Because the temperature distribution in the SLS process is nonuniform, a non-isothermal recrystallization kinetic model is used here to describe the degree of crystallization [84–86]:

$$\chi(t) = 1 - \exp\left(-\left(\int_0^t K(T)\mathrm{d}\tau \right)^n \right) \tag{7.25}$$

where χ is the crystallization degree, n is the Avrami exponent, and $K(T)$ is the non-isothermal recrystallization rate expressed as $K(T) = \ln(2)^{1/n}/t_{1/2}$ in which $t_{1/2}$ is the time when the half of crystallization is reached:

$$\frac{1}{t_{1/2}} = K_0 \exp\left(-\frac{U}{R(T - T_\infty)} \right) \exp\left(-\frac{K_g\left(T + T_{\text{m}}^0\right)}{2T^2\left(T_{\text{m}}^0 - T\right)} \right) \tag{7.26}$$

where K_0 is the growth rate, U is the activation energy, R is the universal gas constant, $T_\infty = T_g - 30$ K is the temperature limit below which the crystallization stops, K_g is the nucleation rate constant, and T_{m}^0 is the equilibrium melting temperature. The time differential form of this equation is derived as:

$$\dot{\chi} = n\frac{\ln(2)^{1/n}}{t_{1/2}}(1 - \chi)\left(\ln\frac{1}{1 - \chi} \right)^{(n-1)/n} \tag{7.27}$$

The recrystallization process causes volumetric shrinkage. It is reported that the volumetric shrinkage induced by full recrystallization was about 4.7% for

PA12 [17, 78]. The recrystallization-induced strain is dependent on the crystallization degree:

$$\varepsilon_{\rm re} = -\varepsilon_{\rm ref}\chi \cdot \boldsymbol{I}, \quad \text{for } T_{\rm re,f} < T < T_{\rm re,s} \tag{7.28}$$

where $\varepsilon_{\rm ref}$ is the strain under full crystallization, and $T_{\rm re,\,s}$ and $T_{\rm re,\,f}$ are the start and end temperature of the recrystallization process, respectively. The minus symbol in the equation indicates that it is a contraction strain. It is also assumed that the recrystallization causes an isotropic shrinkage.

7.3.1.5 Finite Element Simulation

To simulate the SLS process, a 3D finite element model is developed in the software package ABAQUS. Both one-layer and multi-layer printing processes on a substrate can be simulated. For multi-layer printing, the element birth and death method should be used, i.e. the current layer is activated by the birth method and the upper layers are deactivated by the death method. The fine meshes are established for the current layer to accurately predict the temperature, strain, and stress fields. The other regions are coarsely meshed to save the computational cost.

Before the simulation, the thermal and mechanical properties of polymer materials should be provided. The detailed method to calibrate these material parameters can be referred to literature [35, 80]. In the simulation, the process parameters such as the laser power, scanning speed, printing strategy, layer thickness, and hatch space are needed. The effects of these process parameters on the temperature, deformation, and stress can be evaluated by the simulation.

There are two simulation procedures for the thermomechanical analysis, i.e. fully coupling and sequential coupling. For a fully coupling simulation, the temperature, stress, and strain results are calculated simultaneously. The coupling relationship between the stress and the temperature is considered in each time increment step of the simulation. For sequential coupling simulation, the temperature result is treated as a predefined thermal condition for the calculation of the stress and strain results. This sequential procedure is reasonable if the coupling relationship is not strong. For SLS process modeling, the fully coupling procedure is more realistic to predict the temperature, strain, and stress of the printed part.

The laser heat source model is implemented in the user subroutine DFLUX in ABAQUS. The laser power P is directly used in the subroutine. The movement of the laser beam is described through the scanning speed and time. The distance from the current laser beam center is given as:

$$r = \sqrt{(x - x_0 - v_x t)^2 + (y - y_0 - v_y t)^2} \tag{7.29}$$

where the point (x, y) represents one position on the layer surface, the point (x_0, y_0) represents the initial position of the center of the laser beam, v_x and v_y are the components of the laser scanning speed in x and y directions, respectively, and t is the scanning time. In the DFLUX subroutine, the heat flux at each point (x, y, z) is determined according to Eqs. (7.14) and (7.29). The DFLUX subroutine is called at every integration point in each time increment. The transient heat transfer model is implicitly solved using a modified Newton method to obtain the temperature field.

7.3.2 Numerical Method of the Thermo-Elasto-Viscoplastic Model

The thermo-elasto-viscoplastic model and the recrystallization model are implemented in the user subroutine UMAT in ABAQUS. The UMAT subroutine is called at every integration point in each time increment. The thermal strain and the recrystallization-induced strain are determined before the stress calculation. The thermal strain is determined by the variation of the temperature and the temperature-dependent CTE. The non-isothermal recrystallization kinetic model is solved using the Newton method. The recrystallization-induced strain can be determined by the recrystallization degree. The elastic strain, viscoplastic strain, and stress are then calculated through an incremental method [87, 88].

Assuming that the results of the four strain components and stress at the time increment i are known, results at the time increment $i+1$ could be obtained with [79, 87, 89]:

$$\Delta\boldsymbol{\varepsilon} = \Delta\boldsymbol{\varepsilon}_{\rm e} + \Delta\boldsymbol{\varepsilon}_{\rm vp} + \Delta\boldsymbol{\varepsilon}_{\rm th} + \Delta\boldsymbol{\varepsilon}_{\rm re} \tag{7.30}$$

$$\boldsymbol{\varepsilon}_{{\rm e},i+1} = \boldsymbol{\varepsilon}_{{\rm e},i} + \Delta\boldsymbol{\varepsilon}_{\rm e} \tag{7.31}$$

$$\boldsymbol{\sigma}_{i+1} = \boldsymbol{C}(T_{i+1}) : \boldsymbol{\varepsilon}_{{\rm e},i+1} \tag{7.32}$$

$$\boldsymbol{\varepsilon}_{{\rm vp},i+1} = \boldsymbol{\varepsilon}_{{\rm vp},i} + \dot{\boldsymbol{\varepsilon}}_{{\rm vp},i+1}\Delta t \tag{7.33}$$

The increments of the thermal strain and the recrystallization-induced strain, $\Delta\boldsymbol{\varepsilon}_{\rm th}$ and $\Delta\boldsymbol{\varepsilon}_{\rm re}$, are determined before the calculation of stress. The total increment of the strain $\Delta\boldsymbol{\varepsilon}$ is also known at the beginning of the time increment $i+1$. The key component of the incremental method is to calculate the viscoplastic strain whose rate at the time increment $i+1$ is written as:

$$\dot{\boldsymbol{\varepsilon}}_{{\rm vp},i+1} = \dot{\lambda}_0\left(\frac{\sqrt{3\boldsymbol{s}_{i+1} : \boldsymbol{s}_{i+1}/2}}{\sigma_y(T_{i+1})(1+\delta-\delta\exp(-p_{i+1}/p_{\rm y}))}\right)^m \frac{3\boldsymbol{s}_{i+1}}{2\sqrt{3\boldsymbol{s}_{i+1} : \boldsymbol{s}_{i+1}/2}} \tag{7.34}$$

The right side of the equation is a function of the variables $\boldsymbol{s}_{i+1}$ and p_{i+1}, which are related to the stress $\boldsymbol{\sigma}_{i+1}$ and the viscoplastic strain $\boldsymbol{\varepsilon}_{{\rm vp},\,i+1}$. According to Eqs. (7.30–7.32), the stress can be described as a function of the increment in the viscoplastic strain:

$$\boldsymbol{\sigma}_{i+1} = \boldsymbol{C}(T_{i+1}) : [\boldsymbol{\varepsilon}_{{\rm e},i} + \Delta\boldsymbol{\varepsilon} - (\boldsymbol{\varepsilon}_{{\rm vp},i+1} - \boldsymbol{\varepsilon}_{{\rm vp},i}) - \Delta\boldsymbol{\varepsilon}_{\rm th} - \Delta\boldsymbol{\varepsilon}_{\rm re}] \tag{7.35}$$

Therefore, the right side of Eq. (7.34) is a function of the viscoplastic strain $\boldsymbol{\varepsilon}_{{\rm vp},\,i+1}$ and the left side of the equation is the time differentiation of the viscoplastic strain. After the determination of the viscoplastic strain, the other three strain components and the stress are determined.

The consistent tangential modulus at the end of the time increment $i+1$, $\boldsymbol{C}_{a\lg,\,i+1}$, should be updated based on the results of the strain and the stress. The differentiations of Eqs. (7.30, 7.32, 7.33, 7.24, 7.28) are written as [89]:

$$\mathrm{d}\boldsymbol{\varepsilon} = \mathrm{d}\boldsymbol{\varepsilon}_{\rm e} + \mathrm{d}\boldsymbol{\varepsilon}_{\rm vp} + \mathrm{d}\boldsymbol{\varepsilon}_{\rm th} + \mathrm{d}\boldsymbol{\varepsilon}_{\rm re} \tag{7.36}$$

$$d\sigma = \frac{\partial \boldsymbol{C}}{\partial T} dT : \boldsymbol{\varepsilon}_e + \boldsymbol{C} : d\boldsymbol{\varepsilon}_e \tag{7.37}$$

$$d\boldsymbol{\varepsilon}_{vp} = \frac{\partial \dot{\boldsymbol{\varepsilon}}_{vp}}{\partial \boldsymbol{\sigma}} : d\boldsymbol{\sigma}\Delta t + \frac{\partial \dot{\boldsymbol{\varepsilon}}_{vp}}{\partial T} dT\Delta t \tag{7.38}$$

$$d\boldsymbol{\varepsilon}_{th} = \left(\alpha dT + \frac{\partial \alpha}{\partial T} dT(T_{i+1} - T)_i\right) \cdot \boldsymbol{I} \tag{7.39}$$

$$d\boldsymbol{\varepsilon}_{re} = -\varepsilon_{ref} \frac{\partial \chi}{\partial T} dT \cdot \boldsymbol{I} \tag{7.40}$$

The subscript $i+1$ is omitted for clarity. Because the temperature is also a variable, the full differentiation includes the contribution from the varied temperature. The viscoplastic strain rate relates to the stress and the temperature according to Eq. (7.34). Therefore, the differentiation form of the viscoplastic strain has two components. Substituting Eqs. (7.36 and 7.38) into Eq. (7.37), the differentiation of the stress is written as:

$$\begin{aligned} d\boldsymbol{\sigma} &= \frac{\partial \boldsymbol{C}}{\partial T} dT : \boldsymbol{\varepsilon}_e + \boldsymbol{C} : (d\boldsymbol{\varepsilon} - d\boldsymbol{\varepsilon}_{vp} - d\boldsymbol{\varepsilon}_{th} - d\boldsymbol{\varepsilon}_{re}) \\ &= \frac{\partial \boldsymbol{C}}{\partial T} dT : \boldsymbol{\varepsilon}_e + \boldsymbol{C} : d\boldsymbol{\varepsilon} - \boldsymbol{C} : \left(\frac{\partial \dot{\boldsymbol{\varepsilon}}_{vp}}{\partial \boldsymbol{\sigma}} : d\boldsymbol{\sigma}\Delta t + \frac{\partial \dot{\boldsymbol{\varepsilon}}_{vp}}{\partial T} dT\Delta t\right) \\ &\quad - \boldsymbol{C} : d\boldsymbol{\varepsilon}_{th} - \boldsymbol{C} : d\boldsymbol{\varepsilon}_{re} \end{aligned} \tag{7.41}$$

Regrouping the aforementioned equation leads to

$$\begin{aligned} \left(\boldsymbol{I} + \boldsymbol{C} : \frac{\partial \dot{\boldsymbol{\varepsilon}}_{vp}}{\partial \boldsymbol{\sigma}} \boldsymbol{I}\Delta t\right) d\boldsymbol{\sigma} &= \frac{\partial \boldsymbol{C}}{\partial T} dT : \boldsymbol{\varepsilon}_e + \boldsymbol{C} : d\boldsymbol{\varepsilon} \\ &\quad - \boldsymbol{C} : \frac{\partial \dot{\boldsymbol{\varepsilon}}_{vp}}{\partial T} dT\Delta t - \boldsymbol{C} : d\boldsymbol{\varepsilon}_{th} - \boldsymbol{C} : d\boldsymbol{\varepsilon}_{re} \end{aligned} \tag{7.42}$$

Therefore, the consistent tangential modulus is determined as:

$$\boldsymbol{C}_{a\lg,i+1} = \frac{\partial \boldsymbol{\sigma}}{\partial \boldsymbol{\varepsilon}} = \left[\boldsymbol{C}^{-1} + \frac{\partial \dot{\boldsymbol{\varepsilon}}_{vp}}{\partial \boldsymbol{\sigma}} \Delta t\right]^{-1} \tag{7.43}$$

The variance of the stress with respect to the temperature is also obtained as:

$$\begin{aligned} \frac{\partial \boldsymbol{\sigma}}{\partial T} &= -\left[\boldsymbol{C}^{-1} + \frac{\partial \dot{\boldsymbol{\varepsilon}}_{vp}}{\partial \boldsymbol{\sigma}} \Delta t\right]^{-1} : \\ &\quad \left(-\boldsymbol{C}^{-1} \frac{\partial \boldsymbol{C}}{\partial T} : \boldsymbol{\varepsilon}_e + \frac{\partial \dot{\boldsymbol{\varepsilon}}_{vp}}{\partial T} \Delta t + \alpha \boldsymbol{I} + \frac{\partial \alpha}{\partial T}(T_{i+1} - T_i)\boldsymbol{I} - \varepsilon_{ref} \frac{\partial \chi}{\partial T} \boldsymbol{I}\right) \end{aligned} \tag{7.44}$$

The updated consistent tangential modulus is used to calculate the increment of the total strain for the next time increment.

The update of the stress, strain, and consistent tangential modulus is coded in the subroutine UMAT using Fortran. Because there are several variables updated in the calculation, a group of the solution-dependent variables (SDVs) in UMAT is used to store these variables. The SDVs include the stress and these four strain components. Furthermore, several intermediate variables such as the recrystallization degree and the equivalent viscoplastic strain are also stored in SDVs. These SDVs are used for

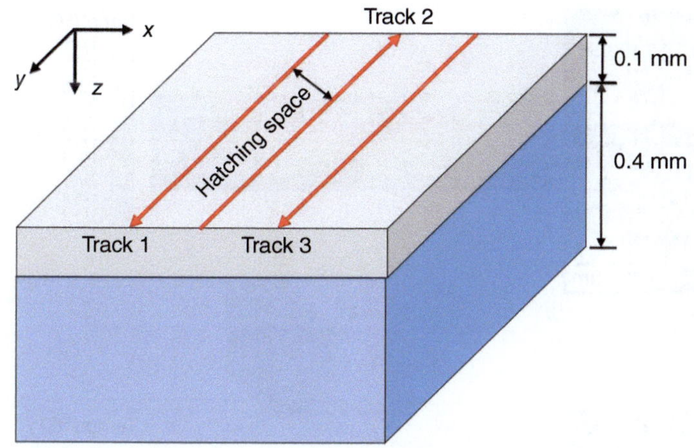

Figure 7.18 Simulation model with three scanning tracks on the powder layer. Source: Shen et al. [35]/With permission of Elsevier.

stress and strain updates and are also recorded in the analysis result file. The distribution and evolution of these SDVs can be obtained for postprocessing.

7.3.3 Temperature Distribution

It is of great significance to provide an accurate prediction of the temperature distribution during the SLS process because the maximum temperature and the dimension of the melt pool are important for the accurate evaluation of the printing quality. The SLS process is a layer-by-layer printing technique with each powder layer of a certain thickness. The depth of the melt pool should be larger than the layer thickness so that parts of the previously sintered layers are melted again, resulting in strong bonding between adjacent printed layers. If the depth of the melt pool is less than the layer thickness, unmelted regions are preserved in the printed layer and thus the bonding is weak. Another requirement of the processing temperature is that the maximum temperature should be less than the degradation temperature of the material.

The thermal simulation of the SLS process for polymer composite powder is conducted using the transient heat transfer model [35]. The powder used is PA12 reinforced by carbon nanotubes (CNTs) with a weight fraction of 0.5%. Figure 7.18 shows the simulation model with a bulk sintered part and one powder layer. The thickness of the powder layer is set at 0.1 mm. The scanning process of three tracks on the powder layer is simulated to evaluate the temperature distribution and the dimension of the melt pool. The results of the melt pool depth under two sets of process parameters were validated by experimental results. Figure 7.19a shows the microstructure of the sintered specimen on the x–z plane under the process parameters of $P = 10\,\text{W}$, $s = 3000\,\text{mm s}^{-1}$, and $h = 0.2\,\text{mm}$. The microstructure is nonuniform with horizontal dark and bright strips. The dark strips indicate the fused regions and the bright ones represent the unmelted regions. The average fusion depth of each layer is about

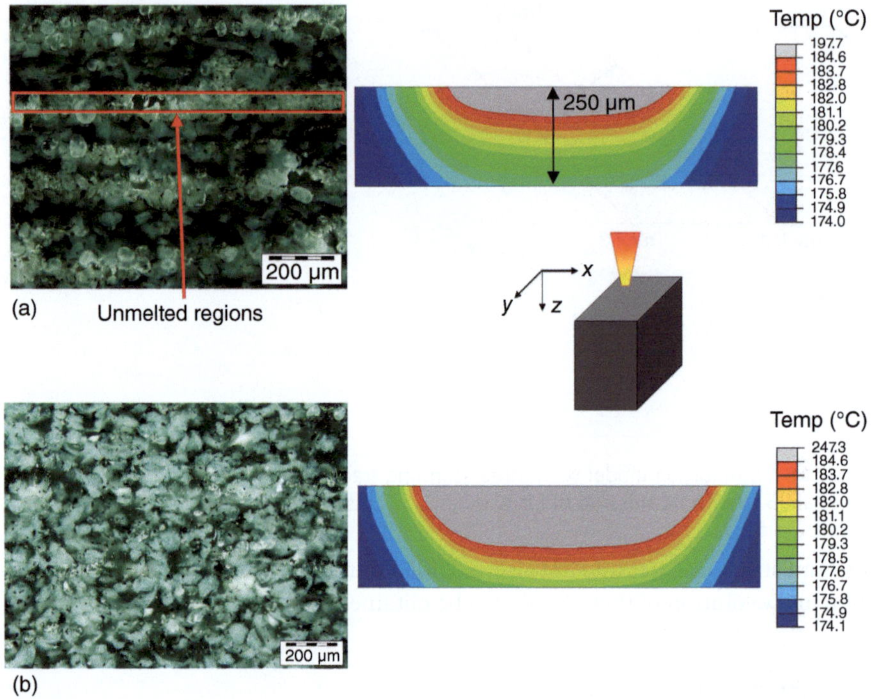

Figure 7.19 Microstructures of the sintered specimen on the x–z plane and the fusion zones after the scanning of three tracks under the process parameters: (a) $P = 10$ W, $s = 3000$ mm s^{-1}, and $h = 0.2$ mm, and (b) $P = 20$ W, $s = 3000$ mm s^{-1}, and $h = 0.2$ mm. Source: Shen et al. [35]/with permission of Elsevier.

72 μm. The simulation result shown in Figure 7.19 is the temperature distribution on the x–z plane of the powder layer. The gray zone indicates the melt pool where the temperature is higher than the melting point of the CNT/PA12 composite powder. The maximum depth of the predicted melt pool is about 82 μm which is less than the layer thickness. The experimental depth of the melt pool validates the numerical prediction. Figure 7.19b shows the microstructure and temperature distribution results under the process parameters of $P = 20$ W, $s = 3000$ mm s^{-1}, and $h = 0.2$ mm. Under the higher laser power, the obtained microstructure is uniform, and the bonding of adjacent layers is strong. The predicted depth of the melt pool is about 160 μm, which is larger than the layer thickness. The results in both simulation cases validate that the thermal model can provide a reasonable prediction of the temperature distribution of the SLS process.

The maximum SLS processing temperature for PEKK powder under different process parameters was numerically investigated through the transient heat transfer model [20]. Figure 7.20 shows the 3D finite element model to calculate the temperature distribution. The thermal radiation boundary condition is considered on the top surface in addition to the heat convection. It is also assumed that no heat flux crosses the side and bottom surfaces. The scanning speed is 600 mm s^{-1}, and the bed

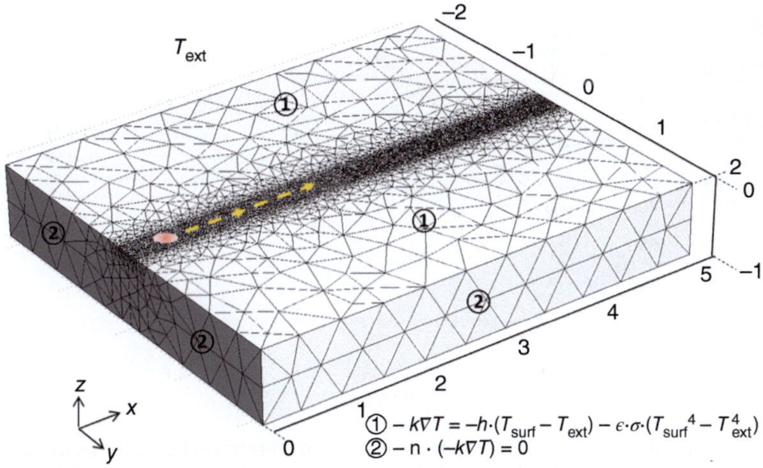

Figure 7.20 3D finite element model of the SLS process for PEKK powder. Source: Peyre et al. [20]/With permission of Elsevier.

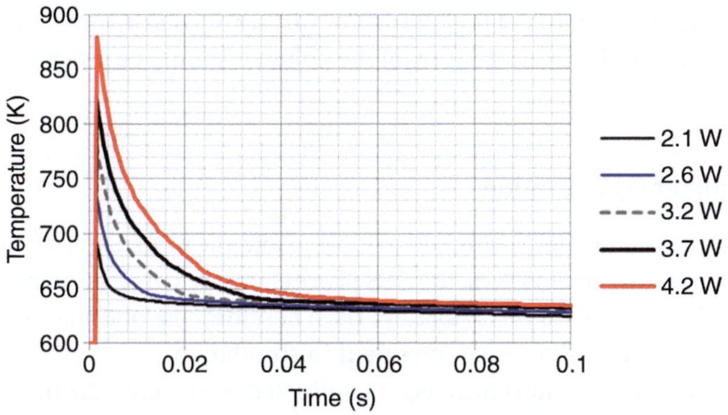

Figure 7.21 Temperature evolution on the powder bed under different laser powers. Source: Peyre et al. [20]/With permission of Elsevier.

temperature is set at 600 K. Figure 7.21 shows the evolution of the maximum temperature during the SLS process for PEKK. The temperature rapidly increases at the beginning of the process and reaches a maximum of 690 K under the laser power of 2.1 W. A larger laser power is observed to induce a higher maximum temperature. Figure 7.22 shows a detailed diagram of the maximum temperature under different sets of laser powers and scanning speeds. Under a certain scanning speed, the maximum temperature experiences a linear relationship with the laser power. By using the morphology and the color of the printed specimen, the temperature threshold of the thermal degradation is determined to be 760 K. No black spots are shown on the surface of the printed specimen, indicating the absence of degradation. Large laser power and high scanning speed are recommended for the SLS process for PEKK.

The next stage is partial degradation. Some black spots indicate polymer degradation. When the threshold temperature in this stage ranges from 760 to 825 K, several

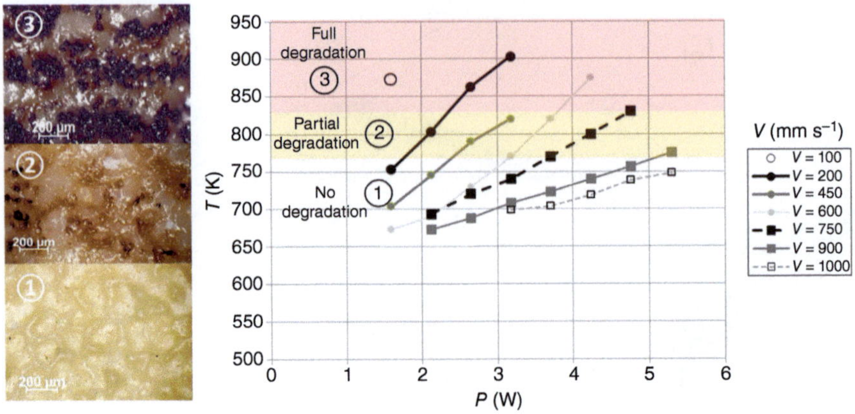

Figure 7.22 Diagram of the maximum temperature under different sets of laser power and scanning speed. Three zones are divided: no degradation, partial degradation, and full degradation. Source: Peyre et al. [20]/With permission of Elsevier.

black spots indicate the partial degradation of the polymer. If the maximum temperature is higher than 825 K, the majority of regions on the surface are black, indicating full degradation. The partial and full degradation of the polymer should be prevented in the SLS process.

7.3.4 Process Parameter Optimization

The process parameters affect the temperature distribution and the dimension of the melt pool. Under inappropriate parameters, the SLS process induces defects such as unmelted regions and degradation in the printed parts. Therefore, the optimization of SLS process parameters is an important component in simulation.

Extensive experimental investigations have been conducted to evaluate the influence of process parameters on print quality. The task is tremendous to study the effect of individual process parameters on the quality of printed parts. Therefore, the energy density E_d is calculated to represent the combined effect of the laser power, scanning speed, and laser diameter [90, 91]:

$$E_\mathrm{d} = \frac{P}{s\Phi} \tag{7.45}$$

where Φ is the laser diameter. The energy density is the average energy deposited on the powder surface per unit area. Caulfield et al. [19] studied the effects of the energy density on the density, yield stress, Young's modulus, fracture strength, and elongation of the printed PA12 specimens. Results showed that these physical and mechanical properties improved with the increase in the energy density, while decreased as the energy density exceeded a threshold of about 0.023 J mm^{-2}. Based on the experimental data, the neural network approach [24] and the variance analysis method [25, 92] are used to predict the effects of the process parameters on the quality of printed parts and optimize the process parameters. However, the trial and error process is too time-consuming and costly. A numerical method can be used together with the experimental method to speed up the process optimization. After the validation of the numerical method, it can produce necessary data that can aid in the process optimization.

The common objectives of process optimization are to achieve excellent physical and/or mechanical properties of SLS-printed parts. However, the finite element modeling of the SLS process has limitations in reproducing the micro-defects (such as unmelted zones, miro-voids, and microcracks) in printed parts. Therefore, printing productivity and energy consumption efficiency are used as the objectives for process optimization. The printing productivity P_V signifies the volumetric incremental rate of the melt pool and is defined as the volume of printed parts in unit time:

$$P_V = swd \tag{7.46}$$

where s is the scanning speed, and w and d are the width and depth of the melt pool, respectively. The width and the depth of the melt pool depend on the process parameters. The energy consumption efficiency is the ratio of the effective laser power to the total laser power. Because the laser energy also increases the temperature of the surrounding powder particles, only a part of the laser energy is used to melt the powder particles. Effective laser energy is defined as the energy to produce the melt pool with a volumetric incremental rate P_V. The energy consumption efficiency α is thus defined as:

$$\alpha = \frac{P_V E_V}{P} \tag{7.47}$$

where P is the laser power and E_V is the energy required to melt the powder particles per unit volume. The parameter E_V can be obtained through the DSC analysis of polymer powder. For the polymer PA12 reinforced by CNTs with a weight fraction of 0.5% (CNTs/PA12), $E_V = 0.064\,\text{J mm}^{-3}$ [22].

There are two constraint conditions of the process parameter optimization. The first one is that the maximum temperature, T_{max}, should be higher than the melting point of the polymer, T_m, and lower than its degradation point, T_d:

$$T_m < T_{max} < T_d \tag{7.48}$$

For the CNTs/PA12 composite, $T_m = 184.6\,°\text{C}$ and $T_d = 300\,°\text{C}$, i.e. $184.6 < T_{max} < 300\,°\text{C}$. The second constraint is that the depth of the melt pool must be larger than the layer thickness and less than twice the layer thickness. For the CNTs/PA12 composite, the thickness is set at 100 μm. Therefore, the second constraint condition is as follows:

$$100 < d < 200\ \mu\text{m} \tag{7.49}$$

The relationship between the objectives and the process parameters (laser power, scanning speed, and hatching space) should be established for optimization. The energy density is used here to combine these parameters. Furthermore, the relationship between the maximum temperature, melt pool depth, and energy density can also be developed. Extensive numerical simulations are conducted to get the results

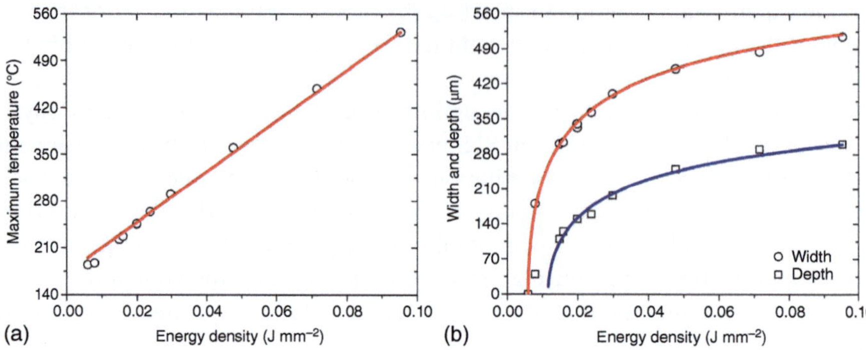

Figure 7.23 Relationship between (a) the maximum temperature, and (b) the width and depth of the melt pool and the energy density. Source: Shen et al. [35]/With permission of Elsevier.

of maximum temperature and melt pool dimension under varied process parameters. Figure 7.23a shows that the maximum temperature has a linear relationship with the energy density:

$$T_{\text{max}} = T_{\text{b}} + 3784.8E_{\text{D}} \tag{7.50}$$

A wide range of energy density from 0.01 to 0.1 J mm^{-2} is used in the modeling. The relationship between the melt pool dimension and the energy density is not linear as shown in Figure 7.23b. Two equations are obtained through a nonlinear curve fitting method:

$$\begin{cases} w = 312\sqrt{\ln(168E_{\text{D}})} \\ d = 206\sqrt{\ln(86.1E_{\text{D}})} \end{cases} \tag{7.51}$$

These three equations provide an accurate prediction of the relationship between the objectives and process parameters. According to the constraint conditions and these equations, the available ranges of the energy density can be determined. The available zone in a diagram with the laser power as the horizontal axis and the scanning speed as the vertical axis is shown in Figure 7.24a. The blue and red regions indicate that the process parameters in these regions meet the constraint conditions of the maximum temperature (Eq. (7.48)) and the melt pool depth (Eq. (7.49)), respectively. Therefore, the overlapped region marked by two red lines is the available region for the optimization of the process parameters. The suggested energy density is about 0.015–0.04 J mm^{-2}.

The two objectives are functions of the energy density, and their values can be also shown in the diagram of the laser power and the scanning speed. Figure 7.24b,c show the distribution of the printing productivity and the energy consumption efficiency. The printing productivity is high when the laser power and scanning speed are large. However, the energy consumption efficiency can also reach a high value when the laser power and the scanning speed are small. Through these two plots, the optimized process parameters are $P = 40$ W, $s = 3800$ mm s. For the hatching space, a value $h = 0.3$ mm is suggested for the SLS process of CNTs/PA12 composite powder.

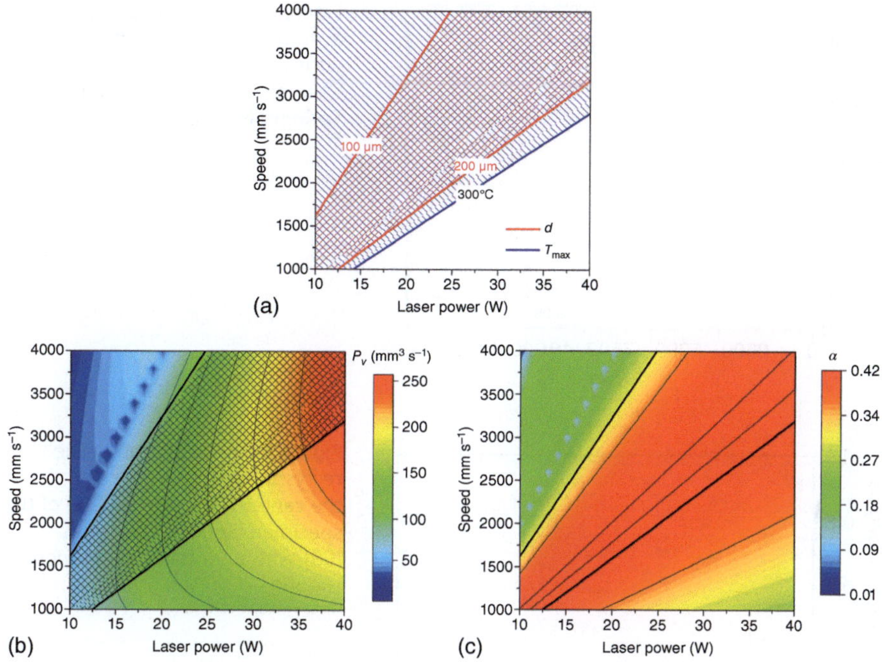

Figure 7.24 Diagram of (a) the available regions determined by two constraint conditions, (b) the printing productivity, and (c) the energy consumption efficiency. Source: Shen et al. [35]/With permission of Elsevier.

7.3.5 Recrystallization, Strain, and Stress Results

Deformation, recrystallization, strain, and stress results are obtained through the developed thermomechanical model. The temperature evolution during the cooling process is the fundamental information for stress and strain calculation. Figure 7.25 shows the temperature evolution of the printed PA12 part during the cooling process. The process parameters are $P = 35$ W, $s = 4000$ mm s^{-1}, and $h = 0.25$ mm. The whole evolution process can be divided into three stages. In the first stage, the temperature rapidly decreases to the onset point of recrystallization within about 500 seconds, where the liquid polymer changes to the solid state. The second stage occurs during the recrystallization process. For the PA12 polymer, the start and end temperatures of the recrystallization process are 168 and 160 °C, respectively. The rate of decrease in temperature is slower than that in the first stage because the recrystallization process is exothermic. The recrystallization process lasts for about 700 seconds. In the last stage, the temperature decreases from 160 to 30 °C. Therefore, the average cooling rate of the whole cooling process is about 0.6 °C min^{-1}.

The recrystallization degree of PA12 in the cooling process is obtained according to the temperature evolution. The Avrami exponent is determined to be $n = 2.6$ for the PA12 polymer [77]. Figure 7.26 shows the evolution of the recrystallization degree within the first 5000 seconds. The maximum value of the recrystallization degree is 0.999, indicating the completion of the recrystallization process. The process begins

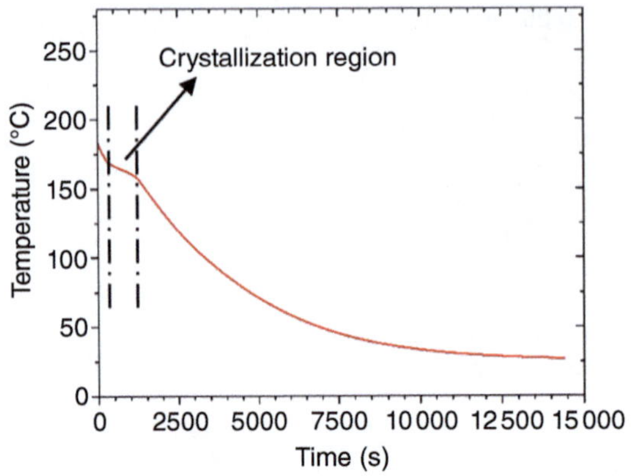

Figure 7.25 Temperature evolution of the PA12 polymer during the cooling process. Source: Shen et al. [93]/with permission of Springer Nature.

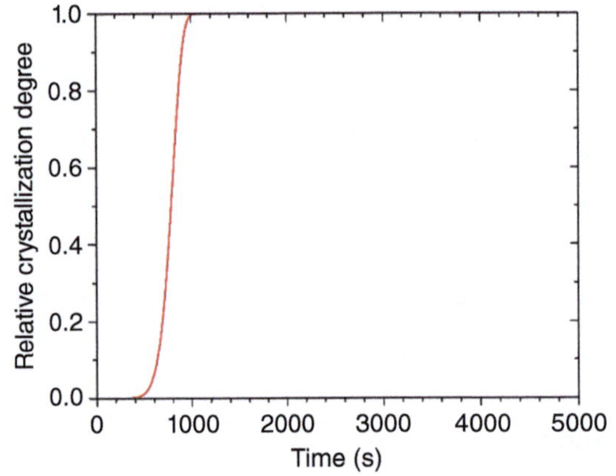

Figure 7.26 Evolution of the recrystallization degree at the center of the top surface of the specimen. Source: Shen et al. [93]/with permission of Springer Nature.

at about 500 seconds, and then the recrystallization degree significantly increases to the maximum value. The value remains unchanged after the end of the recrystallization process.

Figure 7.27 shows the evolution of the recrystallization-induced strain and the equivalent viscoplastic strain. The evolution of the recrystallization-induced strain has a similar shape as that of the recrystallization degree. The maximum recrystallization-induced strain is about 0.0157. There is no equivalent viscoplastic strain before about 500 seconds. In this initial stage, only two strain components exist, i.e. the elastic strain and the thermal strain. Because no external loading is applied on the SLS-printed parts, the thermal contraction strain is accommodated by the elastic strain. The equivalent viscoplastic strain begins to increase after the initial stage. The rate of increase during the recrystallization process is high due to the rapid increase in the recrystallization-induced strain. In the recrystallization process, the thermal contraction strain is small because the temperature change is only 8 °C. Most of the recrystallization-induced strain is accommodated by the equivalent viscoplastic strain. At the end of the recrystallization process, the equivalent viscoplastic strain is about 0.008, which occupies 51% of the recrystallization-induced strain. The rest is accommodated by the elastic strain. After the recrystallization

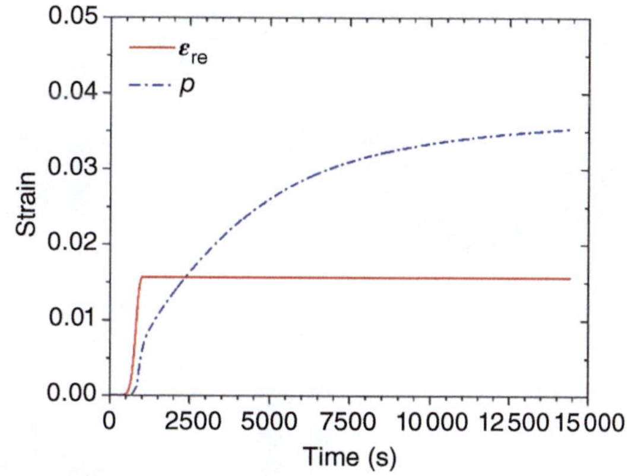

Figure 7.27 Evolution of the recrystallization-induced strain (ε_{re}) and the equivalent viscoplastic strain (p). Source: Shen et al. [93]/with permission of Springer Nature.

Table 7.1 Predicted deformation ratio and experimental deformation ratios of printed parts under the simulation case of $P = 35$ W, $s = 4000$ mm s^{-1}, and $h = 0.25$ mm.

No.	Experimental deformation ratio (%)	Predicted deformation ratio (%)
1	−2.36	−2.43
2	−2.32	
3	−2.28	

Source: Adapted from Shen et al. [93].

process, the equivalent viscoplastic strain still increases because of the thermal contraction strain. However, its rate of increase gradually decreases. The equivalent viscoplastic strain is about 0.035 at the end of the simulation.

The geometric accuracy of the SLS-printed parts is a critical printing quality indicator. The lengths of three SLS-printed specimens are measured and compared with the design length. The deformation ratio is defined as the ratio of the difference between the printed (or predicted) length and the design one, to the designed length. The experimental deformation ratios are calculated by using the realistic lengths of three specimens. The predicted deformation ratio is calculated by using the predicted length. Table 7.1 shows the comparison between both deformation ratios. Negative deformation ratios indicate that the SLS-printed specimen is shorter than its design length. The average deformation ratios of three printed specimens under the process parameters of $P = 35$ W, $s = 4000$ mm s^{-1}, and $h = 0.25$ mm is −2.32%. The predicted deformation ratio under the same process parameters is −2.43%. The results show that the developed thermomechanical model can provide accurate predictions of the deformation of the SLS-printed specimens.

Figure 7.28 shows the distributions of the Mises stress and two in-plane stress components, i.e. S11 and S22, at the end of the simulation. The maximum Mises stress occurs at the center of the printed layer. The main reason is that the deformation of the polymer material at the layer center is constrained by the

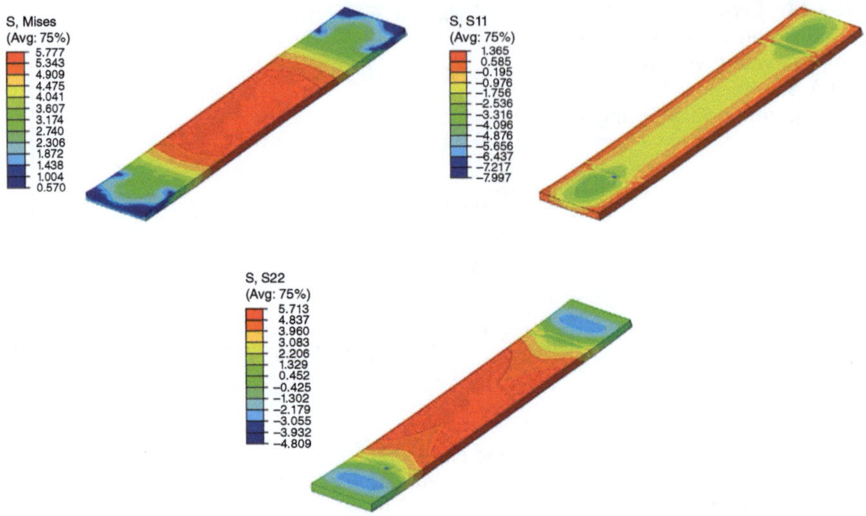

Figure 7.28 Distribution of Mises stress and other two stress components. Source: Shen et al. [93]/with permission of Springer Nature.

surrounding material. The maximum Mises stress is about 5.9 MPa. The stress component along the scanning direction, S22, is larger than S11. Its maximum value also occurs at the layer center. Figure 7.29 illustrates the evolution of the Mises stress at the layer center. Before the recrystallization process, the Mises stress increases linearly to about 0.7 MPa due to the thermal contraction strain. In the recrystallization process, the Mises stress increases significantly from 0.7 to 2.4 MPa. The recrystallization-induced strain leads to the increase of the elastic stress. After the recrystallization process, the Mises stress experiences a slight drop. This can be attributed to two main reasons: no further increase in the recrystallization-induced strain and the small thermal contraction strain. After the drop, the Mises stress gradually increases to 5.9 MPa, which is mainly driven by the thermal contraction strain. Another reason for the increase in the Mises stress could be due to the increase in the material properties such as the elastic modulus. At the end of the

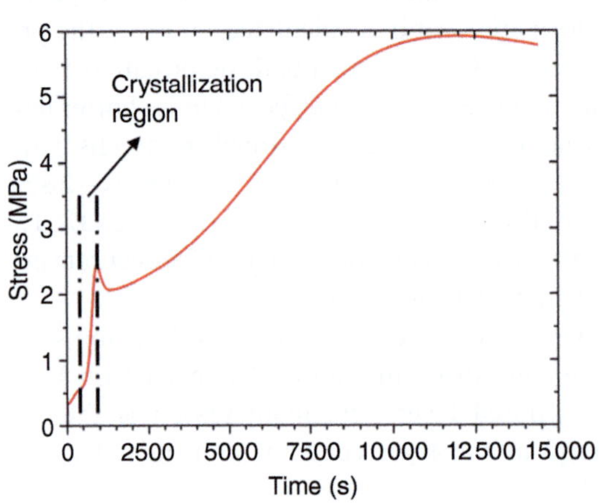

Figure 7.29 Evolution of Mises stress during the cooling process. Source: Shen et al. [93]/with permission of Springer Nature.

simulation, the Mises stress begins to decrease due to the stress relaxation induced by a small temperature variation.

7.4 Summary and Outlook

In this chapter, the recoating and sintering processes of polymer and composite powder particles in powder-based AM are simulated. Firstly, a DEM model has been presented for simulating the recoating process of polymer or fiber/polymer composite powder. The effects of recoating parameters, including the layer thickness, recoating velocity, fiber loading, and particle shape, on the recoating quality of the powder bed are investigated. Secondly, an finite element model has been developed to study the process efficiency in the SLS process of CNT/PA12 composite powder. The optimized laser power, scanning speed, and powder layer thickness are 40 W, 3800 mm s^{-1}, and 0.3 mm, respectively. Finally, a thermomechanical model has been developed to predict the temperature, recrystallization, strain, and stress in the SLS process of PA12 powder.

Due to the complexity of polymer powder particles, an efficient and versatile approach is still needed to accurately represent the polymer particles for the DEM simulation. Currently, enormous simulation works have been focusing on the study of the recoating process of pure polymer or metal particles. The recoating characteristics of composite powder such as fiber homogeneity and orientation that are of great significance for the mechanical strength and anisotropic properties of printed composites should be explored. Moreover, the powder recoating process in AM of polymer materials should be associated with the subsequent melting and coalescence processes. The modeling of the whole printing process can allow us to investigate the complex multi-physics AM process of polymer materials more accurately and provide more comprehensive guidance for parameter optimization. Particularly for polymer composites, a thorough investigation on the recoating, melting, and coalescence processes of multiple materials is worthwhile. Although the thermomechanical model has been widely used for the process simulation of AM, most of the studies focus on a multitrack or multi-layer printing process. Part-scale modeling that considers the realistic processing conditions of specific products is still needed for process optimization.

References

1 Yuan, S., Shen, F., Chua, C.K., and Zhou, K. (2019). Polymeric composites for powder-based additive manufacturing: materials and applications. *Progress in Polymer Science* 91: 141–168.

2 Tan, L.J., Zhu, W., and Zhou, K. (2020). Recent progress on polymer materials for additive manufacturing. *Advanced Functional Materials* 30 (43): 2003062.

3 Chatham, C.A., Long, T.E., and Williams, C.B. (2019). A review of the process physics and material screening methods for polymer powder bed fusion additive manufacturing. *Progress in Polymer Science* 93: 68–95.

4 Haeri, S. (2017). Optimisation of blade type spreaders for powder bed preparation in additive manufacturing using DEM simulations. *Powder Technology* 321: 94–104.

5 Ziegelmeier, S., Christou, P., Wöllecke, F. et al. (2015). An experimental study into the effects of bulk and flow behaviour of laser sintering polymer powders on resulting part properties. *Journal of Materials Processing Technology* 215: 239–250.

6 Escano, L.I., Parab, N.D., Xiong, L. et al. (2018). Revealing particle-scale powder spreading dynamics in powder-bed-based additive manufacturing process by high-speed x-ray imaging. *Scientific Reports* 8 (1): 1–11.

7 Ruggi, D., Lupo, M., Sofia, D. et al. (2020). Flow properties of polymeric powders for selective laser sintering. *Powder Technology* 370: 288–297.

8 Chen, H., Wei, Q., Zhang, Y. et al. (2019). Powder-spreading mechanisms in powder-bed-based additive manufacturing: experiments and computational modeling. *Acta Materialia* 179: 158–171.

9 Steuben, J.C., Iliopoulos, A.P., and Michopoulos, J.G. (2016). Discrete element modeling of particle-based additive manufacturing processes. *Computer Methods in Applied Mechanics and Engineering* 305: 537–561.

10 Xiang, Z., Yin, M., Deng, Z. et al. (2016). Simulation of forming process of powder bed for additive manufacturing. *Journal of Manufacturing Science and Engineering* 138 (8): 081002.

11 Zohdi, T.I. (2013). Rapid simulation of laser processing of discrete particulate materials. *Archives of Computational Methods in Engineering* 20 (4): 309–325.

12 Parteli, E.J.R. and Pöschel, T. (2016). Particle-based simulation of powder application in additive manufacturing. *Powder Technology* 288: 96–102.

13 Haeri, S., Wang, Y., Ghita, O., and Sun, J. (2017). Discrete element simulation and experimental study of powder spreading process in additive manufacturing. *Powder Technology* 306: 45–54.

14 Lee, Y.S., Nandwana, P., and Zhang, W. (2018). Dynamic simulation of powder packing structure for powder bed additive manufacturing. *The International Journal of Advanced Manufacturing Technology* 96 (1–4): 1507–1520.

15 Meier, C., Weissbach, R., Weinberg, J. et al. (2019). Critical influences of particle size and adhesion on the powder layer uniformity in metal additive manufacturing. *Journal of Materials Processing Technology* 266: 484–501.

16 He, Y., Hassanpour, A., and Bayly, A.E. (2020). Linking particle properties to layer characteristics: discrete element modelling of cohesive fine powder spreading in additive manufacturing. *Additive Manufacturing* 36: 101685.

17 Verbelen, L., Dadbakhsh, S., Van den Eynde, M. et al. (2016). Characterization of polyamide powders for determination of laser sintering processability. *European Polymer Journal* 75: 163–174.

18 Laumer, T., Stichel, T., Nagulin, K., and Schmidt, M. (2016). Optical analysis of polymer powder materials for selective laser sintering. *Polymer Testing* 56: 207–213.

19 Caulfield, B., McHugh, P.E., and Lohfeld, S. (2007). Dependence of mechanical properties of polyamide components on build parameters in the SLS process. *Journal of Materials Processing Technology* 182 (1–3): 477–488.

20 Peyre, P., Rouchausse, Y., Defauchy, D., and Régnier, G. (2015). Experimental and numerical analysis of the selective laser sintering (SLS) of PA12 and PEKK semi-crystalline polymers. *Journal of Materials Processing Technology* 225: 326–336.

21 Yuan, S., Bai, J., Chua, C.K. et al. (2016). Highly enhanced thermal conductivity of thermoplastic nanocomposites with a low mass fraction of MWCNTs by a facilitated latex approach. *Composites Part A: Applied Science and Manufacturing* 90: 699–710.

22 Yuan, S., Bai, J., Chua, C.K. et al. (2016). Material evaluation and process optimization of CNT-coated polymer powders for selective laser sintering. *Polymers (Basel)* 8 (10): 370.

23 Yuan, S., Zheng, Y., Chua, C.K. et al. (2018). Electrical and thermal conductivities of MWCNT/polymer composites fabricated by selective laser sintering. *Composites Part A: Applied Science and Manufacturing* 105: 203–213.

24 Wang, R.-J., Wang, L., Zhao, L., and Liu, Z. (2006). Influence of process parameters on part shrinkage in SLS. *The International Journal of Advanced Manufacturing Technology* 33 (5–6): 498–504.

25 Singh, S., Sachdeva, A., and Sharma, V.S. (2017). Optimization of selective laser sintering process parameters to achieve the maximum density and hardness in polyamide parts. *Progress in Additive Manufacturing* 2 (1–2): 19–30.

26 Rong-Ji, W., Xin-hua, L., Qing-ding, W., and Lingling, W. (2008). Optimizing process parameters for selective laser sintering based on neural network and genetic algorithm. *The International Journal of Advanced Manufacturing Technology* 42 (11–12): 1035–1042.

27 Körner, C., Attar, E., and Heinl, P. (2011). Mesoscopic simulation of selective beam melting processes. *Journal of Materials Processing Technology* 211 (6): 978–987.

28 Körner, C., Bauereiß, A., and Attar, E. (2013). Fundamental consolidation mechanisms during selective beam melting of powders. *Modelling and Simulation in Materials Science and Engineering* 21 (8): 085011.

29 Khairallah, S.A. and Anderson, A. (2014). Mesoscopic simulation model of selective laser melting of stainless steel powder. *Journal of Materials Processing Technology* 214 (11): 2627–2636.

30 Khairallah, S.A., Anderson, A.T., Rubenchik, A., and King, W.E. (2016). Laser powder-bed fusion additive manufacturing: physics of complex melt flow and formation mechanisms of pores, spatter, and denudation zones. *Acta Materialia* 108: 36–45.

31 King, W., Anderson, A.T., Ferencz, R.M. et al. (2014). Overview of modelling and simulation of metal powder bed fusion process at Lawrence Livermore National Laboratory. *Materials Science and Technology* 31 (8): 957–968.

32 Dong, L., Makradi, A., Ahzi, S., and Remond, Y. (2009). Three-dimensional transient finite element analysis of the selective laser sintering process. *Journal of Materials Processing Technology* 209 (2): 700–706.

33 Riedlbauer, D., Drexler, M., Drummer, D. et al. (2014). Modelling, simulation and experimental validation of heat transfer in selective laser melting of the polymeric material PA12. *Computational Materials Science* 93: 239–248.

34 Foroozmehr, A., Badrossamay, M., Foroozmehr, E., and Si, G. (2016). Finite element simulation of selective laser melting process considering optical penetration depth of laser in powder bed. *Materials & Design* 89: 255–263.

35 Shen, F., Yuan, S., Chua, C.K., and Zhou, K. (2018). Development of process efficiency maps for selective laser sintering of polymeric composite powders: modeling and experimental testing. *Journal of Materials Processing Technology* 254: 52–59.

36 Cundall, P.A. and Strack, O.D. (1979). A discrete numerical model for granular assemblies. *Géotechnique* 29 (1): 47–65.

37 Cundall, P.A. (1971). A computer model for simulating progressive, large-scale movement in blocky rock system. Proceedings of the International Symposium on Rock Mechanics.

38 Modenese C. (2013). Numerical study of the mechanical properties of lunar soil by the discrete element method. St Anne's College, Oxford.

39 Tsuji, Y., Tanaka, T., and Ishida, T. (1992). Lagrangian numerical simulation of plug flow of cohesionless particles in a horizontal pipe. *Powder Technology* 71 (3): 239–250.

40 Rao, K.K., Nott, P.R., and Sundaresan, S. (2008). *An Introduction to Granular Flow*. New York: Cambridge University Press.

41 Šmilauer, V., Catalano, E., and Chareyre, B. (2019). *Yade Documentation*.

42 Thornton, C., Cummins, S.J., and Cleary, P.W. (2013). An investigation of the comparative behaviour of alternative contact force models during inelastic collisions. *Powder Technology* 233: 30–46.

43 Johnson, K.L., Kendall, K., and Roberts, A. (1971). Surface energy and the contact of elastic solids. *Proceedings of the Royal Society of London A Mathematical and Physical Sciences* 324 (1558): 301–313.

44 Derjaguin, B.V., Muller, V.M., and Toporov, Y.P. (1975). Effect of contact deformations on the adhesion of particles. *Journal of Colloid and Interface Science* 53 (2): 314–326.

45 Parandoush, P. and Lin, D. (2017). A review on additive manufacturing of polymer-fiber composites. *Composite Structures* 182: 36–53.

46 Shi, Y., Li, Z., Sun, H. et al. (2004). Effect of the properties of the polymer materials on the quality of selective laser sintering parts. *Proceedings of the Institution of Mechanical Engineers, Part L: Journal of Materials: Design and Applications* 218 (3): 247–252.

47 Berretta, S., Ghita, O., and Evans, K.E. (2014). Morphology of polymeric powders in Laser Sintering (LS): from Polyamide to new PEEK powders. *European Polymer Journal* 59: 218–229.

48 Schmid, M. (2018). *Laser Sintering with Plastics: Technology, Processes, and Materials*. Carl Hanser Verlag GmbH Co KG.

49 Nan, W., Pasha, M., Bonakdar, T. et al. (2018). Jamming during particle spreading in additive manufacturing. *Powder Technology* 338: 253–262.

50 Tan, P., Shen, F., Tey, W.S., and Zhou, K. (2021). A numerical study on the packing quality of fibre/polymer composite powder for powder bed fusion additive manufacturing. *Virtual and Physical Prototyping* 16 (sup 1): 1–18.

51 Cai, C., Tey, W.S., Chen, J. et al. (2021). Comparative study on 3D printing of polyamide 12 by selective laser sintering and multi jet fusion. *Journal of Materials Processing Technology* 288: 116882.

52 Schiochet Nasato, D. and Pöschel, T. (2020). Influence of particle shape in additive manufacturing: discrete element simulations of polyamide 11 and polyamide 12. *Additive Manufacturing* 101421.

53 Yuan, S., Chua, C.K., and Zhou, K. (2019). 3D-printed mechanical metamaterials with high energy absorption. *Advanced Materials Technologies* 4 (3): 1800419.

54 Geldart, D., Abdullah, E.C., Hassanpour, A. et al. (2006). Characterization of powder flowability using measurement of angle of repose. *China Particuology* 4 (3–4): 104–107.

55 Carr, R.L. (1965). Classifying flow properties of solids. *Chemical Engineering* 1: 69–72.

56 Beakawi Al-Hashemi, H.M. and Baghabra Al-Amoudi, O.S. (2018). A review on the angle of repose of granular materials. *Powder Technology* 330: 397–417.

57 Pitanga, H.N., Gourc, J.-P., and Vilar, O.M. (2009). Interface shear strength of geosynthetics: evaluation and analysis of inclined plane tests. *Geotextiles and Geomembranes* 27 (6): 435–446.

58 American Society for Testing and Materials. Committee D18 on Soil and Rock (2008). *Standard Test Method for Bulk Solids Characterization by Carr Indices.* ASTM International.

59 Coetzee, C.J. (2017). Review: calibration of the discrete element method. *Powder Technology* 310: 104–142.

60 Liu, X.Y., Specht, E., and Mellmann, J. (2005). Experimental study of the lower and upper angles of repose of granular materials in rotating drums. *Powder Technology* 154 (2–3): 125–131.

61 Meier, C., Weissbach, R., Weinberg, J. et al. (2019). Modeling and characterization of cohesion in fine metal powders with a focus on additive manufacturing process simulations. *Powder Technology* 343: 855–866.

62 Rackl, M. and Hanley, K.J. (2017). A methodical calibration procedure for discrete element models. *Powder Technology* 307: 73–83.

63 Roessler, T. and Katterfeld, A. (2018). Scaling of the angle of repose test and its influence on the calibration of DEM parameters using upscaled particles. *Powder Technology* 330: 58–66.

64 Luo, Z. and Zhao, Y. (2018). A survey of finite element analysis of temperature and thermal stress fields in powder bed fusion Additive Manufacturing. *Additive Manufacturing* 21: 318–332.

65 Childs, T., Berzins, M., Ryder, G., and Tontowi, A. (1999). Selective laser sintering of an amorphous polymer—simulations and experiments. *Proceedings of the Institution of Mechanical Engineers, Part B: Journal of Engineering Manufacture* 213 (4): 333–349.

66 Patil, R.B. and Yadava, V. (2007). Finite element analysis of temperature distribution in single metallic powder layer during metal laser sintering. *International Journal of Machine Tools and Manufacture* 47 (7–8): 1069–1080.

67 Antony, K., Arivazhagan, N., and Senthilkumaran, K. (2014). Numerical and experimental investigations on laser melting of stainless steel 316L metal powders. *Journal of Manufacturing Processes* 16 (3): 345–355.

68 Haque, E. and Hampson, P. (2014). Modelling phase change in a 3D thermal transient analysis. *The International Journal of Multiphysics* 8 (1): 49–68.

69 Brighenti, R., Cosma, M.P., Marsavina, L. et al. (2020). Laser-based additively manufactured polymers: a review on processes and mechanical models. *Journal of Materials Science* 56 (2): 961–998.

70 Soldner, D., Greiner, S., Burkhardt, C. et al. (2021). Numerical and experimental investigation of the isothermal assumption in selective laser sintering of PA12. *Additive Manufacturing* 37.

71 Hussein, A., Hao, L., Yan, C., and Everson, R. (2013). Finite element simulation of the temperature and stress fields in single layers built without-support in selective laser melting. *Materials & Design (1980–2015)* 52: 638–647.

72 Tan, P., Kiran, R., and Zhou, K. (2021). Effects of sub-atmospheric pressure on keyhole dynamics and porosity in products fabricated by selective laser melting. *Journal of Manufacturing Processes* 64: 816–827.

73 Rombouts, M., Froyen, L., Gusarov, A. et al. (2005). Light extinction in metallic powder beds: correlation with powder structure. *Journal of Applied Physics* 98 (1): 013533.

74 Tian, X., Peng, G., Yan, M. et al. (2018). Process prediction of selective laser sintering based on heat transfer analysis for polyamide composite powders. *International Journal of Heat and Mass Transfer* 120: 379–386.

75 Tan, P., Shen, F., Li, B., and Zhou, K. (2019). A thermo-metallurgical-mechanical model for selective laser melting of Ti6Al4V. *Materials & Design* 168: 107642.

76 Sutton, A.T., Kriewall, C.S., Leu, M.C., and Newkirk, J.W. (2016). Powder characterisation techniques and effects of powder characteristics on part properties in powder-bed fusion processes. *Virtual and Physical Prototyping* 12 (1): 3–29.

77 Zhao, M., Wudy, K., and Drummer, D. (2018). Crystallization kinetics of polyamide 12 during selective laser sintering. *Polymers* 10 (2): 168.

78 Benedetti, L., Brulé, B., Decreamer, N. et al. (2019). Shrinkage behaviour of semi-crystalline polymers in laser sintering: PEKK and PA12. *Materials & Design* 181: 107906.

79 de Souza, N.E., Peric, D., and Owen, D. (2008). *Computational Methods for Plasticity: Theory and Applications*. Wiley.

80 Shen, F., Kang, G., Lam, Y.C. et al. (2019). Thermo-elastic-viscoplastic-damage model for self-heating and mechanical behavior of thermoplastic polymers. *International Journal of Plasticity* 121: 227–243.

81 Zerbe, P., Schneider, B., Moosbrugger, E., and Kaliske, M. (2017). A viscoelastic-viscoplastic-damage model for creep and recovery of a semicrystalline thermoplastic. *International Journal of Solids and Structures* 110–111: 340–350.

82 Garcia-Gonzalez, D., Zaera, R., and Arias, A. (2017). A hyperelastic-thermoviscoplastic constitutive model for semi-crystalline polymers: application to PEEK under dynamic loading conditions. *International Journal of Plasticity* 88: 27–52.

83 Lemaitre, J. and Chaboche, J.-L. (1990). *Mechanics of Solid Materials.* Cambridge University Press.

84 Avrami, M. (1939). Kinetics of phase change. I General theory. *The Journal of Chemical Physics* 7 (12): 1103–1112.

85 Avrami, M. (1940). Kinetics of phase change. II transformation-time relations for random distribution of nuclei. *The Journal of Chemical Physics* 8 (2): 212–224.

86 Avrami, M. (1941). Granulation, phase change, and microstructure kinetics of phase change. III. *The Journal of Chemical Physics* 9 (2): 177–184.

87 Shen, F. and Zhou, K. (2019). An elasto-plastic-damage model for initiation and propagation of spalling in rolling bearings. *International Journal of Mechanical Sciences* 161–162: 105058.

88 Shen, F., Ke, L.-L., and Zhou, K. (2021). A debris layer evolution-based model for predicting both fretting wear and fretting fatigue lifetime. *International Journal of Fatigue* 142: 105928.

89 Belytschko, T., Liu, W.K., Moran, B., and Elkhodary, K.I. (2014). *Nonlinear Finite Elements for Continua and Structures.* Wiley.

90 Li, M., Han, Y., Zhou, M. et al. (2020). Experimental investigating and numerical simulations of the thermal behavior and process optimization for selective laser sintering of PA6. *Journal of Manufacturing Processes* 56: 271–279.

91 Franco, A., Lanzetta, M., and Romoli, L. (2010). Experimental analysis of selective laser sintering of polyamide powders: an energy perspective. *Journal of Cleaner Production* 18 (16–17): 1722–1730.

92 Sachdeva, A., Singh, S., and Sharma, V.S. (2012). Investigating surface roughness of parts produced by SLS process. *The International Journal of Advanced Manufacturing Technology* 64 (9–12): 1505–1516.

93 Shen, F., Zhu, W., Zhou, K., and Ke, L.-L. (2021). Modeling the temperature, crystallization, and residual stress for selective laser sintering of polymeric powder. *Acta Mechanica* 232 (9): 3635–3653.

Index

Additive Manufacturing Technology: Design, Optimization, and Modeling, First Edition. Edited by Kun Zhou.

n

S